Classics in the History of Greek Mathematics

BOSTON STUDIES IN THE PHILOSOPHY OF SCIENCE

VOLUME 240

CLASSICS IN THE HISTORY OF GREEK MATHEMATICS

Edited by

JEAN CHRISTIANIDIS
University of Athens, Greece

SPRINGER SCIENCE+BUSINESS MEDIA, B.V.

A C.I.P. Catalogue record for this book is available from the Library of Congress.

DOI 10.1007/978-1-4020-2640-9

Printed on acid-free paper

TABLE OF CONTENTS

PART 3. STUDIES ON PROPORTION THEORY AND INCOMMENSURABILITY

Texts selected and introduced by Ken Saito

PART 4. STUDIES ON GREEK ALGEBRA

Texts selected and introduced by Jacques Sesiano

PART 5. DID THE GREEKS HAVE THE NOTION OF COMMON FRACTION? DID THEY USE IT?
Texts selected and introduced by Jean Christianidis

PART 6. METHODOLOGICAL ISSUES IN THE HISTORIOGRAPHY OF GREEK MATHEMATICS
Texts selected and introduced by Sabetai Unguru

PERMISSIONS

The Editor would like to thank the following publishers for granting permission to reprint the papers included in this volume:

1. Jürgen, M.: Die Entdeckung der Möglichkeit von Missenschaft. Reprinted from *Archive for History of Exact Sciences* 2 (1962-66), 410-435, with permission from Springer-Verlag.
2. Szabó, Á.: Wie ist die Mathematik zu einer deduktiven Wissenschaft geworden?. Reprinted from *Acta Antiqua Academiae Scientiarum Hungaricae* IV (1956), 109-152, with kind permission of Akadémiai Kiadó.
3. Knorr, W.R.: On the early history of axiomatics. The interaction of mathematics and philosophy in Greek antiquity. Reprinted from J. Hintikka, D. Gruender, E. Agazzi (Eds), *Theory change, ancient axiomatics, and Galileo's methodology*, 1981, 145-186, with kind permission of Kluwer Academic Publishers.
4. Knorr, W.R.: Construction as Existence Proof in Ancient Geometry. Reprinted from *Ancient Philosophy* 3 (1983), 125-148, with kind permission of Mathesis Publications, Inc.
5. Saito, K.: Book II of Euclid's Elements in the Light of the Theory of Conic Sections. Reprinted from *Historia Scientiarum* 28 (1985), 31-60, with kind permission of History of Science Society of Japan.
6. Lloyd, G.E.R. The Meno and the Mysteries of Mathematics. Reprinted from *Phronesis* 37 (1992), 166-183, with kind permission of Brill Academic Publishers.
7. Becker, O.: Exdoxos-Studien I: Eine voreudoxische Proportionenlehre und ihre Spuren bei Aristoteles und Euklid. Reprinted from *Quellen und Studien zur Geschichte der Mathematik, Astronomie und Physik*, B.II (1933), 311-330, with permission from Springer-Verlag.
8. Von Fritz, K.: Discovery of Incommensurability by Hippasus of Metapontum. Reprinted from *Annals of Mathematics* 46 (1954), 242-264, with kind permission of Princeton University Press.
9. Freudenthal, H.: Y avait-il une crise des fondements des mathématiques dans l'antiquité? Reprinted from *Bulletin de la Société mathématique de Belgique* 18 (1966), 43-55, with kind permission of Société Mathématique de Belgique.
10. Knorr, W.R.: Impact of Modern Mathematics on Ancient Mathematics. Reprinted from *Revue d'histoire des mathématiques* 7 (2001), 121-135, with kind permission of Société Mathématique de France.
11. Toomer, G.J.: Lost Greek mathematical works in Arabic translation. Reprinted from *Mathematical Intelligencer* 6 (1984), 32-38, with permission from Springer-Verlag.
12. Knorr, W.R.: Techniques of fractions in Ancient Egypt and Greece. Reprinted from *Historia Mathematica* 9 (1982), 133-171, with permission from Elsevier.

13. Fowler, D.H.: Logistic and fractions in early Greek mathematics: a new interpretation. Reprinted from P. Benoit, K. Chemla, J. Ritter (Eds), *Histoire des fractions, fraction d'histoire*, 1992, 133-147, with permission from Birkhäuser Verlag AG.
14. Unguru, S.: On the Need to Rewrite the History of Greek Mathematics. Reprinted from *Archive for History of Exact Sciences* 15 (1975), 67-114, with permission from Springer-Verlag.
15. Van der Waerden, B.L.: Defence of a 'Shocking' Point of View. Reprinted from *Archive for History of Exact Sciences* 15 (1976), 199-210, with permission from Springer-Verlag.
16. Weil, A.: Who Betrayed Euclid? Reprinted from *Archive for History of Exact Sciences* 19 (1978), 91-93, with permission from Springer-Verlag.
17. Unguru, S.: History of Ancient Mathematics: Some Reflections on the State of the Art. Reprinted from *Isis* 70 (1979), 555-565, with permission from The Chicago University Press.

PREFACE

The twentieth century is the period during which the history of Greek mathematics reached its greatest acme. Indeed, it is by no means exaggerated to say that Greek mathematics represents the unique field from the wider domain of the general history of science which was included in the research agenda of so many and so distinguished scholars, from so varied scientific communities (historians of science, historians of philosophy, mathematicians, philologists, philosophers of science, archeologists etc.), while new scholarship of the highest quality continues to be produced.

This volume includes 19 classic papers on the history of Greek mathematics that were published during the entire 20th century and affected significantly the state of the art of this field. It is divided into six self-contained sections, each one with its own editor, who had the responsibility for the selection of the papers that are republished in the section, and who wrote the introduction of the section. It constitutes a kind of a Reader book which is today, one century after the first publications of Tannery, Zeuthen, Heath and the other outstanding figures of the end of the 19th and the beginning of 20th century, rather timely in many respects. First, the inclusion in one volume of a considerable number of papers that had been published for the first time in old, and in certain cases hard to find, scientific journals representing turning-points in the history of the field, constitutes a particularly useful aid for all those working on the history of mathematics. Second, by means of the selected papers and the introductory texts that accompany them, the reader can follow the ways the historiography of Greek mathematics developed – that is to say, the questions which occupied the community of the historians of mathematics in various periods during the development of the field, how these questions were answered, the conflicting tendencies that grew up within the community over this period etc. Finally, the introductory texts that precede each chapter help the reader to approach critically the selected papers and at the same time to get an idea of the issues being further clarified by the new historiographical approaches.

Six well-known historians of Greek mathematics have contributed to the preparation of this volume. I thank my colleagues and friends Hans-Joachim Waschkies (University of Kiel), Reviel Netz (Stanford University), Ken Saito (Osaka Prefecture University), Jacques Sesiano (École Polytechnique Fédérale de Lausanne), and Sabetai Unguru (Tel Aviv University) for their participation in the preparation of this volume. I also thank Kostas Gavroglu and Jurgen Renn for their useful advice during the preparation of the volume for the series *Boston Studies in the Philosophy of Science*.

Jean Christianidis
University of Athens

PART 1

THE BEGINNINGS OF GREEK MATHEMATICS

Texts selected and introduced by Hans-Joachim Waschkies

HANS-JOACHIM WASCHKIES

INTRODUCTION

The question of the origins of Greek mathematics has always been considered to be an extremely difficult one. The amazing achievements of ancient Greek mathematicians have been passed down to us almost exclusively via the works of authors such as Autolycus, Archimedes, Euclid, Apollonius, Theodosius, Menelaus, Diophantus, and Pappus, the earliest of whom wrote in the second half of the 4th century BC. Additionally Aristotle's philosophy of science as put forward in his *Analytica posteriora*, which was written slightly earlier, shows that at his time mathematical proofs were performed according to methods already close to those applied by Euclid. This is strong evidence that there had been a considerable period before the middle of the 4th century BC during which mathematics had been developed by the Greeks, from their very beginnings, to the state of perfection to which the classical works mentioned testify. But, as has been a continual source of regret ever since, there is an almost total lack of sources concerning the early phases of this process. Moreover Theophrastus of Ephesus and Eudemus of Rhodes seem not to have fared much better when they wrote histories of mathematics in the second half of the 4th century BC. Both have been lost (Waschkies, 1998, p. 368), but the work of Eudemus has left some traces in later works. Proclus refers to it repeatedly, and therefore his so called *Survey*, which outlines the history of Greek mathematics from its beginnings up to Euclid (Proclus, 1873, pp. 64-68), is often supposed to be from Eudemus, although Proclus does not refer to him by name there. This *Survey* served as a guide when the exploration of the history of Greek mathematics took a fresh start in 1870 with Carl Anton Bretschneider's *Geometrie und die Geometer vor Eukleides*. As it has been customary it seems appropriate to start by quoting some lines from the *Survey* which refer to the very beginnings of Greek mathematics:

> We say, as have most writers of history, that geometry was first discovered among the Egyptians and originated in the remeasuring of their lands. This was necessary for them because the Nile overflows and obliterates the boundary lines between their properties. It is not surprising that the discovery of this and the other sciences had its origin in the necessity, since every thing in the world of generation proceeds from imperfection to perfection. Thus they would naturally pass from sense – perception to calculation and from calculation to reason. Just as among the Phoenicians the necessities of trade and exchange gave the impetus to the accurate study of number, so also among the Egyptians the invention of geometry came about from the cause mentioned.
>
> Thales, who had traveled to Egypt, was the first to introduce this science to Greece. He made many discoveries himself and taught the principles for many others to his successors, attacking some problems in a general way and others more empirically (Proclus, 1970, pp. 51-52).

This account, although rather vague, looks like a valuable source at first glance; however, there is a rule which is to be obeyed by historians in general (and therefore historians of mathematics in particular) and which advises caution. An iron law which has often been ignored in the history of early Greek philosophy, sciences and mathematics

J. Christianidis (ed.), Classics in the History of Greek Mathematics, 3–18.

demands that all sources considered in the course of an historical investigation should be read initially in the chronological order of the authors whose works contain these sources and not in the order of the authors, scientists or mathematicians cited, reported on, or even only mentioned therein. If this rule is obeyed, reports by late authors may appear in quite a new light. Historical digressions within Proclus' *Commentary on the first book of Euclid's Elements* tell us primarily what Proclus himself thought about the way in which mathematics had developed in the 750 and more years up to his own times, and this obliges every historian of science to compare his notes with earlier or independent sources provided there are some.

Regarding Proclus' claim that geometry was an invention of the Egyptians, earlier sources are indeed known. The oldest of these is to be found in the *Histories* of Herodotus. He refers first to the obliteration of boundary lines between fields in the Nile valley by the annual flood, then to the need of remeasuring the land thereafter, and finally remarks: "This, in my opinion (δοκέει δέ μοι) was the origin of geometry, which then passed into Greek" (II.109). It is evident that Herodotus himself knew nothing about the discovery of geometry in Egypt. His ideas about its origins were mere guesses. Nevertheless his claim that geometry was a child of ancient Egypt was repeated by Plato and Aristotle, who, quite as speculatively as Herodotus, put its birth into a new context. In Plato's *Phaedrus* (274*C*-275*D*) the Egyptian god Theut is said to have invented not only geometry, but also number and reckoning, astronomy and letters, while Aristotle in his *Metaphysics* starts by outlining a process of general intellectual development which led to the invention of sciences. According to his social theory of cultural development it was only after useful arts and arts aiming at giving pleasure had been established that arts, which aim neither at the necessities of life nor at pleasure, were emerging. These demand leisure; and according to Aristotle mathematics were founded by Egyptian priests during their spare time (*Metaphysics* I.1, 981*b*10-25). The three passages just mentioned were combined and slightly altered to fit better in a new context by Hero (*Metrica*, I, p. 2, ed. H. Schoene) and Iamblichus (*De vita Pythagorica liber*, § 12). Evidently another such example is the first item of Proclus' *Survey* cited above. The trustworthiness of this source cannot be enhanced by the (doubtful) assertion that Proclus relies on Eudemus. Even if this were the case, it would only show that Eudemus himself stays within the tradition starting with Herodotus, Plato, and Aristotle who had merely speculated about the beginnings of geometry.

The views held by Herodotus, Aristotle and Proclus concerning the origins of Greek geometry deserve to be considered from another point of view. Especially since the appearance of Thomas S. Kuhn's monograph *The structure of scientific revolutions* in 1962 historians of science are in the habit of asking whether the emergence of a scientific discipline sprang from roots in pre-scientific knowledge, and had therefore been stimulated by internal causes, or had been due to (changes within) the social or cultural environment in question, and thus had been triggered by external causes. As far as the origins of Greek mathematics are concerned Herodotus, Aristotle and Proclus clearly favor an external point of view, a variant of which has often also been fostered by historians of mathematics. Once more a quotation from the *Survey* of Proclus is appropriate:

> [Oenopides of Chios and Anaxagoras of Clazomenae] are mentioned by Plato in the *Erastae* (132A-C) as having got a reputation in mathematics. Following them Hippocrates of Chios, who invented the method of squaring lunules, and Theodorus

> of Cyrene became eminent in geometry. For Hippocrates wrote a book on elements, the first of whom we have any record who did so.
>
> Plato, who appeared after them, greatly advanced mathematics in general and geometry in particular because of his zeal for these studies. It is well known that his writings are thickly sprinkled with mathematical terms and that he everywhere tries to arouse admiration for mathematics among students of philosophy (Proclus, 1970, pp. 53-54).

According to this statement it was the philosophical zeal to which Plato's writings testify (and not his competence as a mathematician) which caused remarkable advances in mathematics; in the wake of Proclus the idea of philosophy as an external cause of Greek mathematics has had its defenders up to the present day. One of them is Árpád Szabó, who puts forward these ideas in a paper which seeks to answer the question "Wie ist die Mathematik zu einer deduktiven Wissenschaft geworden?", while an approach which advocates that both the origins of Greek mathematics and its further development were triggered by difficulties arising within mathematical contexts is represented in this anthology by Wilbur Richard Knorr.

Jürgen Mittelstraß has not been obliged to choose between these positions. He argues that Greek mathematics owed its origins to Thales who, being reckoned to have been both the first Greek philosopher and the first Greek mathematician, was motivated by both internal and external causes when shaping the foundations of Greek mathematics, but, as I shall explain in some length, his point of view differs more fundamentally in another point from that held by Szabó and Knorr. When Paul Tannery published his *Pour l'histoire de la science hellène* in 1887, he pointed out that before Thales not only had there been neither sciences nor mathematics, but further nobody had any idea of what sciences and mathematics were or ought to be (p. 1). In the same sense Jürgen Mittelstraß wrote in 1966 that the question of the origins of Greek mathematics cannot be answered by a reference according to which some pre-Greek scribe or some early Greek knew this or that matter of fact which may be fitted into some mathematical theory of today (or some theory of the times from Aristotle to Pappus). The possibility of doing mathematics had to be discovered like the possibility of using fire or of breeding animals, but as Mittelstraß was forced to admit the question of the nature of the difference between Greek mathematics and the pre-scientific knowledge of Egyptian or Babylonian scribes had not yet been discussed as thoroughly as it should have been. This was of course a serious shortcoming, for without an explicit statement of this difference the question of the origins of Greek mathematics cannot be unambiguously discussed.

To begin with, the most typical feature of Greek mathematics in its developed form is the presentation of areas of mathematical knowledge as theories. These are sets of general statements concerning mathematical objects. A (small) subset of them, called principles, are held to be true while the truth of all other statements, whether theorems or general statements concerning the solution of a problem, is deduced logically from principles and statements belonging to the theory in question which were already proved. A. Szabó and W.R. Knorr both consider that the first appearance of a mathematical theory marks the very beginnings of Greek mathematics, while Mittelstraß argues that the prerequisites necessary for any mathematical theory but not yet known to pre-Greek scribes had to be acquired by the Greeks well before such theories could have been composed. It is the emergence of such prerequisites Mittelstraß is looking for when he tries to date the origins of Greek mathematics.

To assess this point of view, it will be helpful to look at an example which shows the way in which mathematical knowledge was encoded by pre-Greek scribes. According to Szabó (who draws on O. Becker, K. v. Fritz, W. Burkert and B.L. van der Waerden) "the most substantial difference between Greek and Oriental sciences is that the former is an ingenious system of knowledge built up according to the method of logical deduction, whereas the latter is nothing more than a collection of instructions and rules of thumb, often accompanied by *examples*, having to do with how some particular mathematical tasks are to be carried out" (Szabó, 1978, p. 186); citing O. Becker (1957, p. 157) he continues: "It is not even certain that the Babylonians knew how to formulate general theorems. [...] *Proofs* do not appear in any of the ancient Oriental texts known to us" (Szabó, loc. cit.). This characterization of pre-Greek mathematical knowledge may seem to settle all questions regarding its difference from Greek mathematics, but even a short glance at a so called mathematical problem text from the times of ancient Babylon shows that the gap between pre-Greek mathematical knowledge and Greek mathematics is not as wide Szabó supposes.

Problem 20 of the Old Babylonian mathematical cuneiform text *BM 85 194* (Neugebauer, 1973, I, pp. 142-193; King, 1962, plates 8-13) reads in the translation of Jens Høyrup (2002, pp. 272-275):

1, the circle.	$p = 60\ [= \pi \cdot \mathrm{d}]$*
2 NINDAN I have descended,	$a = 2$
the crossbeam [is] what?	$c = ?$
You, {...} 2 make hold, 4 you see.	$2a = 4$
	$[d = p/\pi \approx p/3 = 20]$
4 from 20, the crossbeam, tear out, 16 you see.	$d - 2 \cdot a = s = 16$
20, the crossbeam, make hold, 6,40 you see.	$d^2 = 400$
16 make hold, 4,16 you see.	$s^2 = 256$
4,16 from 6,40 tear out, 2,24 you see.	$c^2 = d^2 - s^2 = 144$
[By] 2,24, what is equalside?	
12 is equalside, the crossbeam.	$c = 12$
Thus the procedure.	

Before starting to analyze this text, I refer to a problem which never seems to have been given due attention by historians of mathematics. Any text is primarily the result of somebody's effort to encode knowledge he has gained and therefore the Old Babylonian clay tablet *BM 85 194* is the result of an ancient Babylonian scribe's effort to encode some details of his mathematical knowledge – just as a German pupil's mathematical exercise or a modern algebra textbook are examples of efforts to encode mathematical knowledge gained by a pupil or a mathematician of our times respectively. In these cases as in all comparable situations a reader – be it the writer himself or a person living some five thousand years after the scribe of clay tablet *BM 85 194* – is confronted with the task of decoding the message in question to regain or to share insight into the knowledge the author had at his disposal when encoding it. This knowledge is quite another thing than the text showing the result of the effort to encode it. The style and type of coding used by an author who tries to inform someone depends heavily on his cultural milieu and therefore texts containing coded mathematical knowledge may seem to transmit something meager because their author did not have at his disposal techniques of coding everybody today is expected to have.

* The equations to the right of the text refer to my interpretation of the text.

Precisely this has happened in the efforts of historians of Greek mathematics to asses the niveau reached by ancient Babylonian and ancient Egyptian scribes.

The most important line of *BM 85 194, Problem 20* and also of many other old Babylonian problem texts is the last one reading: "Thus the procedure" (which, astonishingly, is suppressed in B.L. van der Waerden's citations and reports of comparable texts commented on in his most influential *Science Awakening*, pp. 62-81). As has already been emphasized by Otto Neugebauer and Otto Struve who were the first to interpret *BM 85 194, Problem 20* (Neugebauer and Struve, 1931, pp. 81-92), Old Babylonian mathematical texts such as these unambiguously show that the scribes of those days were not at all interested in the specific results of the individual problems solved on their clay tablets, for otherwise the words "Thus the procedure" at the end of the solution would make no sense. On the contrary, their representation of the steps by which a specific solution of a single problem might be found had the function of teaching by example how to solve problems of some general type, though the general rule to be applied was never explicitly stated. Old Babylonian problem texts were therefore aide-mémoires used by scribes teaching pupils verbally to solve a class of problems (and in some cases filling out their explanations with graphical representations). On the other hand *BM 85 194, Problem 20* like comparable examples from ancient Babylonian and ancient Egyptian mathematical texts, shows that the scribes of those days had not yet discovered the linguistic means by which modern-day mathematicians express the treatment of something general. For example, modern-day mathematicians would state *BM 85 194, Problem 20* something like this:

> Given a circle let a chord be drawn within it. The length of the perimeter and the length of the arrow belonging to the chord being known, find the length of the chord.

The deciphering of *BM 85 194, Problem 20* by Otto Neugebauer and Otto Struve was made (a little) easier by the fact that its text is enriched by a tiny diagram outlining a circle divided by a chord. Next to the circle and between the perimeter and the chord one (resp. two), wedges are impressed, being sexagesimal figures representing the numbers $1[\cdot 60 = 60]$ and 2 (fig. 1). Given this numbered diagram and the words accompanying it, it seems that the Babylonian teacher started his instruction by drawing the perimeter $[p]$ of a circle on the floor (or another surface suitable for this purpose) with a chord $[c]$ inscribed into it, explaining that his "handy" figure represents a circle with a perimeter of sixty NINDAN, just like the tiny circle on clay tablet *BM 85 194*, which is no bigger than a pea. There can be no doubt that he made some such explanation (if he did not think it to be a matter of course), for it seems

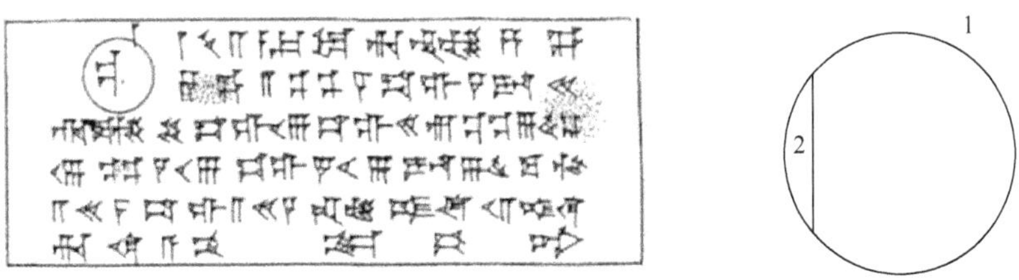

Fig. 1

unlikely that he tried to draw a circle with a perimeter of 60 NINDAN, a NINDAN being equivalent to ca. 6 m. There is a passage form Aristotle's *Posterior Analytics* alluding to a similar situation:

> Nor does the geometer make false hypotheses, as he has been charged with doing, when he says the line he draws is a foot long, or straight, when it is not. He infers nothing from this; his conclusions are only made obvious by this. (*Analytica posteriora* I.10, 76*b*39 – 77*a*3, transl. by W.D. Ross).

The reader of Aristotle was aware that every stroke on a papyrus may be thought of as representing a straight line one foot long just as the Babylonian scribe and his pupils were aware that every shape drawn on a surface whatsoever and being somewhat circular represents the circle with a perimeter of sixty NINDAN mentioned in *BM 85 194, Problem 20*. But what is more, they understood the calculation of the length of the chord belonging to the arrow of length 2 NINDAN in a circle with a perimeter of 60 NINDAN as the encoding of a rule, according to which it is possible to calculate the length of the chord in any circle, provided the length of its perimeter and that of the arrow belonging to the chord are given. Consequently, in *BM 85 194, Problem 20* the numbers 60, 2 and 12 are used to encode "length of the perimeter", "length of the arrow", and "length of the chord", while today letters like p, a, and c in fig. 2 and the equations explaining *BM 85 194, Problem 20* are used to encode them.

In the next stage the scribe of *BM 85 194, Problem 20* drew both perpendiculars from the middle of the chord, extending them until they met the perimeter of the circle (fig. 2). One of them is the arrow [a], which together with the other makes a diameter [d] of the circle. Now the scribe proceeded to draw perpendiculars from the ends of chord [c] pointing into the interior of the circle. These were extended too until they met the perimeter. Then he joined the ends of the perpendiculars just mentioned, whereby a rectangle with the sides [s] and [c] was inscribed into the circle, and by a last preparatory step this rectangle was bisected by one of its diagonals [of length d].

A glance at figure 2 together with a look at the text of *BM 85 194, Problem 20* shows how the scribe may have taught the solution of the problem in question. Led by considerations of symmetry he argued that one side [s] of the rectangle may be determined by calculating the difference [$s = d - 2a$] between the length of the diameter [d] and twice the length of the arrow [a]. Before this could be done he (and his pupils) had performed a deduction which is "hidden" in the text. There the scribe silently passes from 60, the given length [p] of the perimeter to the length 20 of the diameter [d] by applying the relation $D = P/\pi$ between the length P of the perimeter and the

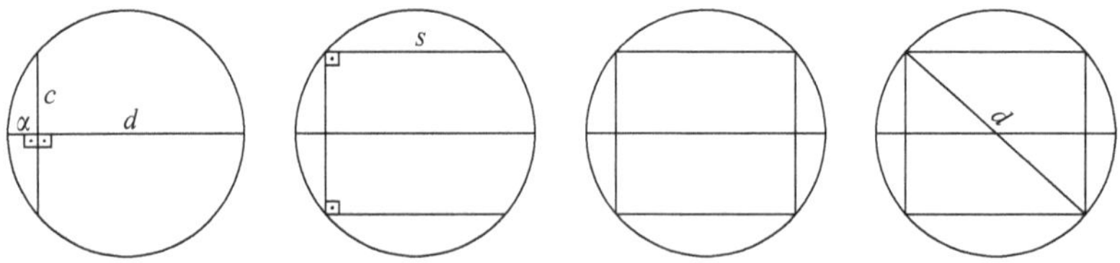

Fig. 2

length D of the diameter of any circle to the special circle given with perimeter 60 and taking π to be 3. This crude approximation of π is often referred to as evidence that Babylonian scribes applied empirically discovered rules of thumb which matched practical purposes quite well, but few claims pertaining to ancient Babylonian mathematical knowledge are more misleading than this one. The point is not that the constant π (which, being irrational, as shown by J.L. Lambert in 1770 and indeed transcendental as discovered by F. Lindemann in 1882, cannot in fact be calculated exactly) was approximated in those times with more or less accuracy. Teaching by example how to determine the length of a chord in a circle, given its perimeter and the length of the arrow pertaining to the chord would not have been altered in the slightest if the scribe had used an approximation closer to π than 3, and I therefore only mention in passing that the Babylonian scribes knew the more refined approximation $\pi \approx 3\frac{1}{8}$ (Bruins and Rutten, 1961, p. 26 & p. 33; Neugebauer and Sachs, 1945, p. 59). Much more revealing is their awareness of a constant of proportionality enabling them to determine the diameter of any circle whatsoever provided its perimeter is known. What is more, they even knew how to derive the constant to be applied in the calculation of the area of a circle given its diameter [using a relation equivalent to our formula $F = \pi \cdot (D/2)^2$] from the constant they used when calculating the length of its diameter from its perimeter (see Waschkies, 1993, pp. 47-61).

Having calculated the length of the side [s using the relation $s = d - 2 \cdot a$] the scribe goes on to determine the length [c] of the chord by first calculating its square [via $c^2 = d^2 - s^2$]. "Behind" this step of the solution stands the general premise that the square on the diameter of a rectangle is equal to the sum of the squares on its sides, which the ancient Babylonians seem to have justified by a reference to fig. 3 (Waschkies, 1993, p. 46). This premise is equivalent, though not identical, to the so called theorem of Pythagoras, for it is to be kept in mind that pre-Greek mathematicians did not have the notion of an angle at their disposal (Thureau-Dangin, 1938, pp. xvii–xviii; Bruins, 1955, pp. 44-49) nor did they ever use a premise corresponding to statements about triangles in general, and in a last step the length of the chord [c] is calculated by extraction of the root [of c^2].

I may seem to be trying to reduce the difference between pre-Greek mathematical knowledge and Greek mathematics too much or even to deny that there was one. On the contrary I have commented on *BM 85 194, Problem 20* to illustrate the tremendous gap which separates pre-Greek mathematical knowledge and Greek mathematics. As further evidence to make this clear I shall compare the use made of the so-called theorem of

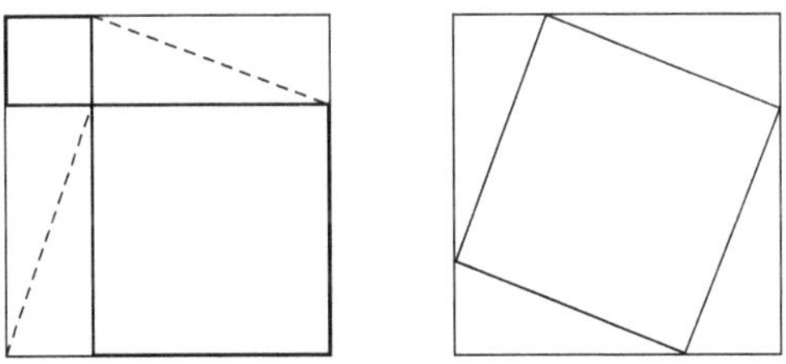

Fig. 3

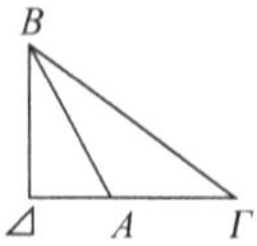

Fig. 4

Pythagoras in the solution of *BM 85 194, Problem 20* with the use made of it by Euclid when proving Prop. II.12 in the *Elements* (transl. by T. L. Heath):

> In obtuse-angled triangles the square on the side subtending the obtuse angel is greater than the squares on the sides containing the obtuse angel by twice the rectangle contained by one of the sides about the obtuse angle, namely that on which the perpendicular falls, and the straight line cut off outside by the perpendicular towards the obtuse angle.

This statement is followed by the so called exposition [of the premises] and the specification [of the goal] which are stated with reference to a lettered diagram (a notion coined and discussed by Reviel Netz (1999, pp. 12-88), in detail) and therefore with reference to a mere representative of the mathematical objects mentioned (fig. 4) in the statement of Prop. II.12, which are classes of equivalence and cannot be looked at at all:

> Let *ABΓ* be an obtuse-angled triangle having the angle *BAΓ* obtuse, and let *BΔ* be drawn form the point *B* perpendicular to *ΓA* produced. I say, that the square on *BΓ* is greater than the squares on *BA*, *AΓ* by twice the rectangle contained by *ΓA*, *AΔ*.

The style of this sentence corresponds fairly well to that of the first three lines of *BM 85 194, Problem 20* and the "numbered" diagram accompanying them, but the two sources discussed here have something else in common. According to Proclus (1873, p. 203) the exposition (ἔκθεσις) and specification (διορισμός) of every problem or theorem furnished with all its parts should be followed by the auxiliary construction (κατασκευή). For *BM 85 194, Problem 20* this would be at least the completed configuration represented by fig. 2, which is missing from the clay tablet. The comparable figure presupposed by Euclid when proving Prop. II.12 is also missing. This is not at all exceptional, for Euclid often does without a κατασκευή in his *Elements*. Nevertheless he presupposes that his reader has its details in mind [or has himself worked out a suitable auxiliary construction]. This is the case when, continuing his proof of Prop. II.12, he simply states:

> The square on *ΓB* is equal to the squares on *ΓΔ*, *ΔB*, for the angle at *Δ* is right.

As mentioned above *BM 85 194, Problem 20* also does not show any of the auxiliary constructions presupposed by its text. The intelligent ancient Babylonian reader knew how to complete the rudimentary diagram accompanying the texts, just like Euclid's reader, but on another point the difference between them is fundamental. The Babylonian scribe calculates the length of the chord [*c*] by applying without any explanation mathematical knowledge equivalent to that stated generally in Eulid's Theorem I.47. The text of the proof of Prop. II.12 is also silent on this point, but the ensemble of the text of the *Elements* shows that Euclid knew how to prove statements like "the square on *ΓB* is equal to the squares on *ΓΔ*, *ΔB*" by using explicitly stated general premises, although these deductions are passed over in silence in the written proof of Prop. II.12, sure they were "well known" to the reader. If he had been asked to make the proof more explicit Euclid could have proceeded in the following way: From the premise that $B\Delta$ is perpendicular to $A\Gamma$ and Definition I.10, which states that "When a straight line set up on a straight line makes the adjacent angles equal to one another, each of the right angles is right, and the straight line standing on the other is called a perpendicular to that on which it stands", it follows that the angle at Δ is right. This enables Euclid to deduce that the square on the hypotenuse ΓB of the right angled triangle is equal to the squares on the sides $\Gamma\Delta$ and ΔB containing the right angle, using the so-called theorem of Pythagoras I.47 as a generally stated premise which reads as follows (transl. by T. L. Heath): "In right-angled triangles the square on the side subtending the right angle is equal to the squares on the sides containing the right angle".

Babylonian scribes knew how to use mathematical knowledge corresponding to general premises such as the above-mentioned variant of Euclid's theorem I.47 to justify their methods for solving sets of problems by applying them to special configurations. However, they never stated these premises in the form of general propositions nor did they formulate the problems they solved in all generality by applying generally valid methods, as the Greeks later did. It is always hazardous to propose an explanation of why some person or some culture did not cross this or that border, even though the step seems obvious; however a supposition presented by Jens Høyrup deserves mention. Høyrup presented ample evidence showing that the social context of pre-Greek mathematical knowledge was a school in which teachers (themselves being scribes) were teaching future scribes. These were trained to solve problems they would meet after becoming administrators. Therefore, the aim of their training was *to find the right number* specifying weight, measure or prices (Høyrup, 2002, p. 8). Nevertheless, their mathematical exercises included lot of *art pour l'art* (as mathematics in school has done ever since). In this way teachers and pupils could display their mathematical abilities by solving difficult problems. This was done by putting into practice sophisticated procedures using methods generally applicable; however, as a result of the paradigm of their social environment their aims never evolved beyond that of calculating the correct number. Having achieved this end, they never indicated or even stated explicitly the general premises they used. The objective of doing so had first to be discovered, and only in the wake of this discovery could these statements themselves become a set of well-defined elements to be brought into an order induced by logical-deductive dependences. Mittelstraß argues that the crucial step leading out of the impasses of mathematical knowledge of pre-Greek cultures and to the road to axiomatic-deductive mathematics was precisely the discovery of the possibility of expressing generally valid mathematical knowledge verbally by formulating general statements about mathematical objects and relations pertaining to them. Therefore, his paper represents an attempt to mark the origins of Greek mathematics, which is quite different from those by Szabó and Knorr.

Dating the origins of Greek mathematics thus evidently depends on the notion "*beginnings of Greek mathematics*" held by the individual historian of science, but even after this decision there is no easy way to determine the period in which Greek mathematics arose. Jürgen Mittelstraß takes the 'Thales-fragments' to be reliable sources and argues that according to his concept of the origins of Greek mathematics, we owe them to Thales, but this needs further discussion. As mentioned above the notion of an angle has not been found in any pre-Greek source. In Greece it seems also have been introduced rather late, for there is no evidence for it antedating Herodotus' *Histories* (I.51) written about 425 BC, and even there it is used only to describe the shape of a corner in a temple. Any fragment which attributes a mathematical insight depending on the notion of an angle to a person much earlier than Herodotus should therefore be doubted, and indeed such fragments have been used as cornerstones by historians trying to reconstruct the contributions of Thales to Greek mathematics. One of these is from the *Vitae philosophorum* of Diogenes Laertios (mentioning as his source Pamphile [I.24-25], a contemporary of Nero) while all the others have come down to us via Proclus' *Commentary on the first Book of Euclid's Elements* (who in turn sometimes refers to Eudemus of Rhodes as his source). As an example I cite the following remark pertaining to Euclid's Prop. I.15: "This theorem, then, proves that, when two straight lines cut one another, the vertical angles (*αἱ κατὰ κορυφὴν γωνίαι*) are equal. It was first discovered by Thales, Eudemus says, but was thought worthy of a scientific demonstration only with the author of the *Elements*" (Proclus, 1970, p. 233).

To be aware of the relation of equality stated in Euclid's Proposition I.15 clearly requires an insight into the notion of an angle, as do two other mathematical discoveries ascribed to Thales who lived about 585 BC. I therefore consider it to be unlikely that Thales had mathematical knowledge of the kind ascribed to him at his disposal, and I may add that D.R. Dicks (1959) had argued convincingly as early as 1959 (using evidences other than the late appearance of the notion of an angle) that all fragments ascribing to Thales some specific knowledge in the realm of mathematics are post-Aristotelian fabrications. Thus, even if one agrees with Mittelstraß that Greek mathematics started with the discovery that mathematical knowledge can be expressed by verbally stating general propositions one still has to look for early sources to get a *terminus ad quem* for their appearance in Greece. This has been done recently by Markus Asper (2001). Greek legal texts inscribed on marble slabs have come down to us from about 650 BC onward. The laws in question are general statements. The structure and the terminology of them are highly standardized. Further unambiguousness of these texts is achieved by avoiding *synomyma* but rarely by introducing definitions of terms used in them. These do not appear before about 450 BC at all. The oldest inscriptions of this kind antedate clearly the times of Thales, and therefore the discovery of linguistic means to state general propositions has probably been achieved by the Greek even earlier than suggested by Mittelstraß.

The problems implicit in Szabó's ideas on the origins of Greek mathematics are by no means easier to deal with than those just mentioned. Early in the 20th century scholars analyzing the works of Aristotle realized that it is possible to separate them into sections which, like the layers of an archeological site, can be brought into a chronological order (and thus indicate an intellectual development of Aristotle). In 1936 Oskar Becker showed that the use of this method to establish a relative chronology could also be applied *mutatis mutandis* to Euclid's *Elements*. He argued that Propositions IX.21-36 together with some arithmetical definitions from Book VII (including definitions VII.6-7 fixing the notions of even and odd numbers) and an appendix to Book IX stating that it is impossible to generate a square number by doubling a square number constitute the remnants of an arithmetical theory originally established independently of Euclid's arithmetical theory of proportions as exposed in Books VII-XI. Let me add that this old core of the theory contains also Propositions VII.31-32 which state that any number either is prime or divided by some prime number, Theorem IX.20 which states that there are more prime numbers than any assigned multitude of prime numbers and the so-called "Sieve of Erathostenes" (Waschkies, 1989, pp. 280-301). Becker was also aware of a fragment from a comedy by Epicharmus implying that spectators of about 480 BC knew that addition or subtraction of unity to or from an even number produces an odd number. Becker took this to be evidence that the whole theory of even and odd as reestablished by him was known about 480 BC too, but this is not as certain as one might wish.

The 'Epicharmus-fragment' has come down via the *Vitae philosophorum* (III. 9-11) of Diogenes Laertius who himself refers to a book written by Alkimos probably at the end of the 4th century BC. These are late sources but as far as I know this 'Epicharmus-fragment' is believed to be authentic (although other 'Epicharmus-fragments' are not). This problem having been settled a far more difficult question to answer is whether one may draw as far-reaching a conclusion from Epicharmus' text as Becker did. The crucial lines of it read:

> If someone adds to an odd, or if you wish, to an even number one ψῆφος or if he takes one of those [you just see laid out] away, do you think the [number represented there] is still the same?
> Not at all!
> And further, if a [small] length is added to a tape measure one cubit long or if there is some [small length] cut off from it, will it be still the same measure?
> Of course not!

While the second example doubtless relates to everyday experience the first one is a little more demanding. As Becker noticed Epicharmus did not simply mention that the addition or subtraction of a unit alters every number. He was evidently alluding to the fact that the addition or subtraction of a unit to an even number produces an odd number. This is stated in Euclid's Def. VII.7, but one might doubt whether or not Epicharmus was referring to an explicit definition of even and odd numbers as part of a more or less extensive deductive theory founded in definitions, for it seems possible that he was alluding only to everyday experience gained by his audience – for example in conjunction with some game of chance. Ca. 480 BC as *terminus ante quem* for the theory of even and odd is therefore anything but easy to confirm, and this is of importance for Szabó's ideas concerning the beginnings of Greek mathematics.

As mentioned above Árpád Szabó advocates an external point of view when trying to explain the origin and early phases of Greek mathematics. Strictly speaking he states that in their very beginnings Greek mathematics were a consequence of Eleatic philosophy. On the other hand he adopts the thesis of Oskar Becker that the theory of even and odd was a very early (if not the very first) mathematical theory developed by the Greeks, and therefore Szabó was obliged to supply arguments showing it to be Eleatic. When he wrote the paper reprinted in this anthology Szabó still adopted Becker's thesis that the theory of even and odd had been composed by the Pythagoreans. He therefore tried to show that they were able to do so by applying methods of proof developed by Parmenides and his sect, while he tried to show in his later papers (and especially in his monograph *The beginnings of Greek mathematics*), that the ensemble of the theory of even and odd had emerged within the Eleatic school. Szabó tries to reach this goal via two roads. One of his arguments depends crucially on the premise that Euclid's Def. VII.1 reading "An unit is that by virtue of which each of the things that exist is called one (transl. by T.L. Heath)" was part of the theory of even and odd from the very beginning. To assert this Szabó, after having unfolded a long series of considerations, states that it seems to be "fair to say, therefore, that the Euclidean definition of 'unit' is nothing but a concise summary of the Eleatic doctrine of 'Being'" (Szabó, 1978, p. 261), passing a little later to the conclusion, that his investigation of Euclid's "definition of 'unit' has led us straight to Eleatic philosophy" (loc. cit.). Putting aside the question of whether the route from Euclid's Def. VII.1 to Eleatic philosophy is as straightforward as Szabó pretends, the definition itself looks much more Platonic than Eleatic. It is strongly reminiscent of Plato's doctrine of *μέθεξις* and seems to have been coined by a Platonic or even post-Platonic mathematician having in mind *Phaedo*, 101*B-C*. Szabó's conclusion from Euclid's Def. VII.1 that there was an Eleatic origin of the theory of even and odd seems therefore rather doubtful, while the second strategy he proposes to argue for an Eleatic origin of Greek mathematics (already exposed in his paper "Wie ist die Mathematik zu einer deduktiven Wissenschaft geworden?" and repeated in all his later papers discussing the origins of Greek mathematics) seems more promising. It takes as its starting point the assumption that the proof of the impossibility to generate a square number by doubling a square number, which is found today at the end of Euclid's book IX (or some essentially equivalent proof) was part of the old theory of even and odd. Proofs of this type require a *reductio ad absurdum* which has been thought of as a hallmark of Greek mathematics ever since. There is no pre-Greek mathematical text showing traces of the use of this method to justify statements, but neither is there any source telling us anything about its origin. Szabó simply states "that neither anti-empiricism nor the method of indirect proof could have arisen spontaneously in mathematics [for he does] not believe, for example, that mathematicians were prompted solely by their dealing with numbers and geometrical figures to change radically their way of thinking and to adopt an interesting new method of proof, they must have been subjected to some influence from outside mathematics. [.... . His] view is that these two features of Greek mathematics, its rejection of empiricism and its characteristic use of indirect proof, are attributable to the decisive influence of the *Eleatic* school of philosophy (loc. cit., p. 217)", but unfortunately there is no source to support this claim. Nobody will deny that Eleatic philosophy is anti-empirical nor that it relies heavily on indirect proofs, but this observation does not prove the thesis of Szabó that Greek mathematics was guided in its initial development by the Eleats.

Wilbur Richard Knorr is a historian of science convinced that the development of Greek mathematics was induced and guided by internal stimuli from its very beginning. He surmises that the insights gained when Greek mathematicians examined the consequences of their discovery that there are incommensurable pairs of magnitudes caused them to develop mathematics as a deductive discipline starting from principles among which definitions had a key role. What is more Knorr argues that even pre-Greek scribes were accustomed to unconsciously use the method of proof by *reductio ad absurdum*. Most of their problems "are accompanied by a 'proof', or, as we might say, a 'check'. [...] We thus see the germ of indirect reasoning" (Knorr, 1981, pp. 147-148). I might add that the attempt to generate a square number by doubling a square number by applying a method much used by the so-called *ψῆφοι*-arithmeticians (to whom, incidentally, according to Becker, we owe the old theory of even and odd), changes quite unintentionally into a *reductio ad absurdum*. To them to generate a square number by doubling a square number meant to rearrange two squares, each composed

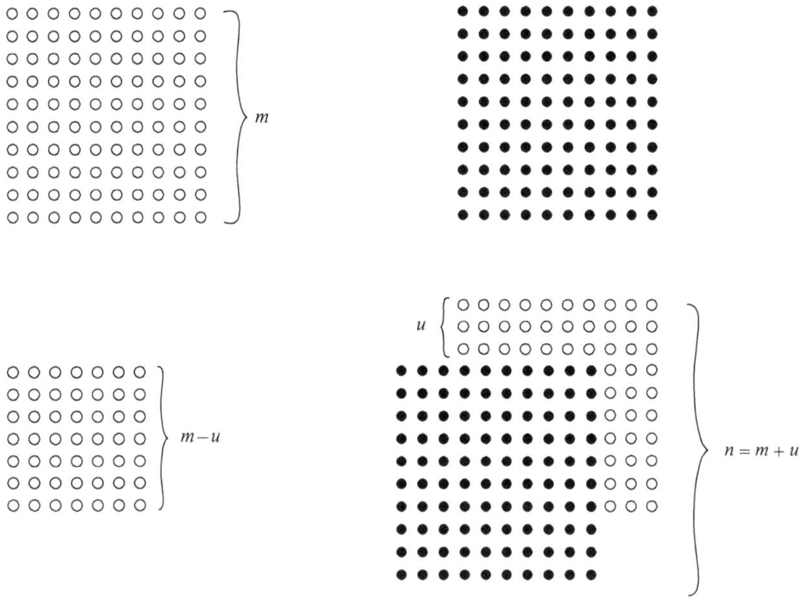

Fig. 5

of m^2 pebbles, into a larger square containing n^2 pebbles (fig. 5). If this is possible then $m < n < 2m$ will hold. Therefore the difference $n - m$ will be a number u $(1 < u < m)$ with $2m^2 = (m + u)^2$. A ψῆφοι-arithmetician would therefore start by laying out two squares containing m^2 pebbles each. Then he would proceed by separating from one of them a so-called gnomon of width u and adding it immediately to the other square of m^2 pebbles. There is now a square containing $(m - u)^2$ pebbles left. According to the assumption that $2m^2 = (m + u)^2$, this square must fit exactly into the two quadratic 'holes' able to receive u^2 pebbles each. Therefore, if there is a square number m^2 which generates a square number n^2 by being doubled, then there is a smaller square number u^2 too which, on being doubled, generates a square number $(m - u)^2$. Iterating one reaches the conclusion that there are infinitely many square numbers smaller than n, which, on being doubled, produce a square number. Therefore it is impossible to construct square numbers in this way (Waschkies, 1989, pp. 272-275). This discovery easily leads to the insight that the side and the diagonal of a single square are a pair of incommensurable magnitudes, and therefore the method of proof by *reductio ad absurdum* might be an invention by early Greek mathematicians (as well as by the Eleatic philosophers) if it had to be invented at all (Mueller, 1969, pp. 297-298). This is of course no proof showing that Szabó is wrong and Knorr is right for there are no sources enabling us to answer this question definitely, but there is something more in Knorr's paper that needs further discussion.

Knorr argues that the first mathematical theories shaped as deductive disciplines starting from principles did not emerge before the times of Theaetetus. This seems rather late if we consider that Hippocates of Chios, who is said to have observed a comet about 430 BC, was according to Proclus (1873, p. 66) the first Greek mathematician to write a book on elements. Proclus' *Commentary* is a late source, and the 'Elements' of Hippocrates are totally lost, but there is a long passage in the *Commentary* of Simplicius *to Aristotle's Physics* that relates how Hippocrates managed to square three types of lunules (Rudio, 1907). Simplicius, drawing on Eudemus, claims to have enriched the latter's text with explanatory remarks and a series of later additions have indeed been spotted by historians of mathematics. The old core, which such emerged, is primarily a source from Eudemus' times, but there is evidence that it is essentially at least pre-Aristotelian. In the passages in question the terms *στιγμή* and *σημεῖον* are missing. These nouns denoting 'the point in general' have not been found in any source antedating Aristotle while ample use is made of them from Aristotle onward (Waschkies, 2000). In pre-Aristotelian texts points are denoted by 'proper names' like *κέντρον*, *πέρας*, etc., and therefore the old core of the 'Hippocrates-fragment' may well be authentic. The logical deductions worked out there are good evidence that Hippocrates wrote something like a first 'edition' of Euclid's Elements presenting the elementary mathematics of his time shaped into theories, and therefore the date for the origins of Greek mathematics as a deductive discipline founded in principles given by Becker, Szabó and Mittelstraß may be nearer to the truth than that given by Knorr.

The reader may wish to know something about my own view concerning the origins of Greek mathematics. First of all I tend to regard its development into a deductive discipline starting from beginnings such as those described by Jürgen Mittelstraß as a process caused intrinsically and driven by mathematically minded people not (too much) influenced by philosophical schools. First sets of theorems expounded in an order induced by stating some of them to be principles and deducing all others logically may have been composed from about 475 BC onward, but this is a guess (although something more than a mere guess) just like any other idea concerning the origins of Greek mathematics, and finally I should mention that I do not know of sources permitting the attribution of special mathematical achievements with some confidence to a Greek mathematician earlier than Hippocrates of Chios.

Gino Loria in Volume I of his *Le scienze esatte nell' antica Grecia* dealing with *I geometri greci precursori di Euclide* in 1893 emphasized that sources supporting his views are rare and often also of questionable value. He therefore asked his reader to enter into a dialogue with him to discuss the ideas put forward in his book and just in the same way the reader is invited to enter into a dialogue with Mittelstraß, Szabó, Knorr and me too.

I thank Miss Catriona Maclean, Université Pierre et Marie Curie (Paris 6), Institut de Mathématiques, for rutting my essay into English.

University of Kiel

REFERENCES

Aristotele: *Prior and posterior analytics*. A revised text with introduction and commentary by W. D. Ross. Reprinted lithographically from corrected sheets of the first edition. Oxford, 1965 [[1]1949].

Aristotele: *Metaphysics*, vol. I-II. A revised text with introduction and commentary by W.D. Ross. Reprinted lithographically from corrected sheets of the first edition. Oxford, 1966 [[1]1924].

Asper, Markus: "Stoicheia und Gesetze. Spekulationen zur Entstehung mathematischer Textformen in Griechenland". Althoff, Jochen, Bernhard Herzhoff und Georg Wöhrle (Herausgeber): *Antike Naturwissenschaft und ihre Rezeption*. Bd. XI. Trier, 2001, pp. 73-106.

Becker, Oskar: "Die Lehre vom Geraden und Ungeraden im Neunten Buch der Euklidischen Elemente. (Versuch einer Wiederherstellung in der ursprünglichen Gestalt)", *Quellen und Studien zur Geschichte der Mathematik, Astronomie und Physik*. Abt. B: Studien, Bd. 3 (1936), pp. 533-553.

Becker, Oskar: "Frühgriechische Mathematik und Musiklehre", *Archiv für Musikwissenschaft*, Bd. 14 (1957), pp. 156-164.

Bretschneider, Carl Anton: *Geometrie und die Geometer vor Euklides. Ein historischer Versuch*. Leipzig, 1870.

Bruins, Evert Marie: "On the system of Babylonian geometry", *Sumer*, vol. 11 (1955), pp. 44-49.

Bruins, Evert Marie and Rutten, Marguerite: *Textes mathématiques de Suse* [Mémoires de la Mission Archéologique en Iran, t. 34]. Paris, 1961.

Burkert, Walter: *Weisheit und Wissenschaft. Studien zu Pythagoras, Philolaos und Platon*. Nürnberg, 1962. [English edition (revised): *Lore and science in ancient Pythagoreanism*. Translated by E.L. Minar jr. Cambridge (Mass.), 1972.]

Dicks, D.R.: "Thales", *The Classical Quarterly*, vol. 53 (1959), pp. 294-309.

Diogenes Laertios: *Vitae philosophorum*, t. I-II. Recognovit brevique adnotatione critica instruxit H.S. Long. Oxford, 1966 (first edition 1964).

Euclid: *The thirteen books of Euclid's Elements*, vol. I-III. Translated with introduction and commentary by Sir Thomas Little Heath. Cambridge, 1926 [first edition 1908].

von Fritz, Kurt: "Die APXAI in der griechischen Mathematik", *Archiv für Begriffsgeschichte*, Bd. 1 (1955), pp. 13-103.

Hero Alexandrinus: *Opera quae supersunt omnia*, vol. III: *Rationes dimetiendi et commentatio dioptrica – Vermessungslehre und Dioptra*. Griechisch und deutsch herausgegeben von Hermann Schoene. Leipzig, 1903 (reprint edition 1976).

Herodotus: *Historiarum libri IX*. Edidit Heinrich Rudolph Dietsch. Editio altera curavit H. Kallenberg. Vol. I-II. Leipzig, 1899-1901 [Editio stereotypa, [2]1885]

Høyrup, Jens: *Lengths – widths – surfaces. A portrait of Old Babylonian algebra and its kin*. New York/Berlin/Heidelberg, 2002.

Iamblichus: *De vita Pythagorica liber*. Edidit L. Deubner. Stuttgart, 1975 [first edition, Leipzig 1937].

King, Leonard William: *Cuneiform texts from Babylonian tablets in the British Museum*, vol. IX. London, 1962 [first edition 1900].

Knorr, Wilbur Richard: "On the early history of axiomatics. The interaction of mathematics and philosophy in Greek antiquity". In Hintikka, Jaakko, David Gruender and Evandro Agazzi (Eds.): *Proceedings of the 1978 Pisa Conference on the History and Philosophy of Science*, vol. I: *Theory change, ancient axiomatics, and Galileo's methodology*. Dordrecht/Boston/London, 1981, pp. 145-186.

Kuhn, Thomas S.: *The structure of scientific revolutions*. Chicago, 1962 [[2]1970 (with a postscript)].

Lambert, Johann Heinrich: "Vorläufige Kenntnisse für die, so die Quadratur und Rectification des Circuls suchen". In Lambert, J.H.: *Beyträge zum Gebrauche der Mathematik und deren Anwendung*. Teil II.1, 5. Abhandlung, pp. 140-169. Berlin, 1770.

Lindemann, Ferdinand: "Über die Zahl π", *Mathematische Annalen*, Bd. 20 (1882), pp. 213-225.

Loria, Gino: *Le scienze esatte nell' antica Grecia*, vol. I-V. Modena, 1893-1902.

Mittelstraß, Jürgen: "Die Entdeckung der Möglichkeit von Wissenschaft", *Archive for History of Exact Sciences*, vol. 2 (1962-66), pp. 410-435.

Mueller, Ian: "Euclid's 'Elements' and the axiomatic method", *British Journal for the Philosophy of Science*, vol. 20 (1969), pp. 289-309.

Netz, Reviel: *The shaping of deduction in Greek mathematics. A study in cognitive history*. Cambridge/New York/Melbourne, 1999.

Neugebauer, Otto and W. [Vasilij Vasil'evič] Struve: "Über die Geometrie des Kreises in Babylonien", *Quellen und Studien zur Geschichte der Mathematik, Astronomie und Physik*. Abt. B: Studien, Bd. 1 (1931), pp. 81-92.

Neugebauer, Otto: *Mathematische Keilschrift – Texte*, Bd. I-III. Berlin/New York/Heidelberg, 1973 (first edition 1935-1937).

Neugebauer, Otto and Sachs, Abraham: *Mathematical cuneiform texts*. With a chapter "The Akkadian dialects of the Old Babylonian mathematical texts" by A. Goetze. New Haven (Connecticut), 1945.

Plato: *Opera*. Recognovit brevique adnotatione critica instruxit Ioannes Burnet. Vol. I: *Eutyphro, Apologia Socratis, Crito, Phaedo, Cratylus, Theaetetus, Sophista, Politicus*. Vol. II: *Parmenides, Philebus, Symposium, Phaedrus, Alcibiades I, Alcibiades II, Hipparchus, Amatores*. Oxford, 1899-1900.

Proclus Diadochus: *In primum Euclidis elementorum librum commentarii*. Ex recognitione Godofredi Friedlein. Leipzig, 1873 [reproduction Hildesheim, 1967].

Proclus Diadochus: *A commentary on the first book of Euclid's Elements*. Translated with introduction and notes by Glenn R. Morrow. Princeton (New Jersey), 1970.

Rudio, Ferdinand: *Der Bericht des Simplicius über die Quadraturen des Antiphon und des Hippokrates (griechisch und deutsch)*. Leipzig, 1907.

Szabó, Árpád: "Wie ist die Mathematik zu einer deduktiven Wissenschaft geworden?", *Acta Antiqua Academiae Scientiarum Hungaricae*, t. IV (1956), pp. 109-151.

Szabó, Árpád: *The beginnings of Greek mathematics*. English translation by A.M. Ungar. Dordrecht and Boston, 1978 [first German edition 1969].

Tannery, Paul: *La géométrie grecque, comment son histoire nous est parvenue et ce que nous en savons*. I. Partie: *Histoire générale del la géométrie élémentaire*. Paris, 1887 [^{2}Paris (under the title *Pour l'histoire de la science hellène. De Thalès à Empédocle*), 1930].

Thureau-Dangin, François: *Textes mathématiques babyloniens*. Transcrits et traduits par François Thureau-Dangin. Leiden, 1938.

van der Waerden, Bartel Leendert: *Science awakening*. English translation by Arnold Dresden. Groningen, 1954 [first Dutch edition 1950].

Waschkies, Hans-Joachim: *Anfänge der Arithmetik im Alten Orient und bei den Griechen*. Amsterdam, 1989.

Waschkies, Hans-Joachim: "Mündliche, graphische und schriftliche Vermittlung von geometrischem Wissen im Alten Orient und bei den Griechen". In Kullmann, Wolfgang and Jochen Althoff (Herausgeber): *Vemittlung und Tradierung von Wissen in der griechischen Kultur*. Tübingen, 1993, pp. 39-70.

Waschkies, Hans-Joachim: "Mathematische Schriftsteller". In *Grundriss der Geschichte der Philosophie*. Begründet von Friedrich Überweg. Völlig neubearbeitete Ausgabe. Redaktion Wolfgang Rother. *Die Philosophie der Antike*. Herausgegeben von Hellmut Flashar. Bd. II/1: *Sophistik. Sokrates. Sokratik. Mathematik. Medizin*. Basel, 1998, pp. 365-453.

Waschkies, Hans-Joachim: "Indizien zur Datierung der geometrischen Termini *στιγμή* und *σημεῖον* als Bezeichnungen für den 'Punkt im allgemeinen' auf die Zeit von Eudoxos und Aristoteles". Althoff, Jochen, Bernhard Herzhoff und Georg Wöhrle (Herausgeber): *Antike Naturwissenschaft und ihre Rezeption*. Bd. X. Trier, 2000, pp. 43-81.

JÜRGEN MITTELSTRASS

DIE ENTDECKUNG DER MÖGLICHKEIT VON WISSENSCHAFT

Meinem verehrten Lehrer WILHELM KAMLAH zum 60. Geburtstag
Vorgelegt von
O. FLECKENSTEIN

I

Wenn wir heute von Wissenschaft sprechen, dann denken wir zumeist an die neuzeitliche Wissenschaft, wie sie seit GALILEI und NEWTON entstanden ist; wir denken dabei also insbesondere an die *exakten* Wissenschaften, wobei exakt hier heißen soll, daß gewisse Wissenschaften sich in ihrem methodischen Aufbau der Mathematik und der formalen Logik bedienen. In exemplarischer Form ist dies bei der Physik der Fall, und insofern diesen Aufbau begonnen zu haben, das Verdienst eines einzelnen Mannes, nämlich GALILEIS ist, scheint es darum auch gerechtfertigt zu sein, mit dem Anfang des 17. Jahrhunderts einen absoluten Anfang in unserer Wissenschaftsgeschichte, zumindest aber in der Geschichte des Selbstverständnisses der Wissenschaft zu setzen. Gleichwohl glauben wir nun aber zu wissen, daß nicht erst die Neuzeit seit 1600, sondern weit früher schon das griechische Denken die Wissenschaft in die Welt brachte, und von einem Anfang der Wissenschaft zu reden, nur Sinn hat, wenn man bedenkt, welche Rolle die Griechen in der Geschichte der wissenschaftlichen Bemühungen des Menschen gespielt haben. Tatsächlich wird denn auch zu Beginn des 17. Jahrhunderts nicht die Wissenschaft überhaupt entdeckt, sondern lediglich die Möglichkeit der Physik als Wissenschaft[1]. Und nur weil die Physik alsbald zur exemplarischen Wissenschaft schlechthin wurde und darüber hinaus bis heute diejenige Wissenschaft geblieben ist, die auf dem Umweg über die Technik in einzigartiger Weise verändernd in die Welt und in das Leben des einzelnen eingreift, kann der Eindruck entstehen, es handele sich bei der Entstehung der neuzeitlichen Wissenschaft um das Entstehen von Wissenschaft überhaupt.

Nun könnte man meinen, die griechische Wissenschaft stelle lediglich die Vorgeschichte jener Wissenschaft dar, wie sie recht eigentlich erst im 17. Jahrhundert beginnt; man würde also einräumen, daß es auch vorher schon Wissenschaft gab, dann jedoch glauben, daß es sich hierbei lediglich um mehr oder weniger triviale Vorstufen handele, die man getrost wieder vergessen kann, nachdem man sich der Ziele der

[1] Gemeint ist hier eine Physik, die sich als eines Beweismittels des Experimentes bedient. Daher kann die Physik des ARCHIMEDES, in der nicht Experimente, sondern nur geometrische Überlegungen zur Beweisführung zugelassen sind, an dieser Stelle außer Betracht bleiben. Die sachliche Frage nach dem Verhältnis dieser beiden Möglichkeiten, Physik zu treiben, ist damit noch nicht erörtert und soll hier auch nicht erörtert werden.

J. Christianidis (ed.), Classics in the History of Greek Mathematics, 19–44.

neuzeitlichen Wissenschaft versichert hat. Aber diese Betrachtungsweise greift zu
411 kurz. Sie greift zu kurz nicht nur, weil gewisse Erkenntnisse | z.B. der griechischen Mathematik und der griechischen Astronomie keineswegs trivial sind, sondern weil es, wenn wir nach dem Begriff der Wissenschaft fragen, gar nicht auf den sogenannten Stand der Forschung — z.B. also den Stand der antiken Forschung um — 400 im Vergleich zu dem der neuzeitlichen Forschung, sagen wir im Jahre 1687 (Erscheinungsjahr von NEWTONS „Philosophiae naturalis principia mathematica") — ankommt, sondern auf die *Idee* der Wissenschaft, wie sie sich Tag für Tag in der wissenschaftlichen Arbeit realisiert und wie sie irgendwann einmal bewußt erfaßt worden sein mußte.

Wissenschaft treiben ist ja eben nicht so etwas Selbstverständliches wie z.B. Essen, Trinken und Schlafen, also etwas, das der Mensch immer schon tut und tun muß, um existieren zu können. Auch daß der Mensch nicht nur Ackerbau und Handel, sondern ebenso Wissenschaft treibt, ist keineswegs selbstverständlich und kann nur jenem selbstverständlich erscheinen, der bereits in einer langen wissenschaftlichen Tradition steht, die selbst die Frage nach ihrem Ursprung nicht mehr stellt. Das heißt aber: die *Möglichkeit* der Wissenschaft mußte ausdrücklich erst *entdeckt* werden; und dies — die Entdeckung der Möglichkeit von Wissenschaft überhaupt — ist es, was wir den Griechen als geniale Tat zuschreiben. Nicht das Verdienst also, einiges schon erkannt und damit der neuzeitlichen Wissenschaft wertvolle Ansätze oder, wie in der Mathematik und in der Logik, die charakteristischen Hilfsmittel zu ihrer Konstituierung als exakter Wissenschaft an die Hand gegeben zu haben, ist das Entscheidende, sondern die Einsicht in eine Möglichkeit des Menschen, an die bislang niemand gedacht hatte, deren Realisierung aber dann das Leben des Menschen bis auf den heutigen Tag bestimmen sollte.

Wenn somit im folgenden von dem Begriff der Wissenschaft in der Antike die Rede sein soll, dann nicht von irgendeinem beliebigen Begriff, wie man etwa auch nach dem Wissenschaftsbegriff der römischen Kaiserzeit oder der Spätscholastik fragen könnte, sondern von der Entdeckung der Möglichkeit von Wissenschaft überhaupt und davon, wie die Griechen mit dieser ihrer großartigen Entdeckung „fertiggeworden" sind, was sie mit ihr anfingen und wie sie selbst über sie dachten.

II

Wollte man sich darauf beschränken, aus der Sicht dessen, der auch die auf die Antike folgende Wissenschaftsgeschichte einigermaßen überblickt, festzustellen, was die besondere Stellung der Griechen in dieser Wissenschaftsgeschichte ausmacht, so wird man sagen dürfen, daß sie als erste *Theorien* ausgebildet haben. Unter einer Theorie verstehen wir eine Reihe von Sätzen, die z.B. in bestimmter geordneter Weise voneinander logisch abhängig sind und auf diese Weise einen Satzzusammenhang oder ein *Satzsystem* darstellen. *Bewiesen* werden Sätze eines solchen Satzsystems in der Regel, indem man gewisse Sätze als logische Folgerungen anderer, bereits *gesicherter* Sätze aufweist. Wie man in diesem Zusammenhang zu *ersten* gesicherten Sätzen kommt, ist dabei eine spezielle Frage, die im Augenblick noch außer Betracht bleiben kann. Worauf es hier zunächst lediglich ankommt, ist die Feststellung, daß man mit einer solchen Überlegung zwar das charakteristische und für die Folgezeit so bedeutende Moment des griechischen wissenschaftlichen Denkens erfaßt haben

dürfte, daß | man damit aber noch nicht verstanden hat, wie es zu dieser speziell 412
griechischen Theorienbildung kam und wieso damit eigentlich etwas entscheidend Neues in die Geschichte des Menschen trat. Schon die Verwendung solcher Wörter wie „Theorie", „Satzsystem", „beweisen" und „logisch folgern" erfolgt ja im Rahmen einer Fachsprache, die wir zu sprechen gewohnt sind, ohne daß uns normalerweise einfiele, hinter diesen Wörten „Entdeckungen" zu vermuten, die keinesfalls selbstverständlich sind. Daß man es in der Wissenschaft mit *Sätzen* zu tun hat, daß man diese Sätze *beweisen* kann, das sind z.B. solche Entdeckungen, von denen der Wissenschaftler zwar täglich Gebrauch macht, deren Bedeutung aber nur selten richtig eingeschätzt wird.

Der erste, der nach griechischer Überlieferung Sätze aufgestellt und bewiesen haben soll, war THALES von MILET. Im einzelnen werden ihm bekanntlich die folgenden elementaren geometrischen Sätze zugeschrieben: (1) Der Kreis wird durch jeden seiner Durchmesser halbiert[2], (2) die Scheitelwinkel sich schneidender Geraden sind gleich[3], (3) die Basiswinkel im gleichschenkligen Dreieck sind gleich[4], (4) zwei Dreiecke, die in einer Seite und den anliegenden Winkeln übereinstimmen, stimmen in allen Stücken überein[5], und (5) der Peripheriewinkel im Halbkreis ist ein rechter[6]. Die Überlieferung darf, mit Ausnahme des fünften Satzes[7], als einigermaßen gesichert
gelten[8], zumal EUDEM, auf den sich PROKLOS | in seiner Wiedergabe der Sätze bezieht, 413

---

[2] PROCLI DIADOCHI in primum Euclidis elementorum librum commentarii, ed. G. FRIEDLEIN, Leipzig 1873, S. 157, 10–13.

[3] PROCL. in Eucl. 299, 1–5 FRIEDLEIN (EUDEM Fr. 135, ed. F. WEHRLI, Die Schule des Aristoteles VIII, Basel 1955).

[4] PROCL. in Eucl. 250, 20–251, 2 FRIEDLEIN.

[5] PROCL. in Eucl. 352, 14–18 FRIEDLEIN (EUDEM Fr. 134 WEHRLI). Man darf im übrigen vermuten, daß auch die Sätze (1) und (3), bei deren Wiedergabe sich PROKLOS nicht auf EUDEM beruft, auf dessen Autorität hin angeführt werden. Vgl. TH. HEATH, The thirteen books of Euclid's Elements, 3 Bde, Dover Publications 1956 (Nachdruck der 2. Aufl. Cambridge 1926), I, S. 36.

[6] DIOG. LAERT. I, 24–25.

[7] DIOGENES LAERTIUS vermag sich in seinem Referat lediglich auf eine Mitteilung der im 1. vorchristlichen Jahrhundert lebenden Geschichtsschreiberin PAMPHILE zu stützen. EUKLID beweist diesen Satz (III, 31) mit Hilfe des Satzes von der Winkelsumme im Dreieck (I, 32), dessen Entdeckung EUDEM wiederum erst den *Pythagoreern* zuschreibt (PROCL. in Eucl. 379, 2–16 FRIEDLEIN; EUDEM Fr. 136 WEHRLI). Dies scheint gegen die Annahme, man habe hier ebenfalls einen thaletischen Satz vor sich, zu sprechen, doch hat TH. HEATH nachweisen können, daß sich dieser Satz auch ohne den Winkelsummensatz aufstellen läßt und somit jedenfalls der Sache nach nichts gegen PAMPHILES Zeugnis spricht. TH. HEATH, A history of Greek mathematics, 2 Bde, Oxford 1921, I, S. 136f.

[8] Vgl. P. TANNERY, Pour l'histoire de la science hellène, Paris 1887, S. 52ff.; TH. HEATH, a.a.O., S. 128ff.; K. v. FRITZ, Die *APXAI* in der griechischen Mathematik, Archiv für Begriffsgeschichte I (1955), S. 77ff.; im Anschluß an K. v. FRITZ: Á. SZABÓ, Wie ist die Mathematik zu einer deduktiven Wissenschaft geworden?, Acta Antiqua IV (1956), S. 130ff.; B. L. VAN DER WAERDEN, Erwachende Wissenschaft, Basel-Stuttgart 1956, S. 143ff.; O. BECKER, Das mathematische Denken der Antike, Göttingen 1957, S. 37ff. Scharfen Zweifel an der Überlieferung äußern J. BURNET, Early Greek philosophy, 4. Aufl. London 1930, S. 45f., und D. R. DICKS, Thales, The Classical Quarterly 53 (1959), S. 301 ff. Dabei beschränkt sich BURNET jedoch auf wenige skeptische Bemerkungen, während DICKS von der irrtümlichen Annahme ausgeht, thaletische Geometrie müsse, wenn es sie wirklich gegeben haben sollte, in ihrem Aufbau bereits in der Weise der euklidischen Geometrie verfahren sein. Ansätze zu einem solchen Aufbau werden nun in der Tat erst in der zweiten Hälfte des 5. Jahrhunderts faßbar, doch läßt sich — wie im

selbst Originales von Erschlossenem trennt[9] und beim dritten Satz entgegen der damals üblichen Bezeichnung *ἴσαι* für gleiche Winkel das altertümliche *ὅμοιαι* bringt[10]. Nun ist die Frage nach der Anzahl wirklich auf THALES zurückgehender Sätze in diesem Zusammenhang nur von sekundärer Bedeutung; entscheidend ist, daß überhaupt solche Sätze (*θεωρήματα*) von THALES, oder sei es auch einem anderen Griechen um — 600 aufgestellt wurden[11]. Sätze nach Art der fünf genannten hat es vorher, speziell in der babylonischen Mathematik[12], nicht gegeben. Diese vor-
414 griechische Mathe|matik begnügte sich vielmehr mit der Zusammenstellung praktischer Regeln und Verfahren zur Lösung konkreter Aufgaben, wie sie etwa die Praxis der Feldereinteilung ergab, und ging, wie sich heute bereits mit einiger Sicherheit sagen läßt, noch nicht zu einer in dieser Weise „zweckfreien" Betrachtung der Regeln oder Rezepte über. So wurde z.B. praktisch bereits nach dem sogenannten pythagoreischen Lehrsatz gerechnet, eine Feststellung, die das Niveau dieser vorgriechischen

---

folgenden gezeigt werden soll — die thaletische Geometrie eben auch anders verstehen, und zwar ohne die von DICKS empfohlene Auffassung, es handele sich hier wohl um den „empirischen Typ" babylonischer und ägyptischer Geometrie (DICKS, a.a.O., S. 303). Skeptisch, unter Hinweis auf BURNET, äußern sich auch G. S. KIRK / J. E. RAVEN, The presocratic philosophers, Cambridge 1957 (2., geringfügig veränderter Nachdruck 1962), S. 83 f. Ihr, von KIRK formuliertes Urteil ist charakteristisch für diesen Standpunkt; sie vermuten, „that THALES did gain a reputation with his contemporaries for carrying out various far from straightforward empirical feats of mensuration, without necessarily stating the geometry that lay behind them" (84).

9 PROCL. in Eucl. 352, 16–18 FRIEDLEIN. Zu der hier von EUDEM erwähnten thaletischen Methode der Distanzbestimmung zwischen Schiffen auf hoher See vgl. B. L. VAN DER WAERDEN, a.a.O., S. 144 f.

10 PROCL. in Eucl. 251, 1–2 FRIEDLEIN: *ἀρχαϊκώτερον δὲ τὰς ἴσας* (sc. *γωνίας*) *ὁμοίας προσειρηκέναι*. Vgl. dazu K. v. FRITZ, Gleichheit, Kongruenz und Ähnlichkeit in der antiken Mathematik bis auf Euklid, Archiv für Begriffsgeschichte IV (1959), S. 48 f.; H. D. RANKIN, *῞Ομοιος* in a fragment of Thales, Glotta. Zeitschrift für griechische und lateinische Sprache 39 (1961), S. 73–76.

11 Die Thalestradition selbst ist gut bezeugt, sie läßt sich bis ins 5. Jahrhundert v. Chr. zurückverfolgen; vgl. W. BURKERT, Weisheit und Wissenschaft. Studien zu Pythagoras, Philolaos und Platon, Nürnberg 1962, S. 393.

12 Vgl. O. NEUGEBAUER, The exact sciences in antiquity, 2. Aufl., Providence 1957, S. 48. 146; Á. SZABÓ, a.a.O., S. 115; O. BECKER, a.a.O., S. 11; H. GERICKE, Über den Unterschied von griechischer und vorgriechischer Mathematik, Gymnasium 67 (1960), S. 128. Wenn die griechische Tradition immer wieder auf Ägypten als das Ursprungsland speziell der Geometrie hinweist (HERODOT II 109; ARISTOTELES Met. A 1.981 b 23–25; PROCL. in Eucl. 64, 16–65, 7 FRIEDLEIN, unter Berufung auf die vorherrschende Meinung und unter wörtlicher Anspielung auf ARISTOTELES), so kommt darin nur zum Ausdruck, daß sie selbst schon nicht mehr in der Lage war, die geschichtliche Leistung der eigenen Mathematik richtig einzuschätzen. Babylonischer Einfluß wird merkwürdigerweise überhaupt nicht registriert, obgleich er faktisch weitaus größer gewesen sein dürfte als der Einfluß ägyptischer Mathematik. Vgl. dazu TH. HEATH, a.a.O., S. 122 ff.; O. BECKER, Grundlagen der Mathematik in geschichtlicher Entwicklung, Freiburg-München 1954, S. 22; ders. Das mathematische Denken der Antike, S. 9; B. L. VAN DER WAERDEN, a.a.O., S. 23 ff. und 49 ff. Was das eigenartige Zurücktreten der babylonischen Tradition hinter der ägyptischen Tradition in der griechischen Mathematikgeschichtsschreibung betrifft, so glaubt Á. SZABÓ hierfür einen Grund darin sehen zu können, daß bereits im 4. Jahrhundert, als EUDEM die erste Geschichte der Mathematik schrieb, die griechische Mathematik völlig geometrisiert war und man deswegen „schon eine nähere Verwandtschaft mit der ägyptischen Geometrie als mit der babylonischen Algebra fühlen" konnte (a.a.O., S. 130 Anm. 49). Gegen diesen Vorschlag spricht allerdings, daß auch auf dem Gebiete der Geometrie — mit Ausnahme in der Berechnung des Kreisinhaltes, und nur auf dieses Beispiel stützt SZABÓ seinen Vorschlag — die Babylonier den Ägyptern überlegen waren; vgl. O. BECKER, Das mathematische Denken der Antike, S. 9 f.

Mathematik deutlich vor Augen stellt, ohne daß dieser Lehrsatz jedoch unabhängig von bestimmten vorgegebenen Aufgaben jemals ausdrücklich als Satz über Hypothenusenquadrat und Kathetenquadrate formuliert worden wäre[13]. Wenn wir heute sagen, daß die Babylonier den pythagoreischen Lehrsatz praktisch schon benutzt haben, so machen wir hierin eben Gebrauch von einer Ausdrucksweise, die erst mit der griechischen Mathematik, speziell mit der thaletischen Geometrie sinnvoll wurde.

Im Unterschied zu den praktischen Sätzen der vorgriechischen Mathematik, die in Form von Rechenvorschriften und Rezepten ausgesprochen wurden, lassen sich die thaletischen Sätze vielleicht am besten als *theoretische* Sätze bezeichnen. In diesen Sätzen ist von speziellen Aufgaben und Anwendungsmöglichkeiten ganz abgesehen; diese Sätze sind nicht formuliert, um z.B. dieses oder jenes Dreieck konstruieren zu können, sondern um etwas mitzuteilen, was der Konstruktion aller möglichen Dreiecke noch vorausgeht. Vielleicht könnte man sagen, daß diese Allsätze, um solche handelt es sich hier, hervorgegangen sind aus einer Reflexion auf das Funktionieren jener babylonischen Rezepte, sie wären also die griechische Antwort auf ein ebenso griechisches wissenschaftliches Warum. Doch müssen solche Überlegungen notwendig bloße Vermutungen bleiben, in denen zudem im einzelnen immer noch allzu viele Einsichten und Fragestellungen vorausgesetzt werden, auf die zu reflektieren doch erst durch das griechische Denken möglich wurde. Man wird sich also mit der Feststellung zufriedengeben müssen, daß aller Wahrscheinlichkeit nach THALES als erster die Möglichkeit gesehen hat, theoretische Sätze (und das heißt für den Fall der Geometrie: Sätze über *ideale Gegenstände*) zu formulieren. „Sätze über Verhältnisse im Kreis" oder „Sätze über Winkel" aufzustellen und in dieser Aufstellung eine sinnvolle Tätigkeit zu sehen, das ist eine „Entdeckung", die weder notwendig noch einfach naheliegend war, und die gemacht zu haben, das Verdienst eines Denkens ist, das wir, zumeist recht allgemein, als das griechische Denken zu bewundern pflegen.

Wichtiger noch als der Übergang von praktischen zu theoretischen Sätzen aber ist für das Entstehen der Wissenschaft der Umstand, daß THALES diese seine | theoreti- 415
schen Sätze ausdrücklich zu *beweisen* suchte[14]. Auf den Gedanken, seine Sätze zu beweisen oder gar beweisen zu müssen, um sie in Form von Vorschriften bestimmten Aufgaben beifügen zu können, ist kein Babylonier gekommen; wie sich bezeichnenderweise auch nachweisen läßt, daß es hier falsche Rezepte gab, nach

13 O. NEUGEBAUER, Zur geometrischen Algebra (Studien zur Geschichte der antiken Algebra III), Quellen und Studien zur Geschichte der Mathematik Astronomie und Physik B 3, Berlin 1936, S. 257; ders. The exact sciences in antiquity, S. 35 ff.; vgl. B. L. VAN DER WAERDEN, a.a.O., S. 122 ff.; O. BECKER, a.a.O., S. 10. 55f. Explizite Formulierungen dieses Lehrsatzes sind allerdings aus der indischen Sakralgeometrie bezeugt (Āpastamba-Śulva-Sūtra, vgl. O. BECKER / J. E. HOFMANN, Geschichte der Mathematik, Bonn 1951, S. 39), doch ist ihr Alter ungewiß (überliefert sind sie aus dem 5./4. Jahrhundert) und läßt sich ein für die griechische Mathematik erstmals charakteristischer Zusammenhang von Satz und Beweis für sie nicht nachweisen; vgl. TH. HEATH, a.a.O., S. 145 ff.; O. BECKER, a.a.O., S. 11. Zum sakralen Charakter indischer Geometrie vgl. A. SEIDENBERG, The Ritual Origin of Geometry, Archive for history of exact sciences I (1960–1962), S. 488 ff.

14 Die wahrscheinlich auf EUDEM zurückgehende Nachricht über Satz (1) lautet: *Τὸ μὲν οὖν διχοτομεῖσθαι τὸν κύκλον ὑπὸ τῆς διαμέτρου πρῶτον Θαλῆν ἐκεῖνον ἀποδεῖξαί φασιν* (PROCL. in Eucl. 157, 10–11 FRIEDLEIN).

denen gleichwohl, weil sie noch einigermaßen brauchbare Resultate lieferten, fortwährend gerechnet wurde[15]. Die Frage aber, warum THALES gerade auf diesen Gedanken kam, läßt sich noch schwieriger beantworten als die Frage nach der „Herkunft" seiner theoretischen Sätze. Die bisher vorgebrachten Erklärungsversuche sind wenig befriedigend. Wenn z. B. VAN DER WAERDEN erklärt, THALES habe gelegentlich zwischen zwei divergierenden Resultaten, etwa zwischen der babylonischen[16] Kreiszahl $\pi = 3$ und der ägyptischen $\pi = 4 \cdot (\frac{8}{9})^2$, entscheiden müssen, und er habe dies eben „einfach" dadurch getan, „indem er sie bewies"[17], so wird damit nur gesagt, wie man *heute* natürlich vorgehen würde, nachdem die Möglichkeit des Beweisens jedermann geläufig ist, aber nicht erklärt, was es bedeutet, diese Möglichkeit allererst entdeckt zu haben.

Wenn man hier überhaupt von einem Bedürfnis sprechen will, das bei THALES zur Entdeckung der Beweismöglichkeit geführt haben mag, so ließe sich eventuell vermuten, daß hinter dieser Entdeckung die Frage steht, wovon denn in den theoretischen Sätzen überhaupt die Rede ist. Nach PLATON sagen wir, daß diese Sätze Sätze über ideale Gegenstände sind; um aber deutlich zu machen, was damit gemeint ist, muß man über die Formulierung der Sätze hinaus offenkundig noch zusätzlich etwas tun, d. h. man muß Handlungen anführen oder Zusammenhänge aufweisen, die auch jemand anderen in die Lage versetzen, diese Sätze seinerseits zu übernehmen. Gelingt dies, d. h. läßt sich die Wahrheit dieser Sätze jedermann einsichtig machen, und sei es durch Beachtung gewisser Evidenzen, so mag damit dann die zweite thaletische Entdeckung einigermaßen umschrieben und als provoziert durch die erste Entdeckung verstanden sein, wobei es relativ gleichgültig ist, ob man das hier angedeutete Verfahren „Beweisen" nennt oder irgendwie anders.

Glücklicherweise sind wir nun in der Lage, mit einiger Sicherheit sagen zu können, wie THALES seine Sätze bewiesen hat. Unter den Axiomen findet sich bei EUKLID als siebtes das sogenannte Kongruenzaxiom: „Was einander deckt, ist einander gleich" (τὰ ἐφαρμόζοντα ἐπ' ἀλλήλα ἴσα ἀλλήλοις ἐστίν). Von diesem Axiom macht
416 EUKLID dann im folgenden lediglich beim Beweis dreier Sätze, | darunter des ersten Kongruenzsatzes (I, 4), Gebrauch[18], und dies ist um so verwunderlicher, als sich gewisse Ungenauigkeiten in den Gleichheitsdefinitionen ohne weiteres hätten vermeiden lassen, wenn EUKLID häufiger auf dieses Axiom zurückgegriffen hätte[19]. Offenbar sucht er jedoch absichtlich die Verwendung dieses Axioms auf jene Fälle zu

---

[15] Mit falschen Formeln wurde z. B. bei der Berechnung von Kegel- und Pyramidenstumpf gearbeitet; B. L. VAN DER WAERDEN, a.a.O., S. 120 ff. Allerdings läßt sich nach O. NEUGEBAUER auch die exakte Formel für das Volumen eines quadratischen Pyramidenstumpfes nachweisen; Bedenken, die NEUGEBAUER (Vorlesungen über Geschichte der antiken mathematischen Wissenschaften I, Berlin 1934, S. 171) hier gegenüber seinem eigenen Vorschlag geltend machte, zerstreut VAN DER WAERDEN, a.a.O., S. 121 f.

[16] Neuerdings läßt sich in einigen Keilschrifttexten auch der Wert $\pi = 3\frac{1}{8}$ nachweisen; vgl. O. BECKER, a.a.O., S. 10.

[17] B. L. VAN DER WAERDEN, a.a.O., S. 147.

[18] Die beiden anderen Sätze sind I, 8: Wenn in zwei Dreiecken zwei Seiten zwei Seiten entsprechend gleich sind und auch die Grundlinie der Grundlinie gleich ist, dann müssen in ihnen auch die von gleichen Strecken umfaßten Winkel einander gleich sein; III, 24: Ähnliche Kreissegmente über gleichen Strecken sind einander gleich. Hierzu vgl. K. v. FRITZ, a.a.O., S. 25 ff.

[19] Vgl. K. v. FRITZ, Die *APXAI* in der griechischen Mathematik, Arch. f. Begriffsgesch. I, S. 76 f.

beschränken, in denen er, wie beim ersten Kongruenzsatz, keine andere Beweismöglichkeit sieht. Diese Abneigung dem siebten Axiom gegenüber läßt sich wiederum daraus erklären, daß dieses Axiom „Bewegung" benutzt, sich also auf ein Verfahren bezieht, das offenkundig einen stark *empirischen* Charakter besitzt; beim Beweis des ersten Kongruenzsatzes heißt es ausdrücklich: „*Deckt* man nämlich Dreieck ABC *auf* Dreieck DEF (*ἐφαρμοζομένου*) und *legt* dabei (*τιθεμένου*) Punkt A *auf* Punkt B sowie die Gerade AB *auf* DE..."[20]. Dieses „empirische", genauer vielleicht: dynamische Verfahren, das Übereinanderlegen oder Aufeinanderklappen von Figuren, ist EUKLID zweifellos ein Ärgernis; wie man aber aus einigen Mitteilungen, die wir PROKLOS verdanken, entnehmen kann, ist damit nun genau das Vefahren benannt, dessen man sich früher ausgiebig zu Beweiszwecken bediente! Insbesondere erwähnt PROKLOS einen solchen Klappbeweis für den thaletischen Satz, daß der Kreis durch jeden seiner Durchmesser halbiert wird[21], und dies unmittelbar im Anschluß an jene von EUDEM übernommene Bemerkung, daß THALES diesen Satz *bewiesen* habe[22]! Es liegt damit nahe, in der von EUKLID so ängstlich gemiedenen und in neuerer Zeit bei HILBERT durch Kongruenzaxiome ersetzten Methode des Aufeinanderlegens und Aufeinanderklappens das seit THALES übliche Beweisverfahren zu sehen, zumal sich alle THALES zugeschriebenen Sätze tatsächlich ohne weiteres mit Hilfe dieser Methode beweisen lassen[23].

Damit wäre also das thaletische Beweisverfahren rekonstruiert und scheinbar gleichzeitig als ein *empirisches* Verfahren bezeichnet. In der Tat nehmen die meisten der modernen Interpreten an, daß THALES in dieser Weise empirisch vorgegangen ist. Doch hat bereits 1902 BERTRAND RUSSELL in einem Enzyklopädieartikel darauf aufmerksam gemacht, daß auch die im *ἐφαρμόζειν* beschrie|bene Klappmethode 417
schon im Sinne des modernen Kongruenzbegriffes geometrischer Figuren aufgefaßt werden kann; es wird nicht wirklich mit Figuren hantiert, sondern lediglich die Aufmerksamkeit von einer Figur auf eine andere gelenkt, die unter anderem durch gewisse Eigenschaften bestimmt ist, die sie mit der ersteren teilt[24]. Diese Bemerkung RUSSELS, in der K. v. FRITZ explizit eine Ehrenrettung EUKLIDS gegenüber dem Vorwurf, er benutze selbst empirische Hilfsmittel, sehen will[25], läßt sich nun auch zur

---

[20] Opera omnia, ed. J. L. HEIBERG / H. MENGE, 9 Bde, Leipzig 1883–1916, I, S. 16 ff.

[21] PROCL. in Eucl. 157, 17–158, 2 FRIEDLEIN.

[22] PROCL. in Eucl. 157, 11 FRIEDLEIN; vgl. Anm. 14.

[23] Insbesondere K. v. FRITZ und Á. SZABÓ haben in zahlreichen mathematikhistorischen Arbeiten immer wieder auf das *ἐφαρμόζειν*-Verfahren hingewiesen und seine Bedeutung im Zusammenhang mit ihrer Beurteilung der thaletischen Geometrie hervorgehoben. Vgl. K. v. FRITZ, a.a.O., S. 72 ff.; ders. Der gemeinsame Ursprung der Geschichtsschreibung und der exakten Wissenschaften bei den Griechen, Philosophia Naturalis II (1952), S. 206 ff.; ders. Gleichheit, Kongruenz und Ähnlichkeit in der antiken Mathematik bis auf Euklid, Arch. f. Begriffsgesch. IV, S. 7 ff. Á. SZABÓ, a.a.O., S. 131 ff.; ders. Die Grundlagen in der frühgriechischen Mathematik, Studi italiani di filologia classica XXX (1958), S. 16 ff.; ders. *ΔΕΙΚΝΥΜΙ*, als mathematischer Terminus für „beweisen", Maia. Rivista di letterature classiche N.S. X (1958), S. 115 f.

[24] Artikel: Geometry, non-Euclidean, The New Volumes of the Encyclopaedia Britannica constituting in combination with the existing volumes of the 9. edition the 10. edition of that work. The fourth of the new volumes, being volume 28 of the complete work, London 1902, S. 671. Diese Bemerkung wurde wieder aufgegriffen von TH. HEATH, The thirteen books of Euclid's Elements I, S. 227.

[25] K. v. FRITZ, Gleichheit, Kongruenz und Ähnlichkeit in der Mathematik bis auf Euklid, Arch. f. Begriffsgesch. IV, S. 18. RUSSELL geht es an dieser Stelle jedoch weniger um eine Ehrenrettung

Interpretation thaletischer Beweisformen heranziehen, zumal ja keineswegs ausgemacht ist, daß THALES selbst dieses Klappverfahren so empirisch verstanden hat, wie EUKLIDS Ausdrucksweise es nahelegt[26]. Anhand der sogenannten thaletischen Grundfigur, einem Rechteck mit Diagonalen und umschriebenem Kreis, kann man sich vielmehr ohne große Mühe deutlich machen, daß hier als Beweismethode auch einfache *Symmetriebetrachtungen* in Frage kommen können[27]. Mit dem Hinweis auf ein Klappverfahren würde dann nur auf handgreifliche Weise zum Ausdruck gebracht, daß z. B. im Falle des dritten Satzes (Gleichheit der Basiswinkel im gleichschenkligen Dreieck) das gleichschenklige Dreieck in bezug auf die Winkelhalbierende symmetrisch ist. THALES hätte demnach seine Sätze durch Symmetriebetrachtungen bewiesen, die ihrerseits nicht als Sätze auftreten, sondern unmittelbare, am Objekt gewonnene Einsichten — nämlich daß bestimmte Homogenitätsforderungen am Objekt erfüllt sind — darstellen. Erst die Tradition hat diese Betrachtungen dann durch gewisse, dabei stillschweigend immer schon mitbenutzte Sätze ersetzt. So werden, worauf bereits kurz hingewiesen wurde, von HILBERT Kongruenzaxiome ergänzend in die Geometrie eingeführt, die z. B. bei EUKLID noch fehlen[28].

Mit der Realisierung der Möglichkeit theoretischer Sätze und der Möglichkeit des Beweises sind damit am Anfang des griechischen Denkens jene Entdeckungen gemacht, die für den dann faktisch erfolgten Aufbau der Wissenschaft von eminenter
418 Bedeutung sind, ohne daß man behaupten könnte, Wissenschaft sei | schlechterdings auf diese Entdeckungen als die Bedingungen ihrer Möglichkeit angewiesen. Dies ist zwar richtig für jene Form der Wissenschaft, wie sie sich seit den Griechen entwikkelt hat, doch wäre auch eine Weiterführung der babylonischen Verfahren ohne die thaletischen Entdeckungen immerhin denkbar, die nicht unbedingt weniger anspruchsvoll als die griechische und neuzeitliche Wissenschaft hätte ausfallen müssen und die sich natürlich auch als Wissenschaft hätte verstehen können. Nun interessiert an dieser Stelle nicht die Frage, wie Wissenschaft allenfalls auch noch möglich sein könnte, sondern allein, wie sie faktisch möglich wurde. Und in Beantwortung dieser Frage haben sich die beiden genannten Entdeckungen zunächst als die zentralen Voraussetzungen herausgestellt. Allerdings genügen sie so, wie sie bisher in ihrem Ursprung behandelt wurden, noch nicht, um das Entstehen des wissenschaftlichen Denkens, speziell bei den Griechen, gänzlich einsichtig zu machen.

Wie zu Beginn erwähnt wurde, liegt ein wesentliches Charakteristikum der Theorienbildung weiterhin darin, daß die Sätze einer solchen Theorie untereinander

---

EUKLIDS als vielmehr um die Unterscheidung von *axiomatischer* und *physikalischer* Geometrie (Kongruenzaxiome versus Existenz starrer Körper). EUKLIDS mißverständliche Formulierung wird hier für die Kontroverse um diesen Unterschied verantwortlich gemacht.

26 Als selbstverständlich empirisches Verfahren wird es von Á. SZABÓ angesehen: Wie ist die Mathematik zu einer deduktiven Wissenschaft geworden?, Acta Ant. IV, S. 132; Die Grundlagen in der frühgriechischen Mathematik, Stud. ital. di filol. class. XXX, S. 18; *ΔEIKNYMI*, als mathematischer Terminus für „beweisen", Maia N.S. X, S. 116.

27 Darauf machen bereits aufmerksam: O. BECKER / J. E. HOFMANN, a.a.O., S. 45; O. BECKER, Das mathematische Denken der Antike, S. 39.

28 D. HILBERT, Grundlagen der Geometrie, 9. Aufl. Stuttgart 1962. — Eine moderne Darstellung der Geometrie, die nicht die Kongruenz, sondern die Spiegelung als Grundbegriff verwendet (F. BACHMANN, Aufbau der Geometrie aus dem Spiegelungsbegriff, Berlin-Göttingen-Heidelberg 1959), könnte als Wiederaufnahme einer sich Symmetriebetrachtungen bedienenden Geometrie aufgefaßt werden.

durch gewisse *logische Abhängigkeiten* ausgezeichnet werden. Man kann auch sagen, daß überhaupt erst durch die Herstellung solcher Abhängigkeiten eine Theorie entsteht. In der Durchführung führt dies nun in der griechischen Geometrie, wie sie bei EUKLID in beispielhafter Form vorliegt, zu einem axiomatisch-deduktiven Aufbau oder kurz: zu einer *axiomatischen* Theorie, in der aus an den Anfang gestellten nicht bewiesenen Sätzen alle übrigen Sätze erschlossen werden. Dieses Verfahren aber setzt nun zweifellos Logik, zumindest die Kenntnis des *logischen Schließens* voraus, womit eine weitere „Entdeckung" an den Anfang des wissenschaftlichen Denkens bei den Griechen rückt. Blickt man dabei auf den Werdegang der griechischen Geometrie, d. h. auf ihre allmähliche Entwicklung zu einer axiomatischen Theorie, so scheint diese Entdeckung sogar gegenüber den beiden anderen Entdeckungen den Vorrang zu haben. Tatsächlich wird häufig die Meinung vertreten, erst die Logik habe überhaupt die spezifisch griechische Form der Mathematik im Unterschied zur babylonischen Form ermöglicht und es seien insbesondere die Eleaten gewesen, die mit der originalen Entwicklung logischer Beweisformen, vor allem der Form des *indirekten* Beweises, diese Mathematik hervorgebracht hätten[29]. Gegenüber dieser Behauptung hat die Gegenthese, die ebenfalls gelegentlich vertreten wird, daß nämlich die Mathematik es gewesen sei, die historisch gesehen die Logik aus sich entlassen habe, einen schweren Stand, wenn man unter Mathematik hier schon einen axiomatischen Aufbau im Stile EUKLIDS versteht[30]. In der Tat ist es absurd, sich eine axiomatische

---

[29] Dieser These sind vor allem die Arbeiten Á. SZABÓS gewidmet. Neben den bereits mehrfach angeführten Aufsätzen verdienen hier noch zwei frühere der Erwähnung: Zum Verständnis der Eleaten, Acta Antiqua II (1953–54), S. 243–286, und: Eleatica, Acta Antiqua III (1955), S. 67–102. Zum indirekten Beweisverfahren vgl. insbesondere den zuletzt genannten Aufsatz sowie: Anfänge des euklidischen Axiomensystems, Archive for history of exact sciences I (1960–62), S. 55 ff. In diesem Zusammenhang als begriffsgeschichtliche Vorarbeit wichtig: K. v. FRITZ, *ΝΟΥΣ, ΝΟΕΙΝ,* and their derivatives in pre-socratic philosophy (excluding Anaxagoras), Classical Philology XL (1945), S. 223–242; XLI (1946), S. 12–34.

[30] Diese Auffassung findet sich zweifellos bei F. M. CORNFORD, der parmenideisches Philosophieren nur auf dem Hintergrund pythagoreischer Geometrie (!) glaubt verstehen zu können („the method of reasoning he [sc. PARMENIDES] imported into philosophy is the method of geometry"), Principium Sapientiae, A study of the origins of Greek philosophical thought, Cambridge 1952, S. 117; vgl. von demselben Verfasser: Plato and Parmenides, London 1939, S. 29. Ein wenig zurückhaltender: H. CHERNISS, The characteristics and effects of presocratic philosophy, Journal of the history of ideas XII (1951), S. 336 — Vernünftig verstehen läßt sich diese These nur, wenn mit dem Hinweis auf eine beispielhafte Mathematik Arithmetik gemeint ist. Diese ist von Hause aus durchaus keine axiomatische Theorie (als solche wird sie erst Ende des 19. Jahrhunderts von PEANO dargestellt) und kann zunächst sogar völlig logikfrei betrieben werden. An Arithmetik denkt auch K. REIDEMEISTER, wenn er erklärt, daß „aus dem Umgang mit Zahlen das Denken und die Idee des widerspruchsfreien Denkens in Begriffen" entsteht (Das exakte Denken der Griechen, Hamburg 1949, S. 11). Mit dieser Bemerkung schließt REIDEMEISTER seine kurze Darstellung eines Beweises aus der bei EUKLID VII–IX niedergelegten elementaren Zahlentheorie ab, der bei der Rückführung eines Satzes (IX,30; „Geht eine ungerade Zahl als Teiler in einer geraden Zahl auf, so geht sie auch in der Hälfte dieser Zahl auf") auf einen anderen, bereits bewiesenen Satz (IX,29: „Das Produkt zweier ungerader Zahlen ist ungerade") von einem indirekten Schluß Gebrauch macht. Beide Sätze gehören zu der seit O. BECKERS grundlegender Studie „Die Lehre vom Geraden und Ungeraden im Neunten Buch der Euklidischen Elemente" (Qu. u. Stud. z. Gesch. d. Math. Astr. u. Phys. B 3 [1936], S. 533 ff.) als ein genuines Stück altpythagoreischer Arithmetik geltenden „Lehre vom Geraden und Ungeraden" (EUKLID IX, 21–34) (Zur Kritik des BECKERschen Datierungsversuches vgl.

419 Theorie vor|zustellen, die sozusagen logikfrei sich konstituieren sollte, um erst nachträglich zum methodischen Vorbild einer sich langsam entwickelnden Logik zu werden.

Nun gehen diese Überlegungen in der Regel davon aus, daß die griechische Mathematik, speziell die griechische Geometrie, von Anfang an auf eine axiomatische Theorie hin entworfen wurde. Daß sie diesen Weg faktisch beschreiten sollte, darüber kann in der Tat kein Zweifel bestehen, doch ist sehr die Frage, ob man die Behauptung, sie habe das immer schon getan, wirklich aufrechterhalten kann. Eine nüchterne Betrachtung des thaletischen Beweisverfahrens bietet z. B., wie vielleicht deutlich geworden ist, noch keinerlei Anlaß dazu, diese Behauptung aufzustellen oder zu übernehmen. In der bisherigen Literatur wird allerdings auch THALES, wenn man ihn
420 nicht von vornherein als quasi mythische | Gestalt aus der Wissenschaftsgeschichte herausnimmt, schon das methodische Bewußtsein zugesprochen, Sätze ließen sich durch andere Sätze beweisen. So pflegt man insbesondere in der Betrachtung des Satzes über die Gleichheit der Basiswinkel darauf hinzuweisen, daß der Beweis dieses Satzes unter anderem vom ersten thaletischen Satz über die Halbierung des Kreises durch den Durchmesser Gebrauch mache, indem von der Annahme ausgegangen werde, daß die gemischtlinigen Halbkreiswinkel im Kreis einen festen Wert besitzen[31]. Gerade dieser Basiswinkelsatz aber läßt sich unter Hinweis auf das Klappverfahren, wie gezeigt werden konnte, auch durch einfache Symmetriebetrachtungen beweisen, womit zumindest deutlich sein dürfte, daß THALES nicht unbedingt, wenn er diesen Satz überhaupt bewiesen hat[32], auf andere Sätze angewiesen war[33]. Im Gegenteil; da in dem so verstandenen Klappverfahren

---

neuerdings W. BURKERT, a.a.O., S. 410 ff.). REIDEMEISTERS Überlegungen machen dabei deutlich — und dies ist hier das Entscheidende —, daß man auch bei der Entwicklung einer konstruktiven Mathematik logisches Schließen als geeignetes Hilfsmittel entdeckt haben könnte. Gerade weil er dabei nicht an axiomatische Mathematik denkt, sondern die konstruktive Arithmetik im Auge hat, geht darum auch SZABÓs Kritik, in der er REIDEMEISTER auf die These vom Primat der Mathematik gegenüber der Logik festlegen will, fehl (Eleatica, Acta Ant. III, S. 73 f.). SZABÓ kann sich offenbar Mathematik nur als axiomatische Theorie vorstellen („Bekanntlich ist die Widerspruchsfreiheit das einzige Kriterium einer mathematischen Wahrheit. Man kann einen Satz in der Mathematik beweisen, d. h. seine mathematische Wahrheit legitimieren, indem man zeigt, daß er keinen Widerspruch in sich enthält und in keinem Widerspruch zu den allgemein anerkannten Axiomen steht“, a.a.O., S. 75 Anm. 28), und diese Auffassung unterstellt er wie selbstverständlich auch REIDEMEISTER. Die Existenz einer konstruktiven Mathematik ist hierbei nicht einmal zur Kenntnis genommen. Gerade mit ihrem Aufweis aber verliert in unserem Zusammenhang SZABÓs These vom Primat der Logik gegenüber der Mathematik ihre charakteristische Schärfe; es könnte sein, daß sowohl eleatische Argumentationskunst als auch die Kunst des Zählens und Messens durchaus selbständig das logische Schließen als Werkzeug vernünftigen Denkens hervorgebracht haben. Eine solche Betrachtungsweise läßt dann selbstverständlich immer noch den speziellen und, wie es scheint, auch begründeten Nachweis zu, daß eine sich als axiomatische Theorie etablierende Geometrie historisch gesehen von der eleatischen Logik abhängig ist.

31 Vgl. O. BECKER, Das mathematische Denken der Antike, S. 38 f.

32 PROKLOS weiß lediglich zu berichten, daß THALES diesen Satz als erster erkannt (ἐπιστῆσαι) und ausgesprochen habe (εἰπεῖν); in Eucl. 250, 22 FRIEDLEIN.

33 Auch ein bei ARISTOTELES (Analytica Priora A24. 41 b 13–22) überlieferter alter Beweis des Basiswinkelsatzes, der ebenfalls von der Annahme Gebrauch macht, daß die gemischtlinigen Halbkreiswinkel im Kreis einen festen Wert besitzen, dürfte, wie TH. HEATH bemerkt hat (Mathematics in Aristotle, Oxford 1949, S. 23 f.), ursprünglich ein ἐφαρμόζειν-Beweis

gegenüber dem in einem axiomatischen Aufbau verwendeten Beweisverfahren mit Sicherheit ein Stück archaischer Beweismethode greifbar ist, wird man annehmen düfen, daß es auch bei THALES gegenüber anderen Verfahren primär ist. Wir hätten damit entgegen der sonst üblichen Ansicht in der thaletischen Geometrie ein Stück logikfreier Elementargeometrie vor uns, eine Geometrie, in der noch keine logischen Schlüsse verwendet zu werden brauchen. Dieses Ergebnis ist dabei insofern von Bedeutung, als man einerseits jetzt sagen kann, daß zu Beginn der griechischen Geometrie über deren axiomatischen Werdegang noch keineswegs entschieden war, und andererseits auch gegenwärtig Bemühungen im Gange sind, die Axiome der Geometrie womöglich anders als durch bloße Evidenzen zu begründen[34].

Wie es scheint, hat die griechische Geometrie jedoch sehr früh, sozusagen noch in ihrer Gründerzeit, begonnen, axiomatisch zu denken. PROKLOS gibt zweifellos einer bald uneingeschränkt herrschenden Überzeugung Ausdruck, wenn er in seinem Kommentar zu EUKLIDS „Elementen" erklärt: „Da wir behaupten, daß diese Wissenschaft, die Geometrie, auf Voraussetzungen beruhe und von bestimmten Prinzipien aus die abgeleiteten Folgerungen beweise..., so muß unbedingt der Verfasser eines geometrischen Elementarbuches gesondert die Prinzipien der Wissenschaft lehren und gesondert die Folgerungen aus den Prinzipien; von den Prinzipien braucht er nicht Rechenschaft zu geben, wohl aber von den Folgerungen hieraus"[35]. | 421

Zum ersten Male faßbar ist diese Tendenz der griechischen Geometrie, sich als axiomatische Theorie zu etablieren, bei dem durch seine Möndchenquadraturen, also die Quadraturen krummliniger Figuren, berühmt gewordenen Mathematiker HIPPOKRATES von CHIOS. In dem im wesentlichen auf EUDEM zurückgehenden Geometerkatalog findet sich bei PROKLOS die Notiz, HIPPOKRATES habe als erster *Στοιχεῖα*, nämlich „Elemente" im Stile EUKLIDS, geschrieben[36], und das ebenfalls aus EUDEM stammende Referat des SIMPLIKIOS über die Möndchenquadraturen hebt gleich zu Beginn ausdrücklich hervor, daß HIPPOKRATES sich eine *ἀρχή* gemacht und zum Beweis eines Satzes andere Sätze herangezogen habe[37]. Interessant ist dabei nun, daß dort, wo zum ersten Male Anfänge einer axiomatischen Theorie sichtbar

---

gewesen sein. Vgl. O. BECKER, a.a.O., S. 38; K. v. FRITZ, Gleichheit, Kongruenz und Ähnlichkeit in der Mathematik bis auf Euklid, Arch. f. Begriffsgesch. IV, S. 51 f. Ausführliche Erörterung dieses voreuklidischen Beweises bei TH. HEATH, The thirteen books of Euclid's Elements I, S. 252 ff.

34 P. LORENZEN, Das Begründungsproblem der Geometrie als Wissenschaft der räumlichen Ordnung, Philosophia Naturalis VI (1960), S. 415–431. Wiederabgedruckt in: Das Problem der Ordnung (Kongreßberichte München 1960), Meisenheim 1962, S. 58–73.

35 *ἐπειδὴ τὴν ἐπιστήμην ταύτην τὴν γεωμετρίαν ἐξ ὑποθέσεως εἶναί φαμεν καὶ ἀπὸ ἀρχῶν ὡρισμένων τὰ ἐφεξῆς ἀποδεικνύναι ... ἀνάγκη δή που τὸν τὴν ἐν γεωμετρίᾳ στοιχειώσιν συντάττοντα χωρὶς μὲν παραδοῦναι τὰς ἀρχὰς τῆς ἐπιστήμης, χωρὶς δέ τὰ ἀπὸ τῶν ἀρχῶν συμπεράσματα, καὶ τῶν μὲν ἀρχῶν μὴ διδόναι λόγον, τῶν δὲ ἑπομένων ταῖς ἀρχαῖς* (PROCL. in Eucl. 75, 6–14 FRIEDLEIN). Übersetzung nach der deutschen Ausgabe von P. L. SCHÖNBERGER/M. STECK, Halle 1945, S. 218.

36 PROCL. in Eucl. 66, 7–8 FRIEDLEIN. Vgl. W. BURKERT, *ΣΤΟΙΧΕΙΟΝ*. Eine semasiologische Studie, Philologus 103 (1959), S. 193 ff.

37 SIMPL. in Arist. Phys. I 2 comment. 61, 5–7 DIELS (Comm. in Arist. Graeca IX, Berlin 1882) (EUDEM Fr. 140 WEHRLI). Vgl. die vorzügliche Separatausgabe von F. RUDIO, Der Bericht des Simplicius über die Quadraturen des Antiphon und des Hippokrates, Leipzig 1907, S. 48.

werden, zugleich eleatischer Einfluß bemerkbar ist. Wenn Hippokrates z. B. jene Fälle unterscheidet, bei denen der äußere Bogen des Möndchens gleich, größer oder kleiner als der Halbkreis ist, um auf diese Weise alle möglichen Fälle zu diskutieren[38], so ist mit Händen zu greifen, daß er darin methodisch dem bereits bei Parmenides selbst in den sogenannten „drei Wegen der Forschung" faßbaren eleatischen Argumentationsschema folgt, das in der Dialektik zunächst die Unterscheidung verschiedener möglicher Behauptungen und schließlich die Eliminierung der falschen
422 Behauptungen verlangt[39]. | Die These, zwischen eleatischer Reflexion auf die Form des Argumentierens und dem Entstehen der Mathematik, speziell der Geometrie, als beweisender Wissenschaft bestehe ein Abhängigkeitsverhältnis derart, daß erst jene Reflexion diese Wissenschaft ermöglicht habe, ist also nur richtig, insofern man dabei an eine Geometrie nach euklidischem Muster denkt, die ja zumindest Vertrautheit mit logischen Schlüssen voraussetzt, und die Möglichkeit thaletischer Geometrie einmal außer acht läßt. Faktisch hat sich denn auch die „Eleatisierung" der griechischen Geometrie so ausgewirkt, daß man jene beinahe 200 Jahre ältere Möglichkeit gänzlich wieder vergaß, die thaletischen Sätze im Sinne einer axiomatischen Theorie verstand, wie sie Proklos in seinen Worten zu rechtfertigen sucht, und diese „neue" Geometrie bald als Musterbeispiel des mit dem eleatischen Denken begonnenen „wissenschaftlichen" Denkens auffaßte. Wo bereits in der Antike über die Wissenschaftlichkeit der Wissenschaft diskutiert wird, geschieht dies ausdrücklich immer im Hinblick auf die Mathematik als exemplarische Wissenschaft.

III

Mit dem Hinweis, daß sich die antike Reflexion über die Wissenschaft an der Mathematik als exemplarisch verstandenen Wissenschaft orientiert, ist nun ein

---

[38] Damit sind natürlich nur alle möglichen Fälle des äußeren Kreisbogens, nicht aber alle möglichen Möndchen erschöpft. Die bereits mit Aristoteles (An. pr. B 25. 69a 30 ff.; Soph. El. 11. 171 b 14 ff.; Phys. A2. 185a 14 ff.) einsetzende Diskussion über das tatsächliche Niveau der hippokratischen Bemühungen wird von der zweifelhaften Vermutung gespeist, Hippokrates habe möglicherweise diesen Unterschied nicht beachtet. Zu dieser Diskussion vgl. K. v. Fritz, Die *ΑΡΧΑΙ* in der griechischen Mathematik, Arch. f. Begriffsgesch. I, S. 93.

[39] Zur eleatischen Methode der Fallunterscheidung vgl. K. Reinhardt, Parmenides und die Geschichte der griechischen Philosophie, Bonn 1916, S. 36 ff., 65 ff. Auf die Verwandtschaft zwischen dem eleatischen Argumentationsschema und dem Vorgehen des Hippokrates macht auch W. Burkert, Weisheit und Wissenschaft, S. 402 aufmerksam. Wo und wann Hippokrates mit dem Eleatismus oder einem bereits von diesem beeinflußten Denken vertraut wurde, läßt sich bei den spärlichen biographischen Zeugnissen nicht entscheiden. Immerhin stimmen diese Zeugnisse darin überein, daß Hippokrates ein vielgereister Mann war und sich längere Zeit auch in Athen aufgehalten hat (Arist. Eth. Eud. H 14. 1247a 17–20; Plutarch Vita Solonis 2; Philop. in Arist. Phys. comment. 31, 3–5 Vitelli [Comm. in Arist. Graeca XVI, Berlin 1897]). Pythagoreischer Einfluß — einmal angenommen, es hätte eine vorhippokratische pythagoreische Geometrie überhaupt gegeben (dazu W. Burkert, a.a.O., S. 425 ff.) — läßt sich nicht nachweisen, Aristoteles macht einen deutlichen Unterschied zwischen den, wie er sich auszudrücken pflegt, „sogenannten Pythagoreern" und Hippokrates (Meteor. A 6.342b 30 ff.) bzw. der Schule des Hippokrates (*οἱ περὶ Ἱπποκράτην*, Meteor. A 7.344 b 15). Daß auch eine leicht mißverstädliche Formulierung bei Iamblich (De communi mathematica scientia, ed. N. Festa, Leipzig 1891, S. 77, 18 ff.) keinen begründeten Hinweis auf eine Verbindung zwischen Hippokrates und Pythagoreern enthält, hat bereits F. Rudio (a.a.O., S. 97 ff.) hervorgekehrt.

weiterer Gesichtspunkt geltend gemacht, der geeignet ist, die hier begonnene Betrachtungsweise zu rechtfertigen. In Anlehnung an aristotelischen Sprachgebrauch erblickt man üblicherweise im „Denken des Allgemeinen", also in generellen Sätzen, die besondere Leistung und wissenschaftsbildende Eigenart des griechischen Geistes. Dieses Denken manifestiert sich nun zweifellos auch in der Aufstellung jener thaletischen theoretischen Sätze (das sind ebenfalls generelle Sätze!), von denen bisher die Rede war, ließe sich aber ebenso gut auch in den spekulativen Äußerungen frühgriechischer Naturphilosophie über das Seiende als Ganzes nachweisen. Wenn von diesen Spekulationen hier ganz abgesehen wird, dann einmal darum, weil — wie bereits hervorgehoben — mit der Theorienbildung und den dabei vorausgesetzten Entdeckungen der Möglichkeit theoretischer Sätze und Beweise für den zukünftigen Gang der Wissenschaft die entscheidenden Schritte getan sind, zum anderen aber, weil eben auch in der Antike schon die Frage, wie Wissenschaft möglich ist und wie sie im einzelnen verfahren soll, am Beispiel der Mathematik erörtert wird. Allerdings tritt diese Frage, wie sich zeigen wird, nicht unmittelbar mit den Anfängen der Mathematik auf. Es ist keineswegs so, daß etwa die griechische Geometrie sofort begonnen hätte, über ihr eigenes Tun, z. B. also über die Auszeichnung gewisser Sätze als erster Sätze bei ihrer Konstituierung als axiomatischer Theorie zu reflektieren. Diese Reflexion wird vielmehr erst bei Platon und Aristoteles in einem ersten Versuch nachgetragen, wobei deutlich wird, daß gewisse Fragen in der griechischen Geometrie bislang entweder gar nicht gestellt oder doch zumindest vernachlässigt wurden.

Vielleicht die wichtigste dieser Fragen ist die nach der Idealität geometrischer Gegenstände. Bei der Erörterung der thaletischen Entdeckung des Beweises war gesagt worden, daß diese Entdeckung möglicherweise durch die Frage erzwungen wurde, wovon denn in den theoretischen Sätzen überhaupt die Rede ist, wobei die beim Beweisen vorgenommenen Handlungen, etwa an der thaletischen Grundfigur, dann dazu dienten, sich der idealen Gegenstände, über die man sprach, zu vergewissern. Dieser Vorgang des Sichvergewisserns wird nun im Verlauf | der Axiomatisierung der Geometrie in zwei 423
Schritte zerlegt; man hat sich jetzt einerseits der Axiome, andererseits der Folgerungen aus diesen Axiomen zu vergewissern. Ist dies nun bei den Axiomen einmal geschehen, was z. B. durchaus im thaletischen Sinne nach wie vor möglich ist, so braucht im folgenden die Frage nach der Idealität geometrischer Gegenstände nicht mehr behandelt zu werden. Es genügt dann in der Regel, für einen gewünschten Beweisgang die Axiome einfach beizubringen, ohne explizit diese Frage noch zu beantworten. Es liegt mithin nahe, die Frage nach der Idealität geometrischer Gegenstände überhaupt nicht mehr als besonders dringlich anzusehen, und tatsächlich scheint man sich auch über den Status dieser Gegenstände, von denen man etwa bei der Formulierung von Sätzen über Winkel spricht, weiter keine Gedanken gemacht zu haben. Um so gefährlicher aber mußte dann gerade darum ein im Grunde so naiver Einwand gegen die Geometrie als Wissenschaft erscheinen, den Protagoras erhoben haben soll, indem er sich unter Berufung auf empirische Feststellungen gegen den geometrischen Satz wandte, daß eine Tangente eine Gerade sei, die den Kreis nur in einem Punkte beühr[40]. Natürlich wußten auch damals

[40] Arist. Met. B 2.998a 2–4. Vgl. H.-G. Gadamer, Dialektik und Sophistik im siebenten platonischen Brief, SB Heidelberg, philos.-hist. Kl. 1964, 2. Abh., S. 18 Anm. 24, der mit Recht unter Hinweis auf diese Episode hervorhebt, „wie schwierig damals noch für einen Mathematiker seine methodische Selbstrechtfertigung war".

die Geometer, daß sie bei der Formulierung dieses Satzes z. B. nicht über einen an einem Rad lehnenden Stock sprachen, woran PROTAGORAS bei seinem Einwand gedacht haben mag; worüber sie nun wirklich sprachen, konnten sie aber offenbar auch nicht sagen.

Der Hinweis darauf, daß es sich empirisch ganz anders verhalte, als gewisse Sätze der Geometrie es formulierten, war aber nicht nur geeignet, Mathematiker, die mit diesen Sätzen umgingen, zu irritieren, ihnen gleichsam bewußt zu machen, sie wüßten nicht, worüber sie sprächen — er hatte vielmehr noch einen weiteren unangenehmen Begleiteffekt, da durch ihn nun auch jene Bemühungen diskreditiert erschienen, die von der Geometrie zu anderen wissenschaftlichen Zwecken Gebrauch machten. Hier ist vor allem an die wahrscheinlich auf EUDOXOS zurückgehende Einführung von geographischen Koordinaten zu denken, die schematische Einteilung der Erdoberfläche in Längen- und Breitengrade, mit deren Hilfe wiederum auf geometrischem Wege die Lage einzelner Orte, ganz unabhängig von bestimmten geographischen Eigenheiten, zueinan-
424 der bestimmbar wurde[41].|Diese künstliche Geometrisierung der Wirklichkeit, wie sie seit EUDOXOS dann auch in der Astronomie praktiziert wird, beruht nun auf der klaren methodischen Einsicht, daß der Mensch von sich aus Gliederungen und Unterscheidungen treffen muß, um der verwirrenden Vielfalt der Erscheinungen beizukommen — sie ist insofern auch dem schlichteren Versuch des HEKATAIOS weit überlegen, der um — 500 wohl zugunsten besserer Überschaubarkeit auf seiner Weltkarte geographische Gegebenheiten durch geometrische Figuren, Quadrate und Trapeze, wiedergab[42] und sich damit den Spott HERODOTS zuzog[43] —, doch dürfte wohl kein Zweifel darüber bestehen, daß dieses methodische Vorgehen empfindlich an

---

[41] Die Datierung dieser „Entdeckung" bereitet einige Schwierigkeiten, da sich eindeutige Zeugnisse nicht anführen lassen. Alles spricht jedoch für EUDOXOS, dessen Planetentheorie von einer bis dahin unbekannten Einsicht in die Leistungsfähigkeit mathematischer Vernunft beim Entwurf kinematischer Modelle für die Erklärung der Phänomene zeugt. Zweifellos hat er auch in seiner Geographie, die erstmals auf eine Erdkugellehre hin entworfen ist, mathematische und astronomische Kenntnisse als Hilfsmittel eingesetzt. Möglicher Vorläufer: der Astronom (STRAB. I,29) BION von ABDERA (DIOG. LAERT. IV, 58; zur Datierung vgl. jetzt W. BURKERT, a.a.O., S. 284). Greifbar ist dieses Programm einer „wissenschaftlichen", sich nicht in reiner Länderkunde erschöpfenden Geographie in EUDOXOS' Behandlung der Zonenlehre (vgl. F. GISINGER, Geographie, RE Suppl. IV 578 ff.). Er greift damit erstmals bei PARMENIDES (POSEIDONIOS bei STRAB. II,94) nachweisbare Versuche auf, die ebenfalls schon von einer Projektion astronomischer Einteilungen auf die Erdoberfläche ausgingen (für PARMENIDES selbst wohl zurecht noch bezweifelt von W. BURKERT, ebd.; ein ausführliches Lehrstück über die astronomische Behandlung der Zonenlehre findet sich bei ARIST. Meteor. B 5.362b 6 ff., dazu H. BERGER, Geschichte der wissenschaftlichen Erdkunde der Griechen, 2. Aufl. Leipzig 1903, S. 301 ff., vgl. auch S. 206ff.). Während es bei den früheren Versuchen jedoch im wesentlichen auf „natürliche" Einteilungen ankam, Bestimmung der Klimagürtel etc., scheint EUDOXOS von vornherein der theoretische Charakter solcher Einteilungen, ihre methodische, von „natürlichen" Gegebenheiten unabhängige Zweckmäßigkeit bewußt gewesen zu sein. Daß er sich um geographische Breitenbestimmungen bemüht hat, läßt sich aus STRAB. II, 119 erschließen; dazu H. BERGER, a.a.O., S. 247 f. Vgl. im übrigen F. GISINGER, Die Erdbeschreibung des EUDOXOS von KNIDOS, Berlin 1921, S. 15 ff. und zu der hier vertretenen Ansicht K. v. FRITZ, Der Beginn universalwissenschaftlicher Bestrebungen und der Primat der Griechen, Studium Generale 14 (1961), S. 555.

[42] Zur Zerlegung des Kartenbildes in schematische „Großräume" vgl. HERODOT II 21. 16. 33–34; IV 8. 36. Eine detaillierte Darstellung gibt F. JACOBY, Hekataios, RE VII, 2.2667 ff. Auf die Abhängigkeit des HEKATAIOS von ANAXIMANDER macht CH. H. KAHN, Anaximander and the origins of Greek cosmology, New York 1960, S. 81 ff., aufmerksam.

[43] IV 36.

Überzeugungskraft einbüßen muß, wenn sich herausstellt, daß die Hilfsmittel, die hier eingesetzt werden, selbst noch nicht hinreichend einsichtig und methodisch begründet sind.

In einem weiteren Sinne ist dabei auch die in mancher Beziehung so unglückliche und in der Geschichte des europäischen Denkens oft verhängnisvolle Unterscheidung zwischen einer „wirklichen" und einer „wahren" Welt Ausdruck der in der Geometrie aufgeworfenen Frage nach dem Verhältnis idealer Gegenstände und denen der alltäglichen Erfahrung. In der Arithmetik kann diese Frage gar nicht erst auftreten, da es sich hier von vornherein um ein schematisches Operieren mit eigens „erfundenen" Symbolen handelt; wo sie aber dennoch, wie in der pythagoreischen Identifizierung von Zahlen und Dingen, diskutiert wurde, ist ihre Fruchtlosigkeit und Hinfälligkeit evident. Dieser „Vorteil" der Arithmetik gegenüber der Geometrie in der Konstituierung und Abgrenzung ihrer Gegenstände ist schon in der Antike klar erkannt worden; Proklos z. B. schreibt: „Daß die Zahlen stoffloser (ἀυλότεροι) und reiner (καθαρώτεροι) als die geometrischen Größen sind, und daß die ἀρχή der Zahlen einfacher (ἁπλουστέρα) als diejenige der geometrischen Größen ist, leuchtet einem jeden ein"[44]. Die kompliziertere Lage in der Geometrie prägt sich, historisch gesehen, dann so aus, daß sie einerseits jener, „wirkliche" und „wahre" Welt unterscheidenden Zweiweltentheorie Vorschub leistet, andererseits aber auch zu einem empiristischen Standpunkt führt, der die Geometrie „aus den Dingen" herzuleiten und ihre Sätze unter Hinweis auf diese Dinge zu beweisen sucht, und den z. B. Protagoras zu teilen scheint. Beide Auffassungen sind in gewisser Weise naiv: erstere muß sich den in der Sophistik erhobenen Vorwurf gefallen lassen, die Rede von einer | „wahren" Welt habe 425
keinen angebbaren Sinn, da man diese Welt, selbst wenn es sie gäbe, nicht von der wirklichen unterscheiden könne, letztere muß darauf verzichten, zu erklären, wie es zur Aufstellung wahrer theoretischer Sätze kommt.

Damit dürfte aber nun vollends deutlich sein, daß es in der Antike sowohl bei der Frage, wie Geometrie als Wissenschaft möglich ist, als auch bei der Diskussion über die Anwendbarkeit geometrischer Verfahren und deren exemplarischen Charakter für jede wissenschaftliche Bemühung des Menschen in entscheidendem Maße auf eine Klärung der latenten Frage ankam, wie man den idealen Charakter der geometrischen Gegenstände zu verstehen habe.

Der erste, der diese Frage bewußt exponiert hat, und dies charakteristischerweise im Zusammenhang mit der Frage nach dem Wesen von Wissenschaft überhaupt, war Platon. Gegen Ende des 6. Buches heißt es an der bekannten Stelle im „Staat", daß die Mathematiker „das Gerade und Ungerade, die Figuren und die drei Sorten von Winkeln" voraussetzen, zu ὑποθέσεις machen, „als ob sie dies schon wüßten", und daß „sie es nicht für nötig halten, sich selbst oder anderen darüber Rechenschaft zu geben". Vielmehr täten sie so, fährt Platon fort, „als sei dies schon jedermann klar", und gingen sogleich von diesen Voraussetzungen aus zur Durchführung (nämlich der Beweise) über, „bis sie schließlich dorthin gelangen, auf dessen Untersuchung sie es abgesehen hatten"[45]. Erstaunlich an dieser berühmten platonischen Kritik mathematischer Verfahrensweisen ist zweifellos, daß sie die Mathematiker nicht, wie man

---

44 Procl. in Eucl. 95, 23–26 Friedlein.

45 Rep. 510C/D.

erwarten sollte, auf ihre Begründungspflicht ersten Sätzen gegenüber aufmerksam zu machen scheint, sondern offenbar nur daran Anstoß nimmt, wie diese Mathematiker von Geradem und Ungeradem, den Figuren und den Winkeln sprechen. Unter „Voraussetzungen" oder „Grundannahmen" (ὑποθέσεις) wären hier demnach gar keine Axiome verstanden, sondern eben jenes allzu selbstverständliche Sprechen von nicht-empirischen Gegenständen; und PLATONS Interesse bestünde mithin in diesem Zusammenhang allein darin, dieses Sprechen auf seine ungeklärten Voraussetzungen aufmerksam zu machen und selbst auf ein sicheres Fundament zu heben. Tatsächlich gipfeln PLATONS Erklärungen an dieser Stelle dann auch in dem Nachweis, daß auch die Mathematiker es mit *Ideen* zu tun hätten, die fraglichen Gegenstände der Geometrie also Ideen seien[46], und es lediglich einer dialektischen Anstrengung bedürfe, um diesen ihren Status einsichtig zu machen. Im Ergebnis hätte man damit PLATONS Antwort auf eine erstmals präzise gestellte Frage nach dem Wesen geometrischer Gegenstände vor Augen und gleichzeitig eine wohl einleuchtende Erklärung dafür, wieso PLATON hier von Figuren und Winkeln und nicht von Sätzen
426 spricht[47]. |

Nun ist mit der Erklärung, bei den geometrischen Gegenständen handele es sich in Wahrheit um Ideen, die der Mathematiker in der Regel nur nicht als solche erkenne, für das Verständnis der platonischen Antwort natürlich noch nicht alles getan, müßte man doch jetzt fragen, was es des näheren mit den Ideen und der Behauptung, etwas sei eine Idee, bei PLATON auf sich hat. Aus der Frage: was hat man unter den geometrischen Gegenständen zu verstehen? wird zunächst nur die Frage: was hat man unter den Ideen zu verstehen? Es sieht also so aus, als ob die Explikation der Ideenlehre die endgültige Antwort PLATONS auf die Frage nach den geometrischen Gegenständen geben könnte. Und doch wäre es ein böser Trugschluß, zu glauben, daß sich die Frage auf diese Weise, nämlich ohne ausdrückliche Berücksichtigung der besonderen Situation der Geometrie, leichter behandeln ließe. Alles spricht vielmehr dafür, daß das Faktum der Geometrie der eigentliche Ausgangspunkt der platonischen Bemühungen um eine Ideenlehre geblieben ist, und darum auch den vielfältigen Schwierigkeiten, die sich nach wie vor der Interpretation dieser Ideenlehre entgegenstellen, vermutlich neue Gesichtspunkte abgewonnen würden, wenn man die besondere Problematik der „geometrischen" Ideen mehr in den Mittelpunkt der Betrachtung rückte[48]. Eine solche Betrachtung würde allerdings der hier interessierenden Frage nach der Entdeckung von Wissenschaft und ihrem Verständnis kaum etwas beitragen und soll darum auch nicht weiter verfolgt werden.

---

46 Rep. 510 D, 511 D; vgl. 527 B.

47 Daß hier nicht Sätze gemeint sind, betonen bereits K. v. FRITZ, Die *ΑΡΧΑΙ* in der griechischen Mathematik, Arch. f. Begriffsgesch. I, S. 38 ff.; E. M. MANASSE, in: Philos. Rundschau, Sonderheft Platonliteratur II, Beiheft 2 (1961), S. 156; H. SCHMITZ, System der Philosophie I (Die Gegenwart), Bonn 1964, S. 77. — An anderen Stellen im Corpus Platonicum kommen natürlich ὑποθέσεις auch als Sätze vor (vgl. Phaid. 100 A etc.), doch sind damit keine speziell mathematischen Sätze, sondern allgemein Annahmen gemeint, über die man sich im Dialog zu einigen versucht oder bereits geeinigt hat (THEAIT. 155 B: ὁμολογήματα). Vgl. den schönen Abschnitt über die ὑποθέσεις bei Á. SZABÓ, Anfänge des euklidischen Axiomensystems. Arch. f. hist. of ex. sc. I, S. 43 ff.

48 „Die Idealität der geometrischen Gebilde *erklärt* geradezu, wie PLATON über SOKRATES hinaus zu seiner Ideenlehre weitergegangen ist", W. KAMLAH, Platons Selbstkritik im Sophistes (Zetemata 33), München 1963, S. 4.

Für eine historische Untersuchung genügt hier die Feststellung, daß PLATON die auf die Idealität ihrer Gegenstände nachdrücklich hingewiesene Geometrie, ohne des näheren noch auf ihre Verfahren einzugehen, in den Verband jener Wissenschaften aufnimmt, die nach Auskunft des „Staates" in besonderem Maße geeignet sein sollen, die Vernunft auf ihren Weg zu bringen[49]. Es handelt sich hierbei bekanntlich um Geometrie, Arithmetik, Astronomie und Musik, also um die später wahrscheinlich unter dem unmittelbaren Bildungseinfluß der Akademie zum sogenannten Quadrivium zusammengeschlossenen „exakten" Wissenschaften[50]. PLATON bezeichnet diese Wissenschaften als *μαθήματα*, als besondere „Lehrstücke" oder „Disziplinen", und macht hierin von einer Terminologie Gebrauch, die sich in der Bildungsbewegung des 5. Jahrhunderts immer mehr durchgesetzt hat, jedoch nicht auf die vier genannten „Fächer" beschränkt war[51]. Ein *μάθημα* ist im geläufigen Sinne
vielmehr jeder beliebige Gegenstand, sofern er sich nur | *lehren* bzw. *erlernen* läßt[52], 427
insbesondere also auch dasjenige, wovon die praktischen Künste, die *τέχναι*, wie etwa Politik und Medizin handeln. Tatsächlich treten auch in den platonischen Dialogen die Ausdrücke *τέχνη* und *μάθημα* nicht streng voneinander geschieden auf, wie im übrigen auch die Ausdrücke *τέχνη* und *ἐπιστήμη*, also jener Ausdruck, den wir mit „Wissenschaft" zu übersetzen pflegen, in der Regel, vor allem in den frühen Dialogen, miteinander vertauschbar sind[53]. Ansätze zu einer strengeren Differenzierung, wie sie dann bei ARISTOTELES vorgenommen wird[54], finden sich bei PLATON eigentlich nur in der Abgrenzung der *τέχνη* gegenüber der *ἐμπειρία*, der bloßen Erfahrung, eine Unterscheidung, die vor SOKRATES und PLATON nicht getroffen wurde[55], und erst in den späten Dialogen macht sich gelegentlich unter Hinweis auf die unterschiedliche Rolle der Vernunft in der Organisation des Wissens auch die Tendenz bemerkbar, *τέχνη* und *ἐπιστήμη* stärker voneinander abzuheben.

---

[49] Rep. 524 D ff.

[50] Zur Entstehungsgeschichte des Quadriviums vgl. PH. MERLAN, From Platonism to Neoplatonism, Den Haag 1953, S. 78 ff. Für pythagoreischen Ursprung: B. SNELL, Die Ausdrücke für den Begriff des Wissens in der vorplatonischen Philosophie (Philologische Untersuchungen 29), Berlin 1924, S. 77 ff.; ähnlich K. v. FRITZ, Mathematiker und Akusmatiker bei den alten Pythagoreern, SB München, philos.-hist. Kl. 1960, 11. Abh., S. 20f. Dagegen betont W. BURKERT im Rahmen seiner Destruktion „pythagoreischer" Wissenschaft den fächerbildenden Einfluß der Akademie (a.a.O., S. 399f.). Die Tradition spricht im wesentlichen vom „pythagoreischen" Quadrivium, vgl. PROCL. in Eucl. 35, 21 ff. FRIEDLEIN.

[51] Dazu wieder W. BURKERT, a.a.O., S. 201 Anm. 97.

[52] Vgl. TH. HEATH, A history of Greek mathematics I, S. 10; B. SNELL, a.a.O., S. 72 ff.; K. v. FRITZ, Der gemeinsame Ursprung der Geschichtsschreibung und der exakten Wissenschaften bei den Griechen, Philos. Nat. II, S. 203; ders. Mathematiker und Akusmatiker bei den alten Pythagoreern, SB München, philos.-hist. Kl. 1960, 11. Abh., S. 20.

[53] Vgl. die detaillierte Untersuchung von F. HEINIMANN, Eine vorplatonische Theorie der *τέχνη*, Museum Helveticum 18 (1961), S. 105 ff.; hier auch weitere ausführliche Literaturhinweise. Zur *τέχνη* bei PLATON vgl. auch H. J. KRÄMER, Arete bei Platon und Aristoteles. Zum Wesen und zur Geschichte der platonischen Ontologie, SB Heidelberg, philos.-hist. Kl. 1959, 6. Abh., S. 220 ff.; zu *ἐπιστήμη* K. v. FRITZ, Der Beginn universalwissenschaftlicher Bestrebungen und der Primat der Griechen, Stud. Gen. 14, S. 610.

[54] Met. A1.981 b 25 ff.; Eth. Nic. Z3.1139b 14 ff.; vgl. B. SNELL, a.a.O., S. 87.

[55] Dazu W. CAPPELLE, Zur hippokratischen Frage, Hermes 57 (1922), S. 262 ff.; F. HEINIMANN, a.a.O., S. 115 f.

Die besondere Rolle der Vernunft in den genannten μαθήματα: Geometrie, Arithmetik, Astronomie und Musik (gemeint ist hier eine rationale Harmonielehre) ist es nun auch, die diesen Wissenschaften in der platonischen Hierarchie des Wissens ihren bevorzugten Platz einbringt. Kriterium für die Wissenschaftlichkeit einer Wissenschaft ist nach PLATON einerseits die Existenz eines methodischen Begründungszusammenhangs, den er z. B. in der mathematischen Praxis durch die Unklarheit über gewisse Grundbegriffe empfindlich gestört sah, und andererseits die prinzipielle Unabhängigkeit von der Erfahrung, das Fehlen jeglicher empirischer Elemente. Beide Gesichtspunkte reduzieren aber Wissenschaft auf die Rolle, welche die Vernunft in ihnen spielt, und sofern diese seit THALES im griechischen Denken die Tendenz hat, sich insbesondere in Geometrie und Arithmetik zu verwirklichen, wird Wissenschaft selbst, mit einem Worte GERHARD KRÜGERS, „als grundsätzliche Besinnung auf die Möglichkeiten des Vernünftigen, Mathematik"[56]. Genau diese Eigenart, sich als mathematische Theorie darzustellen, aber zeigen auch die im „Staat" aufgeführten Disziplinen der Astronomie und Musik — jedenfalls so, wie sie sich PLATON vorstellt[57]; man könnte also meinen, daß sich für PLATON Wissenschaft im methodischen Aufbau aller vier Disziplinen, zusätzlich der als fünften noch genann-
428 ten Stereometrie, erschöpft, | oder diese Disziplinen doch zumindest als sozusagen reine Formen von Wissenschaft aufgefaßt werden. De facto stellen sie jedoch den Worten PLATONS nach lediglich die „Vorschule" zu einer quasi Über-Wissenschaft, der Dialektik, und der Erkenntnis des Guten dar.

Es ist zunächst schwer zu sagen, warum PLATON in einer Diskussion über Wissenschaft den genannten vier Wissenschaften lediglich diesen propädeutischen Charakter zuerkennt, und was die Gründe sind, die ihn zu dieser, allem Anschein nach in seinen Augen sogar notwendigen Unterscheidung veranlaßt haben mögen. Ein Methodenmangel der mathematischen Fächer kann jedenfalls kaum der Grund gewesen sein, da sich nicht einsehen läßt, warum die Mathematiker sich nicht PLATONS Kritik in dieser Richtung zu Herzen nehmen könnten, den Stein des Anstoßes entfernten und fortan Mathematik nach besten platonischen Maßstäben betrieben. Da PLATON aber auch weniger einen gegenwärtigen Zustand kritisieren will, als vielmehr ein anscheinend notwendiges Zurückbleiben der Mathematik hinter der Dialektik konstatieren möchte, bleibt nur übrig, den Grund in der Verschiedenartigkeit der *Objektbereiche*, den Zahlen und Figuren einerseits, den Ideen andererseits zu sehen. Nun wurde bereits hervorgehoben, daß es PLATON gerade darauf ankam, zu zeigen, daß es auch die Mathematiker mit Ideen zu tun haben; also könnte es sich bei der Verschiedenartigkeit der Objektbereiche nur um zwei verschiedene Sorten von Ideen handeln. In der Tat kommt die Idee des Guten, um die es PLATON in erster Linie geht, unter den mathematischen Ideen nicht vor. Sie allein aber ermöglicht, wie PLATON darzustellen sucht, vernünftige Aussagen über eine richtige Lebensführung. Diese Aussagen wiederum sucht PLATON in den Mittelpunkt seines Philosophierens zu rücken; seiner Überzeugung nach ist es die eigentliche Aufgabe der Vernunft,

---

[56] G. KRÜGER, Grundfragen der Philosophie. Geschichte Wahrheit Wissenschaft, Frankfurt/M 1958, S. 170.

[57] Vgl. Verf. Die Rettung der Phänomene. Ursprung und Geschichte eines antiken Forschungsprinzips, Berlin 1962, S. 117 ff.

„praktisch" zu sein. Und unter dieser Auszeichnung der praktischen gegenüber der theoretischen Vernunft muß sich allerdings dann die Mathematik treibende Vernunft mit jener bloß propädeutischen Funktion begnügen, wobei man hinzufügen darf, daß auch innerhalb der platonischen Philosophie immer wieder Schwierigkeiten dadurch entstehen, daß die ins Auge gefaßte „praktische Philosophie" quasi als eine theoretische Lehre aufgestellt werden soll. Man wird jedenfalls sagen können, daß es PLATON an der hier interessierenden Stelle im „Staat" — vielleicht auch wieder aus dem eben angeführten Grunde — schwerlich gelungen ist, die Relevanz jener Unterscheidung zwischen zwei verschiedenen Sorten von Ideen hinsichtlich seines Wissenschaftsbegriffes wirklich einsichtig zu machen. Von den Schwierigkeiten, die sich ihm hier offenbar entgegengestellt haben, zeugen nicht zuletzt auch seine keineswegs klaren Mitteilungen, die in Verbindung mit der aristotelischen Behauptung, PLATON habe die mathematischen Gegenstände als eine dritte Klasse von Seiendem zwischen den Dingen und den Ideen angesehen[58], in der Platonforschung zu einer kaum mehr überschaubaren Dis|kussion darüber geführt haben, was PLATON 429
nun wirklich gemeint haben könnte[59].

Wie immer man sich aber auch angesichts dieser Diskussion entscheiden mag, so viel dürfte wenigstens sicher sein, daß die Mathematik für PLATON bei der Bestimmung der wahren Wissenschaft eine Orientierungsfunktion ausübt, ohne selbst jedoch den Status einer solchen Wissenschaft zuerkannt zu bekommen. Dies liegt aber gerade daran, daß der platonische Wissenschaftsbegriff im Grunde ein *normativer* Begriff ist, den sozusagen an der Sache zu begründen, PLATON zwar nicht recht gelingt, dessen historische Wirksamkeit man jedoch nicht zuletzt daran erkennen kann, daß bis hin zu GALILEI — jedenfalls in der platonischen Tradition — die in ihrer Bindung an die Empirie sogar noch unter der Mathematik stehende Naturerkenntnis keinen wissenschaftlichen Rang besaß und auch auf dem Umweg über die aristotelische spekulative Physik als Wissenschaft nicht möglich wurde.

Nun hat es schon in der Antike nicht an Versuchen gefehlt, über diesen normativen Wissenschaftsbegriff hinauszukommen. Den ersten Versuch in dieser Richtung unternimmt bereits ARISTOTELES, indem er genauer noch als PLATON den Aufbau einer, wie er jetzt sagt, „beweisenden Wissenschaft" (ἀποδεικτικὴ ἐπιστήμη) untersucht und dieser Wissenschaft die Dialektik bewußt unterordnet. Allerdings handelt es sich hierbei um eine von vornherein um die platonische Ideenlehre verkürzte Dialektik, in der

---

58 Met. A 6.987b 14–18; B1.995b 15–18; B2.997a 35–b 3; K1.1059b 3–8 und öfter. Vgl. H. CHERNISS, The riddle of the early Academy, Berkeley 1945, S. 75 f.; dort auch weitere Stellenangaben. Zur recht „aristotelisch" klingenden Begründung (μαθηματικά zwar „ewig" und „unbewegt", aber „zahlreich") vgl. Met. A6.987b 14–18; B 6.1002b 14–16. Zur Beurteilung dieser These vgl. W. D. ROSS, Aristotle's Metaphysics, a revised text with introduction and commentary, Oxford 1924, I, S. 166 ff. K. GAISER, der im Anschluß an H. J. KRÄMER (Arete bei Platon und Aristoteles) den höchst zweifelhaften Versuch unternimmt, die gesamte, nun als „exoterisch" apostrophierte Philosophie der platonischen Dialoge auf eine „esoterische" Zahlenspekulation, wie sie bei SPEUSIPP, XENOKRATES und PHILIPPUS von OPUS greifbar wird, hin zu interpretieren, übenimmt die aristotelische These als gesichert (Platons ungeschriebene Lehre. Studien zur systematischen und geschichtlichen Begründung der Wissenschaften in der Platonischen Schule, Stuttgart 1963, S. 91 ff.). Zu GAISERS Buch vgl. die ausführlichen Besprechungen von K.-H. ILTING in: Gnomon 37 (1965), S. 131 ff., Verf. in: Philosophische Rundschau 13 (1965).

59 Vgl. Literaturbericht H. CHERNISS in: Lustrum 5 (1960), S. 388 ff.

es nicht, um mit PLATON zu reden, darum geht, sich des „wahrhaft Seienden" zu vergewissern, sondern darauf ankommt, geeignete Prämissen zu finden, die, sofern sie der Dialogpartner akzeptiert, diesen schließlich zur Übernahme eines zuvor behaupteten Satzes zwingen. Wichtig im Sinne der anfangs aufgestellten These, daß die antike Wissenschaftsdiskussion immer am Leitfaden der Mathematik als exemplarischer Wissenschaft geführt wird, ist dabei, daß auch ARISTOTELES das Wesen der von ihm intendierten „beweisenden" Wissenschaft am Beispiel der Mathematik deutlich zu machen sucht. Und wie sich bereits PLATON keineswegs damit zufriedengab, lediglich auf den wissenschaftlichen Gang der Mathematik hinzuweisen, sondern selbst in der Frage nach dem Status mathematischer Gegenstände in die interne Fachdiskussion eingriff, so beschränkt sich nun auch ARISTOTELES nicht auf einen solchen Hinweis. Er unternimmt es vielmehr im Rahmen seiner Beschreibung der beweisenden Wissenschaft, den Aufbau der Geometrie methodisch neu zu formulieren, indem er diese erstmals konsequent als umfassende axiomatische Theorie darzustellen versucht.

In gewisser Weise könnte man sagen, daß PLATON und ARISTOTELES dabei, entgegen der sonst vorherrschenden Meinung, die in den aristotelischen Bemerkungen zur Geometrie einen Rückschritt gegenüber PLATON sehen will, Hand in Hand arbeiten. PLATON hatte nach den Gegenständen gefragt, um die es in der Geometrie geht, und hierbei speziell die Unterscheidung zwischen Axiomen und Folgerungen dahingestellt
430 sein lassen; ARISTOTELES dagegen will sich nun gerade | um diese Unterscheidung, den Aufbau der Sätze in der Mathematik kümmern, indem er zugleich PLATONs Korrektur, zumindest ihrer Tendenz nach, akzeptiert[60]. Seine berühmte Abstraktionstheorie, nach der wir die geometrischen Figuren aus den Dingen selbst gewinnen, indem wir von deren materieller Bedingtheit absehen und lediglich ihre Form betrachten[61], läßt sich nämlich, mit ein wenig gutem Willen, auch als bloße Variation des platonischen Standpunktes verstehen. Nur die Akzente wären demnach verschieden gesetzt: wenn PLATON erklärt, daß es sich bei den mathematischen Gegenständen um Ideen handele, so kommt es ihm vor allem auf deren *nicht-empirischen* Charakter an; ARISTOTELES wiederum betont auf dem Hintergrund seines Begriffssystems, in dem die Begriffe Form (μορφή) und Materie (ὕλη) die entscheidende Rolle spielen, daß diese Gegenstände die *mögliche*, aber nie restlos *realisierte* Form *wirklicher* Dinge darstellen. Gemeint ist im großen und ganzen dasselbe, gleichviel, ob man nun lieber mit PLATON von Ideen oder mit ARISTOTELES von reinen Formen sprechen will.

Wenn es dennoch eine aristotelische Kritik der Position PLATONs in dieser Frage gibt, so beruht diese auf dem Verdacht, in der platonischen Sprechweise von Ideen könnte es sich um Existenzbehauptungen im Sinne einer naiven Zweiweltentheorie handeln. Die Reflexion, und das wird gerade in dieser Diskussion besonders deutlich,

---

60 K. v. FRITZ glaubt allerdings in An. post. A 10.76 b 35–39 eine direkte Kritik an PLATONS Stellungnahme im „Staat" sehen zu können (Die *APXAI* in der griechischen Mathematik, Arch. f. Begriffsgesch. I, S. 38 ff.). Doch selbst wenn ARISTOTELES in seiner Erklärung, unter ὅροι dürfe man keine ὑποθέσεις verstehen, tatsächlich auf entsprechende Partien im „Staat" anspielt, könnte es sich doch lediglich um eine terminologische Korrektur handeln.

61 Phys. B 2.193 b 22–194 a 12; de cael. *Γ* 1.299a 15–17; de an. A 1.403 b 11–16; *Γ* 7.431 b 13–16; Eth. Nic. Z 9.1142a 11–20. Vgl. K. REIDEMEISTER, a.a.O., S. 84 f.; O. BECKER, Grundlagen der Mathematik in geschichtlicher Entwicklung, S. 118 ff.

entbehrt eben noch der methodischen Sicherheit, die dazu nötig ist, um sozusagen jenseits von Empirismus und Hypostasierung, oder auch nur der Befürchtung, in eines dieser Extreme zu fallen, im Sprechen über die Idealität geometrischer Gegenstände *Homogenitätsforderungen* an wirkliche Gegenstände zu sehen. So jedenfalls lautet der Vorschlag, den Hugo Dingler wiederholt vorgetragen und den in unseren Tagen Paul Lorenzen wieder aufgegriffen und weitergeführt hat. Durch diesen Vorschlag läßt sich das eigentümliche Verhältnis, das die idealen geometrischen Gegenstände den wirklichen Gegens tänden gegenüber haben und das bei einem nur axiomatischen Aufbau der Geometrie unberücksichtigt bleibt, näher bestimmen und erstmals wirklich einsichtig machen. Um ein Beispiel zu geben: Ein Oberflächenstück heißt *eben*, wenn sowohl die Punkte auf diesem ebenen Stück als auch die beiden Seiten oberhalb und unterhalb des Stückes in bezug auf dieses Oberflächenstück ununterscheidbar sind. Die Punkte der Ebene heißen dabei ununterscheidbar, wenn jede Aussage über einen Punkt auch über jeden anderen Punkt gilt, und die beiden Seiten der Ebene heißen in bezug auf die Ebene ununterscheidbar, wenn jede Aussage über die Ebene und einen Punkt oberhalb von ihr auch über jeden Punkt unterhalb von ihr gilt. Die Ebene verhält sich, euklidisch gesprochen, zu allen ihren Punkten und zu ihren beiden Seiten „gleichartig“[62]. Auf ähnliche Weise, | nämlich durch *Invarianzforderungen*, 431
geht man etwa auch von den „wirklichen“ Wörtern zu den „idealen“ Begriffen über: Über den *Begriff* „rot“ spricht man, wenn man über das *Wort* „rot“ spricht, allerdings nur, wenn man eben auf *invariante* Weise spricht, d. h. so, daß die Sätze über „rot“ wahr bleiben, wenn man „rot“ durch gewisse andere Wöter, etwa „red“ oder „rouge“ ersetzt[63].

Wichtiger als die sogenannte Abstraktionstheorie, mit der Aristoteles auf Platons Klärungsversuch antwortet, ist jedoch im Sinne des eingangs hervorgehobenen Gesichtspunktes der Theorienbildung, wie er für die griechische Idee der Wissenschaft charakteristisch ist, jene Beschreibung einer beweisenden Wissenschaft, die Aristoteles in den ersten Kapiteln der „Zweiten Analytiken“ gibt. Diese Wissenschaft, die gleichsam das Modell jeder zukünftigen strengen Wissenschaft sein soll, stellt sich als ein System von Sätzen dar, die untereinander durch logische Abhängigkeiten ausgezeichnet sind, deren formales Beweismittel der Syllogismus

---

[62] Vgl. H. Dingler, Die Grundlagen der Physik. Synthetische Prinzipien der mathematischen Naturphilosophie, 2. Aufl. Berlin-Leipzig 1923, S. 151 ff.; ders. Das Experiment. Sein Wesen und seine Geschichte, München 1928, S. 56 ff. („Wir denken uns in unserer Realität eine Gestalt derart, daß wir sie von zwei Seiten aus betrachten können, eine Gestalt der Art, die wir eine ‚Fläche‘nennen, und bestimmen sie näher dadurch, daß es unmöglich sein soll, sowohl im ganzen als an irgendeiner Stelle ihre beiden Seiten an sich selbst (abgesehen von der Umrandung) zu unterscheiden“ [57]...„Sie ist zunächst für sich betrachtet lediglich eine ‚Forderung‘ oder ein ‚Verfahren‘, um eine vorher in der Realität nur der Möglichkeit nach vorhandene Form in dieser zu bestimmen“ [58]); ders. Aufbau der exakten Fundamentalwissenschaft, aus dem Nachlaß hrsg. v. P. Lorenzen, München 1964, S. 175 ff. — P. Lorenzen, Das Begründungsproblem der Geometrie als Wissenschaft der räumlichen Ordnung, Philos. Nat. VI, S. 425 ff.

[63] Vgl. K. Lorenz / J. Mittelstrass, Die Hintergehbarkeit der Sprache, Archiv für Philosophie 13 (1965). Es ist dabei sinnvoll, diese Invarianzforderungen an Aussagen über *Wörter* von jenen Homogenitätsforderungen an wirkliche Gegenstände — und das sind ebenfalls Invarianzforderungen an Aussagen, aber nun an Aussagen über *wirkliche Gegenstände* — zu unterscheiden, weil die Erfüllbarkeit dieser Forderungen im ersten Fall allein von sprachlichen Vereinbarungen, im zweiten Fall jedoch von technischen Verfahren abhängt.

ist[64], und von denen es einige gibt, die, durch Syllogismen nicht mehr beweisbar, unbewiesen als erste Sätze am Anfang des Systems stehen[65]. Zweifellos orientiert sich ARISTOTELES hierbei an dem axiomatischen Aufbau der Geometrie, wie er in Ansätzen zuerst bei HIPPOKRATES greifbar ist, befaßt sich nun aber explizit insbesondere mit der Form jener unbewiesenen Sätze, um diesen Aufbau erstmals wirklich zu rechtfertigen und gleichzeitig rein durchzuführen. Er schaltet sich damit in eine Diskussion ein, die seinen eigenen Worten nach bisher wenig erfolgreich geführt und in der Antithese steckengeblieben war, entweder den Gedanken einer streng beweisenden Wissenschaft überhaupt aufzugeben, weil die Beweispflicht offenbar in einen unendlichen Regreß führe, oder aber den Zirkelbeweis zuzulassen[66]. Dagegen behauptet ARISTOTELES nun, daß es durchaus sinnvoll sei, von unbeweisbaren Sätzen zu sprechen, und im übrigen auch die anspruchsvollste Form der Wissenschaft nicht auf solche Anfangssätze verzichten könne. Diese Sätze müssen allerdings folgende Bedingungen erfüllen: sie sollen wahr (ἀληθεῖς), elementar (πρῶται) und unvermittelt (ἀμέσοι), einleuchtender
432 (γνωριμώτεραι) und fundamentaler (πρότεραι) als die aus ihnen | abgeleiteten Sätze und schließlich Gründe (αἴτιαι) der abgeleiteten Sätze sein[67]. Als Beispiel eines Satzes, der diese Bedingungen erfüllt, führt ARISTOTELES mit besonderer Vorliebe den bei EUKLID als 3. Axiom aufgenommenen Satz an, daß Gleiches von Gleichem abgezogen Gleiches ergibt[68]. Dieser Satz gilt nach ARISTOTELES κατ᾽ ἀναλογίαν in allen Wissenschaften als ein Axiom[69], er ist insofern schlechterdings allgemeingültig und noch von jenen ersten Sätzen deutlich unterschieden, die — wie der Satz, daß alle rechten Winkel einander gleich sind, in der Geometrie (EUKLID Post. 4) — in speziellen Wissenschaften unbewiesen und durch Syllogismen unbeweisbar am Anfang stehen[70].

Mit der Behauptung, jede strenge Wissenschaft müsse im beweisenden Aufbau ihrer Sätze irgendwo sozusagen einen Beweisverzicht leisten und eine Fundierung in allgemeinen und speziellen Axiomen hinnehmen, ist nun eine Position erreicht, die für die neuzeitliche Wissenschaftstheorie charakteristisch werden sollte. Es scheint insbesondere auch jener „konventionalistische", in der Geometrie z. B. von HILBERT vertretene Gedanke nicht mehr allzu fern zu liegen, nach dem die Wahl solcher Axiome mehr oder weniger in das Belieben des einzelnen gestellt ist, prinzipiell immer auch andere Sätze als Axiome ausgezeichnet werden könnten. Schon die Forderungen, die ARISTOTELES mit der Aufstellung solcher erster Sätze verknüpft, zeigen jedoch, daß dies definitiv nicht seine Meinung ist, sich vielmehr die beweisende Wissenschaft gerade darin von der Dialektik in seinem Sinne

---

64 An. pr. A 4.25 b 30; An. post. A 2.71 b 17–18 und öfter.

65 Vgl K. v. FRITZ, a.a.O., S. 19 ff.; G. PATZIG, Die aristotelische Syllogistik. Logisch-philologische Untersuchungen über das Buch A der „Ersten Analytiken", Göttingen 1959, S. 137 f.

66 An. post. A 3.72 b 5 ff.

67 An. post. A 2.71b 19–22; vgl. An. post. A4.73 a 21 ff.; Top. A 1.100 a 25–29; dazu TH. HEATH, A history of Greek mathematics I, S. 336 f.; ders. The thirteen books of Euclid's Elements I, S. 117 ff.

68 Vgl. An. post. A 10.76 a 41; A 11.77 a 30–31; Met. K 4.1061 b 19–25.

69 An. post. A 11.77 a 26 ff.

70 Zur Unterscheidung von τὰ κοινά und τὰ ἴδια, d. i. von, euklidisch gesprochen, Axiomen und Postulaten: An. post. A 10.76 a 37 ff. Vgl. TH. HEATH, The thirteen books of Euclid's Elements I, S. 119 ff.; K. v. FRITZ, a.a.O., S. 25 ff.

unterscheiden soll, daß sie es nicht wie diese mit bloß geeigneten Prämissen zu tun hat, die sich von Fall zu Fall, wenn es die Dialogsituation erfordert, auch austauschen lassen, sondern mit *wahren* und *notwendigen* Prämissen[71]. Diese Behauptung verpflichtet ihn nun wiederum, noch etwas zu tun, wovon sich konventionalistische Standpunkte gleich dispensiert wissen, nämlich zu *begründen*, warum bestimmte Sätze „notwendig“ erste Sätze sind und wie man zu ihnen kommt. ARISTOTELES hat sich dieser Verpflichtung auch nicht entzogen, er betont, daß eine solche Begründung durch *ἐπαγωγή*, durch Induktion, wie wir übersetzen, vorgenommen werden kann[72], und seine wiederholten Erklärungen, daß man etwas entweder durch Beweis (*ἀπόδειξις*) oder durch Induktion (*ἐπαγωγή*) wisse, machen deutlich, daß er dabei an eine vollständige Disjunktion beider Methoden denkt[73]. | 433

Hier treten nun leicht Mißverständnisse darüber auf, was ARISTOTELES mit seinem Hinweis auf die *ἐπαγωγή* gemeint haben mag. Sie beruhen zumeist darin, daß man unter dieser *ἐπαγωγή* Induktion im modernen Sinne versteht und damit zugleich den *empiristischen* Grundzug in der aristotelischen Philosophie, speziell in der aristotelischen mathematischen Grundlagenforschung bestätigt sieht. Nun sollen aber erstens nach ARISTOTELES die Axiome gar nicht *bewiesen* werden, indem man etwa hingeht und seine Sätze der Kontrolle der Erfahrung unterwirft. Vielmehr sollen diese Sätze prinzipiell so beschaffen sein, daß man jemand anderen „dahinführen“ kann, sie als wahr, elementar, unvermittelt etc. einzusehen und selbst zu übernehmen; ihre *Annahme* soll also gerechtfertigt werden. Zweitens denkt ARISTOTELES in diesem Zusammenhang nicht an eine möglichst umfangreiche Aufzählung zur Begründung geeigneter Fälle, schon gar nicht an eine vollständige Induktion, wie es die moderne Terminologie nahezulegen scheint, sondern sieht die Annahme eines Axioms bereits gerechtfertigt, wenn aufgrund eines einzigen Falles dessen Sinn unmittelbar einsichtig wird[74]. Die Instanz, an die hier appelliert wird, ist der *νοῦς*, die intuitive Einsicht, und nicht etwa schon die Wahrnehmung (*αἴσθησις*) selbst oder die Erfahrung (*ἐμπειρία*), wie sich ARISTOTELES gelegentlich leicht mißverständlich ausdrückt[75], ohne daß man ihm deswegen gleich einen Vorwurf machen müßte. Für ihn besitzen ja diese

---

[71] Allerdings sieht O. BECKER (Arch. f. Begriffsgesch. IV, S. 220 f.) im Anschluß an K. v. FRITZ (Die *APXAI* in der griechischen Mathematik, Arch. f. Begriffsgesch. I, S. 24 etc.) in einigen Bemerkungen bei ARCHIMEDES bereits erste Schritte in Richtung auf eine konventionalistische Auffassung der Axiome, die sich deutlich von der aristotelischen Auffassung unterscheide. Er bezieht sich dabei auf eine Stelle aus der Vorrede zur „Quadratur der Parabel“, in der ARCHIMEDES betont, daß es ihm bei dem nicht-evidenten „archimedischen Axiom“ nur auf die Evidenz der mit ihm gewonnenen Resultate ankomme (ARCHIMEDIS Opera omnia, ed. J. L. HEIBERG, 3 Bde, 2. Aufl. Leipzig 1910–15, II, S. 264, 22–26.

[72] An. post. A 1.71 a 6; B 19.100 b 3–5 etc.

[73] Vgl. An. post. A 18.81 a 38–81 b 1; An. pr. B 23.68 b 13–14 (hier *συλλογισμός* für *ἀπόδειξις*). G. PATZIG (a.a.O., S. 138) weist in diesem Zusammenhang besonders auf Eth. Nic. Z 3. 1139 b 26–31: *διδασκαλία δι᾽ ἐπαγωγῆς — διδασκαλία δι᾽ συλλογισμῷ* hin, eine Unterscheidung, die seiner Meinung nach die spätere Einteilung in „induktive“ und „deduktive“ Wissenschaften inspiriert haben dürfte. Für ARISTOTELES selbst ist dagegen die *ἐπαγωγή* noch genuiner Bestandteil, nämlich „Anfang“, einer „deduktiven“ Wissenschaft, der *ἀποδεικτικὴ ἐπιστήμη* (1139 b 29–31).

[74] Dies hat kürzlich erst K. v. FRITZ in überzeugender Weise hervorgekehrt: Die *ἐπαγωγή* bei Aristoteles, SB München, philos.-hist. Kl. 1964, 3. Abh.; vgl. bes. S. 32 ff.

[75] Vgl. An. pr. A 30.46 a 17–18 (*ἐμπειρία* vermittelt die *ἀρχαί*); An. post. B 19.100 a 3ff. etc.

Ausdrücke noch nicht jenen fatalen empiristischen Klang, den sie bei demjenigen hervorrufen müssen, der in der neuzeitlichen Auseinandersetzung zwischen Empirismus und Rationalismus auf der Seite KANTS steht. Evidenzen, und auf die allein kommt es ARISTOTELES in Wahrheit an, sind nichts „Empirisches", auch dann nicht, wenn sie an Beispielen, z. B. also aufgrund von Handlungen erzeugt werden, die man an der thaletischen Grundfigur vornimmt.

ARISTOTELES hat selbst nicht versucht, die methodischen Grundsätze der „Zweiten Analytiken" auf seine Philosophie und seinen Entwurf einzelner wissenschaftlicher Disziplinen anzuwenden; dies sollte, wenn auch unter anderen Voraussetzungen, DESCARTES vorbehalten bleiben, der am Leitfaden der neuzeitlichen Physik mit dem Aufbau einer auf evidenten Grundannahmen und aus diesen abgeleiteten Deduktionen beruhenden Universalwissenschaft begann, ohne dabei jedoch aus spekulativen, wenig überzeugenden Ansätzen herauszufinden. Für ARISTOTELES genügt es, gezeigt zu haben, wie eine Wissenschaft prinzipiell aufgebaut sein muß, wenn sie wahrhaft als Wissenschaft auftreten will, wobei aus gelegentlichen Bemerkungen deutlich wird, daß er wohl selbst diesen Aufbau nur in der Mathematik für möglich hielt und glaubte, sich in anderen Wissenschaften mit weniger strengen Maßstäben begnügen zu müssen[76]. Sein eigentliches philosophisches Interesse bezieht sich denn auch nicht so sehr auf den methodischen Fortgang dieser Wissenschaften, die von nun an unter der
434 Bedin|gung der ἕξις ἀποδεικτική, des *Beweisvollzuges*, wie man vielleicht am besten übersetzen sollte, gestellt sind[77], sondern auf jene ersten Sätze, die am Anfang jeder Theorie stehen und von denen es einige gibt, die allen Theorien oder allen Wissenschaften gemeinsam sind. Und hierbei macht sich nun sogleich wieder ein *platonischer* Zug im aristotelischen Denken geltend. Zwar geht ARISTOTELES nicht so weit wie PLATON, der alle die Grundlagen einer Wissenschaft betreffenden Fragen der Kompetenz der Fachwissenschaftler entziehen wollte, doch erlaubt die Unterscheidung zwischen allgemeineren und spezielleren Axiomen es auch ihm, hier einen Bereich, nämlich den der allgemeineren Axiome, auszuklammern, der nicht mehr von den Wissenschaften selbst verantwortet werden muß, sondern einer der platonischen Dialektik durchaus noch vergleichbaren Über-Wissenschaft zufallen soll. Gemeint ist jene „erste Philosophie" (πρώτη φιλοσοφία), die „gesuchte Wissenschaft" (ἡ ζητουμένη ἐπιστήμη) der ersten Metaphysikbücher, in der von den Prinzipien im allgemeinen die Rede sein soll[78], die ARISTOTELES an anderer Stelle aber auch selbst nach „Disziplinen", nämlich in Ontologie (Theologie), Mathematik und Physik einteilt. Diese drei Disziplinen als Formen der *theoretischen* Philosophie werden dann zusätzlich noch von einer *praktischen* und einer *poietischen* Philosophie unterschieden[79].

Die Problematik einer solchen Fächerung, in der der Akzent von rein methodologischen Betrachtungsweisen wieder mehr auf die inhaltliche Formierung von Wissenschaften rückt, mag hier außer Betracht bleiben, zumal sich auch ARISTOTELES selbst nicht an diese Einteilung hält, sie also, seiner Vorliebe für systematische Unterscheidungen entsprechend, wohl ad hoc, zur vorläufigen Orientierung im

---

[76] Vgl. Eth. Nic. A 1.1094 b 11–14; B 2.1103 b 34–1104 a 3.
[77] Eth. Nic. Z 3.1139 b 31–32.
[78] Vgl. Met. B 2.996 b 26 ff. TH. HEATH, a.a.O., S. 121.
[79] Met. K 7.1064 b 1–3; Met. E 1.1025 b 25; Top. Z 6.145 a 15–16.

Wissenschaftskosmos, getroffen haben dürfte[80]. Wesentlich dagegen ist, daß im Rahmen jener „ersten Philosophie", d. h. cum grano salis in der „Metaphysik", jetzt Fragen erörtert werden, die mit der Konstituierung einer beweisenden Wissenschaft unmittelbar kaum mehr etwas zu tun haben, und damit auch die Frage nach der Wissenschaftlichkeit der Wissenschaft langsam wieder aus dem Blickfeld gerät. Für den Fortgang griechischen Denkens hat dies schwerwiegende Folgen. Einmal wird in Fortführung sowohl der aristotelischen „metaphysischen" Überlegungen als auch der rein empirischen Forschung die Wissenschaftstheorie der „Zweiten Analytiken" nahezu vergessen — schon EUKLID baut sie für den Bereich der Geometrie allem Anschein nach unabhängig von ARISTOTELES auf —, zum anderen tritt gegenüber dieser, schon aristotelischen Interessenverlagerung bald eine Reaktion hervor, in deren Verlauf mit der „Metaphysik" der gesamte *βίος θεωρητικός*, die theoretische Einstellung, und also auch die Wissenschaftstheorie, sofern sie noch nicht vergessen ist, in Mißkredit zu geraten beginnt. So wendet sich etwa der Peripatetiker DIKAIARCH aus MESSENE ausdrücklich im Namen einer praktischen Philosophie gegen die herrschende Philosophie (und damit natürlich insbesondere gegen PLATON und ARISTOTELES), in deren theoretischer Einstellung er das abschreckende Beispiel „nutzloser" Gelehrsamkeit zu erkennen glaubt[81]. | 435

Dieses Urteil ist in seiner Totalität sicher nicht gerechtfertigt, da gerade für PLATON und ARISTOTELES jene „griechische" Verknüpfung der Frage nach den Bedingungen des Wissens mit der nach den Bedingungen des richtigen Lebens und richtigen Handelns, also die Verbindung von theoretischer und praktischer Philosophie, charakteristisch ist[82], doch macht es überaus deutlich, wie sich bereits unmittelbar nach ARISTOTELES in der Antike die ursprüngliche Einheit der philosophischen Bemühung aufzulösen beginnt. Die (stoische) Flucht in die praktische Philosophie oder die der Kommentatoren in die Philologie, aus Gründen, die hier nicht erörtert werden sollen, bedeutet jedenfalls auch das Ende jenes großartigen Versuches, nach der Entdeckung der Möglichkeit von Wissenschaft sogleich auch Klarheit über das Wesen dieser Wissenschaft, ihre Leistungen und ihre Methoden, zu erlangen. Immerhin hat dieser Versuch vor allem bei ARISTOTELES bereits zu Ergebnissen geführt, die es erlauben, die eingangs hervorgehobene Theorienbildung als entscheidenden Schritt anzusehen; entscheidend nämlich im Rahmen der Reflexion darauf, was man in der Wissenschaft

---

80 Vgl. W. JAEGER, Aristoteles. Grundlegung einer Geschichte seiner Entwicklung, Berlin 1923, S. 399.

81 CICERO ad Att. II 16, 3. Vgl. PORPHYRIUS, De abstinentia IV, 2 (Opuscula, ed. A. NAUCK, 2. Aufl. Leipzig 1886, S. 228 f.).

82 Die „praktische" Seite gerade auch der aristotelischen theoretischen Bemühungen (von den platonischen Bemühungen in diesem Zusammenhang war ja bereits die Rede) hat J. H. RANDALL Jr. in seinem schönen Buch über ARISTOTELES (Aristotle, New York 1960) meisterhaft dargestellt. So heißt es in einem einleitenden Kapitel "The Aristotelian approach to understanding: living, knowing, and talking": „ARISTOTLE is convinced that no way of understanding the world, no scheme of ‚science', is worth its salt unless it provides the means for understanding living processes in general, and the process of human living in particular. That is, for ARISTOTLE the categories of ‚life' in general, and of ‚human life' in particular, are the most fundamental in any enterprise of understanding — because we who are trying to understand are neither angels nor electrons, but living men, and it is ourselves we are ultimately trying to know and understand. Like SOCRATES and like the Seven Wise Men of Greece, ARISTOTLE makes the precept ‚Know Thyself' central in his vision of a world understood" (S. 4).

tun soll. Daß PLATON und ARISTOTELES dabei noch die Begründung selbst der ersten Annahmen oder der ersten Sätze einer axiomatischen Wissenschaft forderten, zeugt von einer Unbefangenheit und kritischen Nüchternheit dem eigenen Tun gegenüber, die später bedauerlicherweise sogar wieder verlorenging, nachdem man sich in der Neuzeit daran gewöhnt hatte, von „der Wissenschaft" schlechthin zu sprechen und sie in Form der neuen Physik mehr oder weniger fraglos hinzunehmen. Im Grunde fragt erst KANT wieder ausdrücklich nach den Bedingungen der Möglichkeit von Wissenschaft und nimmt damit eine Frage auf, die seit den Griechen in dieser Schärfe nicht mehr gestellt worden war.

Überarbeiteter Text eines auf Einladung des Philosophischen Seminars der Freien Universität Berlin am 15. 7. 1964 unter dem Titel „Der Begriff der Wissenschaft in der Antike" gehaltenen Vortrages.

Á. SZABÓ

WIE IST DIE MATHEMATIK ZU EINER DEDUKTIVEN WISSENSCHAFT GEWORDEN?

Wir vertreten in dieser Arbeit die folgenden drei Thesen: 1. die griechische Mathematik ist als deduktive Wissenschaft spätestens in der ersten Hälfte des 5. Jahrhunderts unter dem Einfluss der eleatischen Philosophie entstanden, 2. die Eleaten waren es, die schon *vor* dieser entscheidenden Wandlung zum ersten Male in der Geschichte des europäischen Denkens die grundlegenden Prinzipien der Logik klar formulierten, und 3. die deduktive Mathematik ist solange überhaupt nicht möglich, bis der Mathematiker die Begründung seiner Sätze nicht auf eine schon vorhandene und bewusst angewandte Logik bauen kann. – Um diese Auffassung begründen zu können, gliedert sich die vorliegende Untersuchung in *vier* Kapitel. *Im ersten Kapitel* wollen wir die wichtigsten jener Fragen der griechischen Mathematikgeschichte mindestens kurz erwähnen, die eben dadurch gestellt worden sind, dass man die Mathematik der vorgriechischen Völker des alten Orients besser kennengelernt hatte; *im zweiten* besprechen wir die neueren Erklärungsversuche über das Zustandekommen der griechischen Mathematik; *im dritten* fassen wir jene wichtigsten antiken Angaben und daran anknüpfenden modernen Erklärungen zusammen, auf die sich unsere eigene Theorie baut, und schliesslich *im vierten* entwickeln und begründen wir dieselbe Theorie über das Entstehen der griechischen exakten Wissenschaft in einer ausführlicheren Behandlung des mathematischen indirekten Beweisverfahrens.

I

Im Laufe der letzten Jahrzehnte beschäftigten sich mehrere bedeutende wissenschaftsgeschichtliche Arbeiten mit den mathematischen Kenntnissen der alten vorgriechischen Völker von Aegypten und Babylon.[1] Als Ergebnis der Forschungen auf diesem Gebiete darf nicht allein die Tatsache gelten, dass man die Mathematik der
vorgriechischen Kulturen besser kennen|gelernt hatte, sondern auch der Zustand 110
selbst, dass sich auch unser Bild vom Griechentum und von den Anfängen der Wissenschaft im Lichte unserer neuen Kenntnisse weitgehend veränderte. Solange man nichts von der ägyptischen und babylonischen Mathematik wusste, konnten die Griechen mit mehr oder weniger Recht als die allerersten Schöpfer und Begründer dieser Wissenschaft gelten. Aber es veränderte sich plötzlich die Lage, als es sich herausstellte, dass manche wichtigen mathematischen Erkenntnisse, die in der griechischen Überlieferung auf das 6. oder 5. Jahrhundert v. u. Z. datiert werden, in den

---

[1] Man findet die wichtigsten Ergebnisse dieser Forschungen zusammengefasst in den ersten drei Kapiteln des Werkes B. L. v. D. WAERDEN: Science awakening. Groningen 1954.

J. Christianidis (ed.), Classics in the History of Greek Mathematics, 45–80.

vorgriechischen Kulturen schon Jahrhunderte früher bekannt waren. Das Leben des Pythagoras wird z. B. nach der griechischen Überlieferung auf das 6. Jahrhundert v. u. Z. gesetzt, und dementsprechend könnte aus demselben Jahrhundert der sog. Satz des Pythagoras stammen, der von den Späteren ihm zugeschrieben wird. Manche Forscher haben jedoch früher die traditionelle Zuschreibung dieses Lehrsatzes an Pythagoras in Zweifel gezogen, weil man sich nicht hat erklären können, wie die Erkenntnis dieses Satzes in einem so frühen Stadium der Wissenschaft möglich gewesen wäre.[2] Aber diese Skepsis der antiken Überlieferung gegenüber bekam einen völlig neuen Sinn, als es sich herausstellte, dass die praktische Anwendung des pythagoreischen Lehrsatzes den Babyloniern schon im 2. Jahrtausend v. u. Z. bekannt war.[3] Man musste ähnlicherweise zur Kenntnis nehmen, dass auch der grösste Teil jenes mathematischen Wissens, welches durch Euklid in systematischer Ordnung behandelt wird, mindestens als eine Summe von empirischen Kenntnissen schon lange vor den Griechen in der babylonischen Kultur geschlossen vorlag.[4]

Dadurch, dass man die Vorgeschichte der griechischen Wissenschaft kennenlernte, schienen auch die Griechen selbst jenen alten Ruhm, den sie früher in der Geschichte der Wissenschaft genossen, beinahe zu verlieren. Wie O. Neugebauer, der hervorragende Kenner der älteren Mathematikgeschichte schreibt: seitdem wir nicht nur jene zweieinhalb tausend Jahre Geschichte kennen, die seit dem klassischen Zeitalter verfloss, sondern seitdem wir auch jene anderen zweieinhalb tausend Jahre mehr oder weniger überblicken können, die dem griechischen Altertum voraufgingen, ist es nicht
111 mehr möglich, in den Griechen die allerersten Schöpfer und Begründer | der Wissenschaft zu erblicken.[5] Wie wir heute sehen, stehen die Griechen nicht mehr am allerersten Anfang der Geschichte der Wissenschaft, sondern irgendwo in ihrer Mitte.[6] Ja, Neugebauer hat gerade in Hinsicht auf die babylonische Mathematik sogar die Frage aufgeworfen, ob es überhaupt recht und billig wäre, die Errungenschaften der griechischen Mathematik unter dem Gesichtspunkt der Wissenschaftsgeschichte eindeutig und *nur* bejahend zu bewerten? – Denn die Griechen haben ja schliesslich, anstatt dessen, dass sie das babylonische Positionssystem der Zahlen in ein bewusstes Positionssystem der Basis 10 oder 12 verwandelt hätten, die positionelle Bezeichnung in Zahlbuchstaben modifiziert, – was selbstverständlich ein folgenschwerer Rückschritt war.[7] Ebenso wurde in der griechischen Mathematik «die Einsicht in das

---

2 Vgl. B. L. v. d. Waerden: Die Arithmetik der Pythagoreer I. Math. Ann. 120 (1947/49) 127–153; besonders auf S. 132.

3 Vgl. O. Neugebauer: Vorlesungen über die Geschichte der antiken math. Wissenschaften. Berlin 1934. S. 168 und K. Reidemeister: Das exakte Denken der Griechen. Hamburg 1949. S. 51.

4 Vgl. O. Neugebauer: Studien zur Geschichte der antiken Algebra III (Quellen und Studien zur Gesch. der Math. Abt. B. Bd. 3 [1936] 245–259): «sowohl im Bereich der elementaren Geometrie, wie im Bereich der elementaren Proportionenlehre, wie schliesslich im Bereich der Gleichungslehre liegt in der babylonischen Mathematik das gesamte *inhaltliche* Material geschlossen vor, auf dem die griechische Mathematik aufbaut. Der Anschluss ist in allen Punkten lückenlos herzustellen.»

5 O. Neugebauer: o. c. S. 259.

6 B. L. v. d. Waerden: Math. Ann. 120 (1947/49) S. 132.

7 A. Frenkian: Études de mathématiques suméro-akkadiennes, égyptiennes et grecques (in der «Revue de l'Université de Bucarest» 1953, 9–20) schreibt (im französischen Auszug seiner rumänisch verfassten Arbeit): «Rien n'indique que les mathématiciens grecs aient connu le système de notation de position relative des peuples de la Mésopotamie. Le système grec de notation des

Wesen der Irrationalzahlen erkauft mit dem abrupten Abbrechen eines bereits zu einem algebraischen Formalismus gelangten | Systems, das sich in allen Punkten direkt in die 112
Algebra der Renaissance hätte fortentwickeln können; ohne die tiefsten mathematischen Leistungen der Griechen wären vielleicht 2000 Jahre zu 'gewinnen' gewesen.»[8]

Selbstverständlich ist sich auch Neugebauer dessen wohl bewusst, dass die mathematischen Leistungen der Griechen historisch keineswegs bloss als ein «Rückschritt» im wahrsten Sinne des Wortes angesehen werden können. Er beruft sich ja auf die genannten Fälle, nur um zu illustrieren, wie die Bewertung historischer Prozesse nach dem Prinzip der einfachen Linearität fehlschlagen muss. Die babylonische Mathematik besass in der Tat auch solche Ansätze, die durch die späteren Griechen *nicht* ausgenutzt wurden; ja, es wurde sogar durch die Tatsache, dass die Griechen die Entwicklung der Mathematik in eine bestimmte Richtung lenkten, eine ziemlich lange Zeit hindurch – eigentlich bis zur Zeit der europäischen Renaissance – verhindert, dass die entwicklungsfähigen Ansätze der älteren babylonischen Algebra zur Entfaltung kämen.[9]

---

nombres est des plus défectueux, *incapable d'aider le calcul* (von mir herausgehoben – Á. SZABÓ) inférieur même à celui des Égyptiens, qui avait un signe particulier pour chaque ordre d'unités décimales, qu'ils répétaient autant de fois qu'il était nécessaire pour écrire un nombre donné. Donc, ou bien les Grecs n'ont pas connu le système de notation des Babyloniens, ou bien *l'esprit de leur mathématique étant orienté dans une autre direction, ils n'ont pas adopté ce système, si toutefois ils l'ont connu*» (von mir herausgehoben – Á. SZABÓ). – Im weiteren versucht dann A. FRENKIAN noch zu erklären, warum eigentlich die Griechen nicht ein besseres Bezeichnungssystem für die Zahlen entwickelt hatten, und er kommt zu der folgenden, bemerkenswerten Vermutung: «Les Hellènes avaient deux sciences qui s'occupaient des nombres: *l'arithmétique* qui était la science théorique des nombres, très étudiée et honorée, et la *logistique* qui était lié à la pratique et qu'on avait abandonné à des spécialistes considérés comme des artisans, appelés, avec un certain mépris, du nom de *banausoi*. C'est ainsi que la science théorique qui est l'arithmétique a fait chez eux de grands progrès, sans faire profiter la logistique de ces progrès. Celle ci a continué de travailler avec les anciennes méthodes venues de l'Égypte: à savoir, la multiplication par duplications successives dont parle un scholie au dialogue Charmide de Platon et le calcul fractionnaire seulement avec des fractions ayant l'unité au numérateur, méthode qui fut employé jusqu' aux temps des Byzantins. – Les mathématiques suméro-akkadiennes ont été beaucoup plus liées à la pratique, comme on le voit par les problèmes qu'elles ont à résoudre dans les textes qui nous sont parvenus. Ensuite, la base de 60 pour le système de numération était très grande et c'est pourquoi les tenants de la civilisation mésopotamienne ont eu recours à la notation de position relative, etc.» – Diese Vermutung passt ausgezeichnet zu jener Auffassung, die wir in dieser Arbeit vertreten: nicht nur die grossartigen Errungenschaften der griechischen Mathematik, sondern auch ihre relativen *Rückstände* stellen nur die Folge derselben grundlegenden Tendenz der antiken klassischen Wissenschaft dar. – Man muss zu der im ganzen wohl richtigen Theorie von A. FRENKIAN nur die folgende Korrektur hinzufügen: es ist irreführend einfach nur eine «*theoretische* Arithmetik» der Griechen mit einer «*prakti*schen Logistik» zu konfrontieren. Denn Platon stellt z. B. im Staat und im Philebos der «praktischen» Arithmetik *und* der «praktischen» Logistik die entsprechenden «theoretischen» Disziplinen entgegen. Es können also beide Namen – Arithmetik und Logistik – sowohl theoretische, wie auch banausische, praktische Kenntnisse bezeichnen. Man vgl. zu dieser Frage die gründliche Arbeit von J. KLEIN: Die griechische Logistik und die Entstehung der Algebra I u. II, Quellen und Studien z. Gesch. d. Math. Abt. B. Bd. 3 (1936) S. 18 ff. und 122 ff.

8 Sieh Anmerkung 5.

9 Es ist übrigens interessant, wie dieselbe griechische Entwicklung durch V. D. WAERDEN beurteilt wird. Er schreibt nämlich über die logischen Konsequenzen, die sich aus der Entdeckung der Irrationalität ergeben (dass nämlich Strecken nicht universell durch Zahlen darstellbar sind und daher auch nicht ohne weiteres wie Zahlen behandelt werden dürfen), und vergleicht die Leistung der Griechen auf diesem Gebiete mit der Stellungnahme der Vertreter der europäischen

Wir beurteilen also die Griechen, seitdem wir die vorgriechische Wissenschaft einigermassen besser kennen, schon völlig anders als früher. Ja, wir stellen uns seit derselben Zeit auch die Anfänge der Mathematik schon anders vor, als etwa noch vor achtzig Jahren. Man dachte früher z. B., dass die Anfänge der Mathematik «geometrischen» Charakter haben müssten, da ja die Geometrie «weniger abstrakt» und «viel anschaulicher» als die Algebra sei. Man merkte es kaum, dass diese Anschauung in letzter Linie nur ein Effekt der ungeheueren Wirkung der Sprechweise von Euklids Elementen und der anschliessenden Form der griechischen Mathematik ist, die auch unsere ganze Erziehung entscheidend beeinflusst hat. Man dachte, dass die älteste mathematische Disziplin die Geometrie sei, eigentlich nur deswegen, weil bei den Griechen, dem damals so gut wie einzig bekannten Kulturvolk des Altertums, die Mathematik beinahe völlig in der Geometrie aufging. Und man vergass im Banne dieser Betrachtungsart, dass die tatsächliche Entwicklung dem Postulat von der geschichtlichen Priorität des Geometrischen auf Schritt und Tritt widerspricht. Die grossen Fortschritte der Geometrie sind in allen Phasen immer unlösbar mit der
113 Entwicklung anderer Diszi|plinen verknüpft,[10] so dass das Geometrische an sich immer erst nachträglich wieder aus dieser Verknüpfung gelöst werden musste. Wie uns die neueren Forschungen gelehrt haben, ist diese Beobachtung auch auf die Anfänge der Wissenschaft zutreffend. Für die Frühgeschichte der Mathematik ist eine «reine» («synthetische») Geometrie viel zu schwierig. Das primäre Hilfsmittel ist hier die Verknüpfung mit dem Bereich der (rationalen) Zahlen, und ein wesentlicher Fortschritt der Geometrie ist erst möglich, wenn die ungeometrischen Hilfsmittel weit genug entwickelt sind. Darum ist in jenen 2000 Jahren der Entwicklung, die dem griechischen Zeitalter vorangehen, alles 'geometrische' nur ein *sekundäres* Objekt zunächst der Rechentechnik mit den rationalen Zahlen in beiden vorgriechischen Kulturen, dann der Algebra in Babylonien.[11] Die Geometrie wird erst viel später, nur bei den Griechen in den Vordergrund des mathematischen Interesses gerückt.

Durch diese Entdeckung wurde auch das Verständnis einer solchen Erscheinung innerhalb der griechischen Mathematik selbst erleichtert, die man zwar auch früher schon bemerkt hatte, aber kaum erklären konnte. Es fiel nämlich schon H. G. Zeuthen auf,[12] dass es sich besonders im II. und im VI. Buch der Euklidischen Elemente vorwiegend um algebraische Probleme in geometrischer Form handelt. Aber man konnte früher kaum befriedigend erklären: was eigentlich die Griechen veranlasst haben mag, um die Algebra zu geometrisieren? Man konnte sich früher in diesem Zusammenhang höchstens auf die grössere «Anschaulichkeit» der Geometrie berufen. Dagegen kann

---

Wissenschaft folgendermassen: «die meisten Vertreter der abendländischen Wissenschaft haben die Darstellbarkeit von geometrischen Grössen durch Zahlen nie bezweifelt, obwohl sie mit der Existenz von irrationalen Verhältnissen bekannt waren, und obwohl man vor Dedekind und Cantor nicht über den exakten modernen Begriff der reellen Zahl verfügte. *Die griechische Kultur ist meines Wissens die einzige, die diese logische Konsequenz wirklich vollzogen hat*» (von mir herausgehoben – Á. Szabó).

10 O. Neugebauer o. c. S. 246 verweist in diesem Zusammenhang auf die folgenden historischen Verknüpfungen: analytische Geometrie und elementare Algebra, Differentialgeometrie und Analysis, Topologie und Riemannsche Flächen + abstrakte Algebra.

11 O. Neugebauer: o. c. S. 247.

12 An diese frühere Erkenntnis von H. G. Zeuthen (Die Mathematik im Alterum und Mittelalter) erinnern O. Neugebauer o. c. 249 und B. L. v. d. Waerden: Math. Ann. 117 (1940) S. 158.

man heute die Antwort auf die Frage nach der geschichtlichen Ursache der gesamten «geometrischen Algebra» vollständig geben: «sie liegt einerseits in der aus der Entdeckung der irrationalen Grössen folgenden Forderung der Griechen, der Mathematik ihre Allgemeingültigkeit zu sichern durch Übergang vom Bereich der rationalen Zahlen zum Bereich der allgemeinen Grössenverhältnisse, andererseits in der daraus resultierenden Notwendigkeit, *auch die Ergebnisse der vorgriechischen 'algebraischen' Algebra in eine 'geometrische' Algebra zu übersetzen*».[13] In dieser Beleuchtung wurde es auf einmal auch klar, warum am Anfang des 4. Jahrhunderts *Archytas* die Arithmetik – d. h. also die Algebra der Griechen – der Geometrie noch vorziehen konnte;[14] er hat nämlich die logischen Konsequenzen aus der | Erkenntnis 114
der Irrationalität noch nicht so weitgehend zur Geltung gebracht, wie bald nach ihm Theaitetos und Platon es taten.[15] Mit anderen Worten heisst es auch so viel, dass wir heute schon genau jene Zeitspanne kennen, zu welcher bei den Griechen die ältere Arithmetik durch die Geometrie verdrängt wurde;[16] es ist auch bekannt, dass jenes Interesse, mit welchem man sich jetzt erneut der Geometrie zuwandte, im Grunde durch eine logische Erkenntnis, nämlich durch die Entdeckung der irrationalen Zahlenverhältnisse erweckt wurde.[17] – Es musste also jene ältere Auffassung, welche die «Anschaulichkeit» der griechischen Geometrie betonte, überprüft werden; ja, es

---

13 O. Neugebauer: o. c. S. 250.

14 Archytas B 4 (Diels): «Und die Logistik hat, wie es scheint, in Bezug auf Wissenschaft vor den anderen Künsten einen recht beträchtlichen Vorrang; besonders auch vor der Geometrie, da sie deutlicher als diese behandeln kann was sie will ... und wo die Geometrie versagt, bringt die Logistik Beweise zustande ...» Von «Logistik» redet Archytas in einem Arithmetik und Logistik umspannenden Sinne; vgl. J. Klein: Die griechische Logistik und die Entstehung der Algebra, Quellen und Studien z. Gesch. d. Math. Abt. B. Bd. 3 (1936) S. 32 A. 1. Verstünde Archytas unter «Logistik» nicht die theoretische Mathematik, so könnte er gar nicht behaupten, dass die Logistik «Beweise zustande bringe»; nicht die banausische Techne, sondern Platons *theoretische* Logistik und *theoretische* Arithmetik zusammen, also ein Zweig der «Mathemata», stellen eine beweisende (apodeiktische), deduktive Wissenschaft dar. – Es ist übrigens interessant, dass der wahre Sinn dieses Archytas-Fragmentes solange überhaupt kaum erklärt werden konnte, bis die zitierten Worte durch O. Neugebauer (o. c. S. 245 ff.) nicht in die richtige historische Beleuchtung gestellt worden sind; Neugebauer schreibt nämlich darüber: «Die Prägnanz dieses Ausspruchs ist umso interessanter, als er ja nur um wenige Jahre älter ist, als die Geometrisierung der griechischen Mathematik, die wir als ihre *klassische* Form anzusehen gewöhnt sind.» H. Diels bemerkte noch zu den Worten des Archytas: «*Sinn und Herstellung des Fragments unsicher* ...» – Richtig behandelt wird dagegen das Archytas-Fragment – auch von O. Neugebauer unabhängig – bei A. M. Frenkian: Le postulat chez Euclide et chez les modernes. Paris 1940. S. 20, 1.

15 B. L. v. d. Waerden (Zenon und die Grundlagenkrise der griech. Mathematik, Math. Ann. 117 [1940] 141 ff.) schreibt im Zusammenhang mit den vorigen Archytas-Worten: «Wenige Jahrzehnte später hat sich das Blatt bereits gewendet: Theaitetos entwickelt seine Klassifikation der irrationalen Strecken, und bei Platon ist das Verhältnis zwischen Logistik und Geometrie vollständig umgekehrt. Die bisherige Logistik ist als Wissenschaft verpönt, die geometrischen Schlüsse sind die wahren Vorbilder exakter Beweisführung. Bei Euklid ist die Algebra vollends aus dem Bereich der offiziellen Geometrie verbannt und darf nur in geometrischem Gewande, als Flächenrechnung oder geometrische Algebra ihr Dasein fristen.» (Zum Terminus «Logistik» dieses Zitates vgl. man oben auch die Anm. 7 und 14!)

16 Sieh die beiden vorigen Anmerkungen!

17 Vgl. B. L. v. d. Waerden: Science awakening, Groningen 1954. S. 126: «It is therefore *logical necessity*, not the mere delight in the visible, which compelled the Pythagoreans to transmute their algebra into a geometric form.»

fragte sich sogar, angesichts der neuen historischen Perspektive, die sich erschloss: ob überhaupt das wesentlichste Merkmal der griechischen Mathematik ihre Anschaulichkeit sei?[18] Man musste auf einmal jenem Platon Recht geben, der über die griechische Mathematik schon in der ersten Hälfte des 4. Jahrhunderts betonte, dass die Anschaulichkeit der Geometrie keineswegs eine *konkrete* Anschaulichkeit sei; denn sie *erinnere* nur an etwas, was anders als auf gedanklichem Wege gar nicht
115 zugänglich wäre.[19] |

Man ersieht also aus den angeführten Beispielen, dass die gründlichere Erforschung der vorgriechischen Mathematik in der Tat auch unsere Kenntnisse über die Griechen veränderte und vertiefte. Heute können zwar die alten Griechen nicht mehr in demselben Sinne für die ersten Schöpfer und Begründer der mathematischen Wissenschaft gelten, wie man es früher von ihnen dachte, aber umso konkreter kann man heute die historische Bedeutung dessen unterstreichen, was sie in der Tat in der Mathematik geleistet hatten. Es hat sich z. B. eindeutig herausgestellt, dass in der vorgriechischen Mathematik solche Begriffe, wie *Satz, Beweis, Definition, Postulat* und *Axiom* noch gar nicht existierten.[20] Diese ältere Mathematik war eigentlich nur eine Summe von empirischen Kenntnissen; man stellte nur Regeln zusammen, wie man gewisse Aufgaben mathematischen Inhalts lösen kann, aber nie wurden Sätze in allgemeingültiger Form aufgestellt, und noch weniger wurde es versucht, irgendeinen Beweis für die Sätze zu liefern, man illustrierte höchstens in einer zahlenmässig ausgerechneten Probe die Anwendung der Regel. Zu einer deduktiven Wissenschaft wurde die Mathematik erst bei den Griechen. Diese Umwandlung der Summe von empirischen Kenntnissen in eine exakte Wissenschaft ist natürlich ein Schritt von riesiger Bedeutung für die ganze weitere Entwicklung. Man versteht also das Interesse, mit welchem man sich der Frage zuwendet: *wie, wann* und *warum* bei den Griechen die Mathematik zu einer deduktiven Wissenschaft geworden ist? – Wir wollen zunächst im zweiten Kapitel dieser Arbeit drei interessante Erklärungsversuche auf die letzthin genannten Fragen besprechen.

II

Man vergleicht das Entstehen der deduktiven Mathematik oft mit der Entfaltung der Lehre über die Logik. Unter anderen scheint auch K. v. Fritz dieser Ansicht zu sein,

---

[18] K. Reidemeister: Das exakte Denken der Griechen. Hamburg 1949. S. 51: «Es ist ein weitverbreitetes Vorurteil, das wesentliche Merkmal der griechischen Mathematik sei ihre Anschaulichkeit ... Richtig ist es vielmehr, dass sich in der pythagoreischen Mathematik die Umwendung vom Anschaulichen zum Begrifflichen vollzieht.»

[19] Vgl. Platon, Staat VII 526 und 527. Wenn die Mathematiker über Zahlen sprechen, so verstehen sie unter Zahlen etwas, was nur gedacht werden kann; ähnlich verhält es sich auch in der Geometrie; wenn man nämlich etwas in dieser Wissenschaft veranschaulichen will, so handelt es sich auch hier gar nicht um konkrete Dinge, sondern man will die Aufmerksamkeit auf etwas lenken, was anders als auf gedanklichem Wege gar nicht zugänglich ist.

[20] O. Becker: Grundlagen der Mathematik in geschichtlicher Entwicklung. Freiburg/München 1954. S. 22: «Nicht einmal die Formulierung allgemeiner Sätze ist für Babylonien gesichert. (Dagegen kommen solche in der traditionellen Form des dogmatischen Kurzsatzes [sūtra] in der altindischen Sakralgeometrie vor.) Von Beweisen ist in den erhaltenen altorientalischen Dokumenten erst recht nichts zu finden; höchstens kommen zahlenmässig ausgerechnete Proben vor.»

der zuletzt in einer Arbeit[21] mit Recht den Gedanken vertrat, dass die griechische Mathematik wohl nur infolge ihrer definitorisch-axiomatischen Grundlegung zu einer deduktiven Wissenschaft werden konnte. K. v. Fritz antwortet zwar nicht unmittelbar
auf die Frage, wie, unter welchen | Umständen, und wann eigentlich die definitorisch- 116
axiomatische Grundlegung der griechischen Mathematik erfolgte, aber es lohnt sich dennoch seine Gedankengänge – hie und da auch weiter ergänzt – zu überblicken, denn es gibt darunter auch solche Gesichtspunkte, die unser Problem näher beleuchten können.

Euklid behandelt in seinem klassischen Werk, den «Elementen», die rund um 300 v. u. Z. entstanden, in einer Sondergruppe, gleich am Anfang seiner Erörterungen zusammengestellt, die Definitionen, Postulate und Axiome; erst nach dem Vorausschicken dieser «Prinzipien» geht er auf die Behandlung bzw. auf den Beweis der einzelnen Lehrsätze, der sog. Theoreme hinüber. Man findet bei Euklid selbst gar keine Erklärung dafür, was eigentlich der Sinn dieser Einteilung sei. Nur der Kommentator des 5. Jahrhunderts n. Zw., Proklos erklärt es in den folgenden Worten:

«Da wir behaupten, dass diese Wissenschaft, die Geometrie, auf Voraussetzungen beruhe und von bestimmten Prinzipien aus die abgeleiteten Folgerungen beweise – denn nur eine ist voraussetzungslos, die anderen aber empfangen ihre Prinzipien von dieser –, so muss unbedingt der Verfasser eines geometrischen Elementarbuches gesondert die Prinzipien der Wissenschaft lehren und gesondert die Folgerungen aus den Prinzipien; von den Prinzipien braucht er nicht Rechenschaft zu geben, wohl aber von den Folgerungen hieraus. Denn keine Wissenschaft beweist ihre eigenen Prinzipien und stellt sie zur Diskussion, sondern hält sie für an sich gewiss; sie sind ihr klarer als die Ableitungen; erstere erkennt sie in deren eigenem Licht, die Ableitungen aber durch die Prinzipien ... Wenn aber jemand die Prinzipien und die Ableitungen hiervon in denselben Topf wirft, so richtet er nur Verwirrung an im ganzen Wissensbereich und vermengt, was miteinander nichts zu tun hat. Denn das Prinzip und das davon Abgeleitete sind von Haus aus voneinander gesondert.»[22]

Es wird bei Proklos an anderen Stellen auch genau erklärt, was der Unterschied zwischen Definitionen, Postulaten und Axiomen sei; diese drei Dinge werden bei ihm «Prinzipien», d. h. griechisch: *ἀρχαί* genannt. Besonders wichtig ist für uns jetzt in diesem Zusammenhang, dass nach der Erklärung des Kommentators der Beweis oder die geometrische Begründung im Falle der Prinzipien nicht nötig, ja nicht einmal auch
möglich sei;[23] die | Prinzipien werden als bekannt und als keines Beweises bedürftig 117
vorausgesetzt, und sie gelten als *Gründe* für alle Theoreme, die aus ihnen folgen und gerade deswegen *nach* ihnen behandelt werden.

---

[21] K. v. FRITZ: Die *APXAI* in der griechischen Mathematik, Archiv für Begriffsgeschichte. Bd. I. Bonn 1955. 13–103.

[22] Procli Diadochi in primum Euclidis Elementorum librum commentarii ed. G. FRIEDLEIN, Lipsiae 1873. S. 75.

[23] Proclus (ed. G. FRIEDLEIN) S. 178: «Gemeinsam ist nun den Axiomen und Postulaten, dass sie keiner Begründung und keiner geometrischen Beweise bedürfen, sondern dass sie als bekannt angenommen werden und Prinzipien sind für die Folgenden. (Sie unterscheiden sich aber voneinander ebenso, wie die Lehrsätze von den Aufgaben verschieden sind.)» – Ähnlich heisst es an einer anderen Stelle (S. 76): «Wenn der Hörer das Verständnis einer Behauptung als von sich aus einleuchtend nicht schon in sich hat, die Behauptung aber gleichwohl aufgestellt wird und er die Annahme zugibt, dann handelt es sich um eine *Definition*.» Vgl. auch noch S. 183.

Man versteht in der Tat aus der Erklärung des Proklos sehr gut die Komposition des Euklidischen Werkes. Die Definitionen, Postulate und Axiome werden bei Euklid offenbar wirklich deswegen gleich am Anfang des ersten Buches[24] vor der Behandlung der einzelnen Lehrsätze aufgezählt, weil diese Dinge jene «Prinzipien» sind, aus welchen die Theoreme abgeleitet werden können. Dabei ist es jetzt einerlei, ob die einzelnen Definitionen, Postulate und Axiome *alle* in der Tat von Euklid selber stammen, oder ob er einige von diesen schon fertig übernahm, oder auch: ob nicht andere erst nachträglich in sein Werk hineingefügt worden seien.[25] Der Aufbau des ganzen Euklidischen Werkes ist selber der Beweis dafür, dass in der Tat schon der Verfasser dieser grossartigen Zusammenfassung jene Einteilung des gesamten mathematischen Wissens – in nicht-bewiesene Prinzipien einerseits und in abgeleitete Lehrsätze andererseits –, von welcher Proklos redet, gekannt haben muss. Das heisst aber mit anderen Worten auch so viel, dass Euklid um 300 v. u. Z. schon sehr genau wusste: worin überhaupt der mathematische Beweis besteht, und wie weit er geführt werden kann.

Aristoteles war es, der schon eine Generation früher, als Euklid lebte, im 4. Jahrhundert in einem seiner logischen Werke, den Analytica posteriora die Methoden der sog. beweisenden (apodeiktischen) Wissenschaften einer eingehenden Behandlung unterzog. Die diesbezüglichen Erörterungen des Aristoteles lassen sich im grossen und ganzen damit vereinigen, was Proklos, der Kommentator des Euklidischen Werkes über den mathematischen Beweis entwickelt, obwohl man oft den Eindruck hat, dass Euklid und die antiken Mathematiker nicht über alle Punkte derselben Meinung waren, wie Aristoteles; ja, selbst die Terminologie des Aristoteles ist nicht immer dieselbe, wie diejenige der Mathematiker.[26] Wir können uns jedoch diesmal damit begnügen, dass auch Aristoteles die Methoden der beweisenden (apodeiktischen) Wissenschaft an der Mathematik illustriert, und dass auch er über die Prinzipien ebenso denkt, wie Euklid
118 bzw. Proklos; wie er schreibt: | «Mit dem Namen des Prinzips bezeichne ich in jeder Gattung dasjenige, worüber es sich nicht beweisen lässt, dass es *existiert*, bzw. dass es *gültig ist*.»[27] Kein Zweifel, in demselben Sinne werden die Definitionen, Postulate und Axiome auch durch Euklid und Proklos als «Prinzipien» angesehen.

Aber derselben Meinung war in Bezug auf die mathematischen Prinzipien – oder mindestens in Bezug auf die Definitionen – auch schon Platon um eine Generation früher als Aristoteles, in der ersten Hälfte des 4. Jahrhunderts. Wie der Sokrates des Platonischen Dialoges über den «Staat» entwickelt: «Du weisst doch wohl, dass die Geometer, die Arithmetiker und die übrigen, die sich mit ähnlichen Wissenschaften beschäftigen, allen ihren Untersuchungen *bestimmte Voraussetzungen* zu Grunde legen, wie z. B. die Begriffe 'Gerades' und 'Ungerades', die geometrischen Figuren,

---

[24] Postulate und Axiome werden bei Euklid nur am Anfang des ersten Buches aufgezählt, dagegen findet man Definitionen – vom VIII., IX., XII. und XIII. Buch abgesehen – am Anfang jedes einzelnen Buches. – Da die Axiome und Postulate allgemeingültiger als die Definitionen sind, wird man diese wohl verhältnismässig später gefunden haben als die Definitionen. Die Definitionen sind wohl die am frühesten erkannten mathematischen Prinzipien.

[25] Über den verschiedenartigen Ursprung der Euklidischen Definitionen, Postulate und Axiome siehe P. Tannery: Sur l'authenticité des axiomes d'Euclide (Mém. scient. II. 1912, 48–63) und A. M. Frenkian: Le postulat chez Euclide et chez les modernes. Paris 1940. 11–24.

[26] Treffend bemerkt in diesem Zusammenhang K. v. Fritz: o. c. S. 103: «Die mathematische Entwicklung ist weitgehend an Aristoteles vorbeigegangen.»

[27] Aristoteles, Analytica Posteriora I 10.

die drei Arten von Winkeln und manches ähnliche. Sie nehmen solche Begriffe einfach an, als ob sie sich über diese Dinge schon im klaren wären, und *halten es nicht für nötig, sich und anderen Rechenschaft über etwas zu geben, was einem jeden doch klar sei.* Von dieser Grundlage aus gehen sie dann vorwärts und finden schliesslich in Übereinstimmung mit ihr das, was Gegenstand ihrer Untersuchung war» (VI 510 C–D). «Die Seele ist bei ihren Betrachtungen auf Voraussetzungen angewiesen und geht nicht bis auf den Grund, da sie *über die Voraussetzungen hinaus rückwärts nicht gehen könnte*» (VI 511 A).

Überlegt man sich diese Platon-Zitate, so muss man daraus schliessen, dass die Griechen lange vor Euklid, zu Platons Zeit allerdings, schon sehr wesentliche Dinge über die Art und Weise des mathematischen Beweisverfahrens wissen mussten. Sie wussten nämlich, dass der mathematische Beweis kein unendlicher Prozess sein kann; es gibt in ihm keinen *regressus ad infinitum.* Mit anderen Worten: sie wussten schon, dass die Mathematik auf solche Voraussetzungen gebaut ist, die man nicht weiter beweisen kann. Dieses Wissen ist zu der definitorisch-axiomatischen Grundlegung der Mathematik unerlässlich nötig, und es war zu Platons Zeit allerdings schon vorhanden.

Fragen wir vorläufig noch nicht, auf welchem Wege wohl die Griechen zu der Überzeugung kamen, dass die beweisende (apodeiktische) Wissenschaft auf unbewiesene Voraussetzungen gebaut werden muss. Statt dieser jetzt noch rätselhaften Frage, versuchen wir diesmal eine andere, etwas leichtere zu beantworten: was mag denn überhaupt die Griechen veranlasst haben, eine definitorisch-axiomatische Grundlegung für die Mathematik zu erschaffen? – Diese Frage liesse sich – mindestens provisorisch – etwa folgendermassen beantworten: sie mussten einmal wohl auf den Gedanken kommen, dass es möglich wäre, alle jene Feststellungen – oder sagen wir: Sätze mathematischen Inhalts – zusammenzustellen, die man überhaupt nicht beweisen konnte, die aber auch in sich als evident empfunden wurden, | und auf 119
die man dann im weiteren ihr ganzes mathematisches Wissen aufbauen konnte. (Möge diese Vermutung auch noch so vage und unbestimmt sein, so vermag sie uns dennoch – mindestens als Arbeitshypothese – weiterzuhelfen.)

Das Zustandekommen einer solchen definitorisch-axiomatischen Grundlegung wurde durch K. v. Fritz mit der Entfaltung der Aristotelischen Logik verglichen. Die Aristotelische Logik soll nämlich – wie er schreibt – aus der Dialektik hervorgegangen sein. (Unter «Dialektik» versteht er die Platonische διαλεκτικὴ τέχνη.) Es ist in der dialektischen Auseinandersetzung die Aufgabe des einen Dialogpartners den anderen dazu zu bringen, einen von ihm gewählten Satz zuzugeben. Zu diesem Zweck muss er solche Prämissen finden, die der Partner für richtig hält und daher zugeben wird, und aus denen sich der Endsatz, den der Partner nicht für richtig hält und nicht zugeben will, mit logischer Notwendigkeit ableiten lässt.

Kein Zweifel, dieses Schema lässt sich in der Tat sehr leicht auf die Euklidische Mathematik anwenden. Man kann fast über jeden beliebigen «komplizierten» Euklidischen Satz feststellen, dass er im Beweis auf «einfachere» Sätze zurückgeführt wird; die «einfachen» werden dagegen unmittelbar aus Definitionen, Postulaten oder Axiomen abgeleitet.[28] Die Mathematik heisst ja gerade deswegen «*deduktive*

[28] Es ist bezeichnend für den naiven Empirismus der Griechen, dass sie den Unterschied zwischen «einfacheren» und «komplizierteren» Sätzen für eine *objektive* Tatsache hielten. Sie sind scheinbar

Wissenschaft», weil sie alle ihre Behauptungen (Sätze) aus solchen Prämissen *ableitet*, bzw. die Behauptungen auf allgemein zugegebene Prämissen *zurückführt*.

Lehrreich ist, das Zustandekommen der definitorisch-axiomatischen Grundlegung mit der Entfaltung der Aristotelischen Logik zu vergleichen, auch schon deswegen, weil es uns auf einen sehr wesentlichen Umstand aufmerksam macht. In der dialektischen Auseinandersetzung ist nämlich der Ausgangspunkt desjenigen, der etwas beweisen will, der Endsatz selbst, und er sucht erst *nachträglich* jene Prämissen zu seiner Behauptung, die auch sein Gegner für richtig hält und zugibt; aber es ist gar nicht unbedingt nötig, dass man zu jenem Endsatz, den man beweisen will, in der Tat auf Grund der Kenntnis jener Prämissen gelangt sei, die man im Beweis benutzt. Im Gegenteil, es ist sehr wohl möglich, dass die Teilnehmer der dialektischen Auseinandersetzung die logischen Prämissen irgendeiner wahr empfundenen oder plausiblen Behauptung erst dann erkannten, als der eine von ihnen versuchte den anderen dazu zu bringen, den von ihm gewählten Satz zuzugeben. – Es ist ja klar, dass
120 derselbe Fall auch für einen grossen Teil der mathe|matischen Sätze Euklids gültig ist. In manchen Fällen war der Inhalt dieser Sätze bei den Völkern des alten Orients aus der Praxis schon längst bekannt, als später die Griechen versuchten, jene «einfacheren» Sätze zu finden, aus welchen sich die «komplizierten» ableiten lassen. Denn «nach Euklid dargestellt erscheint zwar die Mathematik als eine systematische deduktive Wissenschaft, aber die Mathematik im Entstehen erscheint als eine experimentelle (induktive) Wissenschaft».[29]

Nachdem im Sinne der versuchten Erklärung die Griechen im Laufe der definitorisch-axiomatischen Grundlegung der Mathematik erst *nachträglich* jene Prämissen suchten, die zum Beweis ihrer Sätze nötig waren, wird es auch verständlich, dass es ihnen kaum gleich am Anfang gelingen konnte, die Frage zu klären: welche sind denn die «einfachsten» Behauptungen mathematischen Inhalts, welche dürfen als allgemein zugegebene Prämissen, als keines Beweises bedürftige Axiome, Postulate oder Definitionen gelten? – Es ist wahrscheinlich, dass alles, was bei Euklid unter den «Prinzipien» zusammengefasst wird, das Ergebnis einer längeren Entwicklung darstellt. Denn im Sinne des obigen Schemas mussten ja die Mathematiker anfangs nur bestrebt gewesen sein, die «komplizierteren» Behauptungen auf «einfachere», auf solche Sätze zurückzuführen, die auch der Gegner für richtig hielt. Soll das wirklich der Weg der Entwicklung gewesen sein, so ist es kaum denkbar, dass man gleich am Anfang, sozusagen mit einem Schlage die «allereinfachsten» Prämissen, die Axiome, Postulate und Definitionen gefunden hätte. Bezeichnend dafür, wie lange Zeit hindurch derartige Versuche angestellt wurden, ist die Erzählung von Proklos, der berichtet,[30] dass auch noch in der Zeit nach Euklid Apollonios von Perge versucht hätte, das erste Euklidische Axiom zu beweisen: «was demselben gleich ist, ist untereinander gleich».[31] Proklos

---

nicht dahintergekommen, wie subjektiver Art jede solche Unterscheidung ist. Allerdings erörtert Aristoteles langwierig, welcher Art jene Prämissen sein müssten, auf die sich die deduktive Wissenschaft bauen kann; vgl. K. v. FRITZ: o. c. S. 23: «Das als Prinzip angenommene muss in sich *einsichtiger* (?), *einfacher* (?) und abstrakter sein als das, was daraus abgeleitet wird.»

29 G. PÓLYA: Schule des Denkens, Vom Lösen mathematischer Probleme. Bern 1949. S. 9.

30 Proclus (ed. G. FRIEDLEIN) 183.

31 Nach der Übersetzung von O. BECKER: Grundlagen der Mathematik. Freiburg/München 1954. S. 90. Dasselbe in der Übersetzung von P. TANNERY (s. oben Anm. 25): *Les choses égales à une même sont aussi égales entre elles.*

behauptet, dass dieser Versuch von Apollonios deswegen fehlschlagen musste, weil er mit «weniger evidenten» Prämissen begründen wollte, was «evidenter» ist, als seine eigenen angeblichen Prämissen.[32]

Wie man sieht, stellt also die versuchte Erklärung – d. h. der Vergleich der definitorisch-axiomatischen Grundlegung der griechischen Mathematik mit der Entfaltung der Aristotelischen Logik – eine Theorie dar, die zwar teilweise *richtig*
jene Umstände beleuchtet, unter welchen die griechische | Wissenschaft zustande 121
kam, aber der Verfasser – K. v. Fritz – versäumt dennoch – wohl auch infolge der anders orientierten Zielsetzung seiner Untersuchung –, die Frage klar und prägnant aufzuwerfen: *wann, wie* und *warum* jene entscheidende Wandlung eintrat, die zur Geburt der deduktiven Wissenschaft führte. Er antwortet auf diese Frage nur nebenbei mit den folgenden drei Behauptungen, – die zwar im Grunde wieder richtig sind, aber in ihren Konturen doch etwas verschwommen bleiben:

1. Er stellt fest, dass man die ersten Schritte zur definitorisch-axiomatischen Grundlegung wohl schon in der Zeit *vor* Aristoteles wird getan haben müssen, denn sonst könnte sich Aristoteles nicht eben auf die Mathematik in jenen Erörterungen berufen, in denen er die Methoden der beweisenden (apodeiktischen) Wissenschaft bespricht.[33]
2. Er weist auch wiederholt darauf hin, dass die ältere griechische Mathematik – besonders diejenige des Thales – allem Anschein nach noch in höherem Masse empirischen Charakters sein musste, und die unmittelbare Evidenz der Anschaulichkeit erstrebte; später – zur Zeit Euklids – wurden der empirische Zug und das Erstreben der Anschaulichkeit in der griechischen Mathematik in den Hintergrund gedrückt; die Mathematik dieses späteren Zeitalters war schon abstrakter.[34]
3. Einmal stellt K. v. Fritz auch die Tatsache fest, dass jene griechische Mathematik, die von den Pythagoreern des 5. Jahrhunderts ausgegangen war, schon einen anderen Charakter hatte, als die Geometrie des Thales im 6. Jahrhundert: sie begnügte sich nämlich nicht mehr mit der Evidenz der Anschaulichkeit.[35]

Ein anderer Mangel dieser Theorie besteht darin, dass sie die folgenden Fragen so gut wie völlig offenlässt: Ob und inwiefern das Zustandekommen der deduktiven Mathematik mit der Entfaltung der Lehre über die Logik zusammenhing? Verliefen die beiden Erscheinungen – die Entwicklung der Logik und diejenige der Mathematik – nur parallel nebeneinander, sich nur stellenweise gegenseitig beeinflussend, oder musste die eine der anderen unbedingt vorangehen? Und welcher gehört dann die Priorität? – Auch wir lassen diese Fragen vorläufig auf sich bestehen, und wir wenden

---

[32] Bemerken wir im Zusammenhang mit diesem Versuch von Apollonios: nach unseren heutigen Kenntnissen wäre die Annahme *nicht* wahrscheinlich, dass Apollonios «bewusst axiomatisiert hätte» – in dem Sinne nämlich, dass er erkannt hätte: Euklids Axiome können in der Tat in einem nicht-euklidischen Axiomen-System als abgeleitete Sätze auftreten.

[33] K. v. FRITZ: o. c. S. 43: «Es ist deutlich zu sehen, dass Aristoteles nicht in der Weise hätte mit konkreten Beispielen aus der Mathematik operieren können, wenn Ansätze zu einer definitorisch-axiomatischen Grundlegung der Mathematik ... nicht schon vor ihm vorhanden gewesen wären.»

[34] Auf die Erörterungen des Verfassers, die die Geometrie des Thales betreffen, kommen wir noch im III. Kapitel dieser Arbeit zurück.

uns statt dessen jener anderen Theorie zu, die sich nicht mehr damit begnügt, das
122 Entstehen der deduktiven Mathematik mit der Entfaltung der Logik zu | vergleichen, sondern auch die Antwort auf die Frage versucht: wodurch eigentlich dieser merkwürdige Entwicklungsprozess veranlasst wurde?

Die ungarischen Verfasser Gy. Alexits und I. Fenyő behandelten zuletzt in einer populärwissenschaftlich orientierten Arbeit die Fragen der Mathematik und des dialektischen Materialismus.[36] Es wird sich trotz des ungebundenen Charakters dieser Schrift dennoch lohnen, einige Ausführungen der Verfasser näher zu besehen, da sie unter anderem auch eine sehr interessante und originelle Vermutung darüber aufstellen, wie die deduktive Mathematik bei den Griechen entstand. Sie schreiben nämlich an einer Stelle ihrer Untersuchung:

«Die wissenschaftliche Mathematik beginnt eigentlich mit dem Entdecken der Notwendigkeit des strengen Beweises (Pythagoras). Das Entdecken der Notwendigkeit des Beweises erklärt sich seinerseits aus den gesellschaftlichen (sozialen) Umständen. Seine Anfänge sind unmittelbar in der gleichzeitigen Philosophie zu suchen. Die Diskussionsart der Sophisten, die jeden Widerspruch ans Licht zu bringen vermochte, erweckte in den griechischen Forschern den Wunsch nach der logischen Beweisführung. Geht man weiter, so wird man feststellen müssen, dass der hohe Entwicklungsstand der Philosophie, und vor allem ihre Verbreitung in weiten Kreisen, nur die Folge der politischen, also der gesellschaftlichen Diskussionen, Polemiken war. Die Auswirkung dieser Diskussionen kam in der Mathematik epochemachend dadurch zur Geltung, dass man die Notwendigkeit des strengen Beweises entdeckte. Den mathematischen Kenntnissen der übrigen Völker des Altertums fehlte eben dieser entscheidend wichtige Zug, darum lassen sich auch diese Kenntnisse mit der echten Wissenschaft der Griechen nicht vergleichen.»[37]

Ehe wir versuchten zu der Betrachtungsart dieses Zitates Stellung zu nehmen, müssen wir noch auf einen anderen Gedanken aufmerksam machen, der die Denkweise der Verfasser weitgehend beeinflusste. Sie wollten nämlich auch im Falle der Griechen die Gültigkeit jener These nachweisen, dass die Entwicklung der Mathematik im allgemeinen mit der Entwicklung der gesamten Produktion Schritt hält, und dass ihre Probleme im Laufe der Geschichte oft einfach aus den Bedürfnissen der Produktion erwuchsen.[38] – Möge aber dieser Gedanke im grossen und ganzen zwar richtig sein, so lässt er sich dennoch nicht verallgemeinern. Im Falle der griechischen Mathematik lässt sich nämlich für die Zeit von Pythagoras bis einschliesslich Euklid kaum irgendeine nähere Verbindung zwischen den Problemen der deduktiven Wissenschaft einerseits und denen der Produktion andrerseits nachweisen.
123 Es wäre auch verkehrt zu vergessen, dass die griechischen Mathematiker in den | meisten Fällen gar nichts davon hören wollten, dass ihre Wissenschaft überhaupt etwas mit der täglichen Praxis zu tun hätte. Sie erblickten in ihrem Wissen keineswegs das

---

35 K. v. FRITZ: o. c. S. 79. – Der Gedanke, dass mit den Pythagoreern des 5. Jahrhunderts eine neue Epoche in der Geschichte der griechischen Mathematik beginnt, stammt eigentlich von K. REIDEMEISTER. Siehe darüber das III. und IV. Kapitel dieser Arbeit.

36 GY. ALEXITS – I. FENYŐ: Matematika és dialektikus materializmus (= Mathematik und dialektischer Materialismus), Budapest 1948.

37 Ebd. S. 39.

38 Daselbst S. 36–37.

Ergebnis irgendwelcher praktischen Tätigkeit, und noch weniger ein Werkzeug der Produktion;[39] im Gegenteil, die Mathematik stellte für sie sozusagen das völlige Sich-Abwenden von der Praxis, die reine Betrachtung, die blosse Gedankentätigkeit dar. – Ohne Rücksicht darauf, ob sie damit auch Recht hatten, darf man diese Ansichten der antiken Mathematiker nicht ausser Acht lassen, denn sonst wird es kaum möglich, jene Umstände zu begreifen, unter denen die deduktive Wissenschaft entstand.

Der Gedanke also, dass die Entwicklung der griechischen Mathematik mit der Entwicklung der Produktion Schritt hielte, lässt sich für die Zeit von Pythagoras bis Euklid kaum rechtfertigen. Noch weniger überzeugt die Gedankenführung des vorigen Zitates, nämlich die versuchte Antwort auf die Frage: wie die Mathematik bei den Griechen zu einer deduktiven Wissenschaft wurde. Die Verfasser möchten nämlich auch in dieser entscheidenden Wandlung den *indirekten Einfluss* der gleichzeitigen Produktion nachweisen, und sie denken folgendermassen:

Die Entwicklung der antiken Produktionsweise ermöglichte zu einer bestimmten Zeit die *Demokratie* der griechischen Sklavenhalter. In der Demokratie herrscht jedoch die *Freiheit der Diskussion*, und im Laufe der Diskussionen bringen die *Sophisten* als geschickte Wortstreitführer die *Widersprüche* der verschiedenen Meinungen ans Licht; dadurch erwacht mit der Zeit *der Wunsch nach der logischen Beweisführung*, und diesen Wunsch erfüllt auch der griechische Mathematiker als er Beweise für seine Behauptungen (Sätze) liefert.

Besieht man die einzelnen Verknüpfungen dieser Gedankenkette genauer, so entdeckt man gleich den chronologischen Fehler der Konstruktion. Man soll nach der dargestellten Denkweise infolge der Tätigkeit der Sophisten auf die Widersprüche der verschiedenen Meinungen aufmerksam geworden sein, und dies soll den Wunsch nach der logischen Beweisführung erweckt haben. Die chronologische Reihenfolge der ineinander knüpfenden Gedankenmotive verläuft also: *entwickelte Produktionsweise – Demokratie – Diskussionsfreiheit – Sophisten – Entdeckung der verwirrenden Widersprüche – Wunsch nach logischer Beweisführung*, und schliesslich: *die Logik selbst.* Ist man einmal auf dieser Bahn bei der Logik angelangt, so ist es schon leicht zu behaupten, dass dieselbe Logik auch in der mathematischen Beweisführung zur Geltung käme. – Aber der auffallende chronologische Fehler dieser | Konstruktion 124
besteht darin, dass man gewöhnlich die Anfänge der deduktiven Mathematik auf das 6. Jahrhundert setzt; Pythagoras, auf den sich auch die Verfasser der behandelten Schrift berufen, lebte ja in diesem Jahrhundert. Es muss also – im Sinne der obigen Konstruktion – zur Zeit des Pythagoras irgendeine Lehre über die Logik schon vorhanden gewesen sein. Es fällt aber die Tätigkeit der Sophisten, die nach dem vorigen Schema die Entfaltung der Logik überhaupt hätte vorbereiten müssen, erst auf das 5. Jahrhundert. – Die dargestellte Konstruktion wäre also nur in dem Falle brauchbar, wenn es erst gelingen sollte nachzuweisen, dass die Logik als Wissenschaft ihren Ursprung in der Tat in den Diskussionen des täglichen Lebens hat,[40] und wenn es

---

[39] Sokrates betont z. B. im Platonischen Dialog über den «Staat», dass ein sehr kurzes Stück Geometrie und ein sehr kleiner Bruchteil Arithmetik vollständig dazu genügte, um die Bedürfnisse des praktischen Lebens zu befriedigen (VII 526 D); der wesentlichere Teil dieser Disziplinen befriedigt nämlich nicht praktische, sondern Bedürfnisse anderer Art.

[40] Der Gedanke, dass die Logik aus den Diskussionen des täglichen Lebens hervorgegangen sei, taucht in der Form einer Vermutung bei O. GIGON auf; er schreibt nämlich im Parmenides-Kapitel

schon bewiesen wäre, dass es eine Logik wirklich auch schon *vor* jener Zeit gab, in welcher sich die ersten Anfänge der deduktiven Wissenschaften melden.

Aber gesetzt, dass der chronologische Fehler sich irgendwie korrigieren liesse, auch so könnte noch die Konstruktion kaum bestehen. Denn fassen wir nur das vorige Zitat genauer ins Auge. – Die Verfasser sprechen von der «Notwendigkeit des strengen mathematischen Beweises», aber sie versäumen, die historische Frage zu klären: worin eigentlich im griechischen Altertum diese «Notwendigkeit» bestanden haben mag? Statt diese Frage zu stellen, erscheint bei ihnen ihre eigene völlig moderne Auffassung von der «Notwendigkeit des mathematischen Beweises» in einer solchen Form, als ob sie in der Tat wirklich eine «Erkenntnis der alten Griechen» gewesen wäre. Es ist nämlich interessant, wie sie die angeblich «*griechische* Entdeckung dieser Notwendigkeit» illustrieren. – Nachdem sie jene alte Approximationsformel der ägyptischen Geometrie erwähnt hatten, dass man die Fläche eines gleichschenkligen Dreiecks bekommt, wenn man die Basis mit der Hälfte der Seite (!) multipliziert,[41]
125 setzen sie fort: |

«Diese Approximation genügte den alten Aegyptern, weil in diesem Lande der Nil meistens solche Gebiete überschwemmte, die in *langgezogene* gleichschenklige Dreiecke aufgeteilt waren, und die Flächen solcher Dreiecke in der Tat auf Grund der gegebenen Formel – für die Genauigkeit der damaligen Messungen! – richtig, d. h. der Erfahrung entsprechend berechnet werden konnten. *Die Griechen wollten jedoch dieselbe Formel für die Lösung architektonischer Aufgaben anwenden, wo sie schon versagte.*»[42]

Wir haben den letzten Satz des Zitates hervorgehoben, weil er in der Tat eine überraschend geistreiche Vermutung ist, die sich aber, leider, kaum mit irgendwelchen Quellenangaben belegen lässt. Haben denn die Griechen wirklich jemals versucht die erwähnte Approximationsformel der Aegypter in der Architektur zu benutzen? Und

---

seines Buches «Der Ursprung der griechischen Philosophie», Basel 1945, S. 251: «Das Verfahren des Parmenides, durch Elimination der Möglichkeiten zum Wahren zu gelangen, setzt voraus, dass Parmenides eine bestimmte formale Methode des Beweisens schon kennt, ehe er sich daran macht, nun das Sein zu beweisen. Die Frage stellt sich dann nach dem Ursprung dieser Methode. Er wird schwerlich in der ionischen Kosmologie oder in der pythagoreischen Verkündigung zu suchen sein. Mit allen Vorbehalten sei bemerkt, dass *eine solche Technik des Beweisens am leichtesten in der Welt der politischen und juristischen Argumentation, in der sogenannten Gerichtsrhetorik sich bilden konnte.* Die konkrete Frage des Advokaten nach einem ungeklärten Tatbestand oder einer ungewissen Täterschaft konnte ohne Zweifel am ehesten zu solchen Beweismethoden führen, wie sie hier Parmenides an einem ganz anderen Objekt übt. Was wir wissen, ist, dass Sizilien zur Zeit des Parmenides die Gerichtsrhetorik geschaffen haben soll. Was uns fehlt, sind die äusseren Beweisstücke, die von der sizilischen Rhetorik zum «Wege der Forschung» des Parmenides hinüberführen. So muss dies lediglich eine nur zögernd angedeutete Hypothese bleiben.»

41 Zu dieser Approximationsformel vgl. man übrigens O. NEUGEBAUER: Vorlesungen über die Geschichte der antiken math. Wissenschaften Bd. 1. Berlin 1934 S. 123: «Eine Anzahl von Feldern (scil.: in Aegypten) sind dreieckig. Die Angabe der Grösse erfolgt dann etwa nach dem folgenden Schema: Die westliche Seite ist *a*, die östliche *b*, die südliche *c*, die nördliche *'nichts'*. Die Fläche ist dann wieder aus $\frac{a+b}{2}\cdot\frac{c}{2}$ zu erhalten. Hier hat man es also immer mit Näherungsrechnungen zu tun, die sich auf ganz bestimmte Felder beziehen und mit einer für praktische Zwecke ausreichenden Genauigkeit die Flächen angeben.»

42 GY. ALEXITS und I. FENYŐ: o. c. S. 37–38.

warum sind eigentlich die Aegypter selber nicht auf die glorreiche Idee gekommen, die oft benutzte Formel für die Lösung ihrer *eigenen* architektonischen Aufgaben zu verwenden? Auch sie kannten ja nicht nur die Feldmessung, sondern auch die Architektur! – Aber man würde das alles noch irgendwie in Kauf nehmen können, wenn die Verfasser ihre kühne Kombination nicht weiterführten:

«Nachdem die eben erwähnte Approximationsformel versagte, und ähnlicherweise auch andere bloss empirisch aufgestellte Sätze, mussten die griechischen Mathematiker zu der Überzeugung gelangen, dass die abstrakte Verallgemeinerung der blossen Erfahrung in sich noch keine hinreichende Sicherheit und Genauigkeit zu gewähren vermag; infolgedessen entdeckten sie die Notwendigkeit des mathematischen Beweises, und damit begründeten sie die mathematische Wissenschaft im heutigen Sinne des Wortes.»[43]

Kein Zweifel, diese Denkweise projiziert die völlig moderne Auffassung von der Notwendigkeit des mathematischen Beweises in das griechische Altertum zurück. Man kann nämlich in der Wirklichkeit gar keinen Beweis dafür angeben, dass die Begründung der deduktiven Wissenschaft in der Tat mit dem Zweck erfolgt wäre, um im Interesse der täglichen Praxis eine grössere Sicherheit und Genauigkeit zu gewinnen, als welche die blosse Erfahrung zu bieten vermag. Im Gegenteil, wie man später sehen wird: die Griechen wandten sich von der Praxis und damit zum Teil auch von der
Erfahrung selbst ab, als sie die theoretische Wissenschaft begründeten. | Vorläufig 126
wollen wir uns jedoch bloss mit der Feststellung begnügen, dass die griechischen Mathematiker – besonders im ersten Jahrhundert der deduktiven Wissenschaft – bestrebt waren, auch sehr viele solche Sätze zu beweisen, deren Wahrheit sie aus der Praxis schon längst gekannt hatten Diese Sätze sind dadurch, dass man sie bewiesen hatte, für die Praxis um gar nichts wertvoller geworden, als früher die blossen empirischen Kenntnisse waren. Man zog aus dem deduktiven Beweis im Altertum gar keinen unmittelbaren praktischen Nutzen. Ja, es scheint sogar, dass die antiken Mathematiker die praktische Brauchbarkeit des deduktiven Beweises nicht einmal geahnt hatten! Nach einer verbreiteten Anekdote liess z. B. Euklid – als ihn einmal ein Schüler fragte, was für einen Nutzen er aus dem Erlernen der mathematischen Ableitungen ziehen könnte – seinen Sklaven kommen: er sollte dem Fragenden einen Obolus schenken, da der betreffende scheinbar unbedingt einen Nutzen davon ziehen müsste, was er erlernt.[44] – Diese alte Anekdote wäre kaum möglich gewesen, wenn die griechischen Mathematiker überhaupt eine Ahnung davon gehabt hätten, wie die exakten mathematischen Beweise in der Praxis unmittelbar brauchbar sind. – Es ist also irreführend zu behaupten, dass die Griechen die Notwendigkeit des strengen mathematischen Beweises entdeckt hätten. Denn es stimmt zwar, dass die griechischen Mathematiker ausserordentlich strenge Beweise erstrebten, aber auf die Frage, wozu eigentlich diese Beweise nötig sind, hätten sie selbst überhaupt nicht antworten können.

Eine andere Schwäche der durch Alexits und Fenyő versuchten Erklärung besteht darin, dass sie den Ursprung des mathematischen Beweisverfahrens aus den Formen

---

43 Daselbst S. 39. – Ich war bestrebt die ungarischen Zitate möglichst genau zu übersetzen; es muss jedoch bemerkt werden, dass die Übersetzung der letzten Stelle nur *dem Inhalt nach* treu ist. Um eine für unsere gegenwärtigen Zwecke völlig nebensächliche Polemik zu vermeiden, habe ich offensichtliche Fehler und Missverständnisse des Textes in der Übersetzung beseitigt.

44 Vgl. G. Sarton: Ancient Science and Modern Civilization, London 1954 p. 20.

des alltäglichen Beweisverfahrens ableiten möchte. Im Sinne ihrer Auffassung wären die Menschen zuerst im Laufe der Diskussionen des alltäglichen Lebens auf die verwirrenden Widersprüche der Meinungen aufmerksam geworden, infolgedessen wäre der Wunsch nach der logischen Beweisführung erwacht, und später hätten auch die Mathematiker diesen Wunsch auf ihrem eigenen Gebiete zu erfüllen versucht. Das mathematische Beweisverfahren wäre also nur eine weiterentwickelte Form des alltäglichen Beweises. – Aber lässt sich in der Tat der Ursprung des exakten mathematischen Beweises aus solchen Formen des Beweisens ableiten, die *in ihrer Qualität* vom mathematischen Beweis grundverschieden sind? Denn die Methoden des Beweisens im alltäglichen Leben besitzen nur eine sehr entfernte Ähnlichkeit mit dem mathematischen Beweis. Der nicht-mathematische Beweis kann meistens nur bestrebt sein, die *Wahrscheinlichkeit* irgendeiner Behauptung nahezulegen; als Kriterium der Wahrheit gelten aber in diesen Fällen immer: die *Praxis*, die *Erfahrung* bzw. die *Tatsachen* selbst. Dagegen mussten die griechischen Mathematiker des Altertums sehr
127 oft | eben solche *Tatsachen* in allgemeingültiger Form beweisen, die man aus der Praxis schon seit Jahrhunderten sehr gut gekannt hatte, und deren Wahrheit man nie bezweifelte. Denn die Mathematik ist ja eben dadurch zu einer Wissenschaft geworden, dass sie sich – abgesehen von den Prinzipien – mit der bloss empirischen Erfahrung der Tatsachen nicht begnügte, sondern erkannte, dass meistens die allerofenbarsten Tatsachen selbst des Beweises bedürftig sind. – Suchen wir also den Ursprung des mathematischen Beweises in solchen Formen des alltäglichen Beweisverfahrens, die in ihren Ansprüchen viel bescheidener als der mathematische Beweis sind, so müssten wir noch erklären können: wieso und warum eigentlich bei den Griechen der mathematische Beweis so ausserordentlich streng und anspruchsvoll geworden ist, – weit über die Möglichkeiten jener Beweisformen hinaus, die im alltäglichen Leben jemals üblich waren? – Die Theorie von Alexits und Fenyő gibt auf diese besonders wichtige Frage gar keine Antwort.

Man findet den dritten Erklärungsversuch über das Entstehen der griechischen exakten Wissenschaft, den wir hier noch erwähnen müssen, bei B. L. v. d. Waerden.[45] Er weist nämlich darauf hin, dass die Griechen viele mathematische Kenntnisse empirischen Ursprungs von den Aegyptern und Babyloniern fertig übernahmen, aber die verschiedenen praktischen Regeln altorientalischer Herkunft nicht immer untereinander im Einklang waren. Die Babylonier berechneten z. B. den Kreisinhalt nach der Formel $3r^2$, dagegen die Aegypter nach der anderen: $(\frac{8}{9} \cdot 2r)^2$. Nun mussten die Griechen, als sie die abweichenden, bloss empirischen und nur für bestimmte praktische Zwecke brauchbaren Regeln kennenlernten, für sich entscheiden, welche von diesen die bessere sei, und wie könnte man den Kreisinhalt noch genauer berechnen. So wären sie langsam auf den Gedanken der exakten Ableitung gekommen.

---

[45] B. L. v. d. Waerden: Science awakening, Groningen 1954. S. 89; vgl. damit O. Beckers Kritik über die holländische Ausgabe desselben Buches (1950) in Gnomon 23 (1951) 297 ff.: «Interessanter als alle Einzelheiten ist die Gesamtauffassung des Verf. von der frühgriechischen Mathematik. Die entscheidende Wendung sieht er mit Recht in dem Auftreten von Theoremen mit Beweisen. Das Motiv liegt nach ihm in der Übernahme einer nicht immer einheitlichen orientalischen Tradition (wie z. B. die verschiedene Bestimmung des Kreisinhalts durch Aegypter und Babylonier) und der sich daraus ergebenden Notwendigkeit einer kritischen Entscheidung zwischen ihnen.»

Diese Theorie ist zweifellos bescheidener als die beiden früheren. Sie sucht den Ursprung der exakten Mathematik nicht irgendwo in der Nähe der Logik, oder auf einem Wege, der der Entfaltung der Logik parallel läuft; auch das mathematische Beweisverfahren will sie nicht auf solche Formen des Beweisens zurückführen, die im alltäglichen Leben üblich waren, sie möchte statt dessen das Streben nach Exaktheit einfach aus solchen Überlegungen ableiten, die rein mathematischer Art sind. – Es ist in der Tat sehr wohl möglich, dass zu der Entfaltung der exakten Wissenschaft – beson|ders am Anfang, im Falle des Thales – auch solche Reflexionen beitrugen. Es 128
ist jedoch ein Mangel dieser an sich wohl treffenden Vermutung, dass sie jene auffallende und bemerkenswerte Erscheinung völlig ausser Acht lässt, welche sonst in einem anderen Zusammenhang auch v. d. Waerden selbst mit Recht betonte, dass nämlich die griechischen Mathematiker schon sehr früh bestrebt waren, möglichst alles, selbst die alleroffenbarsten Tatsachen peinlich exakt zu beweisen. Stellt man die zur Zeit bekannten ältesten mathematischen Beweise der Griechen zusammen, so bekommt man gar nicht den Eindruck, als ob die griechischen Mathematiker die exakte Wissenschaft mit dem Zweck zustande gebracht hätten, um solche früher «strittige Fragen» entscheiden zu können, wie z. B. die Berechnung des Kreisinhalts. Im Gegenteil, die ältesten mathematischen Ableitungen beweisen sozusagen lauter unbestreitbare, beinahe triviale Tatsachen, die aus der empirischen Praxis auch früher schon längst und ebenso tadellos bekannt sein mussten. – Nun was mag aber die ältesten Mathematiker veranlasst haben, um möglichst alles, auch offenbare und wohlbekannte Tatsachen beweisen zu wollen? – B. L. v. d. Waerdens eben erwähnte Theorie antwortet auf diese Frage nicht.

Man sieht also: die hier zusammengestellten drei verschiedenen Erklärungsversuche liefern zwar sehr gute und brauchbare Gesichtspunkte für die weitere Forschung, aber sie lösen das Problem nicht. Die Frage bleibt auch weiterhin offen: wie ist die Mathematik zu einer deduktiven Wissenschaft geworden?

Nachdem die Wissenschaftsgeschichte das Entstehen der griechischen exakten Mathematik bisher nicht beruhigend erklären konnte, wollen wir im nächsten Kapitel mindestens skizzenhaft zusammenfassen, was man über diese Frage auf Grund unserer heutigen Kenntnisse behaupten kann. Zuerst überblicken wir die diesbezüglichen wichtigsten Angaben der antiken überlieferung *(Punkt A)*, dann ergänzen wir diese Angaben mit einigen Feststellungen der gegenwärtigen Forschung *(Punkt B)*.

III

A) Die dreizehn Bücher der Euklidischen Elemente, die rund um 300 v. u. Z. entstanden, stellen die älteste klassisch gewordene Zusammenfassung der antiken Mathematik dar. Euklids Darstellung der Mathematik ist – trotz jener teilweise unbedingt berechtigten Einwände, die seitdem im Laufe der Jahrhunderte gegen sie vielfach erhoben wurden – bis auf den heutigen Tag sozusagen mustergültig geblieben.[46] Man kann also mit | Recht behaupten, dass der exakte mathematische 129
Beweis der Griechen zur Zeit Euklids seine höchste Entwicklungsstufe eigentlich

46 Man vgl. dazu die Worte von G. PÓLYA: o. c. in den Kapiteln «Durchführen eines Planes» (S. 96 ff.) und «Warum Beweise?» (S. 225 ff.).

schon erreicht hatte. Die Ausbildung und Entfaltung der deduktiven Wissenschaft muss also auf die Zeit *vor* Euklid fallen. Was weiss die antike überlieferung über dieses Zeitalter vor Euklid?

Der Verfasser der ältesten griechischen Mathematikgeschichte war Eudemos von Rhodos, ein Schüler von Aristoteles nicht lange vor Euklid, im 4. Jahrhundert. Man findet einen kurzen Auszug seines verlorenen Werkes im Kommentar von Proklos (5. Jahrhundert n. Zw.) zum ersten Buch der Euklidischen Elemente. Das ist das berühmte «Mathematiker-Verzeichnis» von Proklos. Die wichtigsten Feststellungen dieses Verzeichnisses, die uns interessieren, sind die folgenden.

Der älteste Vertreter der griechischen Mathematik, bzw. der Geometrie, Thales im 6. Jahrhundert, soll sein Wissen in Aegypten gesammelt und von dorther diese Disziplin zu den Griechen gebracht haben. Nach dem Text soll er manches selber erfunden, und in manchem seinen Nachfolgern den Weg zu den Prinzipien gezeigt haben, dadurch, dass er einige Fragen allgemeiner, einige aber handgreiflicher (in sinnlich wahrnehmbarerer Form) auffasste (*τοῖς μὲν καθολικώτερον ἐπιβάλλων, τοῖς δὲ αἰσθητικώτερον*).[47]

Die nächste, für uns wichtige Feststellung des Verzeichnisses betrifft jenen Pythagoras, der ebenso im 6. Jahrhundert lebte. Es wird behauptet, dass Pythagoras die Beschäftigung mit der Geometrie – nach dem Wortlaut des Textes: diese «Philosophie» – verändert hätte, indem er ihr eine solche Form gab, die es ermöglichte, dass sie von nun an zu einem Bestandteil der Erziehung des freien Menschen werden konnte. – Mit solchen Ausdrücken wird in unserer Quelle die Behauptung umschrieben, dass die Geometrie, bzw. die Mathematik des Pythagoras nicht mehr eine praktische, sondern schon eine theoretische Wissenschaft war. Nach antiker Auffassung steht nämlich die praktische Tätigkeit unter der Würde des freien Menschen; der Freie darf sich nur mit untätiger Betrachtung, d. h. griechisch: mit *Theorie* beschäftigen. – Man bekommt eine Ahnung davon, wie wesentlich diese Behauptung des Mathematiker-Verzeichnisses ist, erst dann, wenn man daran denkt, dass der mathematische «Satz» griechisch in der Tat *Theorema* heisst. Auch diese Benennung zeigt, dass es sich in der griechischen Mathematik nicht um die Praxis, sondern um die Theorie um ihrer selbst willen handelte. – Pythagoras soll nach dem Wortlaut des Proklos die entscheidende Wandlung dadurch in die Mathematik gebracht haben, dass er die Prinzipien der Geometrie untersuchte, und ihre Sätze vom konkreten Stoff unabhängig (*ἀύλως*) auf rein intellek-
130 tuellem Wege (*νοερῶς*) erforschte. | – Er, Pythagoras soll auch die Theorie der Irrationalen (oder eher: die Theorie der Proportionen?) gefunden[48] und die kosmischen (= regelmässigen) Körper konstruiert haben.

Dann wird im Verzeichnis von Proklos unter anderem noch erzählt, dass schon lange vor Euklid auch andere Männer solche systematische Werke der Mathematik, also «Elemente» geschrieben hätten, wie später Euklid. Der erste Systematiker der Mathematik war Hippokrates von Chios im 5. Jahrhundert, dann Leon in der ersten Hälfte des 4. Jahrhunderts, und Theudios von Magnesia in der zweiten Hälfte des 4. Jahrhunderts.

---

[47] Proclus (ed. G. FRIEDLEIN) 65. – Den ganzen Text des Mathematiker-Verzeichnisses übersetzt und zum Teil kommentiert B. L. v. D. WAERDEN: Science awakening. S. 90 ff.

[48] Nach der Lectio von G. FRIEDLEIN handelt es sich hier um die Theorie der *Irrationalen;* anders gelesen sollte nur von der Theorie der *Proportionen* die Rede sein.

Nun wollen wir jetzt sehen, inwiefern die antike Überlieferung in den genannten Punkten durch die heutige Mathematikgeschichte ergänzt oder modifiziert werden kann.

B) Was Thales betrifft, scheint die moderne historische Forschung die Behauptungen des alten Mathematiker-Verzeichnisses nur zu bestätigen. Man darf vielleicht aus dem Bericht über Thales schliessen, dass sich auch die griechische Überlieferung des orientalischen Ursprungs der empirischen mathematischen Kenntnisse bewusst war; sie spricht ja deswegen von der Reise des Thales in Aegypten, und dass er von dorther diese Disziplin zu den Griechen gebracht hätte.[49] – Es geht zwar aus unserem Text nicht eindeutig hervor, inwiefern eigentlich schon die Geometrie des Thales als exakte Wissenschaft anzusehen sei, aber selbst über diesen Punkt lässt uns der eben zusammengefasste kurze Bericht vielleicht nicht völlig im Stiche. Es heisst nämlich, dass Thales seinen Nachfolgern den Weg zu den Prinzipien gezeigt habe; als wollte der antike Verfasser des Berichtes mit diesen Worten eben den Gedanken zum Ausdruck bringen, dass Thales selber zwar die Prinzipien der Mathematik noch nicht gefunden hätte – seine Geometrie wäre | also noch keine im späteren Sinne des Wortes 131
«definitorisch-axiomatisch begründete Wissenschaft» gewesen –, aber er hätte dennoch die erste Anleitung zu einer solchen Grundlegung erteilt. Und das alles soll dadurch geschehen sein, dass Thales einige Fragen *allgemeiner*, einige jedoch handgreiflicher auffasste. – Man hat beinahe den Eindruck, als liesse sich selbst noch diese Behauptung der Überlieferung mit bekannten historischen Tatsachen belegen. Es ist ja bekannt, dass den Begriff des geometrischen «Winkels» in der Tat Thales, oder mindestens das Zeitalter des Thales einführte; und das war gewiss eine bedeutende wissenschaftliche *Verallgemeinerung.*[50] Das Streben nach Allgemeingültigkeit muss also schon in der Wissenschaft des Thales zur Geltung gekommen sein. – Dabei betont unser Text auch die andere Seite: Thales soll andere Fragen *in handgreiflicher Form*

---

[49] Es ist interessant, dass die antike Überlieferung, wenn sie über die altorientalische Herkunft der rein praktischen, empirischen mathematischen Kenntnisse berichtet, immer nur von Aegypten und nie von Babylon redet, obwohl die Griechen offenbar manches auch von den alten Babyloniern gelernt haben müssen. Vgl. O. BECKER (Grundlagen der Math. S. 22): «Die Griechen übernahmen weithin altorientalisches, besonders wohl *babylonisches* Material, obwohl die griechische Tradition immer nur *von Aegypten* als dem Ursprungslande der Geometrie spricht.» – Was mag wohl der Grund dieser einseitigen, sozusagen schief gewordenen Tradition sein? – Wir glauben, dass man diese Frage mindestens mit einiger Wahrscheinlichkeit beantworten kann. Auch die griechische Tradition scheint nämlich davon zu wissen, wie die Algebra auf dem Ausstrahlungsgebiet der babylonischen Kultur hochentwickelt war. Proklos sagt z. B., dass bei den Phöniziern die Kenntnis der Zahlen weit entwickelt war, eben infolge des Handels und der vielen praktischen Beschäftigung mit der Rechentechnik (FRIEDLEIN p. 65). – In der Zeit jedoch, als Eudemos im 4. Jahrhundert die erste Mathematikgeschichte verfasste, war die griechische Mathematik schon völlig *geometrisiert*. Deswegen konnten die Griechen ihre alten Lehrmeister auf dem Gebiete der Geometrie mit einigem Recht allerdings eher in den Aegyptern als in den Babyloniern vermuten. Denn wir haben ja schon gesehen, dass die Aegypter in der Tat den Kreisinhalt z. B. genauer berechnen konnten, als die Babylonier. Man konnte also in der Zeit der geometrisierten griechischen Mathematik gewissermassen schon eine nähere Verwandtschaft mit der ägyptischen Geometrie als mit der babylonischen Algebra fühlen.

[50] Vgl. O. BECKER: Gnomon. 23 (1951) S. 298: «Im übrigen ist gerade die Einführung des Winkelbegriffs (statt des *śqt*) eine wesentliche neue Errungenschaft der frühgriechischen Geometer (Scheitelwinkel, Basiswinkel, Winkel im Halbkreis), die von weittragenden Folgen war (Winkelsumme im Dreieck, woraus vielerlei abgeleitet werden konnte).»

behandelt haben. – Es ist kein Zufall, dass dieser Zug der thaletischen Wissenschaft in der Überlieferung betont wird. Kaum um einige Zeilen weiter lesen wir in demselben Bericht über Pythagoras gerade das Gegenteil dessen; Pythagoras soll die Sätze der Mathematik schon unabhängig vom konkreten Stoff (ἀύλως), auf rein intellektuellem Wege (νοερῶς) erforscht haben. Diese letztere Behauptung kontrastiert also die Wissenschaft des Pythagoras mit derjenigen des Thales. – Was soll aber heissen, dass Thales einige Fragen der Geometrie «in handgreiflicher Form» behandelte? – Auch diese Frage lässt sich beantworten, wenn wir zunächst die andere genauer prüfen: ob Thales seine Sätze «beweisen» konnte, und in welcher Form? – Die antike Überlieferung berichtet in der Tat von den «Beweisen» des Thales,[51] aber sie gibt, leider, nicht genau an, worin eigentlich der Beweis bestand. Die moderne Interpretation kann zu der Klärung dieses Problems in zwei Punkten beitragen; erstens kann sie nämlich daran erinnern, dass das griechische Wort für «*beweisen*», der Terminus der mathematischen Fachsprache, ἀποδεικνύναι nach seiner ursprünglichen Bedeutung eigentlich nur «*zeigen*» hiess; zweitens kann aber die moderne Wissenschaftsgeschichte auch an eine Beobachtung erinnern, die ermöglicht das Beweisverfahren des Thales mit grosser Wahrscheinlichkeit zu rekonstruieren.[52] – Euklid stellt nämlich am Anfang seines ersten Buches das sog. Axiom der Kongruenz, das ἐφαρμόζειν-Axiom auf: «Dinge, die
132 sich decken (= die aufeinander passen), sind gleich».[53] Das Wort ἐφαρ|μόζειν («auf einander passen») kommt jedoch später bei Euklid nirgends in einem Lehrsatz, nur in diesem Axiom und in den Beweisen einiger Lehrsätze vor. Es ist also auffallend, dass ein Axiom zwar so allgemein formuliert und dann doch eigentlich nur für zwei Sätze gebraucht werde, obwohl sich – wie man bemerkte – manche Unstimmigkeiten in den Gleichheitsdefinitionen bei Euklid leicht hätten vermeiden lassen, wenn von der Deckungsmethode etwas reichlicher Gebrauch gemacht worden wäre.[54] Euklid ist also offensichtlich bestrebt, das Anwenden des genannten Axioms möglichst zu vermeiden. Wir wissen jedoch aus dem Text von Proklos, dass man früher die empirische Methode des Aufeinanderpassens, die ἐφαρμόζειν-Methode auch in solchen Beweisen benutzte, die bei Euklid gar nicht mehr vorkommen. «Diese Methode muss also einmal in viel weiterem Umfang angewendet worden sein, als dies bei Euklid der Fall ist. Sie scheint auf den ersten Anfang der griechischen Mathematik zurückzugehen, und es ist dann nicht leicht, es als einen reinen Zufall zu betrachten, *dass von den fünf Sätzen, die dem Thales in der antiken Überlieferung zugeschrieben werden, vier sich direkt und der fünfte indirekt mit der Deckungsmethode beweisen lassen.*»[55] Geht man noch weiter, so findet man, dass bei Proklos in der Tat ein Beweis mit der empirischen Deckungsmethode auch für einen solchen Satz genannt wird, von dem es früher hiess, dass er schon durch Thales auf irgendeine Weise «bewiesen» worden sei.[56] Man kann

---

51 Sieh Proclus (ed. FRIEDLEIN) 157.

52 Ich fasse im folgenden die Beobachtungen von K. v. FRITZ (s. oben die Anm. 21) – teilweise auch über seine Feststellungen hinausgehend – zusammen.

53 Eucl. I *Κοιναὶ ἔννοιαι* 7: *τὰ ἐφαρμόζοντα ἐπ' ἀλλήλα ἴσα ἀλλήλοις ἐστίν.* «quae inter se congruunt, aequalia sunt» (I. L. HEIBERGS Übersetzung). «Was sich deckt, ist gleich» (A. BECKER). «Les choses qui coincident l'une avec l'autre sont égales entre elles» (P. TANNERY).

54 K. v. FRITZ: o. c. S. 77.

55 K. v. FRITZ: ebd.

56 Proclus (ed. FRIEDLEIN) p. 157. – Es handelt sich hier um den «Satz», der besagt, dass «der Durchmesser den Kreis halbiert». Euklid hat diesen «Satz» in die 17. Definition seines ersten Buches aufgenommen; dadurch wurde jedoch seine Definition überbestimmt, vgl. K. v. FRITZ: o. c.

also auf Grund dieser Beobachtungen annehmen, dass das Beweisverfahren des Thales wohl eben in der Anwendung jener empirischen Deckungsmethode bestand, welche man früher zwar reichlich verwandte, später jedoch möglichst auszuschalten versuchte.[57] Thales hat also seine Sätze wohl dadurch bewiesen, dass er die von ihm behandelten geometrischen Figuren «auf einander passte», und die Wahrheit seiner Behauptungen auf diese Weise *veranschaulichte*, «zeigte». Im ursprünglichen Sinne des Wortes *ἀποδεικνύναι* war es wirklich ein «Beweis» – allerdings noch von einer völlig anschaulichen, empirischen Art. – Es ist wohl möglich, die Worte von Proklos, dass nämlich Thales manches in der Geometrie noch «in handgreiflicher Form» behandelte, in diesem Sinne zu verstehen, | d. h. also, dass für Thales die Evidenz noch die 133
empirische, anschauliche Evidenz war. Zu seiner Zeit erfolgte also noch nicht jene grundsätzliche Wandlung in der griechischen Wissenschaft, die im Mathematiker-Verzeichnis von Proklos der Tätigkeit des Pythagoras zugeschrieben wird.

Was Pythagoras anbelangt, ist die heutige Wissenschaft den Behauptungen des Mathematiker-Verzeichnisses gegenüber sehr skeptisch. Unsere ältesten Quellen über Pythagoras scheinen nämlich noch gar nichts davon zu wissen, dass er ein grosser Mathematiker und Philosoph gewesen wäre.[58] Versucht man die ersten Anfänge der späteren Pythagoras-Legende wiederherzustellen, so stösst man schliesslich auf eine Bewegung, eine Art «pythagoreischer Romantik», die gegen Ende des 5. und am Anfang des 4. Jahrhunderts in den aristokratischen und zugleich spekulativ und religiös ergriffenen Kreisen Unteritaliens und Siziliens sich ausgebreitet hatte.[59] Aller Wahrscheinlichkeit nach ist der «Philosoph und Mathematiker Pythagoras» erst die Schöpfung dieser Zeit und dieser Kreise. Es kann auch kein Zufall sein, dass Platon und Aristoteles meistens nur die Pythagoreer erwähnen, aber so gut wie niemals von Pythagoras selbst reden. Die moderne Forschung hat auch über jene mathematischen Entdeckungen, die in der späteren Überlieferung dem Pythagoras zugeschrieben werden, feststellen können, dass diese zum Teil aus viel älteren Zeiten stammen, zum Teil aber erst späteren Ursprungs als das 6. Jahrhundert v. u. Z. sind.[60] – Nach all dem wird es wohl niemanden mehr wundernehmen, dass man den Bericht des Mathematiker-Verzeichnisses über Pythagoras sehr skeptisch auffasste; man erklärte auch ihn für einen Teil der Pythagoras-Legende. – Es empfiehlt sich jedoch in dieser Beziehung einige Vorsicht. Denn man darf doch nicht vergessen, dass auch sehr zuverlässige antike Quellen hie und da über die arithmetischen Studien der Pythagoreer des 5. Jahrhunderts berichten. Aristoteles behauptet sogar, dass es die Pythagoreer waren,

---

57 K. v. FRITZ: o. c. S. 94: «Dass im Anfang die *anschauliche* Evidenz eine nicht unbeträchtliche Rolle gespielt hat, zeigt die in einem frühen Stadium reichliche Verwendung der Deckungsmethode, während umgekehrt die fortschreitende, wenn auch bis auf Euklid nicht vollständige Ausschaltung dieser Methode und das sichtliche Bestreben Euklids, den letzten Überbleibseln der Methode, die er nicht vermeiden kann, ein axiomatisches Fundament zu geben und sie auch sonst ihres empirischen Charakters so sehr als möglich zu entkleiden, REIDEMEISTER Recht geben: das Charakteristische der griechischen Mathematik ist, dass sich in ihr die Umwendung vom Anschaulichen zum Begrifflichen vollzieht.»

58 K. REINHARDT: Parmenides und die Geschichte der griech. Philosophie, Bonn 1916 S. 233 f. – In demselben Sinne spricht über Pythagoras auch E. FRANK: Plato und die sogenannten Pythagoreer, Halle/Saale 1923. S. 67 und dazu besonders die Anm. 166 auf S. 356.

59 K. REINHARDT: o. c. S. 232.

60 Vgl. K. REIDEMEISTER: Das exakte Denken der Griechen, Hamburg 1949. S. 20 und 51–52; E. SACHS: Die fünf platonischen Körper, Berlin 1917.

die sich als erste mit μαθήματα befassten.[61] Ebenso war auch nach Platon[62] die grösste und erste von den Wissenschaften der Pythagoreer die Lehre von den Zahlen. Und die moderne Forschung hat in der Tat vermocht – wie man es bald sehen wird –, mindestens einen Teil der pythagoreischen Wissenschaft des 5. Jahrhunderts wieder-
134 herzustellen. Vergleicht man aber die zurückeroberte pythagoreische Mathe|matik damit, was im Verzeichnis von Proklos über Pythagoras selbst berichtet wird, so muss man erstaunt feststellen, dass die Behauptungen der antiken Quelle über Pythagoras, wenn auch nicht auf diese halb legendenhafte Gestalt, so doch mindestens auf die Wissenschaft der Pythagoreer zuzutreffen scheinen. Diese Mathematik ist in der Tat schon auf Prinzipien gebaut, und sie erforscht ihre Sätze wirklich «vom konkreten Stoff unabhängig und auf rein intellektuellem Wege». Deswegen kam auch K. Reidemeister zu dem Schlusse, dass die Begründer der exakten Wissenschaft eigentlich die Pythagoreer des 5. Jahrhunderts gewesen seien.[63] Man hat also beinahe den Eindruck, als hätte das Verzeichnis von Proklos diese Feststellung nur zurückprojiziert auf jenen legendenhaften Pythagoras, den die Sekte der Pythagoreer zu ihrem Namengeber wählte. – Damit ist natürlich das Entstehen der exakten Wissenschaft noch gar nicht erklärt, aber man weiss mindestens von einer merkwürdigen Übereinstimmung zwischen der antiken Überlieferung und der modernen Forschung: die Tradition schreibt die Begründung der exakten Wissenschaft dem Pythagoras zu, während die moderne Rekonstruktion die erste Vertreterin der deduktiven Mathematik in der Arithmetik der Pythagoreer erblickt.

Aus dem kurzen Bericht des Proklos über die ältesten Systematiker der deduktiven Wissenschaft interessiert uns am meisten der Name des Hippokrates. Der grösste und berühmteste Geometer des 5. Jahrhunderts, Hippokrates, von der Insel Chios gebürtig, hat sich nach der Tradition längere Zeit (etwa von 450–430) in Athen aufgehalten. Er soll seinen Lebensunterhalt hier durch «Unterricht in Geometrie» verdient haben.[64] Nach Proklos war Hippokrates der erste, der «Elemente» zusammengestellt hatte. Die heutige Forschung bezweifelt die Glaubwürdigkeit dieses Berichts überhaupt nicht. Denn wir können uns ja ein sehr zuverlässiges Bild vom mathematischen Wissen des Hippokrates verschaffen. Es ist nämlich der wörtliche Bericht des Eudemos von Rhodos über die sog. «Quadratur der Möndchen» (μηνίσκοι, lunulae) des Hippokrates glücklicherweise erhalten geblieben,[65] und auf Grund dieser kostbaren «Inkunabel» der voreuklidischen Geometrie erscheint uns die Wissenschaft des Hippokrates in einem solchen Licht, dass man es für möglich hält: Hippokrates hat allerdings schon versuchen können die mathematischen Kenntnisse seiner Zeit in eine streng aufgebaute logische Ordnung zusammenzufassen. – Dabei wird die Angabe des

---

61 Aristoteles, Met. A 5.985 b 23–24. – Die Pythagoreer verstanden unter «Mathema» ein geordnetes System mit Beweisen; zur Deutung des Wortes vgl. K. Reidemeister: o. c. S 52: μάθημα «eine Zusammenstellung mathematischer Sätze und *Beweise*».

62 Platon, Epinomis 990 C.

63 K. Reidemeister: o. c. S. 52. – Wir wollen uns mit dieser Feststellung von K. Reidemeister im letzten Kapitel der vorliegenden Arbeit ausführlicher beschäftigen.

64 Zur allgemeinen Orientierung sowohl über Hippokrates, wie auch über die Frühgeschichte der griechischen Wissenschaft vgl. man das sehr unterhaltende Büchlein von G. Hauser: Die Geometrie der Griechen von Thales bis Euklid. Luzern 1955.

65 Man findet diesen Bericht des Eudemos bei Simplikios, dem Aristoteles-Kommentator des 6. Jahrhunderts n. Zw.; vgl. dazu O. Becker: Grundlagen der Mathematik, S. 29 ff.

Proklos über diese erste systematische Darstellung der griechischen | Mathematik 135
schon im 5. Jahrhundert v. u. Z. merkwürdigerweise auch von einer anderen Seite her, durch die moderne Forschung sozusagen bestätigt. Die neueste Forschung ist nämlich völlig unabhängig von diesem Bericht über die «Elemente des Hippokrates» zur Vermutung gekommen, dass es schon im 5. Jahrhundert, also in der Zeit vor 400, auch einen schriftlich fixierten Lehrgang der Zahlentheorie geben musste, welcher später ohne erhebliche Änderungen, evtl. nur verkürzt in das VII. Buch der Euklidischen Elemente übernommen wurde.[66] Die Elemente des Hippokrates und dieser vermutete «Lehrgang der Zahlentheorie» müssen aller Wahrscheinlichkeit nach mathematische Werke derselben Art gewesen sein. – Lehrreich ist für uns der Bericht des Proklos über Hippokrates darum, weil man auf Grund dessen eine sehr wichtige Vermutung aufstellen kann. Hat Hippokrates in der Mitte oder in der zweiten Hälfte des 5. Jahrhunderts schon versuchen können, das mathematische Wissen seiner Zeit in systematischer Ordnung zusammenzufassen, so müssen die Anfänge der deduktiven Wissenschaft allerdings auf die Zeit *vor* Hippokrates fallen.

Die allzu knappen Angaben des Mathematiker-Verzeichnisses lassen sich einigermassen damit ergänzen, was die moderne historische Forschung aus der voreuklidischen Mathematik der Griechen rekonstruieren konnte. Am wichtigsten sind für uns in diesem Zusammenhang zwei Arbeiten; die eine von O. Becker aus dem Jahre 1936, und die andere von B. L. v. d. Waerden aus 1947/49.

O. Becker hat nämlich aus dem IX. Buch der Euklidischen Elemente die sog. altpythagoreische Lehre vom Geraden und Ungeraden aussondern können.[67] Es ist ihm nachzuweisen gelungen, dass die letzten sechzehn Sätze dieses Buches bei Euklid eigentlich nur einen Anhang darstellen, den entweder der Verfasser der Elemente selbst, oder mindestens einer seiner ältesten Herausgeber noch im Altertum der Aufbewahrung am Ende der Rolle aus Pietätsgründen für würdig befand. Mit diesen sechzehn Sätzen hängt bei Euklid am Ende des X. Buches der 27. Appendix (ed. Heiberg) zusammen. Diese insgesamt 17 Sätze bilden die sozusagen vollständige Lehre vom Geraden und Ungeraden, deren Entstehungszeit durch O. Becker auf die Mitte oder auf die erste Hälfte des 5. Jahrhunderts gesetzt wird. Die Datierung ist zwar nur vermutlich, aber diese Sätze stellen allerdings das älteste zur Zeit bekannte griechische «Mathema» dar.

Ähnlicherweise konnte B. L. v. d. Waerden feststellen, dass die ersten 36 Sätze des
VII. Euklidischen Buches noch aus der Zeit vor 400 stammen.[68] | Bei Boetius (5. 136
Jahrhundert n. Zw.) ist nämlich ein mathematischer Beweis des Archytas (430–360) erhalten geblieben.[69] In seinem peinlich exakten Beweis setzt nun Archytas die Kenntnis solcher mathematischer Sätze voraus, die man im VII. und VIII. Buch bei Euklid wiederfindet. Wir müssen also annehmen, dass die durch Archytas als bekannt vorausgesetzten mathematischen Sätze zu seiner Zeit in der Tat schon in irgendeinem mathematischen Handbuch schriftlich fixiert waren. Prüft man aber die logische Aufeinanderfolge sämtlicher Sätze, die nach dem Zeugnis des erhaltenen

---

66 Vgl. B. L. v. D. WAERDEN: Math. Ann. 120 (1947/49) 145–146.

67 O. BECKER: Die Lehre vom Geraden und Ungeraden im neunten Buch der Euklidischen Elemente, Quellen und Studien zur Gesch. der Math., Abt B. Bd. 3 (1936) 533–553.

68 B. L. v. D. WAERDEN: Die Arithmetik der Pythagoreer I., Math. Ann. 120 (1947/49) 127–153.

69 Boetius, De inst. mus. III 11 (ed. G. FRIEDLEIN) 1867. 285.

Archytas-Beweises diesem alten Mathematiker schon als schriftlich fertig vorliegendes Gedankengut bekannt sein mussten, so bekommt man ein geschlossenes Ganzes, nämlich die ersten 36 Sätze des VII. Euklidischen Buches. Dieses kompakte Ganze ist nach v. d. Waerden ein pythagoreischer Lehrgang der Zahlentheorie aus dem 5. Jahrhundert, oder mindestens ein Teil von einem solchen Lehrgang.

Überblickt man nun diese Ergebnisse der modernen Forschung und jene alte Überlieferung, die wir eben besprochen hatten, so lassen sich die wichtigsten Angaben, die uns interessieren, in den folgenden fünf Punkten zusammenfassen:

1. Der erste griechische Mathematiker des 6. Jahrhunderts, Thales hat in seinen «Beweisen» nur noch die Evidenz der Anschaulichkeit erstrebt. Seine Wissenschaft war noch vorwiegend empirischer Art.
2. Entweder in der Mitte oder noch in der ersten Hälfte des 5. Jahrhunderts entstand die pythagoreische Lehre vom Geraden und Ungeraden (Eucl. IX 21–36 und X App. 27), das älteste zur Zeit bekannte deduktive Lehrstück der Griechen.
3. Um die Mitte des 5. Jahrhunderts war Hippokrates von Chios, der Verfasser der Möndchen-Quadratur, in Athen tätig; er war der erste, der schon Elemente zusammengestellt hatte.[70]
4. Man hat noch im 5. Jahrhundert jenen Lehrgang der pythagoreischen Zahlentheorie zusammengestellt und schriftlich fixiert, der für uns bei Euklid in den Sätzen 1–36 des VII. Buches erhalten blieb.
5. Zwischen 430 und 360 v. u. Z. lebte Archytas, der noch jene alte Auffassung vertrat, nach welcher die Arithmetik der Geometrie vorzuziehen sei. In der Zeit nach ihm erfolgte die Geometrisierung der griechischen Arithmetik.

Dies sind die wichtigsten Angaben, die man bei der Beantwortung der Frage, wie die deduktive Mathematik entstand, berücksichtigen muss.

Was die Beweise, die Deduktionsart dieser alten (in den Punkten 2, 3, 4 und 5
137 zusammengefassten) griechischen Mathematik betrifft, kann man | folgendes feststellen. Bereits sehr früh waren in der pythagoreischen Mathematik die Ansprüche an die Strenge der Beweise sehr hoch. Das sehen wir etwa an der Lehre vom Geraden und Ungeraden im IX. Buch der Elemente. Sätze, wie Eucl. IX 21–29, sind jedem Rechner selbstverständlich, werden aber ausdrücklich formuliert und bewiesen. Andere Sätze, wie IX 30–32 sind sofort auch in sich einleuchtend, werden aber dennoch peinlich exakt aus noch einfacheren Tatsachen abgeleitet. – In dem Hippokrates-Fragment über die Quadratur der Möndchen werden Ungleichungen, die man ohne weiteres aus einer Figur hätte entnehmen können, auf das sorgfältigste bewiesen. (Man begnügt sich also nicht mehr mit der bloss anschaulichen Evidenz!) – Dasselbe gilt für sämtliche Sätze und Beweise des VII. Buches bei Euklid. – Bei Archytas wird die genaue Ausführung aller einzelnen Beweisschritte sogar bis zur Pedanterie übertrieben.[71]

Man sieht also, dass es so gut wie unmöglich ist, irgendeinen Übergang zu beobachten zwischen dem Beweisverfahren des Thales und jenem anderen, welches

---

[70] Die mathematischen Sätze, die Hippokrates von Chios schon kennen musste, werden bei G. HAUSER: o. c. 105 ff, zusammengestellt.

[71] Die Charakterisierung dieser Sätze siehe bei v. D. WAERDEN: Math. Ann. 120 139–140, bzw. bei G. HAUSER: o. c. S. 107 (über Hippokrates).

die Pythagoreer vertreten. Es kann auch gar nicht davon die Rede sein, dass der strenge deduktive Beweis der Mathematik irgendwie langsam und tastend sich ausgebildet hätte. Die neue Art der Beweisführung ist in ihrer vollen Strenge auf einmal plötzlich da bei den Pythagoreern, ohne dass man es sich vorläufig erklären könnte, wie sie eigentlich zustande kam. Wir müssen eben einfach feststellen, dass in der verhältnismässig kurzen Zeitspanne zwischen Thales und den Pythagoreern die deduktive Wissenschaft irgendwie zur Welt gekommen ist. – Es bliebe nur noch zu erklären, wie und warum eigentlich diese plötzliche und überraschende Wandlung eintrat?

IV

Die bisherigen Betrachtungen haben die Vermutung nahegelegt, dass die deduktive Mathematik eigentlich die Schöpfung der Pythagoreer gewesen sei. Denn sie haben jene ältesten mathematischen Sätze und Beweise zusammengestellt, auf Grund welcher man schon auf das Vorhandensein einer deduktiven Wissenschaft schliessen kann. Aber darüber haben wir noch gar nichts sagen können: was eigentlich die Pythagoreer zu diesem bedeutenden Schritt veranlasst haben mag? Wie kamen sie überhaupt auf den Gedanken, dass eine Behauptung mathematischen Inhalts sich auch in solcher Form beweisen lässt, nicht nur in der Art, wie es zu seiner Zeit Thales noch getan hatte? Und wie ist es möglich, dass sie zu einer so frühen Zeit schon so strenge
Beweise lieferten, dass man sie auch später noch kaum über|treffen konnte? Denn 138
bemerken wir sogleich, dass das allererstaunlichste an der Wissenschaft der Pythagoreer gerade die Strenge ihrer Beweise ist.[72]

Die frühere Forschung scheint nur die Tatsache selbst festgelegt zu haben – nämlich in der Behauptung, dass die wissenschaftliche Mathematik der Griechen mit den Pythagoreern beginnt –; aber sie hat eine Antwort auf die vorhin genannten Fragen noch überhaupt nicht versucht. K. Reidemeister vertrat z. B. den eben erwähnten Gedanken, dass also die Pythagoreer die Begründer der wissenschaftlichen Mathematik sind, mit den folgenden Worten:

«Die Pythagoreer entdeckten die Möglichkeit, mathematische Tatbestände auf Hypothesen zurückzuführen, aus denen diese Tatbestände durch *Denken* gefolgert werden können. Damit entdeckten sie aber zugleich einen Weg, der aus dem Anschaulichen heraus zu geometrischen Tatsachen führt, die *nur* dem Denken zugänglich sind.»[73]

---

72 Die logische Strenge der pythagoreischen Mathematik überrascht selbst die besten Kenner dieses Zeitalters. O. Becker bemerkt z. B. in seiner Kritik über v. d. Waerdens Buch (Science awakening) zu jener Vermutung, dass Euklids VII. Buch noch aus der Zeit vor 400 stammt: «die vollendete Form von 7 könnte durch eine spätere Überarbeitung, etwa durch die Mathematiker der Akademie zustande gekommen sein». – Man versteht, wodurch dieser leichte Zweifel des Kritikers ausgelöst wurde. Die Komposition des VII. Euklidischen Buches ist nämlich so geschlossen kompakt, dass selbst Becker sich fragen musste: ob man doch so etwas schon im 5. Jahrhundert zustande bringen konnte? – Aber kaum ist der Zweifel aufgetaucht, so musste er schon vor dem stärkeren Argument weichen: «Indessen macht der Verfasser (d. h. v. d. Waerden) dagegen das gewichtige Argument geltend, dass bei einer streng logischen Untersuchung, wie sie die Begründung der Zahlentheorie in 7 darstellt, eine Unterscheidung von Inhalt und Form untunlich ist.» – O. Beckers Worte findet man in Gnomon 23 1951 S. 299.

73 K. Reidemeister: o. c. S. 52.

Überlegt man sich diese Worte, so wird man in der Tat kaum etwas an ihnen auszusetzen haben, denn sie beschreiben ja tadellos die Leistung der Pythagoreer. Sie haben wirklich die mathematischen Tatbestände (die auch sinnlich wahrnehmbar sind!) auf Hypothesen (also auf rein gedankliche Elemente) zurückgeführt. Kein Wunder, dass dieselben mathematischen Tatbestände aus den Hypothesen durch Denken gefolgert werden konnten. Der Weg führte nicht nur aus dem Anschaulichen, sinnlich Wahrnehmbaren heraus zu dem rein Gedanklichen, sondern auch umgekehrt, vom Gedanklichen zurück zu dem sinnlich Wahrnehmbaren. Das überraschende war eher, dass man von den Hypothesen auch zu solchen anderen Tatsachen gelangen konnte, die nicht mehr anschaulich, *nur* dem Denken zugänglich sind. K. Reidemeister versäumt es auch nicht diese Errungenschaft der Pythagoreer an einem Beispiel zu illustrieren. Wie er schreibt: «Die pythagoreische Lehre vom Geraden und Ungeraden gipfelt nämlich in dem Nachweis, dass die Diagonale *d* eines Quadrates und die Quadratseite *s* desselben nicht mit derselben Einheitsstrecke *e* gemessen werden kön-
139 nen. (Das heisst in der Sprache der Mathematik: die *Inkommensurabilität* | der Quadratdiagonale zur Seite.) Das lässt sich nicht veranschaulichen (man kann die Inkommensurabilität wirklich nicht mehr sinnlich wahrnehmen) – wie sich etwa der Satz des Pythagoras veranschaulichen lässt –, nur denken und erschliessen.»[74] – Natürlich kann man mit diesen Feststellungen einverstanden sein, es fragt sich nur, ob wir daraus auch verstehen, wieso eigentlich die Pythagoreer auf den rätselhaften Einfall kamen, die sinnlich wahrnehmbaren mathematischen Tatbestände auf Hypothesen zurückzuführen? Was wollten sie überhaupt mit ihren Hypothesen erreichen? Denn es ist ja kaum denkbar, dass sie im voraus geahnt hätten: auf dem Wege der Hypothesen werden sie einmal auch so etwas entdecken, was man gar nicht mehr veranschaulichen, nur denken und erschliessen kann, wie z. B. die eben erwähnte Inkommensurabilität. – Mit einem Wort: die vorigen Zitate von Reidemeister beschreiben zwar die Tatsachen genau, und sie erklären, was eigentlich in der pythagoreischen Mathematik vor sich ging, aber sie geben dennoch keine *historische* Erklärung dafür. Man versteht aus ihnen nicht wie das alles stattfinden konnte.

Will man das Entstehen der deduktiven Wissenschaft historisch erklären, so muss man vor allem die Methoden des Beweisens im Falle jener Sätze genauer untersuchen, die nach unserem heutigen Wissen aus der ältesten Zeit der deduktiven Mathematik stammen. O. Becker, der die Lehre vom Geraden und Ungeraden in ihrer ursprünglichen Form wiederherstellte, hat selber gezeigt, wie man eine solche Untersuchung anstellen kann, und sein Beispiel überzeugt auch davon, dass eine solche Prüfung manches Licht auf diese frühe Epoche der Wissenschaft werfen kann. Er konnte z. B. den archaischen Charakter jener Züge nachweisen, die für die altpythagoreische Lehre kennzeichnend sind. Ähnlich wollen wir jetzt das allgemeine Beweisverfahren der ältesten mathematischen Sätze untersuchen.

Vor allem interessiert uns die Frage, ob man jene Erklärung bestätigen könnte, die K. v. Fritz versuchte (vgl. das II. Kapitel dieser Arbeit), als er die Entfaltung des mathematischen Beweisverfahrens mit derjenigen der Logik verglich? Wie steht es im Falle der ältesten mathematischen Sätze? Ist es wirklich so, dass schon im 5. Jahrhundert die komplizierten Sätze im Beweis auf einfachere Prämissen zurückgeführt wurden? Nehmen wir z. B. einen Satz aus der Lehre vom Geraden und Ungeraden.

---

74 Ebd. – Das Zitat ist mit meinen eigenen Worten (in Klammern) ergänzt.

Der Satz Eucl. IX 22 besagt: *Setzt man beliebigviele ungerade Zahlen zusammen und ist ihre Anzahl gerade, so muss die Summe gerade sein.* – Dieser Satz wird in seinem Beweis in der Tat auf den noch einfacheren IX 21. zurückgeführt: *Setzt man beliebigviele gerade Zahlen zusammen, so ist die Summe gerade.* Der Gedankengang des Beweises für den Satz IX 22. ist der folgende: Nachdem wir beliebigviele ungerade Zahlen zu addieren haben, | deren Anzahl jedoch gerade ist, können wir zuerst 140
aus jeder ungeraden Zahl die Einheit subtrahieren. Dadurch werden die gegebenen ungeraden Zahlen zu geraden Zahlen, da im Sinne der 7. Definition des VII. Buches *die ungerade Zahl sich eben um die Einheit von einer geraden Zahl unterscheidet.* Die subtrahierten Einheiten zusammen bilden auch eine gerade Zahl, da ja ursprünglich ungerade Zahlen *in gerader Anzahl* gegeben waren. Wir haben also eigentlich nur gerade Zahlen zu addieren, für welche der Satz IX 21. gültig ist. – Der Beweis besteht also in diesem Fall tatsächlich daraus, dass der «kompliziertere» Satz (IX 22.) auf die einfachere Prämisse (Satz IX 21.) zurückgeführt wird. Ähnlich wird auch der Satz IX 21. in seinem Beweis auf die noch einfachere Definition der geraden Zahl (VII Def. 6), als auf seine Prämisse zurückgeführt.

Man wäre also geneigt, auf Grund solcher direkter Beweise, die in der Tat auch schon in der ältesten Zeit benutzt wurden, jene Theorie restlos zu bejahen, welche den Ursprung der mathematischen Deduktion eben in dieser Art von Beweis erblickt. – Man wird jedoch diese Frage etwas vorsichtiger behandeln wollen, wenn man daran denkt, dass neben dem direkten Beweis schon in der ältesten Zeit häufig auch eine völlig andere Art des Beweisverfahrens benutzt wurde. Es lohnt sich vor allem eine kurze Statistik der pythagoreischen Beweise zu überblicken.

Die Lehre vom Geraden und Ungeraden besteht insgesamt aus 17 Sätzen. Von diesen Sätzen sind die Beweise einiger nicht in ihrer ursprünglichen Form überliefert. Euklid hat nämlich die alten Beweise hie und da überarbeitet, weil er dadurch die ganze Lehre mindestens äusserlich enger in das Gefüge seines Werkes hineinbauen wollte. O. Becker hat allerdings gezeigt, dass man selbst in diesen Fällen die alte Form des Beweises ziemlich leicht wiederherstellen kann. – Nun ist es aber interessant, dass von diesen 17 Sätzen 5 mit indirekter Methode bewiesen werden; in einem 6. Fall ist der Beweis nicht nur indirekt, sondern darüber hinausgehend: eine *deductio ad absurdum.* Zieht man auch die Rekonstruktionen von Becker in Betracht, so steigt die Anzahl der indirekten Beweise noch höher, nämlich auf 8. Beinahe zur Hälfte werden also die Sätze über das Gerade und Ungerade indirekt bewiesen. Diese verhältnismässig häufige Anwendung des indirekten Beweises fällt selbstverständlich auf. Man fragt sich unwillkürlich, ob dem indirekten Beweis nicht eine ganz besondere Bedeutung zukäme, nachdem er ja doch so oft benutzt wurde?

Man wird in dieser Vermutung noch weiter bestärkt, wenn man den Beweis jenes Archytas-Satzes prüft, welcher für v. d. Waerden die Gelegenheit bot, um die pythagoreische Herkunft des VII. Euklidischen Buches nachzuweisen. Sowohl der Satz des Archytas, wie auch die meisten jener Sätze, die zu seinem Stammbaum gehören, werden indirekt bewiesen. Von den ersten 36 Sätzen des VII. Euklidischen Buches sind die Beweise in 15 | Fällen indirekt. Ja, es gibt auch solche Sätze in diesem 141
VII. Buch, die heute zwar auf den ersten Blick so aussehen, als hätten sie einen direkten Beweis, aber prüft man sie genauer, so stellt es sich bald heraus, dass der Gedankengang ihres Beweises eigentlich indirekt war, und nur später oberflächlich auf einen direkten Beweis umgeändert wurde. So z. B. der Satz Eucl. VII 19., der

heute zwar einen direkten Beweis besitzt, aber man sieht an diesem Beweis, dass sein Gedankengang ursprünglich indirekt war.

Wie könnte man nun diese häufige Verwendung des indirekten Beweises erklären? Denn es ist offenbar, dass jene Vermutung, die K. v. Fritz für die Entfaltung des direkten Beweises versuchte, in diesem Falle versagt. Es handelt sich ja bei der Anwendung des indirekten Beweises gar nicht darum, dass man eine «einfachere» Prämisse für den «komplizierteren» Satz sucht. Der indirekte Beweis ist die Frucht einer völlig anderen logischen Überlegung. – Es wird sich vor allem lohnen, die Frage zu stellen, wie etwa heutzutage ein Mathematiker den indirekten Beweis unter dem Gesichtspunkt der Heuristik beurteilt?[75] – B. L. v. d. Waerden erteilt einmal den folgenden heuristischen Rat:[76]

«Wird nicht eine Konstruktion, sondern ein Beweis verlangt, so sucht man oft mit Vorteil einen *indirekten Beweis*. Man nimmt an, die Behauptung sei falsch, man zieht auch noch das Gegebene heran, und man schliesst so lange weiter, bis man auf einen Widerspruch kommt. Oft gelingt es nachher, den Beweis direkt zu führen, aber zu finden ist der indirekte Beweis manchmal leichter, weil man dabei mehr voraussetzen kann, nämlich das Gegebene und die Falschheit des Behaupteten. Von beiden Seiten aus schliessend, hat man eine Chance, sich in der Mitte zu treffen.»

Interessant sind diese Worte des modernen Mathematikers für uns nicht nur deswegen, weil sie den psychologischen Vorgang des mathematischen Entdeckens einigermassen beleuchten, sondern auch darum, weil sie eine völlig neue Perspektive vor der historischen Frage eröffnen: wie man einst wohl auf den Gedanken des deduktiven Beweises kam. Der Mathematiker behauptet, dass wenn man einen Beweis liefern will, oft der Versuch eines indirekten Beweises vorteilhaft sei. Wohl gelingt es nachher, densel-
142 ben Beweis auch direkt zu führen, aber zu finden ist der indirekte Beweis | manchmal doch leichter. Es stehen uns ja im Falle eines solchen Beweisverfahrens mehr Möglichkeiten zur Verfügung. – Liest man diese Worte aufmerksam genug, und vergleicht man sie mit der Tatsache, dass nach unserer Statistik der indirekte Beweis in der Wissenschaft der Pythagoreer, also schon in der ältesten Zeit der deduktiven Mathematik, so oft verwandt wurde, so muss man sich unwillkürlich fragen: *ob der indirekte Beweis nicht von Anfang an eines der allerwesentlichsten Werkzeuge der deduktiven Mathematik war?* – Um diese Frage beantworten zu können, prüfen wir genauer einige Fälle des indirekten Beweisverfahrens an Beispielen der ältesten griechischen Mathematik.[77]

---

[75] Wir verzichten also einstweilen darauf, die Frage des indirekten Beweises auch über den Bereich der blossen mathematischen Heuristik hinausgehend ausführlicher zu erörten. Bekanntlich ist die Schule des mathematischen Intuitionismus bestrebt, die Verwendung dieser Beweisform einzuschräken. Wir sind zwar der Meinung, dass es wirklich nützlich wäre, die Gesichtspunkte des Intuitionismus auch in der Erforschung der griechischen Mathematikgeschichte zu berücksichtigen, wie es z. B. O. BECKER in seinen Eudoxos-Studien getan hatte. Eine noch eingehendere Untersuchung könnte wohl nachweisen, dass das Prinzip vom ausgeschlossenen Dritten anders im 5. und anders im 4. Jahrhundert verwandt wurde. Aber wir wollten mit dieser Frage das Problem vorläufig nicht weiter komplizieren.

[76] «Einfall und Überlegung in der Mathematik», Elemente der Mathematik, Bd. VIII. Basel 1953, S. 123.

[77] Zum folgenden vgl. man auch meine frühere Arbeit «Eleatica» in Acta Antiqua III 67–103. Die Ergebnisse meiner früheren Untersuchungen (Acta Ant. Hung. 1 [1950] 377–410, II 17–62 und 243–289) werden selbstverständlich auch in der vorliegenden Arbeit immer benutzt, auch wenn ich nicht jedesmal auf sie ausdrücklich hinweise.

Den ältesten bekannten indirekten Beweis liefert Euklid in der Lehre vom Geraden und Ungeraden für den Satz IX 30. *Eine ungerade Zahl muss, wenn sie eine gerade Zahl misst, auch deren Hälfte messen.* Dieser Satz wird in seinem Beweis auf den vorangehenden IX 29. zurückgeführt, aber *nicht* direkt, sondern durch einen indirekten Schluss. (Der vorangehende Satz IX 29. besagt, dass «*das Produkt zweier ungerader Zahlen ungerade ist*».) Zu dem Beweis unseres Satzes (IX 30.) muss man vor allem nachweisen, dass der Quotient einer geraden und einer anderen ungeraden Zahl *nur* gerade sein kann. Ist nämlich dieser Nachweis erbracht, so folgt daraus schon selbstverständlich der Satz selbst: *Eine ungerade Zahl muss, wenn sie eine gerade Zahl misst, auch deren Hälfte messen.* – Nun besteht der indirekte Beweis dessen, dass der Quotient einer geraden und einer anderen ungeraden Zahl *nur* gerade sein kann, aus dem folgenden Gedankengang:

Bekannt sind im Sinne des Satzes selbst: der Dividend (eine gerade Zahl) und der Teiler (eine ungerade Zahl). Ob der Quotient wirklich eine gerade Zahl ist, wissen wir noch nicht, wir wollen es erst beweisen. Ehe wir noch den Beweis versuchten, erinnern wir uns daran, dass das Produkt des Teilers und des Quotienten selbstverständlich den Dividenden ergibt. In unserem Fall ergibt also das Produkt des Teilers (einer ungeraden Zahl) mit dem Quotienten (einer anderen Zahl, von der wir noch nicht wissen, ob sie wirklich gerade ist) den Dividenden, der doch eine gerade Zahl ist. Wie könnte man nun beweisen, dass der Quotient in der Tat eine gerade Zahl ist? – Nehmen wir das *Gegenteil* dessen an, was wir beweisen wollen: sei der Quotient eine «ungerade» Zahl; und nun prüfen wir, was aus dieser Annahme folgt. Ist der Quotient ebenso eine «ungerade» Zahl, wie der Teiler, so muss auch das Produkt der beiden eine «ungerade» Zahl sein, da ja im Sinne des vorangehenden Satzes (IX 29.): *das Produkt zweier ungerader Zahlen ungerade ist.* In unserem Fall ist jedoch das Produkt
des Teilers und des Quotienten (der | Dividend) im Sinne des geprüften Satzes selbst: 143
eine gerade Zahl. Wir sind also bei einem *Widerspruch* angelangt, zum Zeichen dessen, dass unsere Denkweise falsch war. Es ist also *nicht möglich*, dass der Quotient in dem gegebenen Fall eine «ungerade Zahl» sei, sie muss gerade sein. Und damit ist der gewünschte Beweis erbracht.

Wir haben den Beweis, den wir prüfen wollen, nicht wörtlich zitiert, statt dessen nur den Euklidischen Gedankengang *haargenau* – dabei selbstverständlich auch mit Interpretation ergänzt! – wiedergegeben, weil eigentlich nur auf diese Weise die Kontrolle der einzelnen Gedankenschritte möglich wird. Nun wollen wir jetzt sehen, was sich von dem erbrachten indirekten Beweis feststellen lässt.

Man sieht vor allem, dass das indirekte Verfahren daraus besteht, dass man eigentlich *nicht* den fraglichen Satz selbst *beweist*, sondern statt dessen das Gegenteil des Satzes *widerlegt.* Wir wollten ja nachweisen, dass die geprüfte Zahl (der Quotient) eine gerade ist, aber statt des direkten Beweises dieser Behauptung haben wir die gegenteilige Behauptung – «die geprüfte Zahl wäre ungerade» – widerlegt. Diese Form des Beweisens hat überhaupt deswegen den Namen «indirekt» bekommen, weil sie eigentlich kein Beweis, sondern eine Widerlegung ist. Nicht der geprüfte Satz wird bewiesen, sondern sein Gegenteil widerlegt. Man wird also eine Behauptung durch indirekten Schluss so beweisen können, dass man zuerst das Gegenteil der Behauptung aufstellt, und dann zeigt man, dass diese gegenteilige Behauptung falsch ist.

Schon auf Grund dieser völlig einfachen Feststellung über die Tatsache selbst, worin eigentlich das indirekte Beweisverfahren besteht, lassen sich sehr wesentliche

Vermutungen auch darüber aufstellen: was alles eigentlich zur Handhabung des indirekten Beweises unerlässlich notwendig ist? Um einen indirekten Beweis führen zu können, muss man vor allem davon überzeugt sein, dass eine Behauptung entweder *wahr* oder *nicht wahr* ist; ausser diesen beiden Möglichkeiten gibt es eine dritte überhaupt nicht. (Entweder ist ein Ding *A*, oder *Non-A; tertium non datur.*) Ja, man muss diese grundlegende Erkenntnis jeder Logik sich schon so felsenfest angeeignet haben, dass man auch davon überzeugt sei, dass der Beweis irgendeiner Behauptung *(A)* der Widerlegung ihres Gegenteils *(Non-A)* äquivalent ist. Solange man das alles nicht genau und unerschütterlich fest weiss, wird man unter keinen Umständen auf den Gedanken kommen, einen solchen indirekten Beweis, wie der vorige ist, zu führen.

Betrachtet man jedoch den eben dargestellten indirekten Beweis genauer, so entdeckt man bald, dass sich auch noch ein anderer sehr wesentlicher Zug an diesem logischen Verfahren beobachten lässt. Es ist nämlich auffallend, wie man eigentlich die Falschheit einer Behauptung im Laufe des Beweisvorganges entdeckt. – Wir haben
144 die Behauptung aufgestellt, | dass der Quotient – nicht wie es richtig: *gerade*, sondern wie es eigentlich falsch ist, dass er nämlich – eine «ungerade» Zahl wäre. Als wir diese Behauptung aufstellten, konnten wir natürlich im voraus noch nicht wissen, dass sie sich als falsch erweisen wird; das hat sich erst später herausgestellt. Denn wir haben aus unserer (falschen) Behauptung Schlüsse gezogen. Wir dachten nämlich: wenn der Quotient eine «ungerade Zahl» ist, dann muss – im Sinne unserer früheren und schon gesicherten Kenntnis, nämlich im Sinne des vorangehenden Satzes (IX 29.) – auch der Dividend eine «ungerade Zahl» sein. Damit sind wir aber bei einem *Widerspruch* angelangt – denn wir wissen ja, dass der Dividend im gegebenen Falle nicht eine «ungerade», sondern eine gerade Zahl ist –, und der *Widerspruch* war das Zeichen dessen, dass unsere Denkweise falsch war. – Ehe wir noch weitergingen, müssen wir genau verstehen, was es eigentlich heisst, dass wir im Laufe der Gedankenführung *bei einem Widerspruch angelangt sind?* Worin bestand denn der «Widerspruch»? – Diese auf den ersten Blick keineswegs völlig durchsichtige Redeweise heisst diesmal nur folgendes: Wir sind im Laufe unserer Gedanken zu der Behauptung gekommen: «*Der Dividend ist eine ungerade Zahl.*» Am Anfang jedoch, als wir den Satz hörten, den wir zu beweisen hatten, hiess es: «*Der Dividend ist eine gerade Zahl.*» Die beiden Behauptungen – «er ist ungerade Zahl» und «er ist eine gerade Zahl» – widersprechen sich. Die beiden Sätze können natürlich nicht auf einmal (gleichzeitig) wahr sein, denn entweder ist ein Ding *A*, oder *Non-A, tertium non datur*, und ausserdem kann von zwei solchen gegenteiligen Behauptungen immer nur die eine *wahr* sein, die andere ist notwendigerweise *nicht wahr*. – Die Redeweise also, dass wir «bei einem Widerspruch angelangt sind», heisst eigentlich so viel, dass wir zu einer Behauptung gekommen sind, welche einer anderen, schon früher als wahr erkannten Behauptung widerspricht. Die beiden Behauptungen können nicht vereinigt werden, weil ihre Vereinigung einen *inneren Widerspruch* des Gedankens («dieselbe Zahl ist gerade *und* ungerade») erzeugte. Darum ist das Auftauchen des *Widerspruches* ein Zeichen dafür, dass die Gedankenführung irgendwo falsch war.

Sehr bezeichnend ist also für den behandelten indirekten Beweis auch die Auffassung, dass der «innere Widerspruch» des Gedankens ein Zeichen für seine Falschheit ist. Der Gedanke, der sich selbst widerspricht, kann nicht wahr sein. Daraus folgt natürlich auch soviel: wahr ist nur der Gedanke, der sich selbst nicht widerspricht, also: *der widerspruchsfreie Gedanke.* Die Widerspruchsfreiheit als einziges logisches Kriterium[78] für die Wahrheit | irgendwelcher Behauptung muss selbstverständlich demjenigen, der einen indirekten Beweis verfasst, bekannt sein. Um diese These genauer zu illustrieren, wollen wir – ehe wir noch weitergingen und die Konsequenzen aus den Feststellungen über den indirekten Beweis im allgemeinen zögen – mindestens noch an einem Beispiel aus der ältesten griechischen Mathematik das indirekte Beweisverfahren ausführlicher untersuchen. 145

Der Satz Eucl. VII 31., dessen pythagoreischer Ursprung aus dem 5. Jahrhundert durch v. d. Waerden nachgewiesen wurde, besagt: *Jede zusammengesetzte Zahl wird von irgendeiner Primzahl gemessen.* Zu dem Euklidischen Beweis dieses Satzes muss man die folgenden Definitionen des VII. Buches kennen:

Def. 2: *Zahl ist die (endliche) Menge von Einheiten.*
Def. 11: *Primzahl ist, die nur die Einheit zum Teiler hat.*
Def. 13: *Zusammengesetzte Zahl (= Nicht-Primzahl) ist diejenige, die irgendeine andere Zahl zum Teiler hat.*

Der Beweis des Satzes (VII 31.) lautet bei Euklid folgendermassen. – Sei *a* eine beliebige zusammengesetzte Zahl. Wir wollen beweisen, dass diese zusammengesetzte Zahl irgendeinen Primteiler besitzt. Nachdem *a* eine zusammengesetzte Zahl ist, muss sie eine andere Zahl, *b* zum Teiler haben (Def. 13). Diese Zahl *b* kann nur Primzahl oder Nicht-Primzahl (= zusammengesetzte Zahl) sein; eine dritte Möglichkeit gibt es nicht. Ist *b* eine Primzahl, so ist der Satz (VII 31.) bewiesen; ist jedoch *b* eine zusammengesetzte Zahl, so muss sie einen Teiler *c* besitzen (Def. 13), der selbstverständlich auch ein Teiler von *a* ist. Nun kann aber *c* wieder entweder eine Primzahl oder eine zusammengesetzte Zahl sein. Im ersten Fall ist der Satz bewiesen, denn wir haben eine Primzahl *c*, die Teiler von *b* und dadurch auch derjenige von *a* ist. Ist aber *c* eine zusammengesetzte Zahl (= Nicht-Primzahl), so prüft man weiter ihren Teiler *d* usw. – Der Beweis betont, dass man auf diese Weise schliesslich einen Primteiler der zusammengesetzten Zahl *a* finden wird. Sollte man nämlich nie den gesuchten Primteiler finden, und wären die Teiler von *a* lauter zusammengesetzte Zahlen, so hiesse es, dass die Zahl *a* unendlich viele immer kleiner werdende Teiler besitzt, was jedoch im Bereiche der Zahlen unmöglich ist.

Überblickt man die eben dargestellte Gedankenkette des Beweisvorganges, so muss man zugeben, dass die einzelnen Glieder der Behauptung | immer durch das Motiv der 146
«Widerspruchfreiheit» aneinander gereiht werden. Die Zahl *b* muss z. B. deswegen

[78] Im täglichen Leben ist man gewohnt eine Behauptung für wahr zu halten, wenn sie mit der praktischen Erfahrung übereinstimmt. Dagegen gibt es ein *logisches Kriterium* für die Wahrheit einer Behauptung *ausser der Widerpruchsfreiheit* eigentlich kaum. Man muss jede Behauptung – natürlich einzig und allein von dem Gesichtspunkte der Logik aus betrachtet – solange für wahr halten, bis es nicht gelingt, irgendeinen Widerspruch in ihr nachzuweisen. Man kann den Widerspruch in einer Behauptung, wie bekannt, dadurch nachweisen, dass man Schlüsse (Konsequenzen) aus der Behauptung zieht, und dass man das Verhältnis dieser Konsequenzen zu anderen, schon für wahr geltenden Sätzen prüft. Gerät man auf diese Weise in Widerspruch mit einem schon wahr anerkannten Satz, so kann die anfangs aufgestellte Behauptung nicht mehr als wahr gelten, denn man hat ihren widerspruchsvollen Charakter erkannt. (Allerdings müsste man einmal noch nachweisen können, dass es auch beim *richtigen* Schlüsse Ziehen aus irgendeiner Behauptung auf das *Vermeiden des Widerspruches* ankommt!)

entweder Primzahl oder Nicht-Primzahl (= zusammengesetzte Zahl) sein, weil der dritte Fall («sie ist Primzahl *und* Nicht-Primzahl») innerer Widerspruch des Gedankens wäre, also unmöglich ist. Sollte der erste Fall zutreffen – «*b* ist Primzahl» –, so braucht man nicht weiterzugehen, da die Aufgabe gelöst ist; im zweiten Fall jedoch – «*b* ist Nicht-Primzahl» – kann das *dreiteilige Kettenglied des Gedankens* auch für ihren Teiler wiederholt werden: *Primzahl – Nicht-Primzahl – dritter Fall unmöglich;* erster Fall – gelöst; zweiter Fall – man wiederholt von vorne den Gedanken. – Dass aber wirklich die indirekt nachgewiesene Widerspruchsfreiheit des Gedankens sozusagen die Grundlage des ganzen Beweises ist, ersieht man besonders aus dem letzten Schritt. Die Behauptung «A», die man diesmal indirekt begründet, heisst: «*das dargestellte Vorgehen des Suchens ist ein endlicher Prozess, man findet am Schluss die gesuchte Primzahl*». Der indirekte Beweis dieser Behauptung besteht darin, dass man ihr Gegenteil, «Non-A» aufstellt, um es zu widerlegen: «*das dargestellte Vorgehen des Suchens ist ein unendlicher Prozess, man findet die gesuchte Primzahl nie*». Um die Falschheit dieser letzteren Behauptung nachweisen zu können, zieht man aus ihr solange Schlüsse, bis der innere Widerspruch des Gedankens offenbar wird. – Hört der Prozess des Suchens «nie» auf, dann heisst es soviel, dass wir immer nur solche Teiler der Zahl *a* finden, die auch selber zusammengesetzte Zahlen sind. Jene Zahlen aber, die sich im Laufe des Suchens als Teiler von *a* erweisen, sind nicht nur alle kleiner als *a*, sondern sie werden ausserdem mit dem Vorwärtsgehen des Prozesses immer auch noch kleiner. Hört also der Prozess nie auf, so besitzt die Zahl *a* unendlich viele immer kleiner werdende Teiler, die auch selber alle zusammengesetzte Zahlen sind. Aber diese letztere Behauptung lässt sich ja nicht vereinigen mit der Definition 2: *Zahl ist die (endliche) Menge von Einheiten*, sie kann also nicht aus unendlich vielen zusammengesetzten Zahlen bestehen. Wir sind also auf den gesuchten «Widerspruch» gestossen, zum Zeichen dessen, dass die Behauptung «Non-A» nicht wahr sein kann, ihr Gegenteil, die Behauptung «A» ist wahr.

Lehrreich ist der letzte indirekte Beweis auch darum, weil er zufälligerweise auch noch das Problem beleuchten kann: wie man wohl überhaupt auf den Gedanken der mathematischen *Definition* kam. Prüft man nämlich den Satz Eucl. VII 31. und seinen eben behandelten Beweis genauer, so entdeckt man gleich einen merkwürdigen historischen Zusammenhang. Dieser Beweis betont ja, dass keine auch noch so grosse zusammengesetzte Zahl *unendlich viele immer kleiner werdende Teiler haben kann.* Aber wer hat im 5. Jahrhundert v. u. Z. das «Gegenteil dieser Behauptung» aufgestellt, oder mindestens etwas, was so aussieht, als wäre es das «Gegenteil dieser Behauptung»? – Bekanntlich war es der Eleate Zenon, der
147 behauptete, | dass jede Strecke AB *die unendlich vielen immer kleiner werdenden Teile* $\frac{1}{2}, \frac{1}{4}, \frac{1}{8} \ldots$ besitzt.[79] Der Verfasser unseres Beweises will Zenons Behauptung auch gar nicht widerlegen; er betont nur, dass der ähnliche Prozess «im Bereiche der Zahlen» unmöglich ist, weil er im Widerspruch mit der Definition der Zahl stünde. Man hat also – mindestens in diesem Fall – die mathematische Definition so formuliert, dass

[79] Vgl. Aristoteles phys. Z 9. 239 b 9 ff. und 2.233 a 21.

sie die Grundlage für die Widerspruchsfreiheit jener Behauptungen (Sätze) sei, die man auf sie baute.

Schliessen wir diese kurze Betrachtung über das Motiv der Widerspruchsfreiheit mit einer Vermutung über das indirekte Beweisverfahren, die prinzipiell wichtig sein kann. Ist nämlich die Widerspruchsfreiheit das einzige logische Kriterium für die Wahrheit einer Behauptung, so muss der indirekte Beweis überhaupt sozusagen *die primäre logische Beweisart* sein. Ihm gegenüber kann der direkte Beweis nur *sekundärer logischer Art* sein. Allerdings müssen wir noch zum Verständnis dieser prinzipiell wichtigen Vermutung erklären, was wir eigentlich in dieser Beziehung unter «primär» und «sekundär» verstehen. Fangen wir mit dem leichteren Terminus «sekundär» an. Wie soll man die Behauptung verstehen, dass der direkte Beweis «sekundärer logischer Art» ist? – Die Wahrheit einer Behauptung kann auf dem Wege der Logik mit direkter Methode nur so gezeigt werden, dass man jene Behauptung, deren Wahrheit nachzuweisen ist, mit einer anderen solchen Behauptung verknüpft, deren Wahrheit ihrerseits schon von früher her erkannt ist. Die Behauptung also, die zu beweisen war, wird darum als wahr erscheinen, weil man erkennt, dass sie irgendwie aus einer anderen schon früher als wahr anerkannten Behauptung folgt. Wohl können also auch durch den direkten Beweis die wahren Behauptungen in eine sozusagen unendliche Kette der «Wahrheiten» hineingefügt werden. Aber mit direkter Methode können wir nie eine solche letzte Behauptung finden, deren Wahrheit oder Unwahrheit wir bloss mit logischen Mitteln feststellen könnten. Deswegen sagen wir, dass der direkte Beweis «sekundärer logischer Art» ist. – Dagegen ist der indirekte Beweis «primärer logischer Art». Denn es gibt in der Tat mindestens *eine* solche Behauptung, deren Unwahrheit man bloss mit logischen Mitteln erkennen kann. Die Behauptung nämlich, die sich selbst widerspricht, kann, ohne Rücksicht auf ihren konkreten Inhalt, unmöglich wahr sein. Der indirekte Beweis baut eben immer auf diese *primäre logische Erkenntnis*. Man zieht solange immer wieder Schlüsse aus der geprüften Behauptung, d. h. man überlegt sich alle Konsequenzen des aufgestellten Satzes, bis man bei dem offenbaren Selbstwiderspruch des Gedankens anlangt. Nur so kann ein indirekter Beweis gelingen. – Ist aber der indirekte Beweis dem direkten gegenüber *logisch primär*, so muss diese Art des Beweisverfahrens auch historisch
ursprünglicher sein, als der direkte Beweis. In der Tat beginnt | die historische 148
Entfaltung der Logik – wie wir bald daran erinnern müssen – mit dem Erscheinen des indirekten Beweises.

Nun kommen wir aber zu den Schlüssen zurück, die sich aus der Betrachtung des indirekten Beweisverfahrens für die Geschichte der griechischen Mathematik ergeben. Das Vorhandensein, ja das häufige Verwenden dieser Art des Beweises in der pythagoreischen Mathematik des 5. Jahrhunderts zeigt, dass zu dieser Zeit den griechischen Mathematikern eine hochentwickelte Logik schon sehr gut bekannt sein musste. Ohne eine schon vorhandene und bewusst angewandte Logik ist ja die deduktive Mathematik gar nicht möglich, denn die mathematische Deduktion ist im Grunde überhaupt nichts anderes, als die bewusste Anwendung der Logik auf Behauptungen mathematischen Inhalts. Es fragt sich nur: woher eigentlich diese Logik kommt? Haben die Mathematiker sie fertig von anderen übernommen, oder haben sie sie etwa selber ausgebildet, indem sie sich mit ihrem eigenen Forschungsobjekt beschäftigten? – Wir wollen zunächst beide Möglichkeiten offen ins Auge fassen ohne dass wir uns dabei schon im voraus auf bekannte historische Tatsachen beriefen.

Gesetzt, dass die Pythagoreer diese Logik selber gefunden hätten, indem sie sich mit jenen empirischen Kenntnissen mathematischen Inhalts intensiv beschäftigten, welche sie fertig vorgefunden hatten, so fragt es sich: was hat sie denn zu dieser über ein solches Objekt bis dahin allerdings völlig unversuchten Gedankentätigkeit veranlasst? Dass sie damit irgendwelche praktische Zwecke gehabt hätten, ist nach all dem, was wir oben entwickelten, völlig unwahrscheinlich. Es bleibt also die andere Möglichkeit: sie müssen diese Beschäftigung aus rein intellektuellem Interesse betrieben haben. Es interessierte sie also nicht so sehr das tatsächliche Material ihrer Forschung – die früheren empirischen Kenntnisse mathematischen Inhalts –, als eher das Gedankliche, welches sie an diesem Material erprobten, also die Logik. Aber wie soll dann die Logik doch aus den empirischen Kenntnissen der früheren Mathematik erwachsen sein? – Auf diese Frage gibt es gar keine Antwort. Sollen also die pythagoreischen Mathematiker selber die Logik erfunden haben, so bleibt es unerklärt, wie sie es eigentlich fertig brachten.

Wenden wir uns nun jetzt der anderen Möglichkeit zu, dass nämlich die Pythagoreer die Logik nicht gefunden, sondern übernommen, und ursprünglich sie nur auf ihr speziales Gebiet, nämlich auf die empirischen Kenntnisse der früheren Mathematik angewandt hätten, so eröffnen sich gleich wahre historische Perspektiven vor uns. Wir können uns nämlich sofort auf jene Eleaten berufen, die am Ende des 6. und am Anfang des 5. Jahrhunderts gerade in Süditalien sehr nahe bei der Heimat der Pythagoreer tätig waren. – Wir brauchen uns wohl nicht zu wiederholen und noch einmal
149 umständlich zeigen, wie die Eleaten die Logik entdeckt hatten.[80] Statt dessen wird | es vielleicht genügen, nur kurz anzuzeigen, wie in der Tat alle für die pythagoreische Mathematik kennzeichnenden Züge schon früher in der Logik der Eleaten vorhanden waren.

Fangen wir damit an, dass das älteste Beispiel eines griechischen indirekten Beweisverfahrens eben aus dem Lehrgedicht des Parmenides bekannt ist.[81] Parmenides verwendet ja dasselbe Schema, welches wir oben in der Behandlung des indirekten Beweises für den Satz Eucl. VII 31. ausführlicher entwickelten. Nur besitzt bei Parmenides «das dreiteilige Kettenglied des Gedankens» noch die einfachere Form: 1. «das Seiende ist», 2. «das Seiende ist nicht» und 3. «das Seiende ist und ist auch nicht» *(tertium exclusum)*. Er, Parmenides war es ja, der zum ersten Male in der Geschichte des europäischen Denkens die These vertrat, dass das einzige Kriterium für die Wahrheit die Widerspruchsfreiheit ist. Wie könnte aber die Widerspruchsfreiheit, dieses Negativum nachgewiesen werden? – Am leichtestens durch die Widerlegung der gegenteiligen Behauptung, durch das Herausstellen ihres inneren Widerspruches. Deswegen hat Parmenides nie etwas «bewiesen», nur das Gegenteil seiner Behauptung widerlegt. Das ist die Art der Eleaten: durch Nachweis des inneren Widerspruches die gegenteilige Behauptung zu widerlegen.

Leitet man die von den Pythagoreern angewandte Logik von den Eleaten her ab, so wird auch jene grundsätzliche Wandlung auf einmal verständlich, die plötzlich in der Mathematik eintrat. – Wir haben schon betont, dass die Pythagoreer sich nicht mehr mit der empirischen, anschaulichen Evidenz begnügten. Sie wollten auch die

---

[80] Vgl. Acta Ant. Hung. 2 (1953) 17–62 und 243–289.

[81] Vgl. A. GIGON: Der Ursprung der griechischen Philosophie. Basel 1945. S. 251: «die These wird (bei Parmenides) durch Widerlegung des Gegenteils bewiesen».

alleroffenbarsten Tatsachen der täglichen Erfahrung «deduktiv» beweisen, ja sie standen beinahe feindlich der Praxis gegenüber. Man kann das alles gar nicht verstehen und erklären, wenn man die Logik einfach aus der Beschäftigung mit den praktischen mathematischen Kenntnissen ableiten will. Ja das ist beinahe sinnwidrig: warum sollte man auf einmal gar nichts mehr von der praktischen Erfahrung hören, gerade im Fall solcher Kenntnisse, die aus der praktischen Erfahrung stammten und für den praktischen Gebrauch bestimmt waren? – Aber es wird alles sofort klar und verständlich, wenn man bedenkt, dass diese Wandlung in der Mathematik eigentlich nur *eine Erbschaft der Eleaten-Logik* war. Die Eleaten haben nämlich jede sinnliche Erfahrung verworfen, weil sie auf rein spekulativem Wege den «inneren Widerspruch» in den sinnlichen Erfahrungen entdeckten. Denn unsere Sinnesorgane täuschen uns ja vor, als wären die widerspruchsvollen Erscheinungen, wie «Entstehen», «Vergehen», «Sich-Verändern», «Sich-Bewegen» usw. alle wahr. Aber wie könnte etwas wahr sein, was sich selbst widerspricht? Nach Parmenides steckt nämlich in allen diesen Dingen der
Widerspruch des Seins und Nicht-Seins. Deswegen ver|kündet er das neue 150
Erkenntnisprogramm: «Lass dich nicht durch die vielerfahrene Gewohnheit auf diesen Weg zwingen, deinen Blick den ziellosen, dein Gehör das brausende und deine Zunge walten zu lassen; nein, *mit dem Verstande* bringe die vielumstrittene Prüfung, die ich dir riet, zur Entscheidung!» – Die pythagoreischen Mathematiker des 5. Jahrhunderts haben sich eben an dieses Programm gehalten, deswegen verachteten sie das bloss Anschauliche, und deswegen erstrebten sie rein gedankliche Beweise. Die Widerspruchsfreiheit des Gedankens ist also eigentlich infolge des eleatischen Erkenntnisprogramms zum einzigen Kriterium der mathematischen Wahrheit geworden.

*

Die Mathematik war in der vorgriechischen Zeit lediglich eine Sammlung von erfahrungsmässigen, praktischen Kenntnissen. Dadurch, dass die ersten Pythagoreer am Ende des 6. oder spätestens in der ersten Hälfte des 5. Jahrhunderts versuchten, die logische Methode der Eleaten auf diese bis dahin nur empirischen Kenntnisse zu verwenden, ist eine überraschende Entwicklung ermöglicht worden. Die Mathematik ist dasselbe geworden, was wir heute unter diesem Namen verstehen: eine deduktive Wissenschaft. Aber auch umgekehrt: auch die Logik hat dadurch jenes Gebiet gefunden, welches ihr am entsprechendsten war. Von nun an förderte nicht nur die Logik die deduktive Wissenschaft, sondern auch die Mathematik trug ihrerseits zu der weiteren Entfaltung der Logik bei. (Darum steht in mancher Hinsicht die Logik der späteren Mathematiker auf einer höheren Stufe, wie z. B. diejenige des Aristoteles.)

Es wäre jedoch verkehrt zu glauben, dass die Anwendung der Logik auf die mathematischen Kenntnisse schon in der ältesten Zeit der Entwicklung die Wissenschaft nur in positiver Richtung beeinflusste. Zum Teil war diese Wirkung in der alten Zeit auch von negativer Art. Es ist z. B. bekannt, wie weit entwickelt in den vorgriechischen Kulturen das Rechnen mit den Brüchen war. Die deduktive Mathematik der Griechen konnte jedoch mit den Brüchen nichts anfangen; sie musste sie eben im Interesse der logischen «Widerspruchsfreiheit der Mathematik» beiseiteschieben. Wie der Platonische Sokrates es einmal erklärt: «Du weisst doch, dass die Mathematiker lachen würden, wenn man versuchte die Einheit zu zerlegen, und sie liessen es nicht gelten. Wolltest du nämlich die Einheit zerlegen, so würden sie sie statt dessen

vervielfältigen. *Denn sie wollten es ja vermeiden, dass die Einheit jemals etwas anderes als sie selbst sei.* Wenn sie dann jemand fragte: Ihr Wunderlichen, von was für Zahlen sprecht ihr eigentlich? Wo ist denn eine Einheit, wie ihr sie fordert, nämlich etwas in sich völlig Gleiches, Unterschiedsloses, keine Teile Enthaltendes?
151 Was würden sie darauf wohl antworten? – | Dass sie lediglich von *gedachten* Zahlen
sprechen, die man nur erschliessen und nicht sinnlich wahrnehmen kann.»[82]

Kein Wunder, dass auf diese Weise das praktische Rechnen mit Brüchen sich nicht weiterentwickeln konnte; es blieb auch noch in byzantinischer Zeit dasselbe, was es im alten Aegypten war.[83]

---

[82] Platon, Staat VII 525 D–526 A.

[83] Vgl. dazu A. FRENKIANS Worte (zitiert oben in Anmerkung 7). Es ist übrigens bekannt, dass die Griechen unter anderem auch darum die Lehre von den Proportionen so weit entwickelten, weil sie auf diesem Wege die Brüche aus der Mathematik eliminieren konnten.

WILBUR RICHARD KNORR

ON THE EARLY HISTORY OF AXIOMATICS: THE INTERACTION OF MATHEMATICS AND PHILOSOPHY IN GREEK ANTIQUITY

The manner of the origins of deductive method both in philosophy and in mathematics has exercised the thoughts of many notable scholars. Unique to the ancient Greek tradition were conscious efforts to comprehend the process of thinking itself, and these inquiries have since developed into the philosophical subfields of logic, epistemology and axiomatics. At the same time, the pre-Euclidean Greek mathematicians turned to the problem of organizing arithmetic and geometry into axiomatic systems, in effect setting a precedent for the field of mathematical foundations. Did these two movements in the history of thought arise independent of each other, through some extraordinary coincidence? It would appear far more likely that this common adoption of deductive method resulted through interaction between the fields of philosophy and mathematics. But if this is so, to which shall we attribute priority?

The present contribution will examine arguments of the Hungarian philologist, A. Szabó, who has adopted a most provocative position on this issue. In brief, he maintains the pre-Euclidean mathematicians accepted a formal deductive methodology as a direct response to the example of the Eleatic dialecticians, Parmenides and Zeno. Indeed, by describing the fifth-century Pythagorean number theory as "perhaps the greatest and most lasting creation of Eleatic philosophy", Szabó subordinates early mathematics to Eleaticism.[1] He maintains further that the subsequent formal studies in geometry resulted from the effort, only partially successful, to establish there a deductive system comparable to that effected for number theory. In particular, Euclid's *Elements* are interpreted as a direct answer to difficulties raised by the Eleatics.

Szabó thus clearly advocates assigning to philosophy priority in the emergence of deductive method. Yet one need not search long to find equally firm advocates of the opposite view: that mathematics owes no such debt to philosophy, either then, or indeed ever.[2] I shall seek here to establish a position between these extremes. But my intuition is that the latter view is closer to the accurate portrait of early Greek | mathe- 146
matics. In the first part of this paper I shall take up Szabó's arguments and indicate certain major areas where his position is unsatisfactory: notably, in its reading of the early mathematical literature and in its account of the alleged dialectical motives of Euclid.[3] In the second part I shall undertake to describe the interaction of geometric and philosophical studies in antiquity through three examples: the relation of Aristotle and Euclid on the first-principles of mathematics; the contributions of Eudoxus to proportion theory; and the formal style of Archimedes. We shall thus see that the interaction between the two disciplines hardly fits the simplified picture of a unilateral influence; nor did this interaction always contribute to the advancement of research in either field.

J. Christianidis (ed.), Classics in the History of Greek Mathematics, 81–109.

I. THE ORIGINS OF DEDUCTIVE METHOD

Szabó's thesis of the Eleatic influence on the rise of deductive method in early mathematics is reminiscent of certain prominent, but now largely discredited, views. I refer to Tannery's thesis that Zeno's paradoxes had been directed against a naive number-atomism among the early Pythagoreans and to the thesis of the pre-Euclidean "foundations crisis", argued by Hasse and Scholz and others.[4] But Szabó is attempting something more subtle, for the question he addresses is more profound: what were the origins of the deductive method itself and what was the cause of its being adopted into mathematics through the notion of "proof"?

Szabó opposes the view that deductive method came to characterize Greek mathematics through some spontaneous evolution internal to it.[5] Older traditions – the Egyptian and Babylonian – attained the *technical* level of the Euclidean arithmetic and plane geometry; but their methods were always presented in the form of the solution of specific problems – over the course of centuries (indeed, millenia), they showed no sign of developing a formal proof-technique by which their concrete methods might be converted into a theoretical structure. In the same way, there seems small reason to suppose the early Greek practical tradition should have spontaneously evolved into such a theoretical system. By contrast, the early development of Greek natural philosophy soon arrived at the stage where attention had to be paid to the forms of discourse.[6] Being a quest for the principles and causes of things in general, natural
147 philosophy was | fundamentally disputative. In the Eleatics, we see the deliberate application of logical forms in the presentation of criticisms of Milesian cosmology. Thus, we have the earliest uses of logical forms – most important, of the indirect proof-technique, "reduction to the impossible" – as well as clear motives for their introduction, in the context of the emerging dialectic of the early fifth century. Szabó maintains that when the mathematicians were for their own reasons required to apply such deductive forms – as in their studies in number theory and in the theory of incommensurable lines – these were already available through the earlier philosophical debates.

Prima facie this might appear to be an appropriate view. For surely, it is the business of mathematicians to solve mathematical problems, not to pronounce upon the general structure of argumentation. Conversely – if our impressions based on the later development of philosophy do not mislead us – precisely such a question is among the legitimate concerns of philosophers. Nevertheless, we are well reminded by an eminent historian of logic that mathematicians need not depend on prior philosophical studies as the source of the basic logical forms they employ.[7] It is thus important that we clarify the possible alternative modes by which such deductive forms may have entered into mathematical study.

First, what shall we intend by "deductive method"? Certainly, if we seek fully articulated axiomatic systems, we shall not expect to find them among the early pre-Euclidean studies. But if we define our quest as the presentation of propositions as true through the ordered sequence of other propositions, in which each passage from two or more propositions to another (i.e., their "conclusion") is effected according to set rules (i.e., "laws of logic") – then mathematics even in the far more ancient Oriental traditions was already deductive. Babylonian and Egyptian mathematicians had to begin their solutions of problems by first stating the givens, and then through a recognized sequence of steps produce a desired conclusion (the construction or

computation of a required term).[8] The very nature of mathematics compels adherence to some such ordered sequence, in effect, to a logical method. Indeed even the method of indirect reasoning had to figure among the most ancient techniques. For most problems are accompanied by a "proof", or as we might say, a "check". While of course most of the surviving computations do indeed check out, surely countless computations did not; for otherwise, checking would | not have become part of the usual 148
computing procedure. Now, in the event that the "check" did not conform with the stated conditions of the problem, the computist would repeat his work until it did; that is, he would recognize the former solution to be incorrect. We thus see the germ of the method of indirect reasoning.

It is thus clear that the appearance of such forms in mathematical work need not signify their prior study by philosophers. Indeed, we might press this point to argue that the efforts by early mathematicians (say, the fifth-century Pythagoreans) to put in order the body of arithmetic and geometric techniques they were assimilating from the older traditions gave rise to the awareness among philosophers that the same ideas and patterns of reasoning might be applied in their cosmological speculations. Lending credibility to this view is the Pythagoreans' organization as a school, within which the instruction of mathematical subjects might well have occasioned such an effort of compilation. Moreover, it is said that Parmenides had studied the Pythagorean ideas before breaking away. In this way, the Eleatics could have derived their logical forms from the prior example of the mathematicians. Lacking any definitive documentation on this question, we shall not propose this view as more than a possibility. At any rate, Szabó's insistence that the dialecticians *could not* have derived their deductive methods from mathematics[9] is hardly compelling, and thus inadmissible as support for his thesis.

Second, Szabó characterizes Greek mathematics of the formal type as "anti-empirical" and as deliberately in avoidance of "graphical" methods of proof.[10] The latter, as a description of classical Greek geometry may appear surprising; for the construction of diagrams is always a characteristic vehicle for the organization of geometric proofs. In principle, however, the diagram might be dispensed with, since the argument can be effected entirely by means of *verbal* reference to the relevant prior principles. Certainly Euclid and those in the formal tradition never presume to prove a theorem via the actual measurement of the elements in a drawn diagram.[11] Szabó's point about the non-graphical nature of Greek geometry is thus apt. Indeed, the converse emphasis on the visual character of Greek mathematics has led some scholars into absurdities about the meaning and limitations of Greek methods in certain fields, particularly number theory.[12]

But is this avoidance of graphical methods and this recognition of | the *generality* 149
of the theorems in mathematics the same as *anti-empiricism*? The geometer knew that his draftsmanship was not a limiting factor on the truth of his theorems; these are about geometrical entities and are true – or, better, are *known* to be necessarily true – through the correctness of the logical sequences followed in their demonstrations. But what *are* these geometrical entities? Impressed by the mathematicians' ability to draw accurate general conclusions, despite the unavoidably specific and limited nature of any concomitants (e.g., drawn figures, specific numbers) of his arguments, Plato proposed that the reasoning was really founded upon an intuition of abstract entities (the *Ideas*, or *Forms*) existing apart from any physically perceived objects.[13] Now, if the early mathematicians did indeed consciously seek through formal methods to

capture the essence of such entities, as Plato supposes, then we should have to acknowledge the anti-empirical aim of their work and its roots in the Eleatic doctrines.

But this account of the mathematicians' enterprise is not the only one possible. One might, for instance, adopt the view argued by Aristotle: that the things studied in mathematics are *abstracted* from sense-particulars; they do not exist apart from those particulars in actuality (although, in a sense, they might exist conceptually); their special character is their generality – an issue discussed in detail in the *Posterior Analytics*.[14] Under this view the geometer would conceive of the "triangle", for instance, not as some abstract ideal entity, but only as a generalization derived from his inductive experience of sense-perceived triangles. In this way the specific Eleatic tenor of his epistemology disappears altogether.

No one would deny that the Eleatics were a dominant influence on Plato's philosophy of mathematics. But we have no means of determining whether the mathematicians themselves subscribed to this theory, or to the Aristotelian – if, indeed, to any philosophical account at all. Thus, in asserting the anti-empirical character of Greek mathematics and then deducing from this evidence of the influence of the Eleatics, Szabó has lapsed into circular reasoning.

Third, Szabó's detailed investigations into the roots of axiomatic terminology – "postulate", "axiom", "hypothesis", and so on – and their dialectical connotations,[15] as well as his tracing the *content* of the Euclidean first-principles back to the Eleatic arguments on being, plurality, motion and the like,[16] are designed to show that the search
150 | for axiomatic foundations of mathematics had already begun in the fifth century. Thus, if we possessed clear signs of an axiomatic interest within this period of mathematics, we might have to admit some influence from the side of the Eleatics. But in the *mathematical* writings which have survived from the pre-Socratic period, however scant and fragmentary these might be, there is nothing to suggest a concern over such foundational issues. Thus, if the mathematicians themselves manifested no awareness in their work of these philosophical issues, the thesis of a pervasive Eleatic influence dissolves. I find it important, then, briefly to survey the principal mathematical fragments and testimonia to establish this point.

Oenopides of Chios. Proclus, doubtless depending on Eudemus, assigns to Oenopides credit for having "first examined the problem" of constructing the perpendicular to a given line through a given point (*Elements* I, 12).[17] Following Heath and others, Szabó maintains that Oenopides must have initiated the first *theoretical* examination of constructions via compass and straight-edge, in the manner presented by Euclid in Postulates 1–3; for it is hardly conceivable that such elementary results should have remained unknown, that is, through practical techniques, until as late as 440 B.C. or so.[18] The latter point is undoubtedly correct. Yet the former need not follow. For we note that Eudemus is referring here to a *text:* he goes on to say that Oenopides called the drawing of the perpendicular "gnomon-wise in the archaic manner" and that he found this construction useful in astronomy. We easily infer that Eudemus found Oenopides' construction among the materials in a work on astronomy, most likely on the construction of sundials. In the context of such a work, the geometric constructions were certainly secondary to the main interest, so that a deliberate axiomatic approach to them is hardly to be expected. Eudemus's naming Oenopides as the "first" to effect these constructions indicates only that he was unable to find a writing earlier than

Oenopides in which the same had been done. Thus, this testimonium may be useful to assist us in gaining a sense of the degree to which mathematical knowledge was becoming public through writings at that time, as opposed to being transmitted through a personal teacher-student relationship which combined a written and oral tradition. But it hardly justifies assigning to Oenopides the axiomatic presentation of the plane constructions in Euclid's *Elements* I.

| *Hippocrates of Chios.* The most important extant mathematical writing from the 151
pre-Euclidean period is the fragment of Hippocrates's quadrature of lunules.[19] From the technical point of view, it reveals a full mastery of the materials presented in Euclid's Books I, III and VI. Eudemus's account indicates further that the work had a careful deductive structure. Not only are the constructions presented and their quadratures effected in fully systematic fashion, but preliminary theorems are proven on which these quadratures depend; among these are results on the areas of circles and their segments which reappear in Euclid and Archimedes. But even allowing that the form of this presentation is due to Hippocrates, rather than to Eudemus in his paraphrase of Hippocrates, we may still justifiably question whether this work is aptly described as axiomatic.

First, the style of the work is comparable to the geometric treatises of Archimedes:[20] the writing is devoted to the examination and solution of a set of related problems. To the extent that certain materials, notably *theorems*, are required for this, they are established in an introductory section. Many other results are assumed in the course of the work, so that this study entails the application and extension of more elementary work. Certainly, there is a strong sense here of what a proof ought to consist of: a precisely ordered deductive sequence starting from accepted and known results. But this is not the same as an *axiomatic* organization of this material.[21]

Second, where Hippocrates' writing touches closely the items which receive axiomatic treatment in the *Elements*, we find him least satisfactory. For instance, what is the definition of proportion by which he establishes that sectors are in the ratio of the angles which subtend them? This is accompanied by the remark (presumably due to Hippocrates) illustrating what is meant by 'same ratio' by reference to the half, the third, and related proper parts. If this indicates Hippocrates' handling of this theorem, we are still far from the technique of the *Elements* which develops all the needed steps right back to the fundamental definition of proportion (V, def. 5). As another example, how did Hippocrates show that circles are as the squares on their diameters? This appears as *Elements* XII, 2 and requires for its demonstration there the "method of exhaustion" (based on X, 1 and the foundation of proportion theory of Book V). Szabó berates Toeplitz for denying Hippocrates any such fully formalized proof and for proposing a looser alternative, applying a direct limiting | argument.[22] Apparently, Szabó 152
holds that an indirect argument of the Euclidean type was available to Hippocrates. The issue is of importance; for access to such a proof would indicate with Hippocrates a concern for just such formal questions of proof which motivate the axiomatic foundation of the *Elements*. But, contrary to Szabó's claim, Toeplitz's reconstruction is not without support. Precisely such a limiting argument as he presents survives in Antiphon's quadrature of the circle; and the grounds for denying an early dating of the "exhaustion" technique rest on Archimedes' ascription to *Eudoxus* of the *first* proper demonstrations of such theorems as the volumes of the cone and pyramid which

require it.[23] In view of these things, the usually accepted view of assigning to Eudoxus the technique of *Elements* XII appears preferable to the arbitrary dating of these back to Hippocrates some 50 years earlier.

Third, Hippocrates is obscure on just that question about his work which is most interesting from the dialectical point of view: did he maintain that his quadratures amounted in any sense to a quadrature of the circle? Aristotle and the commentators suspect he did, but modern commentators are properly reluctant to charge him with the gross logical error thereby entailed.[24] It seems more probable that Hippocrates established those results which were within his mathematical powers, and left to others the controversial matters. This schism between the mathematicians and sophists will arise again in our discussion of Theodorus; it speaks poorly for any view which highlights the positive influence of dialectic on mathematics at this time.

We may mention two other achievements due to Hippocrates. First, he *reduced* the problem of duplicating the cube to that of constructing two mean proportionals between two given lines.[25] Now, the logical method of "reduction" is known to Aristotle and may be viewed as a precursor of the method of "analysis" (or "analysis and synthesis").[26] Both these methods may be viewed as the mathematicians' technique, a powerful tool in the investigation of advanced problems from the fourth century onward. There is no reason to view its appearance with Hippocrates as a mark of dialectical influence.[27] Indeed, when Plato introduces the method of reasoning by "hypothesis" in the *Meno* (86), he does so with reference to a relatively advanced geometric problem. This appears more to suggest the initiation of the method in mathe-
153 matics before its adoption in philosophy. I could imagine one's | viewing the method of analogous reasoning as a precursor of reduction. But this would certainly be an imprecise comparison. Even Aristotle gives a mathematical example (i.e., Hippocrates' quadrature of lunules) to illustrate this method.[28]

The other work ascribed to Hippocrates was his contribution to the "Elements". Proclus, following Eudemus, says Hippocrates was the first to write a work of this type.[29] In the absence of the work, we cannot pronounce on the thoroughness of its axiomatization, or indeed on whether it was axiomatic at all, properly speaking. It seems to me reasonable to suppose that at Hippocrates' time, with the emergence of the professional teacher, a textbook collecting the materials he was prepared to teach would be a useful item for any mathematician to have at his disposal. The production of such a work, organized from simple materials toward more complex, would receive prescriptions for its format from the pedagogical context in which it would be used. There need thus be no more sophisticated influence derived from philosophical debates, such as those spawned by the Eleatic paradoxes; no self-conscious search for the fundamental unprovable first-principles from which the whole mathematical structure might be developed. Indeed, Plato's efforts to articulate the nature of this search and its importance both for mathematics and dialectic (in the *Republic*) would suggest that in that first effort, Hippocrates' "Elements", the search was barely, if at all, begun.

Theodorus of Cyrene. In my book on pre-Euclidean geometry I argue a major role for Theodorus in the formalization of geometry, particularly of those portions of Books I and II on the "application of areas".[30] From Plato we know Theodorus contributed to the early study of incommensurable magnitudes and earned praise for his precision in proof. Now, the theory of incommensurables represents a special problem for the

geometer: its subject matter cannot be articulated nor its claims established by concrete methods, such as those typifying the older practical metrical tradition to which it is related. One might, for instance, recognize computational difficulties in the treatment of quantities like the square root of 2; but the assertion and proof of its *irrationality* is a matter of quite a different order. Indeed, attaining the level even of *defining* what incommensurability is, not to mention probing the various constructions to establish which of these produce incommensurable lines, cannot be possible without a fine
| control of formal proof methods, especially of the method of indirect proof, as well as 154
a recognition that some very basic concepts require special, non-obvious, definitions.[31] For example, how are we to treat proportions among incommensurable magnitudes? Not at once seeing the way to do this, a geometer like Theodorus might well seek an alternative treatment, as in *Elements* II, dispensing with the techniques of proportions. Subsequently, mathematicians like Theaetetus and Eudoxus and their followers could provide suitable definitions by which these techniques might again be admissible under the more stringent logical conditions operating in proofs about irrationals.

In view of this special character of the study of incommensurables, I set the roots of the formal and axiomatic aspect of the *Elements* in Theodorus's work. I will develop this thesis further in the next section. But it is here important to recognize that Theodorus's own inspiration is being linked with the intrinsic nature of the mathematical subject he studied; it is not the response to external strictures of the dialectical philosophy. Plato provides a portrait of Theodorus in the *Theaetetus* which, if not mere dramatic license on his part, bears on Theodorus's debt to philosophy.[32] In his youth, Theodorus was exposed to the principles of Protagoras's philosophy. The explicitly relativist outlook of this school, prescribing the sharpening of dialectical and rhetorical skills so that one might argue his cause successfully, whether weak or strong, must have been uncomfortable for someone like Theodorus, whose talents lay in mathematics. He soon discontinued these studies. Plato describes him as still uneasy in debate, a teacher given to the lecture-method (rather than the question-and-answer technique so favored by Socrates), impatient with those who presume to set their own standards over against his ("they give me all the trouble in the world"), yet one whose standards of rigor in geometric proof were impeccable. How reasonable is it for us to presume strong and direct influences of dialectic on his methods and geometric outlook? To be sure, he must have acquired no small competence in logical technique through his training under Protagoras, and thus indirectly from the Eleatics. But in the highly contentious political and intellectual environment of mid-fifth-century Athens, this would be unavoidable in any course of higher education.

If this is all Szabó intends by attributing the "transformation of geometry into deductive science" to the influence of the Eleatics, then one could hardly object. But
I suspect the above account is decidedly | weaker than the thesis he advocates. 155
Certainly, it would be odd to view Theodorus's work – which could not but be highly theoretical and deductive, marked by an incipient awareness of the axiomatic approach to geometry – as directly indebted to Eleatic methods, when Theodorus himself seems to have fled the arena of dialectical competition through distress over the absence of firm bases of thought and opinion in that field.

Archytas of Tarentum. Perhaps Archytas' greatest interest for us here is his affiliation with the Pythagorean school.[33] In the modern discussion of pre-Socratic philosophy and mathematics, no move is more common than the assignment to the Pythagoreans of remarkable feats of scientific and mathematical reasoning. Some view the elaboration of fully developed theories in arithmetic and geometry, indeed, even the authorship of whole books of the Euclidean corpus as now extant, as early Pythagorean accomplishments, despite the utter absence of documentary evidence from that period.[34] Thus, it is instructive to consider the work of Archytas, stemming from an advanced stage of the history of that school.

On just such points where Szabó's view would lead us to expect the dialectical influence, Archytas is least satisfactory. In the fragment on epimoric ratios the use of indirect arguments is awkward.[35] Indeed, the last step of the proof, which should properly be cast as an indirect argument, is actually phrased in direct form. Further, a comparison of Archytas' proof, in the Boethian version, with the same theorem proved in the Euclidean *Sectio Canonis* indicates that the number theory available to Archytas contained materials on the least terms of ratios, but not on the more sophisticated notion of relative primes – that is, the content of the first part of *Elements* VII, but not the rest. It is difficult to conceive how formally adequate proofs, let alone a complete axiomatic treatment of number theory, could be assigned to Archytas. The difficulty is of course all the greater in supposing any such theory among the earlier Pythagoreans.

One surviving fragment contains Archytas' definitions of the three means: arithmetic, geometric and harmonic.[36] These reveal a symmetry of phrasing which might be termed aesthetic; but from the viewpoint of any systematic inquiry into the properties of these means they are framed poorly, and later arithmetic authors prescribe the
156 investigation and calculation of means according to alternative | formulas. The results of Archytas' division of the canon have been preserved; but the principles under which the numerical ratios for the various intervals were determined seem not at all clear.

Archytas' solution to the problem of duplicating the cube is extant in a paraphrase by Eutocius of a version from Eudemus.[37] The geometrical insight is uncanny and reflects a persistent strength of the ancient Greek geometric tradition: the profound visual intuition of the relationships of plane and solid figures. But Archytas depends upon the conception of the interpenetration of surfaces of solids generated through the *motion* of planes. By contrast, one of Szabó's critical arguments links the *avoidance* of geometric motions (as in Euclid's definitions) to a sensitivity to the Eleatic rejection of motion and change as logically inconsistent.[38] It would thus appear that this level of awareness had not yet been attained by Archytas. These abstract features of Euclid's definitions and methods appear thus to have entered the tradition of *Elements* later, for instance, in the mid-fourth century as a response to the deutero-Eleaticism of Plato in the *Republic* and the *Parmenides*.

Concerning the alleged formalism of early Pythagorean mathematics, a word on the discovery of the irrational is appropriate.[39] As indicated in our remarks above on Theodorus, the notion of incommensurability and the confirmation that given constructed lines are incommensurable are *intrinsically* theoretical and demand deductive arguments of the indirect kind; no concrete or practical procedure can do more than suggest that such lines evade description in terms of ratios of whole numbers. Hence, if the discovery and verification of incommensurability, even in a single case like the side and diagonal of the square, had been achieved by Pythagoreans toward the beginning of the fifth century, we should have to admit some more or less direct debt to the dialectical forms of the Eleatics. Now, dating this discovery is complicated by

the absence of any relevant and trustworthy documentation. While some scholars advocate an early date, there is no direct evidence to support their claim; indeed, the silence of the pre-Socratic literature on this notion until the close of the fifth century would appear to argue against them. Hoping not to be guilty of circularity myself, I find the later dating for this discovery both preferable and readily understood. Being by nature a theoretical result, the incommensurable requires a preparation in the skills of
deductive reasoning before the anomalous facts emerging from com|putations and con- 157
structions could be articulated in the form of concepts like incommensurability and commensurability and of theorems asserting the incommensurability of specified lines. Those anomalies do not of themselves lead to the conclusion of incommensurability; for instance, the ancient Babylonians and Hindus knew of the computational problems in dealing with certain square roots, yet they appear never to have taken this as more than a practical difficulty, that is, to have recognized that such quantities were *in principle* inexpressible in rational terms. We do well not to underestimate the gap which separates these two stages of awareness. Believing the early Pythagoreans had already bridged that gap, one must accept the implications concerning the formal level of their mathematical tradition at that time.

Szabó rightly chides Becker for inconsistency in maintaining the thesis of "Platonic reform" of mathematics on the one hand, while proposing reconstructions of a deductive theory of the odd and even among the early fifth-century Pythagoreans on the other.[40] But I believe, on the basis of the materials presented in this section, that Szabó has affirmed the wrong side of this contradiction. He has proposed the Eleatic logical methods and idealist outlook had already abetted the conversion of mathematics toward deductive, axiomatic theory in the fifth century, so that no major contribution was left to be made by the Platonic philosophy in the fourth century. By contrast, I argue that the mathematical evidence discourages assigning to fifth-century mathematicians any efforts to axiomatize fields in arithmetic or geometry. The signs of such an interest are first seen in association with Theodorus at the threshold of the fourth century.

We thus see that Szabó's thesis of the Eleatic impact on fifth-century mathematics both reinforces and depends upon a distortion of the nature of that tradition, in imputing to it a formal character it did not have. Similarly, Szabó's view requires us to dismiss the explicit testimonies of Plato and Aristotle on the relation of mathematics and philosophy. It is a remarkable feature of the logical and epistemological writings of both philosophers that they turn again and again to mathematics for examples of concepts and methods. In Plato's *Phaedo*, *Meno*, *Republic*, *Theaetetus* and *Philebus* and in Aristotle's *Analytics*, *Physics* and *Metaphysics* one is reminded of a consistent
theme: that the project of philosophy is to obtain the same | kind of certainty, the same 158
degree of rigor which mathematics has achieved – or in principle *can* achieve. We sense how strongly impressed both were at the success of the geometers of their time in their efforts toward this end. How is it possible that Plato and Aristotle could have been so mistaken in their view of the relation of mathematics and philosophy, if, as Szabó maintains, it was the Eleatic dialectic which had initiated the adoption of deductive method by the mathematicians? Szabó recognizes difficulty in his position, but offers an astonishing explanation: that in the half-century between Zeno and Plato the truth in the matter was simply forgotten.[41] In this way, the substantial corpus of Platonic and Aristotelian writings become irrelevant for understanding the pre-Socratic issues. In effect, then, Szabó can support his thesis only by inventing a

suitably formal pre-Socratic Pythagoreanism and then by directly repudiating the documentary evidence of the fourth-century philosophers. As a method for studying the history of mathematics and philosophy, this is one which few should wish to adopt.

While thus dismissing the writers of the mid-fourth century, Szabó turns to Euclid at the very end of the century for evidences of Eleatic influence. He views the definitions, postulates and axioms ("common notions") which preface Book I of the *Elements* as a direct sign of such influence.[42] Now, even if we *could* detect in Euclid such a sensitivity to the issues which had preoccupied the Eleatics, this need not indicate a *direct* response to Eleaticism, that is, a mathematical reaction dating from the mid-fifth century and then transmitted through some continuous line to Euclid. For the Eleatic problems received a lively examination in both Platonic and Aristotelian circles in the latter part of the fourth century, so that any dialectical influence could well have been exerted at this later time. I will return to this view and its implications in the second part of the paper. But I should like now to consider how useful the hypothesis of Eleatic influence really is for the interpretation of Euclid's procedure.

Szabó has argued that Euclid so framed the basic principles of the *Elements* as to avoid just those difficulties to which the Eleatics had called attention. For instance, the definitions of point, line and so on seem deliberately to avoid reference to sensible things or operations. The "common notions" (known to Aristotle and Proclus as "axioms") articulate such evident principles about equality that one is surprised at the
159 need to state them. Why, for instance, must one state that "the | whole is greater than its part"? Wouldn't that be sufficiently obvious to be admissible in any proof without special comment? (In fact, many less obvious principles *are* introduced into proofs. Why should the truly unquestionable principles require such statement?) Now, it happens that precisely these definitions and axioms have a prior appearance in the history of dialectic. Ultimately, the parts-whole relationship and the essential nature of equality, being and unity were the center of interest for the Eleatics. Szabó would thus maintain that the statements in Euclid follow from the earlier examination of these concepts and principles by the Eleatics.

Let us take a specific example: Euclid adopts a statical conception of the circle – as the plane figure bound by a line such that all the lines drawn from a point within it (i.e., its center) to its edge are equal (Book I, Def. 15); he does not use the intuitively clearer conception of the circle as the figure generated by a finite line segment which rotates about one of its fixed end-points – a dynamical conception preferred by Proclus. Can we fathom Euclid's preference? Might it not be that Euclid knew of the Eleatics' refutation of the possibility of motion and change in true being? Mathematical entities – lines, figures and so on – are such types of abstract being; hence, how can we conceive of their motion?[43] Thus, if we follow Szabó's view, Euclid adopted the static conception as a resolution of this puzzle. Now, this bit of dialectic is not hypothetical. For Proclus presents just such an argument against the motion of mathematical entities and has no small difficulty attempting to reconcile his Platonic notion of geometric being with his dynamical conceptions of lines and figures.[44] Let us go further: among the "common notions" there is one which asserts the equality of figures which can be exactly superimposed one on the other. This is a principle of which Euclid makes remarkably little use. Given that it in essence presumes a physical operation – the motion of one figure until it comes in position directly over the other – might not Euclid have wished to minimize appeal to it, for the same reason of the immobility of geometric being?

Ostensibly plausible,[45] this view is revealed upon closer examination to be a fully inaccurate account of Euclid's procedure. First, on the acceptability of motions for geometric objects, we observe that Euclid *does* introduce and use dynamical conceptions – such solids as the cone, cylinder and sphere are defined in terms of generation by plane figures rotating about a fixed axis (Book XI); why not merely | extend the 160
statical mode already used for the circle, so to avoid the appeal to motion? Again, the "postulates" (1, 2, 3 and 5) all use dynamical operations: the *drawing* of a line segment or of a circle, and the indefinite *extension* of a given line segment – why not phrase these in terms of the mere *existence* of geometrical entities (lines, circles) which possess certain specified properties (a given center and radius; or two given points)? We note that neither before nor after Euclid was there any reservation about the admissibility of motions into purely geometric theorems – at least, as far as the mathematicians were concerned: Archytas solved the duplication of the cube by means of lines and surfaces generated by rotating plane figures; the followers of Eudoxus constructed mechanical devices to solve the same problem; Archimedes defined the spirals in terms of a double motion; later writers recognize several varieties of motion-generated curves and surfaces.[46] Plato and the Eleatics to the contrary notwithstanding, the geometers had no qualms at all about such conceptions.

As for the principle of superposition, were Euclid seeking dialectical approval, he would have been compelled to adopt a rather different mode – for if superposition is *in principle* unacceptable, then even *one* use of it is too many.[47] But Euclid uses it twice in Book I and again in Book III (24) to establish theorems on the congruence of triangles and circular segments, respectively. Might he be making a philosophically significant point: that some resort to this principle, itself based on the conception of geometric motion, is indispensable for the working out of his geometric system? Indeed, invariance of area under the so called "Euclidean" motions of translation and rotation can be adopted as the defining property of the Euclidean metric (for in the non-Euclidean geometries figures do not so remain invariant). But Euclid is not troubled by such concerns. Otherwise, his procedure should have been this: eliminate the superposition axiom altogether; adopt as a *postulate* one of the theorems which require it, for instance, the congruence of triangles of which two sides and the included angle are respectively equal;[48] then prove the others by means of this postulate.[49] Euclid does none of this. Nor do his successors reveal uneasiness over superposition: for instance, Archimedes employs it several times, in his study of equilibrium of planes and his proof that the ellipsoid is bisected by any plane passing through its center (note that Euclid asserts the analogue for circles as part of his *definition* of the circle).[50] Proclus provides alternative | proofs of some of Euclid's 161
theorems using the superposition principle where Euclid does not, and even gives a proof of the fourth postulate (that all right angles are equal) by means of this same principle.[51] In sum, then, the ancient geometers felt quite free to adopt the conceptions of motion and superposition in their proofs, despite the dialectical difficulties therein. However much the Eleatics and Plato after them might shout disapproval, the geometers remained remarkably deaf to their objections here.

There still remains a question about Euclid's procedure: why, then, *does* he appear to reduce to a minimum appeal to the principle of superposition? For it is clear that several of his theorems admit of far simpler proofs than he gives them if the principle is introduced. I believe the answer may lie in Euclid's conception of his geometric system. Theorems are not isolated propositions, to be proved in whatever way. They fit into an *ordered* sequence, and there are considerations under which certain choices of order might be viewed as preferable to others. Aristotle devotes no small attention to this very issue in the *Posterior Analytics*.[52] In proving a theorem, although any number of prior assumptions might serve, there will be one choice which affords us knowledge of the *cause* of the truth of what is asserted, and this is determined by the essence of the thing examined. In effect, geometric proofs develop in the order which reverses the order of the familiarity of these things to us, that is, of our inductive discovery of these facts.

If we may consider Euclid's procedure in this light, the manner of his introduction of the primary undemonstrated premisses becomes clear. On the one hand, if a postulate has been used to demonstrate a theorem, that theorem now becomes admissible as an assumption in the proofs of subsequent theorems; as in the order of the increase of our knowledge of things, each thing learned becomes the basis for learning new things. Hence, we do better to demonstrate by means of what we have just proved, if that is possible, rather than by continually referring back to the primary assumptions. I believe this accounts for Euclid's handling of superposition. On the other hand, if it is possible to prove a theorem without recourse to some basic premiss, one should avoid giving a proof which in fact appeals to that premiss. Here, the finest example lies in Euclid's use of the parallel-postulate. For he introduces it precisely at that point in Book I where its
162 use is necessary (I, 29). The earlier theorem (I, 16) – that the | exterior angle of a triangle exceeds each of the non-adjacent interior angles – might easily be viewed as a corollary of I, 32. But the latter theorem requires Postulate 5, whereas the former does not.

One notes that Euclid's commitment to such considerations of deductive order varies in different parts of the *Elements*: Books I, V and VII, for instance, reveal a strong sense of the sequential ordering of results. By contrast, Books II and IV generally lack a cumulative character, amounting to collections of independently established results. One may interpret this as a sign of the earlier provenance of the two latter books, a view confirmed by other aspects of content and method.

Such considerations, then, discourage acceptance of the thesis of a direct Eleatic responsibility for the rise of deductive mathematics: in particular, it misrepresents the pre-Socratic work in geometry and number theory and it proves a questionable instrument for the understanding of Euclid. We shall instead now seek to portray the rise of the axiomatic approach in geometry as in part a reflection of the dialectical interests of the mid-fourth century.

II. FORMAL ASPECTS OF THE WORKS OF EUDOXUS, EUCLID AND ARCHIMEDES

In Plato's *Republic* the acquisition of knowledge, both in mathematics and in philosophy, is described as a quest for the appropriate first-principle from which the truths of things may be derived. Until such universal insight is achieved, one must adopt "hypotheses" as the basis of reasoning, a provisional expedient which can also assist in the discovery of more basic principles. Plato's characterization is imprecise, and may perhaps have been overestimated as a prescription for the systematization of knowledge. But there appears to be here a sense of the project of axiomatization, in particular of the fields of mathematics, and one may easily suspect that work of this type was then being encouraged in the Academy.

This impression is confirmed by Proclus' summary of the pre-Euclidean work which culminated in the *Elements*. Beginning with Archytas, Leodamus and Theaetetus, thirteen contributors are named, the most notable being Theaetetus and Eudoxus, all having an association of some sort with Plato and the Academy. Moreover, a number of Plato's arguments, as in the *Parmenides*, revive difficulties which the Eleatics had introduced on the theoretical understanding of | basic mathematical concepts (e.g., "unit", "num- 163
ber", "point", "line", "construction", "generation through motion", and the like). Somewhat later, in the *Physics* and other writings, Aristotle examined these issues; he criticized the notion of indivisibles, for instance, and provided synopses and commentaries on Zeno's arguments. The Eleatic methods and viewpoints so remained a topic of debate throughout this period. In view of these facts, together with the absence of comparable work in the fifth century, we may place the major contributions to the organization and axiomatization of mathematics among these fourth-century figures, whose treatises were collected and edited by Euclid at the very beginning of the third century.[53]

To what extent does Euclid's work actually reveal a concern over dialectical difficulties? One should note first that those parts of the *Elements* which appear most sensitive to such difficulties have the least bearing on the mathematics developed in the several Books. For instance, many of the definitions are but formal tokens, with no operational value for the proofs of theorems. The definitions of "point" and "line" before Book I, of "ratio" in Book V and of "unit" and "number" in Book VII have no use for the investigation of theorems; indeed, defining "point" as "that which has no part" is unsatisfactory even from the dialectical side, since both Plato and Aristotle disapprove of defining terms by means of negation. On the other hand, some terms which *are* of importance do not receive definition: for instance, the notion of "measure" is employed in Book VII, but is not there defined. Further, many of the definitions are straightforward and directly applied in the proofs; there is no indication in these instances that dialectical debates affected the introduction or choice of wording for these. As for the completeness of Euclid's postulates and axioms, the modern studies of foundations, such as those by Hilbert, have revealed how many principles are tacitly assumed by Euclid and in how many ways their formulation might be improved. For example, the "Archimedean axiom" appears as a *definition* in Book V, rather than in the form of a *postulate* as required for its application in Books V, X and XII. In view of such shortcomings, despite the long series of studies on the foundations of the "Elements" in the century before Euclid, one critic has recommended viewing Euclid's work not as an axiomatic effort at all, but rather solely as a mathematical treatment of plane constructions.[54]

| But given the intensity of interest in the principles of scientific knowledge among 164
the mid-fourth-century philosophers and the simultaneous effort among geometers to organize geometry in an axiomatic fashion, one can hardly suppose that these two enterprises proceeded in complete isolation from each other. By the end of this period both areas of inquiry were ripe for synthesis: the former in Aristotle's *Posterior Analytics*, the latter in Euclid's *Elements*. I wish briefly to explore the relation between these two works in two examples: the classification of the "first-principles" and the meaning of postulates of mathematical existence.[55]

Euclid divides the "first-principles" into three classes: definitions, postulates and axioms (that is, "common notions"). Proclus offers three rationales for this distinction between postulates and axioms, none of which adequately covers all five postulates and all nine axioms, as given by Euclid.[56] It would appear that through inadvertence,

Euclid, or some earlier editor of the "Elements", created a puzzle for the dialecticians. Now, Proclus notes that several writers made no distinction between postulates and axioms. One might wish to take this view as a mark of the influence of dialectics, for Euclid would seem to be requiring only the acquiescence to the stated principles, and the essentially *open* character of such acquiescence (especially when one recalls the Eleatic challenges to the conceptions stated in the axioms) is noteworthy.[57] However, an important feature separates mathematical from philosophical writers on this point: the former, as cited by Proclus, employ only the term "postulate".[58] Proclus remarks further that writers similarly conflate the terms "theorem" and "problem"; "problems" and "postulates" alike address *construction* of geometric entities, whereas "theorems" and "axioms" entail the *statement* of their properties and relations. In the pre-Euclidean period it was the philosopher Speusippus who classed all propositions as "theorems", while it was the mathematician Menaechmus who classed them all as "problems".[59] This, I propose, underscores a persistent difference in outlook between the mathematicians and the philosophers: the former were primarily occupied with the constructive activity involved in geometric research, while the latter were most interested in the formal description of proofs. Euclid's adoption of the term "postulate" for those assumptions most directly concerned with the propositions on geometric con-
165 struction can be understood as a derivation from his mathematical sources. The | other
principles may then have been formulated by way of satisfying more philosophical considerations and given the label "common notions" to signify their general applicability to *all* fields of mathematics.[60] But as Aristotle used the term "axiom" to designate these same "common" principles, the transferral of this term to Euclid's "common notions" in the later discussion of axiomatics is easily explained. One must admit, however, that the diversity of opinions on the subdivision of the "first-principles", from the time before Euclid right through to the time of Proclus, attest that Euclid's own choices in the *Elements* were neither fully clear nor definitive.

One of the classifications criticized by Proclus was that of Aristotle. Many scholars have since attempted to establish some sort of conformity between the Euclidean and the Aristotelian approaches, but without notable success.[61] I should like here to show how this issue reveals an interaction between pre-Euclidean mathematics and philosophy of a kind different from simple unilateral influence.

Aristotle's division of the principles in the *Posterior Analytics* (I, 2) separates the immediate first-premisses of apodictic syllogisms into (a) *theses* and (b) *axioms* – the former need not be proved or held in the prior knowledge of the learner; but the axioms, as it happens, necessarily are already known. The "theses" are then subdivided into *hypotheses* and *definitions* – the former assert one or the other of the two terms in a logical disjunction; definitions do not do this. Later (I, 10) Aristotle expounds this in a somewhat different manner: among the first-principles one must first assume (c) the *meaning* of the terms used and then (d) the *existence* of the entities so defined; in some instances this existence is merely *assumed*, but in others it is *proved*. Next there are (e) propositions which are *proper* to the subject-matter of each particular science, while there are also (f) propositions which are *common* to many or all of them.[62] As examples of such "common" propositions Aristotle cites the logical principle of the excluded middle and the condition that the subtraction of equals from equals leaves equal remainders. Now, this last class is variously called by him "the common things" (*ta koina*), "common opinions" and "axioms". As Aristotle's second

example appears among Euclid's "common notions", one sees that Aristotle's "axioms" and Euclid's "common notions" refer to the same class of principles. In this same passage (I, 10) Aristotle next introduces the terms *hypothesis* and *postulate* which here seem to explicate (e). These may or may not be suscep|tible of proof, but in the teacher-student 166
context they happen not to be proved. A "hypothesis" appears to be true to the learner – the term is thus not an absolute, but a relative term. When the student has no opinion of the truth of the assertion, or even doubts it, it is called a "postulate".

For Aristotle the terms "hypothesis" and "postulate" are thus embedded in a dialectical context. We shall consider later whether a similar attitude is taken by Euclid in his separation of the "postulates" from the "common notions".[63] But now, to help associate these schemes of first-principles, let us note the striking prominence Aristotle seems to accord to assumptions of *existence*, so to separate "definitions" from "hypotheses". The puzzle is that one fails to find in the mathematical literature from antiquity any explicit analogue to such assumptions and proofs of existence.[64] But this puzzle vanishes if we examine more closely Aristotle's intent. In these passages he is not really interested in prescribing the classes of assumptions preliminary to the construction of a formal axiomatic *system*. Rather, he is setting out what things one must do in order to prove a *theorem* scientifically; what things one must know or admit in order to understand a proof and recognize that it is a proof.

In this light let us regard the structure of any theorem in Euclid's *Elements*.[65] It begins with an assertion – the thing to be proved or to be constructed; to understand this much we already must know the meaning of certain terms, the relevant definitions. In the proof premisses are introduced: some of these are definitions; others are theorems which have already been proved or can be accepted as provable (perhaps deferred for separate proof as lemmas); still others are premisses which appear self-evident, so that they may be admitted without proof.[66] We have thus covered the appearance of definitions, postulates, axioms and prior theorems in the course of a proof, and on these items Aristotle's discussion and Euclid's practice are in agreement. But there is another kind of assumption made by Euclid: his proofs always provide an "exposition" (*ekthesis*) in which the theorem, having been stated in general terms, is now actualized as a particular configuration for the purposes of inquiry.[67] This exposition typically includes expressions like "let *ABC* be a triangle ...", "let the line *AB* join the points ...", and so on. In effect, one here asserts the *existence* of the entities discussed in the theorem; that is, they are brought into existence by a sort of intellectual *fiat*, a procedure | indispensable for the successful management of the proof. 167
Surely, it is this "exposition" which Aristotle had in mind when he spoke of the mathematicians' *assumption* ("hypothesis") that "there are such lines, units, etc." (*not* that "lines, units, etc., exist").[68] In this way Aristotle's discussion of the first-principles of proof becomes a straightforward commentary on the actual procedures of proof employed by Euclid's immediate precursors.

It is often maintained that the first three Euclidean postulates[69] are assumptions of geometric existence and that, in general, constructions serve the function of proofs of existence.[70] But I hold such a view to be anachronistic, assigning to the *Elements* a dialectical motive which it just does not have. First, where the existence of a solution is at issue, Euclid and the other Greek geometers established the result not by a construction, but by a *diorismus* – a theorem specifying the conditions under which the construction is possible. Second, the postulates and other alleged theorems of existence never employ a *language* of existence. For instance, in *Elements* I, 1 Euclid does not propose to prove the existence of equilateral triangles, but rather to show how the equilateral triangle may be constructed whose side is a given line segment.[71] The postulates are stated not as assertions of the *existence* of lines, points and circles under specified conditions, but rather of the *constructibility* of those lines, points and circles. Third, in the ancient geometry those situations comparable to cases in modern foundations of analysis and set theory where the question of existence is most interesting, assume the existence without comment: for instance, the existence of the fourth proportional is assumed throughout *Elements* XII, even though constructible alternatives could have been offered; the existence of intersections of given curves is assumed without a statement of the principle of continuity required.[72] It seems improbable that the geometric constructions were intended to establish existence when the issue of existence earns no place among the axioms.

Thus, there appears to be no dialectical motive behind Euclid's statement of the postulates or his presentation of geometric constructions. Moreover, in setting out the postulates, he did not aim to restrict the whole field of research on constructions to those which can be effected on these assumptions (in practice, to those constructible by means of compass and straightedge alone), nor even to suggest that these means
168 were somehow privileged among the variety | of constructing devices possible. We know of many efforts to solve the problems of angle-trisection, cube-duplication and circle-quadrature by means of constructions and curves effected by other than "Euclidean" means.[73] Indeed, sometimes such means were employed even when a Euclidean construction is possible.[74] Only late in the third century do we meet efforts to classify constructions and to establish an ordering of priority according to the means employed (i.e., "plane", "solid", and "linear"). Studies by Apollonius and Nicomedes, for instance, showed how certain types of constructions could be effected on the Euclidean assumptions, while others required conic sections or special motion-generated curves. Apparently this gave rise to a formal recommendation that only the simplest possible devices be employed, if a given construction is to be judged formally proper. Accordingly, both Apollonius and Archimedes came under criticism for their unnecessary assumption of "solid" means in certain constructions.[75]

In these third-century developments we perceive a way in which formal philosophical prescriptions might intrude on the geometer's research. To be sure, a formal commitment on the "priority" of assumptions might contribute to the tightening of the axiomatic structure of geometry.[76] But one can recognize that such conditions might impose an extraneous burden on the geometer in his search for the solutions of problems. Is the discovery of how to trisect an angle by means of the conchoid or the Archimedean spiral any the less useful for its incorrect application of "linear", rather than "solid" techniques? A related example is Archimedes' "mechanical" method: as geometry is "prior" to mechanics (according to Aristotle), mechanical principles are formally inadmissible into proofs about the purely geometric properties of figures. But does this diminish the heuristic power of the method? More to the point, does one impede the search for new discoveries in geometry by discouraging the application of such methods? Indeed, adhering strictly to such orderings of principles prevented the Greeks from perceiving many of the more general structural relations which connected ostensibly disparate problems. For instance, the area of the parabolic segment, the area of the plane spiral, the volume of the cone, the volume of an oblique section

of a cylinder, the center of gravity of the paraboloid – these problems, all examined by Archimedes, would appear to be quite distinct each from the others; yet seventeenth-century | geometers came to recognize that they could be reduced to the same 169
procedure, the summation of consecutive square numbers, as signified by the expression $\int x^2\,dx$.[76a] Certainly, the axiomatic concerns of the ancients offered no assistance toward the recognition of such general relationships of structure linking diverse portions of the whole field of mathematics.

Let us return to Euclid's postulates. Proclus notes that the remaining two – the fourth asserting the equality of all right angles and the fifth providing a condition satisfied by parallel lines – differ from the former three in that they seem to have more the "theoretical" character of axioms than the "problematical" character of postulates. Indeed, following upon Aristotle's observation on the first-principles, that axioms ought to be indemonstrable, later writers on axiomatics, like Geminus, contended that these "postulates" were actually *theorems* whose proofs should be possible on the basis of the other assumptions. A number of proofs for both postulates have been preserved in Proclus' commentary. It is well known how diligently the quest for such proofs, especially of the parallel-postulate, was pursued until the status of this principle as a true postulate was established by nineteenth-century geometers. I. Toth has detected a variety of passages in the Aristotelian corpus which suggest that the search for such a proof was an active concern of geometers in the mid-fourth century.[77] Indeed, the careful organization of *Elements* I, in which the introduction of this postulate is deferred until precisely that point where its use is unavoidable, indicates the intensity of these studies. Certainly that research failed of its purpose, however, so that the principle retained its place as an unproven postulate within Euclid's theory of parallels.

If research on the "Elements" was strongly influenced by dialectical attitudes, to the extent that postulates and axioms were introduced only as provisional and debatable assumptions, we should expect to find the serious consideration of the postulates alternative to the parallel-postulate as the basis for a different geometric system. In this way, the Greeks should have advanced toward the elaboration of "non-Euclidean" systems, in particular the hyperbolic geometry. Indeed, Toth has argued that many of the theorems of "non-Euclidean" geometry were worked out in Aristotle's time, if only along the way toward a hoped-for indirect proof of the parallel-postulate.
Unfortunately, no such systematic treatments have survived; indeed, | save for the few 170
suggestive, but highly problematic passages from Aristotle, there remain no fragments or even references to titles to indicate that such studies had ever been completed. It thus appears that one avenue of research which a geometer with strong dialectical inclinations might have discerned was never in fact followed out.

An explanation for this failure may be seen from another condition which Aristotle imposes on the formulation of axioms: they must be self-evident – that is, true as statements abstracted from our physical experience. Proclus remarks that one might dispute an axiom for the sake of argument, but he appears to accept Aristotle's condition and names no philosopher or mathematician who advocated that assumptions contravening experience might form the basis of a complete mathematical system.[78]

These considerations have thus revealed how the philosophical discussion of the principles of scientific knowledge drew from mathematical work, but that even the axiomatic aspects of such works as Euclid's *Elements* resulted primarily from mathematical concerns. Dialectical motives played at best but a limited role in the production of mathematical works. Nevertheless, the contributions of several geometers, notably Eudoxus and Archimedes, manifested a keen sensitivity to matters of formal precision. It is thus appropriate to review their work for possible signs of philosophical influence.

Eudoxus, an esteemed colleague of Plato in the Academy of the mid-fourth century, made remarkable contributions to geometry and mathematical astronomy. In the former area he is most noted for his theory of proportion and his method of limits (usually called the method of "exhaustion") used in the measurement of non-rectilinear surfaces and solids; both techniques can be associated with the solution of important difficulties in the foundations of geometry.

Euclid presents two forms of the theory of proportion in the *Elements*. The version in Book VII applies only to ratios of integers, but can be extended without difficulty to cover ratios of commensurable magnitudes. The one in Book V is general, applying both to commensurable and incommensurable magnitudes. While the former theory is sometimes ascribed to early Pythagoreans, I have already called into question the attribution of significant formal work to such fifth-century arithmeticians; rather, I have argued that this Euclidean theory was initiated with the work of Theaetetus after the beginning of the fourth century.[79] The theory in Book V is generally assigned to
171 | Eudoxus; but for reasons which I have recently discovered and presented in detail elsewhere, there are good reasons for doubting this. For an alternative general technique of proportions, based on the same principle of convergence by consecutive bisection which typifies Eudoxus' "exhaustion" method, is applied in works of Archimedes and others and can readily be employed as the basis of a full theory of proportions. In addition to these approaches is a fourth, framed around the so-called "Euclidean" division algorithm (*anthyphairesis*), and attributable to Theaetetus.[80]

A comparison of these theories enables one to appreciate the formal motive which gave rise to them. The inadequacy of the numerical theory to cover incommensurable magnitudes led to the search for a technique which might apply rigorously to these. As the division algorithm was already basic to the number theory within which the study of incommensurables was conducted, it was natural to attempt an extension of the theory of proportion on the basis of this same technique. One can see how problems with the proofs of specific theorems encouraged the search for alternative approaches, first that utilizing the bisection-method (the one I have attributed to Eudoxus), and ultimately that based on equimultiples, the theory in Euclid's Book V. In this whole development, technical considerations appear to dominate. Indeed, the formal character of the theory of incommensurable magnitudes, where the theorems on proportion were to be applied, establishes from the start a thoroughly mathematical context for these formal problems, without reference to such external stimuli as dialectical paradoxes. It is as likely that Plato and Aristotle were spectators of the progress of the mathematical researches, as that they were instigators of them. On the other hand, the reasons for passing from the Eudoxean bisection version to the Euclidean equimultiple version of the theory seem more strictly formal than technical; in fact, the resultant theory, while formally elegant, tends often to obscure the mathematical ideas in a way the bisection theory does not, and in at least one theorem (VI, 33) actually introduces a logical error absent from its predecessor.

The roots of the "exhaustion" technique are in the pre-Euclidean efforts relating to the quadrature of the circle. One recognizes that among geometers and philosophers

alike there were disagreements on the admissibility of certain techniques and concepts.
In some mathematical work, procedures appear involving limits, indivisible | magni- 172
tudes, assumptions and intuitions justified through sense-experience, and the like.[81] These were drawn into dialectical controversies over issues like the relation of perception to abstract thought and the necessary forms of logical reasoning. To claim that a mathematical technique like the method of "exhaustion" was designed to circumvent dialectical difficulties like the Eleatics' paradoxes on the divisibility and mutability of real being surely oversimplifies, if not entirely misrepresents, the context of the mathematical studies. Moreover, it is by no means clear in the fourth century whether the resolutions adopted in the Euclidean *Elements* yet commanded unanimous acceptance among mathematicians. For, as we shall see later, techniques whose formal inadmissibility was unequivocal nevertheless arise in third-century work.

This discussion has intended to show the nature and extent of the influence which dialectical controversies brought to bear on mathematical work, especially in the area of axiomatic studies. The evidence indicates that whatever that influence was, it was exerted in the environment of the fourth-century Academy, rather than earlier; that it affected the expression, rather than the substance, of mathematical concepts and techniques; that the most interesting axiomatic works can be understood as primarily an internal evolution of mathematical thought; and that the mathematicians' work was as much the source for dialectical inquiries as the humble recipient of the strictures of the dialecticians.

The relation of mathematical and philosophical studies in the period following Euclid appears comparably complex. Archimedes produced his major works after the mid-third century, and his earliest contributions dated from no less than three decades after Euclid. Despite this, one perceives that his early works rely on Eudoxean technical models, rather than the Euclidean versions.[82] Among these is *Plane Equilibria I*, the only one of Archimedes' extant works which qualifies as an effort of the axiomatic kind. Unfortunately, its success *as* an axiomatic effort is questionable; as many scholars have argued, the handling of the primary notions – equilibrium and center of gravity – is not uniform and appears to require additional assumptions on the physical properties of balance.[83] Of course, the work is not complete as far as its geometric technique is concerned, requiring many results from plane geometry and the theory of proportions, as well as the technique of "exhaustion".

| The formal precision of Archimedes' proofs is beyond question; these served as 173
a model both in his own time and afterward, well into the period of early modern mathematics, of the rigorous style of geometric demonstration. Nevertheless, he makes no claim of an axiomatic intent in his works, and we should avoid imputing any such motive to him. The structure of his treatises is ideally suited for the accurate presentation of specified geometric results, representing the product of original research. Typically, each develops one or two major results entailing the demonstration of a set of theorems, sometimes half a dozen or more depending on the number of cases involved. Leading up to this will be a sequence of theorems establishing results necessary for those theorems. Opening the work will be a few basic results not as specific for the principal theorems of the work, but useful in the proofs. Such lemmas often are stated only, the proofs being assumed from more elementary works. In similar fashion, steps are frequently elided in the course of proofs of the main body of theorems, sometimes with references to works in which the missing proof can be found; again,

a lengthy demonstration may be abridged when a portion of it is strictly analogous to an argument already presented.

This makes clear that Archimedes has no aim to set forth exhaustively *all* the materials required; in such advanced studies he can assume much from the elementary literature. In another variation on Euclidean practice, he may *defer* the proof of a step until a later point in the writing, as in an appendix. Archimedes does provide definitions of terms, but only where these are necessary for the work and are presumed novel or unfamiliar.[84] Prefacing *Sphere and Cylinder I* are two interesting series of "lemmas" (*axiomata* and *lambanomena*) assumed in the theorems: the former comparable to "definitions" in Euclid, the latter to "postulates". The discrepancy in terminology would appear to indicate either that Archimedes was not familiar with Euclid's edition of the "Elements", or that he was not particularly disposed toward imitating its formal usage. By including among the *lambanomena* postulates on the relative magnitude of convex arcs and surfaces, as well as the "Archimedean axiom" on continuity, Archimedes reveals a profound insight into the formal treatment of this subject matter. But as there appears to have been no precedent for the former postulates in the earlier literature on
174 axiomatics, mathematical or philosophical, we may be assured that his own | experience in working out proofs led him to recognize the appropriate assumptions to make.

Among Archimedes' works two differ from the above pattern in an interesting way. Both the *Quadrature of the Parabola* and the *Method* are devoted in major part to the exposition of a "mechanical method" of research, lacking the full force of geometric demonstration. As noted above, the principal objection a formalist would have doubtless was the introduction of mechanical principles into the technically prior area of geometry. Archimedes justifies his use of this method in the preface to the *Method* addressed to Eratosthenes: by its use one may discover new results and gain insight useful toward their proof, even if the method itself does not amount to proof. It would appear that Archimedes let his guard down, as it were, in the interests of communicating his heuristic approach to a fellow mathematician. The situation of the *Quadrature of the Parabola* is comparable, in that Archimedes was here writing for the first time to an Alexandrian geometer, Dositheus, whom he did not know, but presumably might expect to have use for such a method. In the later writings to Dositheus, however, the formal standards were never again so relaxed; moreover, we receive several indications that Dositheus and his colleagues impressed Archimedes but little as far as their mathematical talents and capacity for creative research were concerned. In other words, the formal precision of these treatises appears to have been devised to command the approval of the Alexandrian professionals, a group whose interest in the formal details of proofs might be termed scholastic.

Pappus has preserved a number of alternative treatments of Archimedean theorems. As I argue elsewhere, one has reason to view these as based on early versions by Archimedes of studies on the sphere, the spiral and other figures which receive a more complete and formal treatment in the writings sent to Dositheus.[85] One may infer also that the earlier versions were addressed to Conon, a geometer much respected by Archimedes, as the prefaces to the formal treatises make clear. Considered then as communications by one geometer to another, the writings preserved by Pappus are of some interest with respect to their formal organization. Those aspects of looseness evidenced to some extent even in the formal treatises are much more prominent here. For instance, the "parts" of a proposition and its proof are not

divided in the manner of the formal Euclidean | and Archimedean works; assertions 175
are made, not in general, but with reference to the particular elements of specific diagrams.[86] Proofs are severely abridged: major steps are assumed, whole theorems may be stated as "manifest", portions of proofs omitted as "similar" to arguments just presented; in the case of the theorems on spirals the entire convergence argument is absent, indeed is not even hinted at – although its lines could readily be supplied by one conversant with Archimedes' form of the "exhaustion" method. In the context of such a communication, the strictures on the formal organization of proofs would just get in the way, obscuring the essence of the mathematical ideas. In view of this, one might well suspect that the highly formal, detailed treatments of theorems in the treatises did not represent Archimedes' natural and preferred choice for the exposition of his findings; that even in the formal versions the omission of some steps need not be justified by the actual appearance of the step in a prior work (of course, sometimes it does), but may indicate only a sense of what is "obvious". In other words, Archimedes has not the intent to extend the axiomatic structure of mathematics. He is interested in communicating new discoveries in geometry in as clear and precise a manner as possible and according to a format which he deems will meet the approval of his professional colleagues.

Thus, far from abetting the progress of mathematical research, the emphasis on axiomatic issues appears to have distracted, perhaps even impeded, the geometers' work. Certainly, some geometers took this position. We have seen that aspects of Archimedes' work could not have satisfied certain formal criteria, as laid down by Aristotle. The *neuses* in *Spiral Lines* and the "mechanical method" in the *Quadrature of the Parabola* and the *Method* violated notions of formal priority of principles; one suspects that other treatments would face similar objections: the cubature of the cylindrical section in the *Method* (effected via reduction to the quadrature of the parabola) and the quadrature of the spiral in Pappus' version (reduced to the cubature of the cone). In his prefaces, Archimedes takes care to distinguish heuristic from demonstrative procedures, to articulate the assumptions he must make (most notably, the "lemma" on continuity), and the like; but these appear to be defensive gestures, aimed to convince his readers that his discoveries match those of Eudoxus and the other fine geometers of past generations.

In fact, Archimedes several times makes explicit his impatience | with the formal- 176
ists. In the prefaces to *Sphere and Cylinder I* and *Spiral Lines* he expresses regret at Conon's death, for only Conon, an extremely able geometer, had been qualified to judge Archimedes' work. That office was now occupied by Alexandrians who kept sending for details of proofs, but never contributed any new results of their own. This complaint is repeated by later geometers. From Apollonius (preface to *Conics*, Book IV) we know that Conon himself had received some abusive criticism from members of the Alexandrian community – although Apollonius, as a geometer, insisted on the merit even of these offending works. Again, Apollonius was criticized on the basis of some works he allowed to circulate in preliminary versions.[87] An echo of this division between the geometers and their more scholastically oriented colleagues is heard in a passage preserved by Pappus:[88] the geometer disclaims all competence or interest in the business of splitting hairs over first-principles; he wishes instead to gain praise through the utility of his findings. His case is made in the presentation of an ingenious method for evaluating areas and volumes by means of centers of gravity. A formalist

might cringe. But how could a geometer not delight in this ancient presentation of the technique now known as "Guldin's rule"?

Were these complaints by the geometers merely a special plea? Did the strictures by formalists actually have any such impeding effects on their work? Certainly, Archimedes and Apollonius possessed techniques of their own useful for the discovery of results, even if these would not survive scrutiny in a formal work. But the justice of their case, I believe, must be decided with reference to the larger field of ancient mathematics. If we take into our view the subsequent development of geometry, using the materials presented by such late commentators as Pappus, Theon, Proclus and Eutocius as indicators, we perceive that among the studies which they take to be advanced work there appears to be not a single contribution which would overreach the abilities of a geometer in the late third century applying only those concepts and techniques then available. Almost all the results they mention could be assigned quite credibly to Archimedes or Apollonius or their immediate followers.[89] Such *stasis* is virtually incomprehensible in comparison with the development of mathematics since the Renaissance. Each interval of fifty or one hundred years has since embraced fundamental changes in the concepts, methods and problems studied by mathematicians.
177 Surely | no small part of the explanation for the stagnation in antiquity lies with the scholastic attitude toward mathematics, emphasizing the narrow investigation of purely formal questions. The impact of this emphasis must have been felt in geometry through the curriculum of higher education, by encouraging students toward this scholastic attitude and by limiting their exposure to heuristically useful methods.

III. SUMMARY

We have traced the relation of mathematics and philosophy in the fifth, fourth and third centuries with reference to the development of logical methods and axiomatics. Contrary to the view of Szabó, I have argued that the deductive procedures evident in fifth-century work can be understood as intrinsic to mathematical study, while any strong sense of axiomatic organization is there missing. Thus, the effect of the Eleatic dialectic leading to the rise of formal and axiomatic studies in arithmetic and geometry could only occur later, in the environment of the fourth-century Academy where such an interest in mathematics was combined with a renewal of the logical and philosophical views espoused by the Eleatics.

In this way, many of the features of the formal Euclidean system, which Szabó reads as evidence of a more or less direct influence by the Eleatics in the fifth century, are now argued to be the response to the inquiries into logic in the circles of Plato and Aristotle. But even here, the influence was largely superficial, affecting the form of the presentation of mathematical results, yet only little the problems and concepts and methods of examining them. Where a mathematical advance *was* closely tied to formal matters – most notably, in Eudoxus' contributions to proportion theory and the method of "exhaustion" – any view of one-sided dialectical influence grossly oversimplifies. The nature of these studies combined philosophical and mathematical problems; and in providing material for the continuing dialectical debates on related issues, the mathematicians were as active in shaping the development of philosophical views as responsive to philosophical recommendations concerning the appropriate methodology.

By the third century both philosophers and geometers had resolved the larger issues on the systematization of knowledge, as, respectively, the Aristotelian theory of science and the Euclidean com|pilation of geometry. The subsequent history of math- 178
ematics indicates that the success of this axiomatizing effort eventually served to discourage the creative forms of research which could have advanced mathematical knowledge. Earlier, the philosophical interest in mathematical questions and the exposure of mathematicians, in the course of their education, to the dialecticians' conscious interest in epistemology and the structure of knowledge, had the stimulating effect of making mathematicians aware of the autonomy of their field as an intellectual discipline, rather than as but an art for the practical solution of problems. Unfortunately, mathematics in later antiquity became subordinate to the objectives of the philosophical curriculum. Students might be trained in the subtleties of the foundations of elementary geometry, but they rarely acquired the techniques needed for pursuing researches in advanced fields.

In tracing the interaction of mathematics and philosophy into the later stages of formal studies in axiomatics, we have progressed far from the difficult question with which we opened: what was the manner of the first introduction of deductive methods into Greek geometry? While I have argued that Szabó's hypothesis of a specific influence by the Eleatics is unpersuasive, I have not yet presumed to offer a substitute. Even so, Szabó's view merely pushes our question further back. For we must still inquire into the infusion of rational and deductive modes into the earlier natural philosophy. It does little to assert that dialectics provided the logical model for mathematics, or conversely that mathematics performed this function for dialectics. For there still remains the more basic issue of how deductive methods came to be established at all as the appropriate basis of intellectual inquiry.

I believe that the answer to this question must be sought in the wider cultural context of sixth and fifth century Greece. The political, economic and social environment was then such as to encourage individual expression and thus to give rise to the problem of arbitrating among conflicting policies and opinions. The associated intellectual climate was critical in spirit, sometimes to excess, as in the doctrines of the Eleatics, of the Sophists and of the sceptics. This environment encouraged all thinkers, among these the mathematicians, to look to the coherence of their basic assumptions. In dialectics the grounds of knowledge became a central issue: what is true? how does one know what is true and distinguish it from the false? how does one com|municate 179
and teach? The importance of these questions for the thought of Socrates, Plato, Aristotle and others is clear. But surely this general environment had an equivalent impact on the mathematicians. Examining their arithmetic and geometric techniques, they began to seek justifications, such being especially important in problematic areas like the study of incommensurables. Once the quest for justification was underway, the nature of mathematics itself would lead to the implementation of deductive forms, as I have argued. Furthermore, the incentive to teach spurred efforts to organize large areas of mathematics into coherent systems. Hippocrates of Chios was but the first of more than a dozen compilers of "Elements" active within the century before Euclid.

We thus argue that the origins and elaboration of formal methods in mathematics, both in the pre-Euclidean period and afterward, are best understood as a development internal to the mathematical discipline. In response to the same intellectual climate, mathematics and philosophy advanced along parallel lines of development. But the areas of direct interaction between them appear to have had only a limited impact, at least on mathematical studies. Indeed, judging from the epistemological views of Plato and Aristotle, one cannot escape the conviction that the influence of mathematics on philosophy was far more significant than any influence in the converse direction.

BIBLIOGRAPHICAL NOTE

The following works are cited by abbreviations:
AGM = A. Szabó: 1969, *Anfänge der griechischen Mathematik*, Budapest and Munich/Vienna.
EEE = W. Knorr: 1975, *The Evolution of the Euclidean Elements*, Dordrecht.
HGM = T. L. Heath: 1921, *A History of Greek Mathematics*, 2 vol., Oxford.

NOTES

1 A. Szabó: 'Transformation of mathematics into deductive science and the beginnings of its foundation on definitions and axioms', *Scripta Mathematics* **27** (1964), 27–48A, 113–139 (p.137).

2 "I do not think we should assume that mathematicians cannot use a logically valid pattern of reasoning in their work until some philosopher has written about it and told them that it is valid. In fact we know that this is not the way in which the two studies, logic and mathematics, are related." – W. C. Kneale in his commentary to A. Szabó, 'Greek dialectic and Euclid's axiomatics' (in I.
180 Lakatos (ed.), *Problems in the Philoso|phy of Mathematics*, Amsterdam, 1967, pp. 1–27), p.9. While acknowledging that the Eleatics were the first to make conscious use of indirect arguments in dialectic, N. Bourbaki deems it most probable that the mathematicians of the same period had already availed themselves of the same method in their own work (*Éléments d'histoire des mathématiques*, Paris, 1969, p. 11). The "internalist" position is advocated strongly by A. Weil: "[the question is] what is and what is not a mathematical idea. As to this, the mathematician is hardly inclined to consult outsiders. ... The views of Greek philosophers about the infinite may be of great interest as such; but are we really to believe that they had great influence on the work of Greek mathematicians? ... Some universities have established chairs for "the history and philosophy of mathematics": it is hard for me to imagine what those two subjects can have in common." – 'History of mathematics: why and how', (pp. 6–7): lecture delivered at the International Congress of Mathematicians held in Helsinki, August 15–23, 1978.

3 In addition to the essay cited in note 1, Szabó has expounded his views on the rise of deductive method in the following: '*Deiknymi* als mathematischer Terminus für *beweisen*', *Maia* **10** (N.S.) (1958), 106–131; 'Anfänge des euklidischen Axiomensystems', *Archive for History of Exact Sciences* **1**, (1960), 37–106; 'Der älteste Versuch einer definitorisch-axiomatischen Grundlegung der Mathematik', *Osiris* **14**, (1962), 308–369. These and other essays have been reworked as the basis of his *Anfänge der griechischen Mathematik*, Budapest/Munich, 1969 (esp. its third part, 'Der Aufbau der systematisch-deduktiven Mathematik'). It is this last-named work to which I will most frequently refer here (to be cited as *AGM*). The first and second parts of *AGM* are concerned with the pre-Euclidean study of incommensurability and the terminology of early proportion theory, respectively. These will not be discussed here, but many points have been examined by me in *The Evolution of the Euclidean Elements*, Dordrecht, 1975.

4 P. Tannery, *Pour l'histoire de la science helléne*, Paris, 1887, pp. 259–260. His view was elaborated by F. M. Cornford (1939) and J. E. Raven (1948), but has been discounted by most scholars since; for references and discussion, see W. Burkert, *Lore and Science in Ancient Pythagoreanism*, Cambridge, Mass., 1972, pp. 41–52, 285–289 and my *EEE*, p. 43. On the pre-Euclidean "foundations crisis", see H. Hasse and H. Scholz, 'Die Grundlagenkrisis der griechischen Mathematik', *Kant-Studien* **33** (1928), 4–34. This view has been much modified and criticized, as by B. L. van der Waerden (1941) and H. Freudenthal (1963); see my *EEE*, pp. 306–313.

5 *AGM*, III.1: 'Der Beweis in der griechischen Mathematik', esp. pp. 244–246.

6 *AGM*, III.3: 'Der Ursprung des Anti-Empirismus und des indirekten Beweisverfahrens.'

7 See the remark by W. C. Kneale, cited in note 2.

8 For texts from the ancient Babylonian tradition, see O. Neugebauer, *Mathematical Cuneiform Texts*, New Haven, 1945. For the Egyptian tradition, see A. B. Chace *et al.*, *The Rhind Mathematical Papyrus*, Oberlin, Ohio, 1927–29.

9 'Transformation' (see note 1), pp. 45–48; *AGM*, pp. 292f.

10 *AGM*, III.3: I render Szabo's *anschaulich* as "graphical"; other possibilities are "illustrative", "visual", "perceptual", or even "intuitive". I will understand him to refer to a kind of demonstration based on concrete perceptual acts, such as setting out numbered objects or constructing suitably illustrative diagrams.

11 | Of course, the ancient Babylonian geometers did not do this either. In this sense they too might 181
be viewed as using "non-graphical" approaches in geometry. As far as the Greek classical tradition is concerned, we should beware pushing this point on the non-graphical or abstract nature of the discipline too far. For within it the production of the appropriate diagram was always an integral part of the proof. Proclus, for instance, includes "exposition" (*ekthesis*) and "construction" (*kataskeue*) as proper parts of any demonstration (see note 65 below) and the term for "diagram" (*diagramma*) could actually serve as a synonym for "theorem" (see my *EEE*, ch. III/II). Indeed, the availability of the diagram must have eased considerably the burden of the awkward geometric notation used by the Greeks and perhaps explains why they never saw fit to overhaul it.

12 See my 'Problems in the interpretation of Greek number theory', *Studies in the History and Philosophy of Science* **7** (1976) 353–368.

13 On this much-discussed aspect of Plato's epistemology, see, for instance, F. M. Cornford, 'Mathematics and dialectics in the *Republic* VI–VII', (1932), repr. in R. E. Allen (ed), *Studies in Plato's Metaphysics*, London, 1965.

14 For a discussion of the relevant passages, see T. L. Heath, *Mathematics in Aristotle*, Oxford, 1949, ch. IV, esp. pp. 64–67 and J. Barnes, *Aristotle's Posterior Analytics*, Oxford, 1975, p. 161.

15 *AGM*, III. 6–9, 13, 17, 21–23, 26.

16 *AGM*, III. 18, 20, 24, 25.

17 Proclus, *In Euclidem*, ed. G. Friedlein, Leipzig, 1873, p. 283.

18 *AGM*, III. 19.

19 The fragment is preserved by Simplicius, *In Aristotelis Physica*, ed. H. Diels, Berlin, 1882, pp. 60–68. See T. L. Heath, *A History of Greek Mathematics*, Oxford, 1921, I, pp. 182–202.

20 Notably, *Quadrature of the Parabola* and *Sphere and Cylinder I*. The same format of exposition is followed in several of the discussions in Pappus' *Collection IV* and *V*.

21 As we shall develop below, neither are the works of Archimedes, save for *Plane Equilibria I*, accurately viewed as efforts at axiomatization.

22 *AGM*, pp. 450–452. Given the great importance of the Hippocrates-fragment, Szabó's discussion of it in *AGM* (part III) is remarkably slender, and what he does say inconsistent. On the one hand, Szabó insists on the deductive, even axiomatic, form of Hippocrates' work: the "Elements" attributed to him by Proclus must surely have had some sort of foundations (pp. 309f, 342). Szabó, of course, wishes to assert the nascent axiomatic form of early Greek geometry as a mark of Eleatic influence. Yet he elsewhere stresses the non-axiomatic nature of Hippocrates' study of the lunules; for instance, Hippocrates' "beginning premisses" are *lemmas*, whose proofs are given or assumed, *not* "first-principles" in the Aristotelian axiomatic sense (pp. 330f). Presumably, if the mathematicians had advanced too far in this direction so early, one might propose them as rivals to the Eleatics in the initiation of foundational inquiries. Similarly, Szabó points out that Hippocrates does not use indirect reasoning in the fragment (p. 331) – without, we should observe, recognizing the inappropriateness of such reasonings for this subject-matter; while he leaves the issue open, for want of explicit documentation, as to whether Hippocrates knew of this method, he clearly wishes to deny to the mathematicians priority in its use. Yet his criticism of Toeplitz' | reconstruction of a direct "inductive" proof of the 182
area of the circle is based on the possibility that Hippocrates could have applied some form of the "exhaustion method". This latter technique, ascribed to Eudoxus and applied throughout *Elements* XII, is especially characterized by its use of indirect reasoning. Szabó's intent here is to attack Toeplitz' thesis of the "Platonic reform" of geometry, for Szabó views the adoption of formal methods in geometry as a direct response to the Eleatics much before Plato. It is clear that Szabo has shifted his positions on the interpretation of Hippocrates' work, without regard to consistency, in order to gain a favorable argumentative stance as context recommends.

23 See Heath, *HGM* I, pp. 221f, 327–329.

24 *Ibid*., pp. 198f.

25 Proclus, *In Euclidem*, p. 213.

26 On the nature of this method and the extensive scholarship on it, see J. Hintikka and U. Remes, *The Method of Analysis*, Dordrecht, 1974, esp. ch. I.

27 I do not here claim that Szabo presumes to maintain this.

28 *Prior Analytics* II, 25.

29 *In Euclidem*, p. 66.

30 *Evolution of the Euclidean Elements*, ch. VI/IV.

31 On the early history of the study of incommensurables, see my *EEE*, ch. II. As far as the earliest discoveries are concerned, I argue a dating not much before 420 B.C. and a method involving a form of "accidental" discovery rather than any deliberate formal proof (e.g., as in the indirect argument based on the odd and even). The possibility – indeed, the likelihood – that these early discoveries were accidental in some such way undercuts any attempt, such as Szabo's, to use the *formal* character of the study of incommensurables as a debt to the Eleatics (see *AGM*, I, 12; III, 2).

32 *EEE*, ch. III/III–IV.

33 On Archytas, see Heath, *HGM*, I, pp. 213–216.

34 This view is prominent in B. L. van der Waerden, *Science Awakening*, Groningen, 1954, ch. 5, esp. p. 115.

35 The fragment is preserved in Latin translation in Boethius, *De institutione musica*; see my discussion in *EEE*, ch. VII/I.

36 Cited by Porphyry, *In Ptolemaei harmonica* (cf. H. Diels and W. Kranz, *Fragmente der Vorsokratiker*, 6. ed., Berlin, 1951, 47*B*2).

37 Eutocius, *In Archimedem*, in the commentary to *Sphere and Cylinder* II, 2; cf. Archimedes, *Opera*, ed. J. L. Heiberg, III, Leipzig, 1915, pp. 84–88 and Heath, *HGM*, I, pp. 246–249.

38 *AGM*, III. 20. We return to the use of motion in the Greek geometry below.

39 See my *EEE*, ch. II and note 31 above.

40 *AGM*, III. 30, esp. pp. 446f. Becker develops his view in 'Die Lehre vom Geraden und Ungeraden ...', *Quellen und Studien*, 1936, 3:B, pp. 533–553. While here subscribing to Becker's reconstruction, Szabo had earlier dismissed it on the grounds that knowledge of certain *results* (such as the properties of odd and even numbers presented in *Elements* IX) does not in itself justify ascribing to the Pythagoreans comparable formal *proofs*. As elsewhere Szabó here adopts inconsistent positions as context demands (cf. note 22 above).

41 *AGM*, III. 10, 13, esp. pp. 329, 341f.

183 42 | *AGM*, III. 17–20 (on the postulates), 21–25 (on the axioms). Heath covers this material in considerable detail in *Euclid's Elements*, 2. ed., Cambridge, 1926, I, ch. IX. We will take up the question of Euclid's relation to Aristotle on the classification of the "first-principles" in the second part of this paper.

43 *AGM*, III. 20, 29.

44 *In Euclidem*, pp. 185–187.

45 Indeed, it is accepted by P. Bernays in his commentary to the paper by Szabó cited in note 2 above. S. Demidov also raises this view in his commentary on the present paper.

46 Archytas: see note 37 above. Eudoxus *et al.*: Heath, *HGM*, I, p. 255 and I. Thomas, *History of Greek Mathematics*, London, 1939, I, pp. 262–266, 388. Archimedes, *Spiral Lines*, preface, definitions (preceding prop. 12), and prop. 1, 2, 12 and 14. Motion-generated curves (e.g., "quadratrix", "conchoid"): Heath, *op. cit.*, pp. 238–240, 260–262. Related to this is the solution of problems by means of *neuses* ("inclinations", that is, constructions involving a sliding ruler); see Heath, *op. cit.*, pp. 235–238, 240f, and my article, 'Archimedes' *Neusis*-Constructions in *Spiral Lines*', *Centaurus* **22** (1978), pp. 77–98.

47 Heath, *Euclid*, I, pp. 224ff.

48 Such a recasting of the proof was recommended by Russell; cf. Heath, *ibid.*, pp. 249ff.

49 For instance, the congruence of circular segments subtending equal arcs in equal circles would follow from an "exhaustion" proof based on the triangular case.

50 Archimedes: *Plane Equilibria I*, Axiom 4 and prop. 9, 10 (*aliter*); *Conoids and Spheroids*, prop. 18. Pappus employs superposition in his proof of the proportionality of arcs and sectors in equal circles (*Collection* V, 12).

51 *In Euclidem*, pp. 188–190, 249f (citing Pappus).

52 *Posterior Analytics*, I, 13.

53 For a review of the *Elements* and its relation to the pre-Euclidean studies, see my *EEE*, ch. IX.

54 A. Seidenberg, 'Did Euclid's *Elements*, Book I, Develop geometry axiomatically?' *Archive for History of Exact Sciences* **14** (1975), 263–295.

55 The literature on the Euclidean and Aristotelian divisions of the first-principles is rather large. In addition to the studies by Heath and Szabó cited in note 42 above and to Barnes' notes on the *Posterior Analytics* (cf. note 14), the following may be considered: H. D. P. Lee, 'Geometrical method and Aristotle's account of first principles', *Classical Quarterly* **29** (1935), 113–124; B. Einarson, 'On certain mathematical terms in Aristotle's logic', *American Journal of Philology* **57** (1936), 33–54, 151–172; K. von Fritz, 'Die APXAI in der griechischen Mathematik', *Archiv für Begriffsgeschichte* **1** (1955), 13–103. J. Hintikka discusses this question in his paper in the present volume. Recent contributions include B. L. van der Waerden, 'Die Postulate und Konstruktionen der frühgriechischen Geometrie', *Archive for History of Exact Sciences* **18** (1978), 343–357. On constructions and existence-proofs, see note 70 below.

56 *In Euclidem*, pp. 178–184.

57 Cf. Szabó, *AGM*, III. 17, 20, cf. 21–26.

58 For instance, opening *Plane Equilibria I*, Archimedes "postulates" properties of | equilibrium 184
and centers of gravity. Proclus would prefer that he have used "axiom" in this context (*In Euclidem*, p. 181).

59 Proclus, *In Euclidem*, pp. 77–78.

60 Surely it is this wide applicability which accounts for the designation of these principles as "common", both by Aristotle and by Euclid; cf. the passages discussed by Heath, *Mathematics in Aristotle*, pp. 53–57, 201–203. Admittedly, Aristotle also describes the "axioms" as impossible to be mistaken about (*Metaphysics* 1005*b*11–20), but yet indemonstrable (*ibid*., 1006*a*5–15). Apparently this has given rise to the alternative view that these "common notions" are "common *to all men*", as suggested in Heiberg's rendering of Euclid's *koinai ennoiai* as *communes animi conceptiones* (Szabó also subscribes to this view; cf. *AGM*, III. 25).

61 See the contributions cited in note 55.

62 See note 60.

63 One may note that the attempt, as by Szabó, to assign Euclid a dialectical motive for both his "postulates" and his "common notions" would impute to him an even more extreme dialectical position than that suggested by Aristotle; for this would make Euclid's *reason* for articulating the "common notions" the desire to oblige those extremist critics who might challenge not only the "postulates" (some of which might, after all, appear to permit of proof) but also the "axioms" which seem self-evident.

64 I will review below the familiar thesis of Zeuthen, that constructions served the role of existence proofs in the ancient geometry. One should note well that the modern conception of proofs of existence in mathematics and logic has been greatly influenced by developments in the fields of analysis and set theory in the late nineteenth century. There is thus a real danger of interpreting early geometry anachronistically in the case of such questions.

65 Proclus gives an account of the formal subdivision of a theorem and its proof: *In Euclidem*, pp. 203–205; cf. Heath, *Euclid*, I, pp. 129–131.

66 Indeed, there are many such steps in the ancient geometry which passed by without being then recognized as tacit assumptions; cf. the discussion by Becker cited in note 72.

67 Typically, auxiliary elements may be introduced in the "construction" (*kataskeue*) later in the proof; cf. Hintikka's discussion of auxiliary constructions in the book cited in note 26.

68 76*a*31–35. I have since come upon the article by A. Gomez-Lobo, 'Aristotle's hypotheses and the Euclidean postulates', *Review of Metaphysics* **30** (1977), 430–439 in which the same view is argued in detail.

69 These postulates assert (1) that the line segment connecting two given points may be drawn; (2) that a given line segment may be indefinitely extended in a straight line; and (3) that the circle of given center and radius may be drawn. The remaining two postulates do not involve construction as such: (4) that all right angles are equal; and (5) that if two lines are cut by a third such that the interior angles made on one side of the third are less than two right angles, then the given lines, extended sufficiently far, will meet on that side of the third.

70 The primary statement of this thesis is by H. G. Zeuthen, 'Geometrische Konstruktion als Existenzbeweis ...', *Mathematische Annalen* **47**, (1896), 272–278. It has since been elaborated
by O. Becker, *Mathematische Existenz*, Halle a. d. S., 1927; A. D. Steele, | 'Ueber die Rolle von 185
Zirkel und Lineal in der griechischen Mathematik', *Quellen und Studien* **3:B**, (1936), 288–369; and E. Niebel, '... die Bedeutung der geometrischen Konstruktion in der Antike', *Kant-Studien*, Ergänzungsheft, **76**, Cologne, 1959. I criticize Zeuthen's view in *EEE*, ch. III/II (cf. also note

64 above). Van der Waerden and Seidenberg also deny the existential sense of Euclid's postulates (cf. notes 54 and 55 above).

71 Even proofs of incommensurability do not assume the form of negative existence theorems (i.e., "there exist no integers which have the same ratio as given magnitudes"). In this regard, the closest one comes to an existential expression is in *Elements* X, 5–8; e.g., "commensurable magnitudes have to each other the ratio which a number has to a number" (X, 5). Cf. also X, Def. 1: "magnitudes are said to be commensurable when they are measured by the same measure, but incommensurable if none can become (*genesthai*) their common measure". In such instances an existential statement is hardly to be avoided. Even so, it is remarkable how these formulations emphasize the *properties* of assumed magnitudes, rather than the *existence* of magnitudes having these properties.

72 These issues are examined by O. Becker, 'Eudoxos-Studien I–IV', *Quellen und Studien* **B:1–3**, 1933–36 (esp. II and III).

73 For a survey of these constructions, see Heath, *HGM*, I, ch. VII and the items cited in note 46 above.

74 For instance, Hippocrates' third lunule (Heath, *ibid.*, p. 196) and Archimedes' construction in *Spiral Lines*, prop. 5, are both effected by *neusis*, even though a "Euclidean" construction is possible; this is discussed in my article, cited in note 46 above.

75 Pappus, *Collection* IV, 36; cf. my article on *neusis* (note 46 above).

76 Aristotle set out a doctrine of the "priority" of principles in the *Posterior Analytics* **I**, 6–7.

76a In the *Method* Archimedes reduces the problem of the volume of cylindrical section to that of the area of the parabolic segment. Doubtless, he knew of the equivalence of the problems of finding the area of the parabola and that of the spiral; but this notion is used in neither of the treatises devoted to these problems and does not seem to have appeared before Cavalieri's *Geometria Indivisibilibus* (1635). The reduction of the area of the spiral to the volume of the cone is employed in Pappus' *Collection* IV, 22 in a treatment which appears to stem from Archimedes (see my article cited in note 85). The center of gravity of the paraboloid was solved by Archimedes in a lost work *On Equilibria* and applied in the extant work *On Floating Bodies*; see my discussion in 'Archimedes' lost treatise on centers of gravity of solids', *Mathematical Intelligencer* **1**, (1978), 102–109. In this instance, it seems that Archimedes used an independent summation-procedure based on the same expressions proved in *Spiral Lines* and *Conoids and Spheroids*.

77 I. Toth, 'Das Parallelproblem im Corpus Aristotelicum', *Archive for History of Exact Sciences* **3**, (1967), 249–422.

78 It is not even clear whether among the Eleatics the arguments on the nature of being were intended as a possible basis for a cosmological system, or whether they were merely a negative device for the criticism of other cosmologies.

186 79 See my *EEE*, ch. VII. Theaetetus' role in the foundation of number theory was | argued by H. G. Zeuthen, 'Sur les connaissances géométriques des Grecs avant la réforme platonicienne', *Oversigt Dansk. Videns. Sels. For.*, 1913, pp. 431–473.

80 I present the reconstructed "Eudoxean" theory in 'Archimedes and the Pre-Euclidean Proportion Theory', *Archives internationales d'histoire des sciences*, **28** (1978), 183–244. The anthyphairetic theory was first proposed by Zeuthen and by Toeplitz, later elaborated by O. Becker, 'Eudoxos-Studien I' (see note 72). I review and modify his reconstruction of this theory in *EEE*, ch. VIII/II-III and Appendix B.

81 See, for instance, the circle-quadratures of Antiphon and Bryson (Heath, *HGM*, I, pp. 220–226). Democritus conceived solids as somehow constituted of parallel indivisible plane sections (*ibid.*, pp. 179f). The latter has been construed (I believe, mistakenly) to be the basis of an ancient infinitesimal analysis; cf. S. Luria, 'Die Infinitesimaltheorie der antiken Atomisten', *Quellen und Studien* **2:B**, (1933), 106–185.

82 On Archimedes' early works and their dependence on pre-Euclidean sources, see my 'Archimedes and the *Elements*', *Archive for History of Exact Sciences* **19**, (1978), 211–290 and the article cited in note 80.

83 For a review of criticisms of this work, see E. J. Dijksterhuis, *Archimedes*, 1957, ch. IX and my articles cited in notes 80 and 82 above.

84 See, for instance, the definitions given in *Spiral Lines* and *Conoids and Spheroids* (prefaces).

85 See my article cited in note 82 above and also 'Archimedes and the spirals', *Historia Mathematica* **5**, (1978), 43–75.

86 Among the Archimedean treatises only *Quadrature of the Parabola* adopts this usage.

87 See Apollonius, *Conics* I, preface and Hypsicles, *Elements* XIV, preface.

88 Pappus, *Collection* VII, ed. Hultsch, pp. 680–682.

89 In certain fields outside the area of formal geometry notable advances *were* made in later antiquity: for instance, plane and spherical trigonometry, mathematical astronomy and number theory (in the Diophantine tradition). Each of these was associated with techniques of practical computation, rather than theoretical geometry, and none was systematized along the lines of Euclid, Archimedes and Apollonius. The non-axiomatic character of much ancient mathematics and mathematical science is examined by P. Suppes in his contribution to this volume and by F. Medvedev in his commentary.

PART 2

STUDIES ON GREEK GEOMETRY

Texts selected and introduced by Reviel Netz

REVIEL NETZ

INTRODUCTION

In the view of this editor, the twentieth century made two major contributions to the study of Greek geometry. One contribution looked back to the nineteenth century, the other, perhaps, looks forward to the twenty-first century.

1. The work begun in the nineteenth century, of editing the works of Greek geometry, was considerably extended.
2. Towards the end of the century, a new approach to the field gradually took over. This may perhaps serve as basis for work in the twenty first century.

Among works of the first kind, one can mention Heiberg's discovery of *The Archimedes Palimpsest* and the consequent publications Heiberg (1907), (1910-1915). A fundamental textual study, once again having to do with the Archimedean tradition, was Clagett (1964-1984). Several Greek geometrical works were re-edited, sometimes reaching higher standards than those of the preceding century: one can mention, for example, Jones (1986). Most important, the field of textual studies of Greek geometry in Arabic has been much further explored. Especially significant in this respect is Toomer (1990). A lot still needs to be done in this particular field. As is obvious, this kind of development is best seen in monographs and editions and therefore is not represented in this selection.

I concentrate instead on the conceptual change introduced into the field during the last quarter-century. As will be seen in several other sections, this was a period of transition. In effect, a new departure was made from an old understanding of the field – one that had lasted for centuries, arguably ever since Late Antiquity itself.

Ever since the revival of Greek geometry as part of liberal education in Late Antiquity, the fundamental interest in the field derived from its connection to the main tradition of Greek philosophy and in particular from its connection to Plato. Thus the philosophical significance of Greek geometry took on a larger role. Perhaps as a consequence, more interest was focused on the relatively more accessible work of Euclid. Further, and especially during the twentieth century – when philosophy of mathematics dealt mostly with the question of mathematical 'Foundations' – the focus of interest was even narrower. The key question came to be Euclid's formulation of the *Elements* as an axiomatic system. Finally, since it was always recognized that Euclid's work must have had (lost) antecedents, another natural route for the historiography was to study the historical *Evolution of the Euclidean Elements* – the title of Knorr (1975), which may be seen as the zenith of this long historiographical tradition.

To sum up: the study of Greek geometry focused upon Euclid, particularly upon his axiomatic method; much attention was paid to the interaction between geometry and philosophy, as well as to Euclid's antecedents.

Over the last few decades, as the history of Greek geometry came to be studied apart from that of Greek philosophy, a new awareness took hold. Euclid is seen as an

J. Christianidis (ed.), Classics in the History of Greek Mathematics, 113–114.

author among many, perhaps not as central as later tradition made him to be. The relationship between geometry and philosophy is seen to be more complicated and indirect. The gap in sources for pre-Euclidean mathematics makes many scholars look not backwards from Euclid but, instead, forwards, trying to understand Greek mathematics through the evidence of the extant authors. The main focus of interest is in how Greek geometrical methods were developed for specifically Greek geometrical purposes. The works mainly studied are now those of Euclid, Apollonius and Archimedes, their minor (or lesser-known) followers, and their commentators. (For a survey of this development, see Saito (1998).)

The three articles selected here exemplify this new approach. Knorr (1983) shows how Greek geometrical constructions should be understood as serving specifically geometrical purposes, rather than serving a 'foundational' purpose motivated by philosophical considerations. Saito (1985) reassesses Euclid and shows how the *Elements* make most sense as serving a specific geometrical purpose, this time seen through the application of Euclidean results in Apollonius' *Conics*. Lloyd (1992) approaches the geometry/philosophy interaction from the philosophical side, showing how a source that was often used for the reconstruction of pre-Euclidean geometry is actually meant to serve a philosophical purpose, throwing doubt on its use as source for the history of geometry itself. Taken together, the three articles suggest what may be the most useful question: what were Greek geometrical methods supposed to achieve in the context of Greek geometry itself?

Stanford University

REFERENCES

Clagett, M.: *Archimedes in the Middle Ages*, 5 vols. Madison WI, Philadelphia PA, 1964-1984.
Heiberg, J.L.: "Eine Neue Archimedesschrift", *Hermes* 42 (1907), pp. 234-303.
Heiberg, J.L.: *Archimedes/Opera*, 3 vols. Leipzig, 1910-1915.
Jones, A.: *Pappus of Alexandria. Book 7 of the Collection*. New York, 1986.
Knorr, W.: *The Evolution of the Euclidean Elements*. Boston, 1975.
Knorr, W.: "Construction as Existence Proof in Ancient Geometry", *Ancient Philosophy*, 3 (1983), pp. 125-148.
Lloyd, G.E.R.: "The 'Meno' and the Mysteries of Mathematics", *Phronesis* 37 (1992), pp. 166-183.
Saito, K.: "Book II of Euclid's Elements in the Light of the Theory of Conic Sections", *Historia Scientiarum*, 28 (1985), pp. 31-60.
Saito, K.: "Mathematical Reconstructions Out, Textual Studies In: 30 Years in the Historiography of Greek Mathematics", *Revue d'histoire des mathématiques*, 4 (1998), pp. 131-142.
Toomer, G.J.: *Apollonius. Conics: Books V to VII. The Arabic Translation of the lost Greek Original in the version of the Banu Musa*. New York, 1990.

WILBUR R. KNORR

CONSTRUCTION AS EXISTENCE PROOF IN ANCIENT GEOMETRY

It is better to be than not to be.
[Aristotle, *Generation and Corruption* II. 10, 336b29]

The title of this essay is borrowed from a modern mathematical historian; its tag line is taken from an ancient philosopher. Their shared interest in questions dealing with existence has given rise to a familiar thesis about ancient geometry: that its constructions were intended to serve as proofs of the existence of the constructed figures. I propose here to examine that thesis, to argue its weakness as a *historical* account of ancient geometry, and to offer an alternative view of the role of problems of construction:[1] that constructions, far from being assigned a specifically existential role, were not even the commonly adopted format for treating of existential issues when these arose: that some central questions relating to existence were handled through postulates or tacit assumptions, rather than through explicit constructions; that, by contrast, when constructions were given, the motive lay in their intrinsic interest for the ancient geometers. On this basis I will maintain that preconceptions based on modern theories have interfered in the modern effort to interpret ancient mathematics, thus attaching to the existential view of constructions a greater credence than the ancient evidence could justify.

THE EXISTENTIAL ROLE OF CONSTRUCTIONS

Questions of existence were a prominent concern among the ancient philosophers, many of whom appreciated how their general views on existence related to discussions of the nature of mathematics. Mathematical examples are frequent in the writings of Plato and Aristotle, and one can retrieve from them the outlines for theories on the nature of the existence of mathematical entities.[2] Earlier, Zeno's notorious puzzles on the nature of magnitude and motion brought these concepts within the purview of dialectic, calling attention to the difficulties inherent in providing a logical account for these and other terms fundamental within mathematical studies.[3] In principle, one seemed able to challenge the validity of even the most obvious existential assertions.
In this spirit, the neo-Platonist commentator Proclus ascribed to | Euclid an existential 126
motive for choosing as the first proposition of Book I of the *Elements* the problem of constructing the equilateral triangle on a given line segment:[4]

> For unless he had previously [*sc.* to the theorem I 4] shown the existence of triangles and their mode of construction, how could he discourse about their essential properties? ... Suppose someone, before these have been constructed, should say: "If two triangles have this attribute, they will necessarily also have that." Would it not be easy for anyone to meet this assertion with, "Do we know whether a triangle can be

J. Christianidis (ed.), Classics in the History of Greek Mathematics, 115–137.

> constructed at all?" ... It is to forestall such objections that the author of the *Elements* has given us the construction of triangles .... These propositions are rightly preliminary [to the theorems about the congruence of triangles.][5]

Proclus thus subscribes to a view familiar today in discussions about ancient geometry: that the solution of problems of construction is intended to establish the existence of configurations satisfying conditions required for the proofs of subsequent theorems. Euclid's motivation for the existence proof is here alleged by Proclus to derive from a concern for rigor rooted in philosophical, indeed dialectical, considerations.

The modern formulation of this view is associated with the Danish mathematician-historian H.G. Zeuthen whose paper on the ancient geometric constructions appeared in 1896 and has been much cited since then.[6] Zeuthen, like Proclus, derived his principal evidence for the existential role of constructions from Euclid's Book I. In Euclid's scheme, problems of construction are introduced as preliminaries to other problems or theorems in which their result is used. For instance, the problem in prop. 2, of no intrinsic interest, supplies a construction needed for the following problem, the construction of the difference of given line segments. Moreover, and perhaps more important for Zeuthen's view, problems serve as auxiliaries to subsequent *theorems*.[7] He maintains, for instance:

> Euclid doesn't dare use the midpoint of a segment in a proof (I 16) before he has demonstrated the existence of this point through its construction (in I 10).[8]

Unlike Proclus, however, Zeuthen explains this caution over existential issues through technical, rather than dialectical reasons. In particular, the Pythagorean discovery of irrationals, that is, the discovery that there exist no ratios of integers which express exactly the ratios of certain geometric terms, like the side and diagonal of the square, served as an object lesson compelling geometers to provide explicit proofs of existence of the terms admitted in their demonstrations. For after this, it would be realized that a simple arithmetic assumption could not secure the existence of such terms; to do this their geometric construction would be necessary.

Further, for some problems it may happen that solutions do not in fact exist; one thus includes a *diorism* (διορισμός) or statement of the condition which must be satisfied by the given terms of the problem in order that it be solvable. In effect, the *diorism* sets out necessary conditions for the solution, while the construction in the problem itself reveals sufficient conditions.[9] Zeuthen cites from the *Elements* the problem in I 22 (construction of a triangle whose sides shall equal three given line segments), where the necessity of an auxiliary condition (namely, that none of the given
127 lines | shall be greater than the sum of the remaining two) has been established in a prior theorem (I 20). Additional examples are cited from Book VI and from Archimedes.[10] Concluding his essay, Zeuthen proposes that attention to existential aspects helps resolve two "puzzles," one relating to the manner of Euclid's formulation of the "parallel postulate,"[11] the other concerning the original motivation behind the construction of the conic sections.[12]

Before examining his argument, a further clarification should be given of the constructive aspect of ancient geometry. The ancients divided geometric propositions into two kinds: theorems and problems. This formal distinction appears in ancient philosophical discussions related to geometry,[13] but it is fixed as well in the terminology used in the technical treatises. Invariably, a *problem* is stated in the form "to construct ... ," "to draw ...," "to find ...," or the like, the project being to produce a configuration for which a specified property related to "given" points, lines, planes, numbers, ratios, and so on, holds. In a *theorem*, the construction of the figure is given by hypothesis, and the project is to establish that a specified property holds in its case; accordingly, theorems will typically, if not always, take on a conditional format: "if ... , then. ..." The distinction is largely artificial, in that any problem can be reformulated as a theorem,[14] and *vice versa*. Nevertheless, the ancient geometers somehow intended a real distinction between these two kinds of proposition, where the format of the problem highlighted the activity of construction. It follows that for Zeuthen, the distinct role of the problem would be for establishing existence. Modern logical notation predisposes one to the same view, seeing that any theorem will be framed in terms of a universal quantifier, whereas a problem will be framed in terms of an existential quantifier. This seems further supported by the fact that the infinitive in the statement of problems frequently is governed by the expression "it is possible" (δυνατόν ἐστιν), or words comparable. Since "it is possible to construct an x such that ..." logically entails that "there exists an x such that ... ,"[15] the existential character of the problem results almost tautologically.

These considerations support Zeuthen's thesis as a valid account of the *essence* of geometric problems. Zeuthen, however, intends it as a *historical* account of the geometers' view of their own technical efforts. Doubtless, the ancients recognized the logical connection between possibility and existence. But this of itself does not entail an existential intent behind their constructions. Indeed, this logical connection complicates the project of testing Zeuthen's view, for in reading the ancient evidence, one might easily be led to impute an existential intent where it need not have been. For instance, by providing *diorisms* – which are not themselves problems, but rather conditions of the form "it is necessary that ..." attached to the statements of problems (where the necessity claimed is known from previous theorems) – the geometers are certainly aware of the need to assure that the problem can be solved; but we beg the question if we take this as evidence of their intent to establish the existence of the solution by solving the problem.

Thus, Zeuthen's point relating to *diorisms* holds, only if the ancient manner of treating *diorisms* indicates in some special way a concern over existence. To discover whether this is the case, we may consider the example cited by Zeuthen: the problem associated with Archimedes' division of the sphere in *Sphere and Cylinder* II 4. In this problem, as throughout this book, and in many other places in the ancient corpus | of 128
geometry, the solution is effected *via* the method of "analysis."[16] In this method, the figure to be produced is initially hypothesized to have been effected ("let it be done," γεγονέτω) and other configurations are derived from this until one has obtained a figure whose construction is already known. In the present case, the principal problem is to divide a given sphere into segments whose volumes are in a given ratio. Archimedes assumes the division to have been effected, and then reduces it to another problem: that of dividing a given line segment $2a$ into segments x, $2a-x$, such that $(2a)^2:(2a-x)^2=(x+a):c$, for $c < a$. The solution of the latter problem is here assumed, on the basis of a separate treatment Archimedes claims to have appended to the main work, but which was somehow lost in its later transmission. The commentator Eutocius, however, was able to locate a copy of the promised solution and includes its full text with his own editorial additions in his commentary on *Sphere and Cylinder*.[17]

As for the main problem, the solution of the derived problem follows the analytic method. Archimedes first generalizes the latter, seeking the division of a given line segment a into segments x, $a-x$, such that $a-x:c=b^2:x^2$, for given segments b, c.[18] Assuming this division to have been effected, he reduces it to a configuration depending on the point of interesection of a given parabola and a given hyperbola. As a result of this reductive sequence, one also perceives the appropriate condition for the *diorism*.[19] Archimedes states this in the form that $b^2 \cdot c \leq (a/3)\cdot(2a/3)^2$, without explaining its derivation; but it follows readily from consideration of whether the two curves actually meet: the limiting case, where they intersect only at a single point of mutual tangency readily yields the condition stated in the *diorism*, for there $x=2a/3$. Archimedes now turns to the actual construction of the solution of the division, which according to the standard procedure is in the form of a "synthesis." Here, one begins with the derived configuration and goes through the reverse sequence until the desired configuration is obtained.[20] The problem ends with the proof of a theorem, to the effect that the figure so constructed has the stated property. Archimedes finally turns to the unfinished work of the *diorism*, proving in an appended theorem that of all the values for the product $x^2 \cdot (a-x)$, that for which $x=2a/3$ is maximal. His procedure here is entirely synthetic, and he does not attempt to explain his manner of deriving the condition.

At first glance, the analytic method would seem to highlight the existential aspect of problems; for it hinges on the hypothesis that the construction has already been effected, that is, one might say, that the solution exists. Pappus and other ancient commentators, in their general accounts of this method, do indeed describe the analytic hypothesis as the assumption "of the sought *as being* (ὡς ὄν)."[21] But in the technical literature, no such existential terms are used; it is not the existence, but rather the properties of the figure which are at issue. That is, the reasoning takes the form, not "supposing x exists, therefore y exists," but rather "supposing x has property A, therefore y has property B." The objective of the analysis is to deduce from the hypothesis a configuration whose construction is known. An existential formulation would be: "... therefore z has property C; and it is known there exists z having property C." But in the ancient terminology these derived terms are never said "to exist," but always "to be given." Thus, the tradition has failed to exploit a ready opportunity for revealing its existential purposes. Further, when Archimedes introduces the *diorism* and here recognizes three cases (the product of the givens being either greater than,
129 equal | to, or less than the maximal product), he could have adopted an existential phrasing, by noting that in the first case, no solution exists; but in the second and third cases, a solution does exist. But Archimedes does not formulate these alternatives in terms of the *existence* of solutions, but rather in terms of whether or not the problem "will be synthesized" (συντεθήσεται). This is the common diction in the technical literature; the fact that the ancient tradition so clearly misses this further opportunity to cast the analytic method in an existential mode again weakens Zeuthen's position.

There is a similar risk of circularity in claiming that Euclid's ordering of problems as adjuncts to theorems reveals the existential role intended for the problems. In those cases where the problem has no evident interest in its own right, but merely justifies the introduction of the term in a subsequent theorem, this would seem to be a plausible view. Such is the case with the problem in I 2, cited by Zeuthen, for its clear design is to complete the construction in I 3. Other instances, not cited by Zeuthen, can be proposed: the problem in XII 16 (to inscribe in a given circle a polygon which does not meet a second circle lying inside the first and concentric with it) is used in XII 17 (to inscribe a polyhedral solid in a given sphere so that it does not meet a second sphere inside the first and concentric with it); the latter in turn is necessary for the construction employed in the proof of the next theorem, XII 18 (that spheres are as the cubes of their diameters). The two auxiliary problems are unlikely to have interest outside the context of the proof of this theorem or others like it. Closely related lemmas may be cited from Archimedean materials reported by Pappus.[22]

But to acknowledge that there exist *some* examples of problems which lend themselves to the existential interpretation does not entail that the same view obtains for *all* problems of construction. The project of examining the entire body of ancient problems is effectively open-ended, since the number of problems in the extant technical literature runs into the hundreds.[23] But even the few examples cited by Zeuthen fail to make his point decisively. The interest which attaches to the problem of bisecting the given line segment (I 10), for instance, seems not restricted to its reappearance in the proof of the theorem which follows six propositions later (I 16: that the exterior angle of a triangle is greater than either of the two nonadjacent interior angles). All sixteen propositions of Euclid's fourth book are problems of construction (e.g., inscribing and circumscribing circles to given triangles; and constructing regular polygons). Similarly, six of the nine propositions in Book II of Archimedes' *Sphere and Cylinder* are solutions to problems. Comparable sequences of problems can be cited from Apollonius' *Conics*.[24] This same geometer is reported to have composed an entire treatise on the solution of plane loci, a form of construction problem, and similar treatises are attributed to Eratosthenes and Aristaeus.[25] A characteristic example of problem-solving procedure is preserved by Pappus: a solution of the problem of trisecting the angle is reduced to another problem, the insertion (νεῦσις) of a given line segment between two given lines so as to incline (νεύειν) toward a given point; the latter problem is reduced in turn to the construction of a hyperbola passing through a given point and having two given lines as asymptotes.[26] The sequence can be continued: for the problem of the hyperbola is reduced to that of constructing a conic section of given parameters, solved in Apollonius' *Conics* (I 54). One can argue that the problems of Euclid's Book I form a similar reductive sequence, ultimately founded on the three postulates of construction. The postulates themselves are stated in the form | of *problems* (e.g., *to draw* the line which passes through two points), without 130
overt indication of an existential intent.[27] In assigning an existential motive to any problem of this type, one thereby assigns the same motive to the entire reductive sequence. But the converse also holds: that if the specific manner of the construction of any one of the problems in such a sequence – in particular, the last – is of intrinsic interest, then the same becomes true of the entire sequence. In the case of such a sequence, the imposition of an additional existential motive would become superfluous and farfetched.[28]

But let us suppose for argument's sake that an existential motive could be imputed to such a sequence of problems. Presumably, then, the intent would be like that supposed by Proclus: to forestall dialectical challenges against the existence of the constructed entities. The problems in Euclid's Book XIII provide an interesting context for testing this approach. In XIII 13-17 Euclid constructs the five regular solids, beginning with the tetrahedron and ending with the dodecahedron.[29] We can readily reconstruct what an existential strategy for the presentation of these constructions would be. First, one would show that the number of possible regular solids is limited to five. This entails considering the ways in which three or more regular polygons can meet at a common vertex such that the sum of the angles around that vertex is less than four right angles. Only five configurations are possible, as is proved by Euclid in an appendix to XIII 18: arrangements at each vertex of three, four or five equilateral triangles, or three squares, or three regular pentagons. Accordingly, we could assign to the five solids the names 3-triangle-hedron, 4-triangle-hedron, 5-triangle-hedron, 3-square-hedron, and 3-pentagon-hedron.[30] The constructions would then indicate that each of these configurations can in fact be realized as a solid. An existential approach would have a certain appropriateness, since, for instance, it seems hardly obvious that the mere possibility of configuring five equilateral triangles about a point could be the principle for a regular solid having all of its vertices of this form.

But Euclid adopts an entirely different strategy in Book XIII. The definitions of the solids are prefaced to Book XI, where they are given the familiar names of pyramid, cube, octahedron, icosahedron and dodecahedron; their definitions are in terms of the character and number of the plane faces which bound them: that is, the four triangles, six squares, eight triangles, twenty triangles or twelve pentagons which form them. Thus, Euclid assumes from the start that we have a working conception of these solid forms. The constructions which follow in Book XIII thus cannot be intended to establish their existence, but rather to indicate how actually to produce them.[31]

These examples make clear that the choice by Euclid and other geometers to treat such matters as problems of construction indicates an intent other than that of proving existence. The form, comprehensiveness and efficiency of the construction are factors of real interest, despite their irrelevance to the existential issue. Although a few examples do lend themselves to an existential interpretation, the overwhelming majority of problems do not; for these, the constructing effort seems to engage the geometer for its intrinsic interest.

WHEN EXISTENCE IS THE ISSUE

Although problems of construction cannot be limited to the role of furnishing exis-
131 | tence proofs, the converse question can be raised: how the ancient geometers addressed issues of existence. Is it possible that they resorted to the specific form of problems when it became necessary to justify the introduction of entities whose existence might somehow be in doubt?

A series of propositions from Euclid's *Optics* is quite interesting in this regard. Prop. 37 asserts that "there is a place (τόπος) where, when the eye remains fixed but the object seen is transposed, the object seen will always appear equal."[32] A similar assertion is made in prop. 38, where the eye is shifted while the object seen remains fixed. In both cases, a geometrical locus is identified (namely, where the object, conceived of as a linear magnitude, or the eye, conceived of as a point, moves along a circle). These propositions are overtly existential, and include constructions within their demonstration. But it is difficult to construe them as problems in form. Indeed, the very next set of propositions, from 39 to 43, are presented as theorems.[33] Euclid might just as easily have adopted the existential format of props. 37 and 38 for this second set. In fact, the analogue of prop. 42 in Theon's recension of the *Optics*

(Theon's prop. 43) does precisely this. Euclid's next set, prop. 44–47, returns to the initial existential format; but it is followed by a problem, 48: "to find places from which the equal magnitude shall appear half or a fourth part or, in general, in the ratio in which also the angle is divided."[34] This problem does not depend on the preceding constructions, nor, clearly, does it furnish an existential support, either for them or for subsequent problems or theorems.[35] In the strict sense, prop. 48 is not even constructive; for it depends on the division of the angle into parts in any ratio.[36] That is, if the given magnitude (e.g., line AB) is seen under the angle K, one requires an angle D such that D is the given part of K, in order to specify the places where AB appears reduced in the same ratio (namely, the circular arc ADB). This difficulty seems more clearly present in prop. 49 which asserts the same as prop. 48, but employs an alternative wording and proof:

> let there be a seen magnitude, AB; I say that AB has places in which, when the eye is set there, the same appears sometimes half, sometimes whole [?], sometimes one-fourth, and in general in the given ratio.[37]

Since both versions of the construction deal only with the case of the half, the non-constructive aspect of the generalization is missed. The difficulty would be removed, if we could understand prop. 48 as referring only to those parts of the given angle which are constructible; or if prop. 49 assumed in its hypothesis two given angles and so produced the corresponding loci from these. But neither version, in the extant wordings, escapes the difficulty.

The variety of formats adopted in these propositions, some as theorems, some as problems, some as existential assertions of locus,[38] prohibits assigning to any one of them a special existential role distinct from the others. Indeed, these overt existential claims, of the form "there exist places ...," in this context may merely be an alternative manner of speaking, without substantive difference from the more usual phrasings of theorems. For the results established in this part of the *Optics* are not applied later in the work. As existence proofs, their intent would presumably be to assist the actual contrivance of situations where these phenomena arise.[39] But such theorems also provide explanations of the phenomena, by setting out their geometric causes; in this | case, the existential format would not be intended literally, and this may help account 132
for Euclid's fluctuation between the different formats in expressing these propositions.

Many problems are not constructible under the restriction of the postulates of Euclid's *Elements*. The most celebrated of such problems are those of the cube-duplication (or, more generally, finding two mean proportionals between two given lines), the angle-trisection and the circle-quadrature. One might wish to account for the persistence of the ancient effort applied toward solving these problems, resulting in some two dozen ingenious and successful methods in the extant ancient literature,[40] as urged by the search for solutions of a certain kind: but one would hardly suppose that any serious geometer ever doubted the existence of the entities sought. In his commentary on Aristotle's *Posterior Analytics* I 9, Philoponus makes a point of distinguishing between the construction of the circle-quadrature and the question of the existence of a solution:

> Those who square the circle did not inquire whether it is possible that a square be equal to the circle, but by supposing that it can exist they thus tried to produce (γεννᾶν) a square equal to the circle.[41]

Similarly, the commentator Eutocius defends Archimedes' procedure in the proof in *Dimension of the Circle* prop. 1, that any circle equals a triangle whose altitude and base equal, respectively, the radius and the circumference of the circle:

> Someone might think that he has used for the proof a fact not yet demonstrated. ... To take a line equal to the circumference of the circle neither has been demonstrated by him yet, nor has it been handed down by anyone else. But one must understand that Archimedes has not written anything beyond what he is entitled to. For it is clear to everyone, I think, that the circumference of the circle is some magnitude, and that this is among those of one dimension; moreover, the line is of the same genus (εἶδος). Thus, even if it seemed in no way possible to produce (πορίσασθαι) a line equal to the circumference of the circle, nevertheless that there really (τῇ φύσει) exists some line equal [to it], this is sought by no one.[42]

Eutocius thus views the existence of the posited line as obvious, even without the provision of its explicit construction. Presumably, Archimedes' willingness to present this theorem, dependent as it is on such an assumption of existence, must commit him to much the same view. This of course does not diminish the interest in discovering a suitable construction; indeed, one emerges in connection with Archimedes' studies of the spiral.[43] But the question of existence is seen to be separate from the finding of a construction.

These nonconstructive assumptions of existence made by Archimedes and the circle-squarers rely ultimately on an intuition of the nature of continuous magnitude. Of the same kind is an assumption characteristic of the limiting technique introduced by Eudoxus, presented in Euclid's Book XII, and extended by Archimedes: in order to prove that a stated proportionality A:B = C:D holds, one hypothesizes the contrary and then posits A:B = C:X, where X will be greater or less than D. The existence of such
133 a fourth proportional term X can be justified by construction when the four terms | are line segments *via Elements* VI 12, and *via* VI 22 when they are four rectilinear areas; further, Euclid sets out the conditions under which the relation is possible for four integers in IX 19. But these results do not amount to a general construction, nor do they cover the cases at issue in the limiting theorems.[44] Although one can modify the technique so that it avoids this nonconstructive assumption, Euclid makes no attempt to do so. Moreover, his procedure is fully representative of the ancient technique of limits, extant in dozens of theorems from Euclid, Archimedes and others. These observations conflict directly with any supposition that the ancient geometers subscribed to a constructivist view of mathematical existence.[45]

An example of this technique has already been mentioned: Euclid's proof of the proportionality of spheres in *Elements* XII 18. Not only does he here assume the existence of the fourth proportional X; that is, A:B = C:X, where A and B are the cubes of the diameters of given spheres C and D, respectively; but also, he conceives X itself to be a sphere. Now, the construction of a sphere X in a given ratio (A:B) to a given sphere C is possible, for it is a corollary to the construction of two mean proportionals between two given lines. The latter problem, however, is not constructible under the restriction of Euclid's postulates; since it would require the introduction of auxiliary conic curves or the equivalent, its construction could not be introduced into the *Elements*. But even with this, the proof that the constructed sphere X satisfies the stated proportionality can be effected only *via* the theorem that spheres are in the ratio of the cubes of their diameters, the very theorem that is being proved in XII 18. It thus becomes clear that any attempt to effect this proof constructively along the lines set out by Euclid would ultimately be circular.

In the Eudoxean and Archimedean techniques, these existential assumptions, although nonconstructive, have a sufficiently obvious character, through intuitions of continuity, that they do not give rise to explicit comment. A conspicuous example of this sort is the assumption made in Euclid's construction of the equilateral triangle in *Elements* I 1 that the two circles drawn in the construction have a point of intersection. Nothing in Euclid's postulates or axioms actually warrants this inference.[46] But any intuition of continuous variation – that a quantity now less than, now greater than a specified magnitude will equal the same within the interval of variation; or that a curve now on one side, now on the other of another curve will intersect it somewhere in between – will render the assumption so obvious, apparently, that it need not even be articulated explicitly as an axiom.[47]

An existential assumption of a less obvious kind plays an important role in the ancient studies of isoperimetric figures. The ancient sources preserve versions of a study by the geometer Zenodorus (c. 200 B.C.) on the properties of isoperimetric figures, that is, of plane figures having perimeters of equal length. The project of his study is to show that the circle is greater than any rectilinear figure which has a perimeter equal to the circumference of the circle.[48] The argument develops in three basic steps: (a) that if two regular polygons have equal perimeters, the one with the greater number of sides has the greater area; (b) that of all rectilinear polygons of a specified number of sides and equal perimeters, the regular polygon has the greatest area; and (c) that the circle is greater in area than any regular polygon having equal perimeter. Within this sequence, steps (a) and (c) are relatively straightforward deductions from
Archimedean theorems on measurement, so that, in my view, this | portion of 134
Zenodorus' argument is likely to have been proposed by Archimedes earlier in the 3rd century B.C.[49] But step (b) relating the polygons of specified number of sides to the regular polygon, demands considerable care, as much for logical as for technical reasons.

Zenodorus' procedure in proving (b) is indirect: he assumes the maximal polygon and then shows that if it is not regular, then one can construct a polygon of greater area having equal perimeter and the same number of sides, so that the hypothesized figure could not be maximal. Zenodorus seems to treat this result as equivalent to the assertion that the regular polygon is the maximal figure. But what he has in fact established is that *if there exists a maximal figure*, it must be regular. This situation is subtly different from that in the Archimedean *diorism* discussed above, in that now the proposed maximal figure is not compared directly with the other cases, since an indirect proof is employed. If, for instance, one could devise a sequence of constructions passing from the arbitrary case to the regular case in a finite number of steps, such that the area was increased at each step, then the existential *proviso* could be removed.[50] But Zenodorus does not do this and, in fact, could not do it under standard constructive restrictions, since the general construction of the regular polygon is a transcendental problem.[51]

There is thus a gap in Zenodorus' argument hinging on the issue of the existence of the maximal figure. While the existence of the maximum might seem intuitively obvious, there are comparable situations where it turns out to be false.[52] The technical details of a formally correct proof of the assumption, to the extent that modern treatments can be taken as representative, would appear to lie beyond the domain of the ancient geometric methods. But that does not alter the fact that Zenodorus did not provide such an existence proof; the logical niceties of his situation, which doubtless he did not fully grasp, did not deter him from pushing through to the conclusion of his isoperimetric study. It is impossible for us to say what changes he might have introduced, had he fully appreciated the need for this existence proof. But contrary to Zeuthen's position, Zenodorus did in fact "dare" to admit terms whose existence could not be established *via* construction.

These examples reveal that the ancient geometers had no single specially designated method for treating existential issues. Where these issues seem to arise, the format of problems is sometimes adopted, but just as frequently a theorem is employed; in some cases, moreover, as in the propositions cited from Euclid's *Optics*, an alternative existential format is used, so that their classification either as problems or as theorems is ambiguous. Subtle questions of existence arise in the Eudoxean analysis, centering on the introduction of the fourth proportional, as well as in the assumption of the maximal polygon in the isoperimetric studies. But in neither case is the assumption explicitly recognized as such at all; certainly no explicit axiom is introduced to cover it. Here, intuitions of continuity seem to have sufficed, and from the purely existential viewpoint, the same would also apply to those many instances where constructions are provided: the mere existence of the constructed entity is assured intuitively. Thus, the intent behind supplying the construction in such cases must lie in the specific manner of the construction, for instance, on the basis of certain postulates or techniques of construction.

135 THE ALLURE OF EXISTENCE

From the viewpoint of the historical evidence, then, the thesis that the ancient constructions were intended as a form of existence proof accounts neither for the geometers' manner of treating problems of construction, nor for their ways of handling issues of existence. When a thesis so historically unpersuasive captures as general an acceptance as this one has, however, there must be some less evident factors weighing in its favor. Here, I believe the nice fit of the thesis with important trends in the modern philosophy of mathematics is such a factor; for it yields a view of considerable beauty, projecting onto the ancient work an awareness, however indistinct, of sophisticated aspects of the nature of mathematics. In particular, as I shall discuss further below, special concerns of the modern field on issues like the nature of mathematical existence are being read into the ancient technical efforts.

One encounters this phenomenon of projection already in ancient discussions of mathematics. Proclus' attribution to Euclid of existential motives in his construction of the equilateral triangle is an example. Indeed, Proclus' commentary on Euclid is rife with Neoplatonist glosses on geometric procedure, of varying credibility as historical reconstructions of Euclid's actual intentions. Comparable examples can be cited from works by other mathematical commentators, such as Pappus.[53]

I believe this same intellectual trick can operate in subtle ways within the philosophical literature. The issue of existence is a basic concern within the major philosophies. There are many aspects of Aristotle's thought where existential questions are the center of interest, not only in his view of the first principles of the sciences, but also in his doctrine of the categories, his conception of substance as matter in association with form, his conception of change as transition from the potential to the actual presence of a given form in a given matter, and so on. Existential premises, such as that being is better than nonbeing, are worked into his teleological accounts of natural phenomena.[54] Thus, when Aristotle includes the postulation of the existence of the entities of the sciences as a fundamental aspect of their organization,[55] one might expect to find some conscious existential motive in the postulates preliminary to such works as Euclid's *Elements*. It has been well noted, however, that the kind of postulate that Aristotle has in mind, e.g., postulates of the existence of points, lines, planes, numbers, and so on, are precisely the sort of premise absent from Euclid.[56] The effort to effect a complete harmonization of the Aristotelian and Euclidean views on first principles will thus inevitably introduce distortions.

A similar conflict seems to mark a noted mathematical passage from Plato's *Meno*, that on the so-called "geometers' hypothesis." The general discussion deals with virtue, and the specific question as to whether it is teachable. In the absence of a viable definition of what virtue is, Socrates proposes a strategy of inquiring "from hypothesis," such as he maintains geometers employ. He cites as example the problem of inscribing a given area as a triangle within a given circle:

> Whenever someone should ask them, for instance, about an area whether it is possible that this very area be inscribed as a triangle in this very circle, he would say, "I don't yet know if this can be (inscribed), but I think I have, as it were, a certain hypothesis of use for the matter, namely this: if this area is such that when stretched
> along [or: applied to, παρατείνειν] its given line, it falls short by | such an area as the 136
> stretched area itself, it seems to me one thing results; but again another thing if it is impossible that these things obtain. Then by hypothesizing I am willing to say to you what results in the case of its inscription in the circle, whether it is impossible or not." [*Meno* 86e-87b]

Dozens of reconstructions have been proposed for the mathematical problems cited in this passage.[57] I will omit them here, since no specific reconstruction is required for considering the issue I wish to call attention to: namely, that Plato seems to have cast into the format of a *problem* what the later technical literature usually handles *via diorism*.[58] For Plato seems to require the statement of a condition under which the stated problem is constructible. As such, any geometer would have said simply, "it is necessary that the given area be less than the equilateral triangle inscribed in the given circle." The fact that Plato adopts such a convoluted manner of expressing what ought to have been so simple might suggest that perhaps at this time the fact of the maximal property of the inscribed equilateral triangle was not yet known. Upon consideration of the elementary proof leading to this fact, however, one is reluctant to adopt this view.[59]

The geometric procedure underlying Plato's passage would more accurately be described as the "reduction" (ἀπαγωγή) of a problem, that is, the replacement of one problem by another, such that the solution of the one will entail the solution of the other. This procedure had been used decades before Plato by Hippocrates of Chios in his examination of the cube-duplication.[60] It is closely related to an important element within the method of analysis, discussed above.

But why should Plato confuse the issue of this construction with the determination of its possibility? Context is likely to contribute to the answer, seeing that Plato's purpose in the dialogue is to elucidate an issue of possibility, that of teaching virtue. But it seems to me that another factor is present. In the geometric terminology, "it is possible to inscribe ..." would express the project of actually constructing the specified figure. Taken literally, however, it only claims to establish the possibility of the construction, for which an explicit construction would not be necessary. This trick of terminology has already been mentioned as a potential source of confusion within the modern discussion of the existential element of problems. But it is difficult to imagine that Plato could have been merely deceived by a misleading technical diction. One thus suspects that this confusion was part of a deliberate move by Plato to elaborate his philosophical position. From the *Republic* (527a) we know that Plato ridiculed the notion of taking literally the terminology of construction employed by geometers:

> For as if engaging in practice (πράττοντες) and making all their words (or: reasonings, λόγοι) for the sake of practice (πράξις), they speak of 'squaring' and 'applying' and 'adding,' and utter everything in this manner, but the fact is, one would suppose, every mathematical subject is studied for the sake of knowledge.

Moreover, this "knowledge" (γνῶσις) is "of what always exists" (τοῦ ἀεὶ ὄντος), rather than of things which "come into being and pass away" (527b). In Plato's view, then, geometric constructions have only epistemological significance: they enable the geometer to learn the true nature of ideal mathematical entities, but he does not thereby bring any true being into existence. With such an emphasis, it would be quite
137 | appropriate for Plato to frame the constructing efforts in the *Meno* passage as directed toward determinations of possibility.

Plato's position may be an entirely feasible basis for understanding the true nature of mathematics. But determining that is not our concern here. His remarks in the *Meno* seem to misconstrue, perhaps deliberately, the meanings actually intended by the geometers. But we have no reason to suppose Plato aimed at a historically accurate presentation of their meaning. The historian, interested in comprehending the geometers' concerns, is thus forewarned of the need to exploit evidence drawn from the philosophical literature with the greatest care.

ANCIENT AND MODERN VIEWS OF EXISTENCE

A survey of evidence from the technical literature has shown that the existential and the constructive aspects in ancient geometry are separate. The examples cited here indicate that only in cases where the assumption of existence is somehow not obvious do the ancients explicitly take up the question, whether in the format of a problem of construction or through a theorem. On the other hand, explicating the actual manner of construction takes precedence over any logical interpretation one might attempt to project onto their constructing efforts. The shifting of emphasis away from the activity of constructing solutions toward the elaboration of the logical requirements of proofs reflects, I believe, a profound difference in outlook and intent separating the ancient and modern views of mathematics.

The issue of mathematical existence has many facets. An important one is the strictly technical concern to establish that given problems permit of solution within stated parameters. In the ancient procedure, this elicits the *diorism*, which asserts the necessary condition for solution to be possible. But this technical aspect has minimal interest for the philosophical interpretation of ancient geometry. Thus, the discussion has centered on other aspects, corresponding to interpretive views which I shall designate as dialectical, intuitionist and formalist. Each position attempts to incorporate the ancient mathematical work within a wider range of philosophical commitments.

The ancient mathematician did not have to look far to encounter dialectical challenges to the basic concepts of his science. Eleatics, Sophists, Epicureans and Skeptics all found cause for objection in the axioms of number theory and geometry, and through their critique effectively displayed the difficulties in setting these disciplines on a sound logical foundation.[61] This lends a certain credibility to an interpretive notion often advocated in the modern historical discussion of this field, that the formal moves characteristic of the ancient geometry – its explicit formulation of definitions and axioms, for instance, or its inclusion of constructive proofs of existence – were intended as a response to such dialectical criticisms.[62] Indeed, even the ancient commentators, like Proclus, sometimes attempt to promote such a view.[63]

What the advocates of this view overlook, however, is that only a *dialectical* response is appropriate to challenges of this kind.[64] Euclid's construction of the midpoint of a line segment, for instance, despite claims made in some ancient scholia to *Elements* I 10,[65] is no answer at all to atomists who insist on the indivisibility of certain minimal (but finite) magnitudes. Here, the ancient commentators correctly observe that the divisibility of every geometric magnitude (whence its unlimited divisibility) | must be taken as a *postulate* by the geometers.[66] Moreover, dialectical 138
objections will have interest principally at the level of fundamental assumptions; they will have little bearing on the carrying out of specific constructions. If indeed geometers were pressed by such dialectical demands, we would expect to find express evidence of this concern chiefly in their formulations of first principles, their manner of deductive format, and the like. Signs of this type have been pointed out, and these may well indicate a certain degree of sensitivity on Euclid's part, for instance, to some philosophical aspects relevant to the basic structure of geometry.[67] But other signs are conspicuous by their absence, for instance, his failure to adopt an overtly existential terminology despite the appropriateness of doing that within a Platonist ontological framework. Although we have seen a few examples of this existential mode, they are extremely rare.[68] The virtually uniform commitment to the constructive formulation for problems must thus reflect an absence of concern within the geometric tradition for the kinds of issues which stimulated the development of the Platonist theories.

An "intuitionist" view of existence can be proposed, framed loosely on the ideas of L. E. J. Brouwer and, before him, L. Kronecker.[69] Like the Platonist view, this would provide an absolute basis for mathematical terms; but this basis lies not in entities external to human minds, but rather in considerations of the thinking process itself. Grounds of conceivability justify the introduction of certain primary concepts and operations, and from these alone can the rest of the science be derived. The effect is restrictive, in that terms not so derivable are not to be admitted into mathematical investigations. Further, a premium is set on constructive procedures as the only acceptable mode of establishing existence. One can hardly imagine that Kronecker's controversial positions on mathematical existence were unknown to Zeuthen, and that these did not influence his interpretation of the ancient mathematical constructions. But we have seen that the ancient geometers admitted terms for which explicit constructions could not be given, and occasionally resorted to nonconstructive arguments if the existence of such terms had to be justified. Thus, if one subscribes to such an intuitionist view of the ancient field, one must also acknowledge the ancients' incomplete understanding of the requirements of their position.

The "formalist" view of existence, growing out of Hilbert's researches on the foundations of geometry, is doubtless the prevalent position among modern mathematicians.[70] Under this view, the entities which exist within a mathematical structure are relative to the postulates which specify that structure: absolutist intuitions of existence are not relevant to the study of such structures, since it is easily possible that terms which exist in one context do not exist in another (e.g., the square root of 2 does not exist within the system of rationals). The introduction of any term into a proof thus requires that the term's existence be established *via* deduction from the postulates which define the system; this deduction can take the form of an explicit construction, although nonconstructive proofs are fully acceptable for Hilbert. Zeuthen seems closest to such a view of existence in his discussion of the construction in *Elements* I 2. The existence of the constructed line is intuitively obvious and its construction is instrumentally trivial.[71] Its inclusion, then, seems for reasons only of demonstrating that the operation is possible within the specific set of constructions postulated as the basis of Euclid's system.

This assigns to Euclid an extremely subtle view of the nature of geometric exis-
139 | tence. But its validity, at best, must be confined to Euclid's elementary works (the *Elements* and the *Data*), for the ancient geometric tradition, as represented by the extant treatises, rarely manifests such conscious concern for fundamental issues. The more advanced fields, dealing with constructions *via neusis*, conic sections, and mechanically generated curves, for instance, seem never to have been effected within systems of explicit postulates, while those ostensibly axiomatic efforts which survive in geometric optics and mechanics are incomplete in their formal execution.[72] Even parts of Euclid's *Elements* are defective in this respect, for we have noted his dependence on certain important assumptions not covered in explicit postulates.[73]

More important, the formalist position as such fails to account for the ancients' decided preference for *constructive* procedures.[74] Thus, even if we could articulate a modified constructivist-formalist view, we would still have to explain why constructions have been chosen as the means for securing claims of existence. This is the crux of the issue, exposing the fundamental differences between the ancient and the modern views of geometry. Zeuthen, for instance, takes it for granted that in geometry constructions are directed toward the purposes of theory, almost to the exclusion of more practical aspects.[75] Whether the ancient geometers would acquiesce in such a portrait of the intentions of geometric research, however, is a matter for our historical sources to determine. Remarkably, Zeuthen ignores the testimony of writers like Diocles, Anthemius and Eutocius who take pains to explicate practical procedures for the constructions of conics and other curves;[76] he overlooks Pappus' view that the early studies of such problems were stymied by the difficulty of drawing these curves;[77] and he fails to consider the prominent mechanical element in the methods for solving the cube-duplication, where the ancient accounts often include details on their physical implementation (e.g., materials, positioning of pivots and beams, shapes of grooves, and so on) and name specific practical situations where solutions of these problems can be applied.[78] The modern mathematician may be remote from such practical applications of his researches, and perhaps may even be disdainful of the notion of their relevance, but this was clearly not the case for the ancients.

I do not wish to deny the basic validity of Zeuthen's contention that the ancient formal treatises were exercises in mathematical *theory*. Their formal features, embracing a commitment to the exposition of proofs in fine detail, cannot be accounted for through any practical considerations. But the concerns of theory do not provide a convincing position from which to account for the strongly constructive element in the ancient studies. The importance of this element is emphasized by Apollonius in his prefaces to the books of his *Conics*, as when he observes that these materials are needed "for obtaining a knowledge of the analysis and determination of problems."[79] Similarly, in Pappus' description, the entire corpus of analysis, which includes the *Conics* and comparable works of advanced geometry, "has been compiled for those wishing to acquire the power to solve geometric problems, and is set down as useful *for this end alone*."[80] Such firm statements of the intrinsic interest in solving problems discourage efforts to assign this activity a role subordinate to the purely formal interests of theory.

Thus, problems of construction do not merely provide a vehicle for the extension of mathematical knowledge; rather, their solution constitutes in effect what the ancients *mean* by mathematical knowledge. In the context of a theorem, a construc|tion 140
enables one to articulate what the entity is whose properties are being investigated. But discovering how to construct the solutions of problems is a major aim of research, even apart from the context of theorems. Indeed, it can often strain one's imagination to conceive how some of the theorems in the *Conics*, for instance, could have interested the ancient geometers, until one perceives the associated problems whose solution becomes possible by their means.[81] In cases like this, it is the corpus of theorems which forms the tool for the inquiry into problems, rather than conversely.

By emphasizing the solution of problems, the ancients do not necessarily intend physical constructions, although the literature includes many examples of this type. But neither do they emphasize the pure aspects of theory to the exclusion of interest in the practical aspects of their results. Even in their advanced theoretical efforts in geometry, the ancients are still sensitive, if only in an indirect way, to the demands of practice. As Zeuthen seems to admit, the training of any geometer, ancient or modern, will begin with exposure to the skills of actual construction.[82] But there was nothing in the ancient experience of geometry comparable to the great advances in algebraic and geometric theory in the 19th century, including the elaboration of projective and non-Euclidean geometries, which could instill that sense of divorce between theory and practice so prominent in Zeuthen's attitude. It seems to me more appropriate to describe the ancient geometers' view of mathematical existence as a form of naive realism not far different from the Aristotelian abstractionist view of the objects of science.[83] But one does well to stress "naive;" for I do not see encouragement for assigning to the ancient geometers an explicit commitment to any one specific philosophical position.

THE ANCIENT MATHEMATICAL PHILOSOPHY

The role of constructions in ancient geometry is far more diverse than the simple existential view can comprehend. If some of the ancient construction problems can be plausibly assigned an existential function, for the overwhelming majority of examples this is not the case. When the geometers explicitly raise the question of existence, they show no clear preference for the format of construction problems to address it. Subtle issues of existence sometimes pass unnoticed, as in the tacit acceptance of the existence of the fourth proportional. Occasionally, a critic will recognize the presence of such unstated assumptions: the commentator Sporus, for instance, objects to certain uses of a special curve, the quadratrix, on the grounds that its terminal point is not well-defined:[84] and Philoponus notes of Apollonius' solution of the cube-duplication that its *neusis* "is taken as a postulate (αἴτημα) without proof."[85] Thus, contrary to Zeuthen's view, the ancient geometers were indeed prepared to admit terms into their proofs as these were called for, even without prior explicit construction, especially if the existence of the terms appears obvious through intuitions of continuity. Far from identifying problems of construction with justifications of existence, the ancients actually distinguish the two, as when Philoponus remarks that the project of research into the circle-quadrature is not to establish the *existence* of the solution (for that is simply assumed), but rather to discover the *manner* of its construction.[86]

Within the body of ancient philosophy, one can detect not only one mathematical philosophy, but many; for each of the major schools of philosophical thought in
141 | general – Platonist, Aristotelian, Stoic, Epicurean, Skeptic – carried with it a distinctive view of the nature of mathematics.[87] But if we seek the geometers' view of their own discipline, we must turn to the mathematical literature. Of the mathematical commentators, only Pappus and Proclus have much at all to say on philosophical aspects of the field, and when they do so, their remarks tend to be straightforward adaptations of stock positions drawn from their favorite philosophies, in particular, Neoplatonism.[88] Their views seem surprisingly little affected by their ample exposure to the technical literature. As for the technical treatises themselves, they are silent on issues of philosophical interest. The austere format of enunciations and proofs provides no occasion for metamathematical commentary, and even in the prefaces by Archimedes, Pappus and others, we obtain only the rarest glimpses of their general views.[89]

To examine questions like the ancient geometers' view of mathematical existence, then, one is left with no alternative strategy other than applying a philosophical view, whether ancient or modern, to aspects of the technical literature where we would expect these issues to arise. But our opportunities to probe these interpretations directly are limited. Euclid never says whether he intends this or that construction to have existential import. Consider the problems of angle-trisection and cube-duplication, omitted from the *Elements*: if we could communicate to Euclid the findings of modern algebraic theory, showing that solutions of these problems are impossible under the restriction of the constructing techniques postulated in the *Elements*, would he conclude that the entities constructed in these problems in fact do not exist? We can well imagine that his response would be quite different: that these entities obviously do exist; but that they merely cannot be constructed on the basis of the techniques he has postulated.[90]

A difficulty inherent in this approach to reconstructing the ancient view is that the tenability of one account does not exclude the possibility of others. Coherent mathematical philosophies can be framed on the basis of Platonic, Aristotelian, and other ancient philosophies, as well as on modern views like the intuitionist, formalist, and so on. Each of these can provide a viable account of what the *essence* of the ancient geometry must be; but that in itself does not make it into an account of what the geometers intended. Moreover, a straightforward technical reading, without interesting philosophical connotations, is possible. In my view, this last alternative has a natural claim best to capture the ancient geometers' position; for it accords with the absence of overt attempts to clarify metamathematical issues within the body of the technical treatises, and finds direct confirmation in the mathematicians' repeated recommendations of the activity of problem solving.[91] Further, efforts to assign to the geometers a conscious philosophical program inevitably portray their efforts as unsuccessful, in that the ostensible inconsistencies of formal execution, evident even in the *Elements*, provide a platform more fit for blaming than for praising the ancient tradition. Surely it is better to ascribe this discrepancy between an idealized philosophical account and the actual testimony of the sources to the inappropriateness of the interpretation.

By imputing an existential motive to constructions, Zeuthen has drawn attention away from what I take to be the truly interesting issue. It is a fact that the ancients employ a nonexistential format for the investigation of constructions, despite the ease
| with which they could have devised such a format, had they chosen to. By contrast, 142
modern accounts readily translate the ancient effort in existential terms; for instance, Euclid's constructive acts, like the postulate of *drawing* lines joining given points, are now commonly paraphrased as assertions of existence. Here, Euclid is representative of the entire ancient geometric tradition. What the historian must explain, therefore, is how the modern field has come to emphasize its existential aspects, to such a thorough extent that they are taken as self-evidently true, rather than as interpretations. The historian's service, I maintain, lies in exposing this real difference in outlook, thereby framing the effort to discover how this transition in mathematical intuition has taken place within the modern field.

NOTES

1 Throughout my discussion here the word "problem" will be used only in the technical sense it has within the ancient mathematical literature: as that form of geometric proposition which seeks the construction of a figure in a specified relation to certain designated, or "given," magnitudes. The formal distinction between problems and theorems will be elaborated below. Note that the term "proposition" in this context has a more restricted sense than it does in logic.

2 An account of the Platonic philosophy of mathematics is provided by A. Wedberg, *Plato's Philosophy of Mathematics* (Stockholm: Almqvist & Wiksell, 1955). A commentary on the mathematical passages from Aristotle's works appears in T.L. Heath, *Mathematics in Aristotle* (Oxford: Clarendon Press, 1949).

3 Perhaps the most ambitious effort to portray ancient mathematics as a response to early dialectical critiques, particularly those of the Eleatics, is by A Szabó, *The Beginnings of Greek Mathematics* (Dordrecht, Neth.: Reidel, 1978 [tr. of the German ed., 1969]); see also note 62 below.

4 Proclus, *In Primum Euclidis Elementorum Librum Commentarii*, G. Friedlein, ed. (Leipzig: Teubner, 1873) 233-235; cf. the translation by G. Morrow, *Proclus: A Commentary on the First Book of Euclid's Elements* (Princeton, N.J.: Princeton University Press, 1970) 182-183.

5 It is possible that Proclus has a more limited pedagogical intent here, rather than the general view of existence I ascribe to him. In either event, he can hardly be representing any view Euclid would have maintained. If the existence of triangles is at issue, as background to the theory of congruence, pedagogically it would be important to direct the student toward the *general* case, rather than to the special case of equilateral triangles. A construction could be effected by assuming any three noncollinear points as given, joining them by lines two by two (*via* post. 1), and then proving that the resulting figure is a triangle (*via* def. 19). Of course, we here depend on the existence of three such points; this is not postulated by Euclid, a salient gap in his axiomatic scheme (cf. I. Mueller, *Philosophy of Mathematics and Deductive Structure in Euclid's Elements* [Cambridge, Mass.: MIT Press, 1981] 15). On the other hand, his construction of the equilateral triangle in

I 1 also depends on implicit postulates, specifically the intersection of two given circles (cf. Mueller, *Philosophy* 27-28); objections of a different sort were lodged against this theorem by ancient critics (cf. Proclus, *In Euclidem* 214-215). Ultimately, then, the existence of triangles is more nearly a matter for postulate than Proclus appreciates.

6 "Die geometrische Construction als 'Existenzbeweis' in der antiken Mathematik" *Mathematische Annalen* 47 (1896) 222-228. A characteristic favorable use is by O. Becker, *Mathematische Existenz* (*Jahrbuch für philosophische und phänomenologische Forschung* 8, Halle a. d. S., 1927) 130-133. Perceptive criticisms of the thesis, however, have been made by Mueller, *Philosophy* 14-16; and by A. Frajese, "Sur la signification des postulats euclidiens" *Archives internationales d'histoire des sciences* 4(15) [1951] 383-392.

7 The ancient distinction between problems and theorems will be discussed below.

8 Zeuthen, "Construction" 223 (my translation).

9 *Ibid.* 224.

143 10 | *Ibid.* 225. VI 27 establishes the necessity of the condition $(a/2)^2 \geq b$ for solving the problem in VI 28 of the "defective" case of the application of areas; in Zeuthen's algebraic formulation, this entails finding x such that $x^2 - a \cdot x + b = 0$ (Euclid's formulation is entirely geometric). The Archimedean example in an appendix to *Sphere and Cylinder* II 4 will be discussed below. It effects a third-order analogue to the construction from Euclid just cited, namely (again in Zeuthen's algebraization) to find x such that $x^3 - a \cdot x^2 + b^2 \cdot c = 0$, requiring the condition that $b^2 \cdot c \leq 4a^3/27$.

11 *Ibid.* 225-226. Zeuthen proposes to explain Euclid's postulate *via* his need to guarantee the existence of points of intersection of lines in certain configurations (cf. I 44). But a simpler account is possible: that in his theorems on the sum of angles of triangles, to secure his theorem on the angles formed by the transversal to two parallel lines (I 29), Euclid recognized the need for precisely this step, and that ultimately unable to prove it as a theorem, he enunciated it as the postulate. A similar route seems to have led to the insertion of post. 6 (that two lines cannot enclose a space), assumed in I 4. The modern editors consider this postulate to be an interpolation (cf. Mueller, *Philosophy* 31-32).

12 *Ibid.* 226-228. Zeuthen's argument that the early restriction to perpendicular sections of right cones enables a direct correspondence between the geometric formation of these curves and their expression *via* second-order coordinate relations is technically neat. I believe, however, that he exaggerates the ancients' reluctance to accept pointwise and mechanical constructions, and that his insight can be applied toward a more plausible view than his of the early studies of the conics; see my *The Ancient Tradition of Geometric Problems* (Basel/Stuttgart/Boston: Birkhäuser) ch. 3 (forthcoming).

13 It is discussed prominently by Proclus, who cites many precedents among philosophical and mathematical writers. For instance, he divides the "things (following) from the first principles" (τὰ ἀπὸ τῶν ἀρχῶν) into problems and theorems (*In Euclidem* 77), and then discusses the views on this distinction advocated by Speusippus, Menaechmus and others (77-81). In his remarks on Posidonius' views (80) he distinguishes between the theoretic and problematic "proposition" (πρότασις).

14 In fact, each problem will consist of a section executing the construction, followed by a section proving that the figure has the stated property. This latter section is, in effect, a theorem, and if the preceding construction is viewed simply as an extended *protasis*, then the problem itself becomes a theorem. The conversion of theorems to problems, while possible, is not always so straightforward. Problems of locus (i.e., to identify the figure whose elements all satisfy a stated property) seem to form an intermediate class, since the proposition requires producing a construction, yet its enunciation is typically in the format of a theorem.

15 In a constructivist view, the assertions of possibility and constructibility would be equivalent.

16 For a wide-ranging discussion of this method, see J. Hintikka and U. Remes, *The Method of Analysis* (Dordrecht, Neth.: Reidel, 1974). It is a prominent interest in my study of the Greek geometric tradition (cited in note 12 above).

17 Eutocius' text and commentary appear in J. L. Heiberg's edition of Archimedes, *Opera Omnia* 2nd ed. (Leipzig: Teubner, 1915) III 130-152. For discussions of the construction, see E. J. Dijksterhuis, *Archimedes* (New York: Humanities Press, 1957) 195-200; and my *Ancient Tradition* ch. 5.

18 I here set in more algebraic form what Archimedes frames in a strictly geometric manner.

19 The problem in its initial formulation is always solvable; but in its more general form there can arise values of the given terms for which a solution cannot be constructed, so that a *diorism* is then called for.

20 The relation of the analysis and the synthesis is usually so close that the latter is sometimes omitted as "obvious" (*φανερά*) or provided only in outline; cf. Diocles' treatment of Archimedes' problem on the division of the sphere in *On Burning Mirrors*, G. Toomer, ed. (Berlin/Heidelberg/New York: Springer-Verlag, 1976) 86; and the problems on the regular solids presented by Pappus in the *Collection* III 48-52 (F. Hultsch, ed. [Berlin: Weidmann, 1876-78] I 144, 146, 148, 154, 162).

21 *Collection* Book VII, preface (ed. Hultsch, II. 636). The passage is discussed in detail by Hintikka and Remes, *Method* ch. II.

22 *Collection* V 29 (ed. Hultsch, I 382): to mark off an arc from a given circle such that the segments of the two tangents drawn from its endpoints shall be less than a given line segment. One of the ancient theories of proportion depends on the lemma: given three homogeneous magnitudes, to find a magnitude which shall be less than the first, greater than the second, and commensurable with the third; for references to the ancient texts and discussion, see my "Archimedes and the Pre-Euclidean Proportion Theory" *Archives internationales d'histoire des sciences* 28 (1978) 183-244.

23 | I would estimate that about 300 geometric problems are extant in the works of 144
Euclid, Archimedes, Apollonius, Pappus and the minor writers and commentators. Inclusion of Euclid's *Data*, which adheres to an alternative problematic format would add 94 more. Further, in the area of arithmetic, *all* of the propositions (amounting to almost 300) in the extant ten of the thirteen books of Diophantus' *Arithmetica* are problems. These figures can be considered only a bare sampling of the scope of the problematic literature in antiquity. We know, for instance, of whole treatises on loci by Euclid, Apollonius, Eratosthenes, and others, which are no longer extant; cf. the preface and commentary by Pappus in *Collection* Book VII, a résumé of which is given by T. L. Heath. *A History of Greek Mathematics* (Oxford: Clarendon Press, 1921) II 399-427.

24 E. g. Book I 54-60; II 44-53; VI 28-33; the constructions of normals in V 58-63, which could easily be set in the form of problems, happen here to be framed as theorems (cf. the edition by H. Balsam. *Des Apollonios von Perga sieben Bücher über Kegelschnitte* [Berlin: Reimer, 1861]); most of the propositions of the lost eighth book were problems associated with the theorems of the seventh (cf. VII, pref.)

25 Cf. Pappus' accounts of their work, cited in note 23 above.

26 *Collection* IV 36-42 (ed. Hultsch. I 272-280). The technique of *neusis*, that is, construction *via* marked-ruler, is prominent among the researches by Archimedes and his successors; cf. my discussion in *Ancient Tradition* chs. 5 and 6.

27 The existential variant is common in modern discussions of axiomatics: cf., for instance, D. Hilbert, *Foundations of Geometry* ch. I (English ed., La Salle, Ill.: Open Court, 1971, based on the 10th German ed.). Mueller contrasts the expressions of Euclid and Hilbert, *Philosophy* 14-15.

28 The pseudo-Euclidean *Catoptrics* prop. 29 provides a striking illustration. It is a problem, "it is possible that there be constructed a mirror such that many faces appear in it, some greater, some smaller, some nearer, some farther, some with the right on the right and the left on the left, others with the left on the right and the right on the left" (*Euclidis Opera*, J. L. Heiberg, ed. [Leipzig: Teubner, 1895] VII 338), which concludes a sequence of theorems (prop. 16-28) on the placement, size and orientations of images seen in plane, convex and concave mirrors. The problem can hardly have an existential role relative to those theorems; instead, it recapitulates their results in a manner suggestive of their practical implementation. Thus, the issue here is not the existence of this mirror, but the actual manner of its construction.

29 Euclid orders the solids in the sequence 4, 8, 6, 20, 12 (relative to the number of faces); presumably, this follows the order of the lengths of the edges of the solids inscribed in the same sphere, where the tetrahedron has the longest edge, the octahedron a shorter one, and so on, until the dodecahedron with the shortest. The comparison of these lengths forms the content of the last theorem in the book (XIII 18), so that this scheme is appropriate. By contrast, other rationales for the ordering – e.g., according to the number of vertices, or according to the relative volumes of the solids inscribed in the same sphere – are not relevant to the subjects actually examined in the book.

30 The Greek technical language could presumably permit the coining of terms like *τριτριγωνόεδρον*. The naming of the Archimedean semiregular solids continues the Euclidean pattern by following the number of faces: e.g., 8-hedron, 14-hedron (three forms), 26-hedron (two forms), 32-hedron (three forms), 38-hedron, 62-hedron (two forms), and 92-hedron; cf. Pappus, *Collection* V 19 (ed. Hultsch, I 352-354).

31 Mueller elucidates these constructions by supplying analyses (*Philosophy* ch. 7). Such a procedure, which doubtless was followed by the ancients in working out these constructions, assumes the prior recognition of their qualitative description (cf. 254-255).

32 *Euclidis Opera* (ed. Heiberg) VII 80.

33 For instance, prop. 39: "if a magnitude is set at right angles to the base plane, ... it shall always be seen equal when transposed according to a disposition parallel to its original one" (ed. Heiberg, VII 84).

34 *Ibid.* 104.

35 Contrast the example of *Cat.* prop. 29 (cited in note 28 above) where the problem serves to summarize and apply the results in the preceding theorems.

36 The problem of dividing an arbitrarily given angle into *n* equal parts in general reduces to solving a relation of order *n*. The problem of division for all *n*, however, is transcendental. The ancients produced constructions utilizing the quadratrix and Archimedean spiral (cf. Pappus, *Collection* IV 45-46), but classed it among the "linear" problems, that is, not solvable *via* the Euclidean techniques or *via* conic sections (ed. Hultsch, 284); cf. 54, 270 for Pappus' remarks on the tripartite division of problems, discussed by Heath, *History* I 218-220 and in my *Ancient Tradition* ch. 8.

145 37 | *Euclidis Opera*, ed. Heiberg, VII 106. The phrase "sometimes whole" *(pote holon)* does not make particularly good sense; we should have expected "sometimes one-third" *(pote triton)* in this context. The phrase is absent from the medieval Latin translation (reproduced by Heiberg, 107), which is usually in complete literal agreement with the Greek. Theon's recension omits an analogue of prop. 49, doubtless for its mere duplication of the result of prop. 48.

38 On the hybrid character of locus problems, see note 14 above.

39 Such appears to be the motive of problems in the pseudo-Euclidean *Catoptrics* (cf. note 28 above), and is patently the case in the Heronian *Catoptrics* prop. 11-18 (in *Heronis Opera*, W. Schmidt, ed. [Leipzig: Teubner, 1900] II 336-364); e.g., prop. 16: "in a certain conveniently located window in a house, to place a mirror in the house through which people will appear as they come from the opposite direction in the streets < and will be seen by those > in a certain given place in the house" (352).

40 For a survey, see Heath, *History* I ch. VII; and my *Ancient Tradition*.

41 *In Aristotelis Analytica Posteriora*, M. Wallies, ed. (*Commentaria in Aristotelem Graeca* XIII, Berlin: Reimer, 1909) 112.

42 *Archimedis Opera*, ed. Heiberg, III 230.

43 In *Spiral Lines* prop. 18, Archimedes shows that the subtangent corresponding to the position of the radius vector terminating one full turn of the spiral equals the circumference of the circle with radius equal to that radius vector. Strictly speaking, this does not rectify the circular arc, since it depends on the construction of the spiral and its tangent. Archimedes nowhere claims this to be a solution of the problems of the rectification and quadrature of the circle, but some of the commentators seem to view it as such; cf. Iamblichus, cited by Simplicius, *In Aristotelis Physica* ed. H. Diels, I 60 (alternative version reproduced in I. Thomas, *Greek Mathematical Works* LCL [Cambridge, Mass.: Harvard University Press, 1939-41] I 334).

44 The same assumption is made in Euclid's proof that a magnitude in a given ratio to a given magnitude is itself given (*Data* prop. 2). The editor, R. Simson, adds the phrase "and if ... a fourth proportional can be found" to the enunciation, both here and in all subsequent appearances, since the proof in fact makes this assumption (*The Elements of Euclid* [Glasgow: Robert and Andrew Foulis, 1762] 360, 454-455.). But neither Euclid nor the writers in the later analytic tradition seem conscious of the need for such a qualification.

45 O. Becker has indicated the discrepancy between these appeals to the assumption of the fourth proportional and views of the ancients' adherence to a constructivist position; see his "Eudoxos Studien II" *Quellen und Studien zur Geschichte der Mathematik, Astronomie und Physik* 2, Abt. B (1933) 369-387.

46 Zeuthen ("Construction" 226) maintains that the existence of this point follows from Euclid's definition of the circle as "a plane figure contained by a single line ..." (Def. 15). But in fact, its existence depends on postulates of continuity as well. A discussion of the use of such postulates, particularly in the form proposed by Dedekind, for establishing incidence relations of lines and circles appears in T. L. Heath, *The Thirteen Books of Euclid's Elements* 2nd ed. (Cambridge: Cambridge University Press, 1926) I 234-240.

47 This suggests that, whatever the apparent axiomatic ideal of the *Elements* might be, in execution Euclid does not articulate *every* assumption introduced into his proofs, but only those whose character is sufficiently nonobvious as to attract attention. I think this applies, for instance, to the case of the parallel postulate; cf. note 11 above.

48 For a résumé of the argument, see Heath, *History* II 206-213.

49 I present support for this view in my "Archimedes and the Elements" *Archive for History of Exact Sciences* 19 (1978) 238-239.

50 This can be done for some theorems of maxima, for instance, the fact that of all triangles inscribed in a given circle, the equilateral triangle has the greatest area (cf. my *Ancient Tradition* ch. 3); this example is related to the mathematical passage in Plato's *Meno*, to be discussed below.

51 Cf. note 36 above.

52 A counterexample is presented by O. Toeplitz and H. Rademacher as background to their demonstration of the isoperimetric theorem for the circle; cf. their *The Enjoyment of Mathematics* (Princeton, N.J.: Princeton University Press, 1966) [transl. from the German edition of 1933] chs. 21-22.

53 A very effective way of retrieving the philosophical background of such writers as Pappus would be to attempt a source analysis of those passages where they adopt specific philosophical positions. In the case of Pappus' account of the method of analysis, for instance, an indirect dependence on Aristotelian | passages can be perceived (cf. my *Ancient Tradition* ch. 8). Little of this 146
sort has been done, however, and the fragmentary nature of the evidence might severely limit the scope of one's findings.

54 Cf. the line quoted from *Generation and Corruption* II 10 at the beginning of this paper. There Aristotle seeks to explain the reproductive cycle in living things: since nature strives for the best, and since being is better than nonbeing, and since individual living things necessarily have finite life spans, the species attains a form of eternal existence through the perpetual cycle of generations.

55 Cf. *Posterior Analytics* I 10 (76a32-37):

> What is denoted by the first (terms) and those derived from them is assumed; but, as regards their existence [lit.: that they are], this must be assumed for the principles but proved for the rest. Thus what a unit is, what the straight (line) is, or what a triangle is (must be assumed); and the existence of [lit.: that it is possible to take] the unit and of magnitude must also be assumed, but the rest must be proved. [from Heath, *Mathematics in Aristotle* 50-51]

56 Cf. Mueller, *Philosophy* 14-15.

57 For a survey, see R. S. Bluck, *Plato's Meno* (Cambridge: Cambridge University Press, 1961). I discuss the technical aspects suggested by the passage in *Ancient Tradition* ch. 3.

58 Heath points out this difficulty (*History* I 303). The passage is labelled as a *"diorismos"* by I. Thomas in his reproduction of the text (*Greek Mathematical Works* I 394; cf. 397n).

59 I reconstruct such a proof in *Ancient Tradition* ch. 3.

60 Cf. Proclus, *In Euclidem* 212-213; so also Eutocius, following Eratosthenes, in his commentary on Archimedes' *Sphere and Cylinder* (*Archimedis Opera*, ed. Heiberg, III 88).

61 The mathematical views contained within these philosophies are investigated by Ian Mueller in his *Coping with Mathematics* (*The Greek Way*) (Chicago: Morris Fishbein Center for the Study of the History of Science and Medicine (Publication No. 2), 1980); this inquiry is continued in his "Geometry and Scepticism" in M. Burnyeat *et al.*, ed., *Scepticism and Science* (Cambridge: Cambridge University Press, 1982).

62 See, in particular, A. Szabó, *Beginnings of Greek Mathematics* (cited in note 3 above) sect. III. I review and criticize some of his positions in my "Early History of Axiomatics" *Pisa Conference (1978) Proceedings*, ed. J. Hintikka *et al.* (Dordrecht, Neth.: Reidel, 1981) I 145-186.

63 *In Euclidem* 233-234; cited in notes 4 and 5 above.

64 This impasse is noted by Mueller in his "Geometry and Scepticism." It may also underlie Aristotle's remark that the geometer *qua* geometer does not have to address questions which violate the principles of his science (*Physics* I 2, 185a14-17; *Sophistical Refutations* 11, 171b7-18).

65 Cf. scholium no. 50 (*Euclidis Opera*, ed. Heiberg, V 135): "it is proved from this that indivisible lines do not exist, if indeed it is possible to bisect the proposed side." The existence of incommensurable magnitudes is cited in scholia to Book X as evidence against the atomists; cf. no. 26 (*ad* X 1): "that there is not a least magnitude, as the Democriteans say, is also proved *via* this theorem, if in fact it is possible to take a magnitude smaller than any given magnitude" (436).

[66] Cf. schol. no. 1 to Book X (415-416), where the infinite divisibility of magnitude is cited as the cause underlying the nature of incommensurable magnitudes. Geminus, cited by Proclus (*In Euclidem* 278), claims, with reference to I 10, that the assumption of the continued divisibility of magnitude underlies the construction of the bisection of the line segment, rather than conversely: "that every magnitude shall be continually divided, and shall never result in an indivisible ... is provable; but it is an axiom (ἀξίωμα) that every continuous (magnitude) is divisible." Eudemus (cited by Simplicius, *In Physica*, ed. Diels, I 55) maintains that it is the principle of the infinite divisibility of magnitude which is violated by Antiphon's fallacious circle-quadrature (cf. Heath, *History* I 222).

[67] One would not wish to deny that Euclid's inclusion of statements of the "common notions" (κοιναὶ ἔννοιαι) prefacing Book I, and perhaps also his formulations of some of the definitions, were influenced by philosophical discussions of the first principles. There are interesting parallels to Aristotelian remarks on "axioms," as cited by Proclus (*In Euclidem* 193-194); for discussion, see Heath, *Euclid's Elements* I 221-222 and *Mathematics in Aristotle* 50-57. But the question of the authenticity of the "common notions" is vexed, and Mueller observes that, in any case, the Euclidean listing seems not even intended to be complete (*Philosophy* 35-36).

[68] The problematic format. "to *find* ...," adopted throughout Euclid's so-called "arithmetic books" (VII-X) differs from that in the other books (where an overtly constructive terminology is used), and may thus reflect a sensitivity to the kinds of ontological issues raised by Plato. This is a point raised by H. Mendell in his current studies of the Platonic and Aristotelian mathematical philosophies.

147 [69] | For a review of the intuitionist position, one may consult M. Black, *The Nature of Mathematics* (repr., Paterson, N.J.: Littlefield, Adams, 1959); and W. and M. Kneale, *The Development of Logic* (Oxford: Clarendon Press, 1962) ch. XI.

[70] The formalist position is reviewed by S. Körner, *The Philosophy of Mathematics* (London: Hutchinson, 1960); cf. also the works by Black and Kneale & Kneale cited in the preceding note. Mueller contrasts Hilbert's view with the Euclidean approach (*Philosophy* ch. 1).

[71] Zeuthen, "Construction" 223. The drawing of a line from a given point and equal to a given line is effected immediately *via* standard compasses; Euclid's construction reveals that his third postulate (i.e., the drawing of a circle of given center and radius) is equivalent to construction *via collapsing* compasses. Thus, one must show that constructions *via* the one instrument are possible also *via* the other. For further remarks, see Mueller, *Philosophy* 15-16, 24-25; and Heath, *Euclid* I 246.

[72] The advanced treatises of Archimedes and Apollonius are not even axiomatic *in form*; although they of course develop their subject matters according to a strict deductive sequence of theorems, they presuppose the entire elementary literature, the "data," and other technical materials. To be sure, postulates are stated at the beginning of Archimedes' *Sphere and Cylinder* Book I, and also Apollonius' *Conics* Book I; but these enunciate only those few special principles applied in the treatise, without attempting to provide a complete listing of all assumptions made therein. Euclid's *Optics* and Archimedes' *Plane Equilibria* Book I list their initial postulates in a manner which at first suggests an axiomatic intent. But the works themselves freely admit assumptions not covered in the lists, so that the formal execution of the project, if indeed it was intended to be axiomatic, is seriously flawed (cf. P. Suppes, "Limitations of the Axiomatic Method in Ancient Greek Mathematical Sciences," *Pisa Conference Proceedings*, ed. J. Hintikka *et al.* [Dordrecht, Neth.: Reidel, 1981] I 197-213).

[73] These include the continuity assumptions, discussed in the section above.

[74] The constructive aspect of geometry seems to have been a basic concern for Zeuthen in his own mathematical studies. See the biography and bibliography prepared by M. Noether in *Mathematische Annalen* 83(1921) 1-23 (esp. 9-11). The constructive, even "inductive," approach advanced by Zeuthen would appear to contrast with the more abstract, deductive position on existence advocated by Hilbert.

[75] For instance, Zeuthen insists that the ancients could not have considered the use of conic sections for solving cubic relations to afford any practical advantage, but that the introduction of conics and other special curves was for their use in extending geometric theory. He does admit, however, that "practical execution [of the simpler constructions] is not ruled out, where these actually are of use" ("Construction" 222-223).

[76] Diocles, *On Burning Mirrors* props. 4-5, 10 and 12 present pointwise constructions of the parabolic and "cissoid" curves (see citation in note 20 above); Diocles' pointwise construction of the

"cissoid" is also paraphrased by Eutocius (*Archimedis Opera*, ed. Heiberg, III 66-68). On Anthemius, see G. Huxley, *Anthemius of Tralles* (Cambridge, Mass.: Harvard University Press, 1959) for the texts, translation and commentary on fragments dealing with the pointwise construction of elliptical and parabolic burning-mirrors.

77 *Collection* III 7 (ed. Hultsch, I 54). An interpolation in Eutocius' commentary on Archimedes mentions a mechanical device invented by Isidore of Miletus for drawing parabolas (*Archimedis Opera*, ed. Heiberg, III 84), and in his commentary on Apollonius, Eutocius indicates the practical utility of pointwise constructions for the conic curves (*Apollonii Opera*, J.L Heiberg [Leipzig: Teubner, 1893] II 230-234).

78 Cf. in particular the methods of Hero, Nicomedes, Pappus and Eratosthenes reported by Eutocius (*Archimedis Opera* III 58-96) and Pappus (*Collection* III 7-10 [ed. Hultsch, I 56-68]). Contexts of their practical application are evident from the treatments by Hero in the *Mechanics* (I, 9; *Heronis Opera*, L. Nix, ed. [Leipzig: Teubner, 1900] II 23) and the *Belopoeica* (*Greek and Roman Military Treatises: Texts*, E. W. Marsden, ed. [Oxford: Oxford University Press, 1971] 40-41), and are stated in the account Eutocius draws from Eratosthenes (*Archimedis Opera* III 90). These include any situations where weights or volumes are to be scaled up in proportion, as in the design of military engines or the building of ships.

79 Preface to Book V; for translations and discussion of the prefaces, see T.L. Heath, *Apollonius of Perga* (Cambridge: Cambridge University Press, 1896) lxix-lxxiv.

80 *Collection* VII, preface (ed. Hultsch, II 634); emphasis mine.

81 Many of the theorems in Book III, for instance, deal with relations of the products of segments of chords and tangents to the conic sections. Such relations turn out to be critical for the solution of certain conic problems, such as the drawing of conics which are to pass through given points or have given lines as tangents.

82 | "Construction" p. 222: 148

> We have inherited geometric construction from the Greeks. In particular, such constructions as can be effected *via* compass and straight edge serve as excellent exercises in our schools and are indispensable for practitioners.

83 On the conception of mathematical entities as "abstracted" from substances, see the passages cited by Heath, in *Mathematics in Aristotle* 64-67.

84 Cited by Pappus in *Collection* IV 31 (ed. Hultsch, I 254).

85 *In Analytica Posteriora*, ed. Wallies, 105; reproduced by Heiberg in *Apollonii Opera* II 106.

86 Cf. the passage on the circle-quadrature cited from Philoponus in note 41 above.

87 Cf. the studies of Mueller, cited in note 61 above.

88 The Neoplatonist element is evident throughout Proclus' work, not only in his commentary on Euclid, but also in his commentaries on Plato. Pappus' philosophical inclinations are most clearly seen in his commentary on Euclid's Book X (edited from the Arabic manuscript by G. Junge and W. Thomson [Cambridge, Mass.: Harvard University Press, 1930]); cf. also my *Ancient Tradition* ch. 8.

89 In the preface to the *Method*, for instance, Archimedes insists on the distinction between heuristic treatments, like that in accordance with his "mechanical method," and formally acceptable geometric proofs. His technique in this work makes prominent use of manipulations of infinitesimal magnitudes (e.g., the line segments which comprise a plane figure). Criticizing difficulties in the concept of indivisible magnitudes was a major issue for Aristotle (as in *Physics* Book VI), who already had a wealth of material to draw from in the debates surrounding the doctrines of the pre-Socratic atomists, and left a legacy for future contention among the Epicureans and Stoics. But in the *Method*, Archimedes has not a single word to contribute to these debates. Apparently, the philosophical coherence of his method was irrelevant for its heuristic efficacy. In this, I take him to be representative of the division between technical and philosophical interests in the ancient mathematical field.

90 In Proclus' view the omission follows from the need to apply higher curves, like the conchoids, quadratrices, spirals, or other such "mixed lines," which lie outside the domain of the *Elements* (*In Euclidem* 272). Pappus ascribes the belated efforts on these problems to the ancients' early unfamiliarity with the theory of the conics and the difficulty of drawing these curves (*Collection*, ed. Hultsch, I 54, 272). In neither case is there any question of the *existence* of the solving entities.

91 See the passages from Pappus and Apollonius cited in notes 79 and 80.

KEN SAITO

BOOK II OF EUCLID'S *ELEMENTS* IN THE LIGHT OF THE THEORY OF CONIC SECTIONS

INTRODUCTION

This paper proposes an alternative to the prevailing interpretation which regards the second book of the *Elements* (hereafter *Elem.* II) as basic part of the "geometric algebra". Chapter I of this paper is dedicated to an examination of the *Conics* of Apollonius. Though the central part of the "geometric algebra" is usually explained as a translation of the Babylonian algebraic techniques, it is not reasonable to attempt a determination of the nature of Book II by inconfirmable conjectures regarding its origin. The significance of *Elem.* II should be sought by studying applications of the propositions there. Thus we should examine how Euclid utilizes his propositions in *Elem.* II, when arguing about this book. The propositions in the *Elements* have been thoroughly examined by Ian Mueller.[1] But Mueller's study is not sufficient for purposes of the present paper, precisely because he limits his study to the *Elements*; other works of Euclid as well should be examined. Hence, the study of the *Conics* is necessary, since compilation of the fundamental part of the theory of conic sections is attributed to Euclid. The examination of the *Conics* to shed light on *Elem.* II can also be justified by the fact that the term "geometric algebra" originates in Zeuthen's study of Apollonius. Throughout my examination, geometric intuition in the *Conics* will be emphasized. In the second chapter, I examine *Elem.* II itself. Overall, this study is a refutation of the common interpretation of *Elem.* II, and an attempt to advance Mueller's study a step further.

CHAPTER I. THE "GEOMETRIC ALGEBRA" IN APOLLONIUS'S CONICS

Apollonius's *Conics* is one of the greatest works in Greek mathematics, well known for its difficulties to modern readers. The *Conics* were thoroughly studied by Zeuthen, and his *Die Lehre von den Kegelschnitten im Altertum*[2] remains the standard work. Zeuthen characterized the argument of Apollonius by its two | major auxiliary 32
methods[3] (Hülfsmittel), namely, the theory of proportions (*Elem.* V and VI) and the use of propositions in *Elem.* II. The latter was "geometric algebra", which is now a

1 I. Mueller, *Philosophy of Mathematics and Deductive Structure in Euclid's Elements* (MIT Press, 1981).

2 H. G. Zeuthen, *Die Lehre von den Kegelschnitten im Altertum* (Kopenhagen, 1886; reprint, Hildesheim, 1966) (hereafter *Die Lehre*).

3 *Die Lehre*, 1er Abschnitt.

J. Christianidis (ed.), Classics in the History of Greek Mathematics, 139–168.

standard appellation for this book. Zeuthen insisted that these auxiliary methods were algebraic in essence, and that the whole argument of Apollonius was equivalent to modern algebraic operations. This view has been widely accepted. For example, van der Waerden states:

> ... Apollonius proves geometrically all the algebraic transformations performed on the equation. The line of thought is mostly purely algebraic and much more "modern" than the abstract geometric formulation would lead one to think. Apollonius is a virtuoso in dealing with geometric algebra and also a virtuoso in hiding his original line of thought. ...[4]

That the Greeks had algebraic modes of thought and hid their original (algebraic) line of thought under the guise of geometric formulations, is a leitmotiv of this interpretation. Is this idea, which can be traced back to the 16th- and 17th-century mathematicians, to Viète and Descartes, and even to Ramus, justified? Are Apollonius's arguments so tortuous that they cannot be understood without attributing to him some other line of thought (viz. algebraic) than that derived from the text itself?

In the following examination of some propositions in the *Conics*, I will make it clear that Apollonius's thought can be better understood if we assume that crucial steps of his argumemts depend on geometric intuitions.

34 My examination concentrates on propositions concerning the interchangeability | of the diameters of the conics (*Conics*, I 42–51), and one of the so-called power propositions (III 17) together with its lemmata (III 1–3),[5] leading me to hope that my choice will not be criticized for being arbitrary. Propositions I 42–51 are recognized as the aim of the first book of the *Conics*,[6] and III 17 as one of the propositions used in the solution of the four line locus,[7] of which Apollonius was so proud in his preface.[8]

Before minute examinations of these propositions, a rough sketch of the contents of the first book is in order. Propositions 1–10 are preliminary, and are followed by the derivation of the symptoms of the three conics (or four, if we count the opposite sections—hyperbola with two branches—as another conic section in the case of the hyperbola).

---

4 Van der Waerden, *Science Awakening*, tr. by A. Dresden (Groningen, 1959), p. 248 (hereafter *SA*).

5 All the propositions concerning the opposite sections (hyperbola with two branches) are omitted in my examination. The reasons for this are twofold: first, the characteristics of Apollonius's argument are sufficiently revealed through those propositions not involving the opposite sections; second, arguments involving the opposite sections are not included in the theory of the conic sections at the time of Euclid, and thus are not adequate for my purpose as stated in the introduction.

6 For example, see *SA*, p. 249.

7 *Die Lehre*, pp. 126–154. T. L. Heath reproduces Zeuthen's arguments in his *Apollonius of Perga* (Cambridge, 1896; reprint, 1961), pp. cxxxviii–cl. (hereafter Heath, *Apollonius*).

8 J. L. Heiberg ed., *Apollonii Pergaei quae graece exstant* (1891; reprint, Stuttgard, 1974), Vol. 1, p. 4. Note that there are sufficient reasons to believe that the propositions chosen here had been known before Apollonius, perhaps by Euclid. Archimedes cites the same proposition as *Conics* III 17 in his *Conoids and Spheroids* prop. 3, and states "This is included in the *Conic Elements*", which is usually identified with the lost work of Euclid. In Apollonius's *Conics*, III 17 is based, through III 1–3, on I 42, 43, which form the substantial part of the theorem on the interchangeability of diameters. As a result, it is likely that I 42, 43 were also included in the *Conic Elements* and thus natural to assume that the whole theorem on the interchangeability of the diameters was also included therein. As for a detailed argument on the contents of the *Conic Elements* of Euclid, see Heath, *Apollonius*, pp. xxxiv–xxxvi.

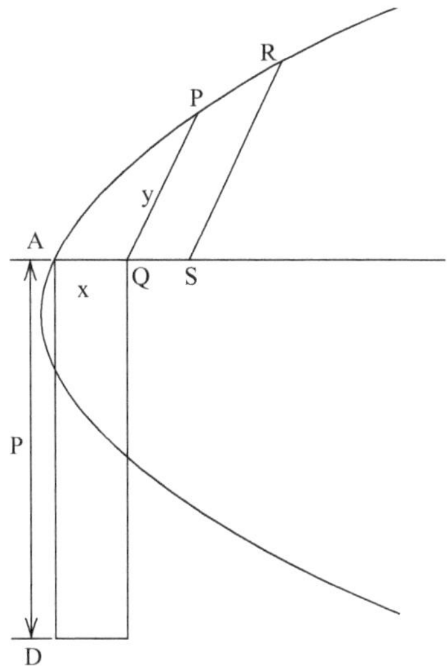

Fig.1

I 11 $\mathbf{T}(PQ) = \mathbf{O}(AD, AQ)$[9] (Fig. 1)

I 12,13 $\mathbf{T}(PQ)$ = rectangle AQEF (Fig. 2, 3 respectively)[10]

AQ is the diameter of the curve and bisects any chord parallel to the ordinate PQ. The point A is called the vertex. Called the *latus rectum*, line AD is drawn perpendicular to the diameter AQ at the vertex A. The length of *latus rectum* is determined to satisfy a certain proportionality for each conic section. A′ is the point where the diameter cuts the cone again. AA′ is called *latus transversum*.[11]

Though Apollonius's symptoms are different from those of his forerunners such as Archimedes, the difference is not substantial. First, Apollonius's symptom allows the diameter and the ordinate to be oblique to each other, while his predecessors made them orthogonal. But this does not mean that Apollonius was the first to discover the oblique symptom. Second, Apollonius's symptoms are stated | as the equalities 35
between the square of the ordinate and the rectangle applied on the *latus rectum*. According as the application is accomplished either without excess and defect, or with excess, or with defect, the conic sections produced are called *parabole, hyperbole*, and *elleipsis*, respectively. Van der Waerden seems to believe that this expression of the symptom is a sign of Apollonius's algebraic line of thought:

> The diagrams of Apollonius, here reproduced, consist of two unequal parts; one could speak of a geometric and an algebraic diagram. The geometric diagram

9 Hereafter I will use **T**(XY) for "the square on XY", and **O**(XY, YZ) for "the rectangle contained by XY and YZ".

10 Proposition I 14, the case of the opposite sections, is omitted.

11 For a detailed description, see Heath, *Apollonius*, pp. 8–12.

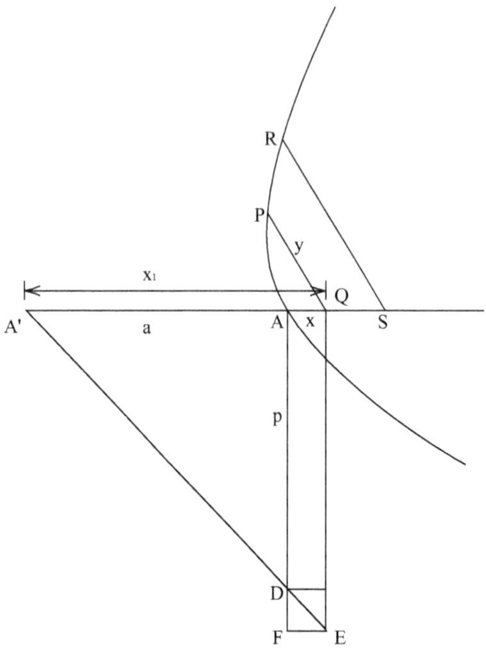

Fig. 2

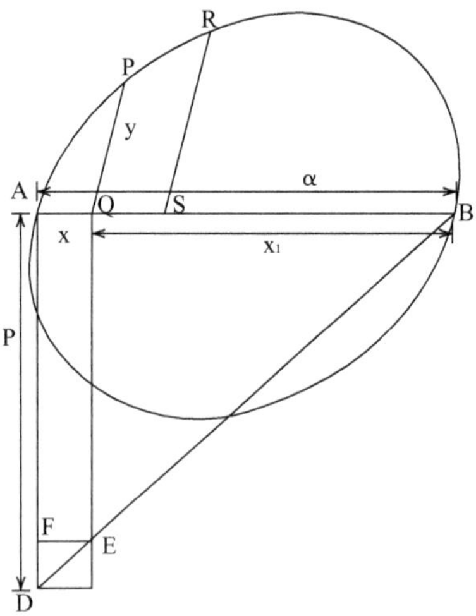

Fig. 3

> consists of the conic section, in which we are actually interested, the oblique axes, the abscissa x and the ordinate y. The algebraic diagram is not oblique, but rectangular; it expresses the algebraic relation between a, p, x, and y. The areas which are here added and substracted correspond to the terms of an equation in modern analytical geometry, ...[12]

Van der Waerden's claim seems, however, deceptive. We should note that these allegedly algebraic symptoms play very little role in the propositions of the *Conics* except in the case of the parabola. In this case the symptom before Apollonius was already stated in virtually the same manner, *i.e.* the equality between the square of the ordinate and the rectangle contained by the abscissa and a line of constant length. In the cases of the hyperbola and the ellipse, Apollonius usually uses other forms of the symptoms which he derives later, namely in

Conics I 20: **T**(PQ): **T**(RS) = QA: AS (Fig. 1. For the parabola)

Conics I 21 (a): **T**(PQ): **O**(AQ, QA′) = *latus rectum: latus transversum* (Fig. 2, 3, for the hyperbola and the ellipse, respectively)

(b): **T**(PQ): **O**(AQ, QA′) = **T**(RS): **O**(AS, SA′)

For the hyperbola and the ellipse, Apollonius almost always uses I 21 (a), which is virtually equivalent to that of Euclid and Archimedes,[13] though Apollonius extends it to oblique conjugations. Thus, *Conics* I 11–14 are stated only as formal definitions of the conic sections, not as substantial bases for the later descriptions. Moreover, it appears that the orthogonal diagram of Apollonius which van der Waerden claims to be algebraic, is in fact a *geometric* expression (or visualization) of the symptom which can be sufficiently stated in terms of proportionality without any additional new diagram, and thus that Apollonius's intention is to visualize the relation expressed by the symptom.

Proposition I 17 states that the line parallel to the ordinate at the vertex is tangent to the section. Propositions I 18–31 deal with the intersections of conic sections and straight lines. I 32 is the converse of I 17, showing the uniqueness of the tangent at the vertex. I 33–34 are concerned with tangents of the conic sections at any point. For the parabola (I 33), the tangent at point C is drawn as follows | (Fig. 4): Draw ordinate CD 36
through C, and take a point E in the extension of the diameter AD, letting EA=AD. EC is thus the tangent of the parabola at the point C. For the hyperbola and the ellipse, point E is taken to satisfy the proportionality BE: EA = BD: DA (Fig. 5, 6), where B is the opposite end of the diameter. This proportionality is transformed in I 37 as follows:

I 37 (a) **O**(DZ, ZE) = **T**(ZA)

I 37 (b) **O**(AD, DB) = **O**(ZD, DE), or

T(CD): **O**(ZD, DE) = *latus rectum: latus transversum*.

---

12 *SA* pp. 247–248. Here van der Waerden describes the symptoms in the following manner. Let AD = p PQ = y, AA′ = a, AQ = x, A′Q = x_1 and the ratio $p: a = \alpha$. Then,

I 11: $y^2 = px$.

I 12, 13: $y^2 = x\alpha x_1 = x(p \pm \alpha x)$. (the + sign applies to the hyperbola and the − sign to the ellipse)

13 See Heath, *Apollonius*, pp. xxxv, 1–li.

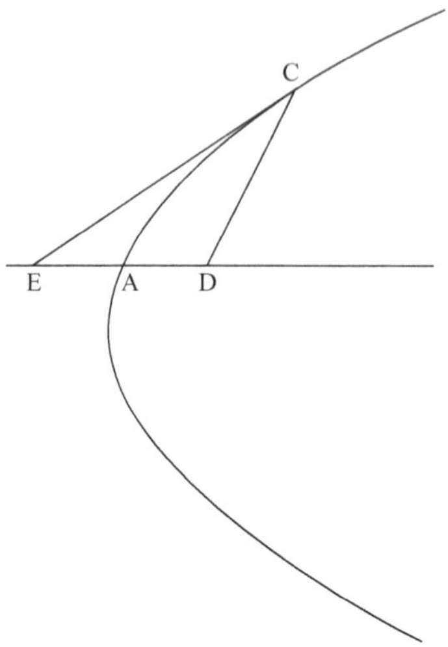

Fig. 4

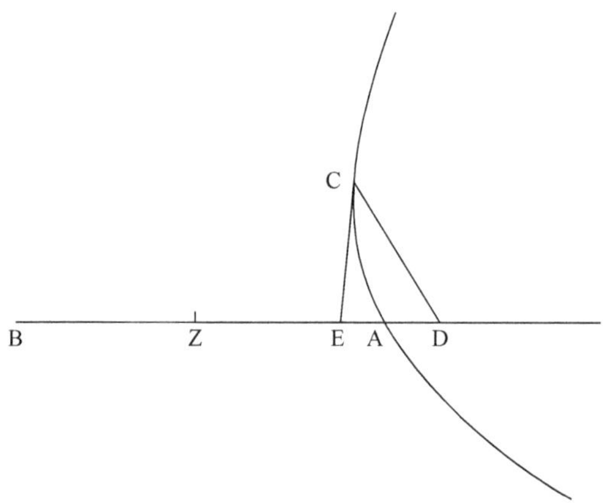

Fig. 5

37 | I 35–36 are the converses of I 33–34 respectively, and state that if CE is the tangent of the conic section, the above equality or proportionality holds. I 39–41 can be regarded as lemmata for the theorems of interchangeability of diameters for the hyperbola and ellipse, while I 42–51 form propositions concerning such interchangeability. I 52–60 demonstrate the existence of a cone and plane generating a conic section given in terms of the symptom. According to these propositions, any curve

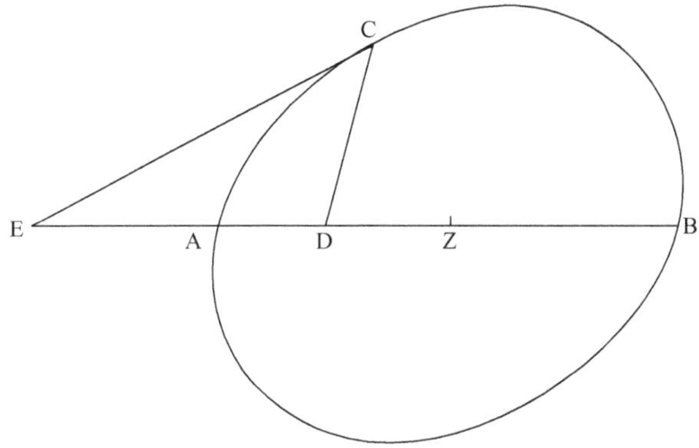

Fig. 6

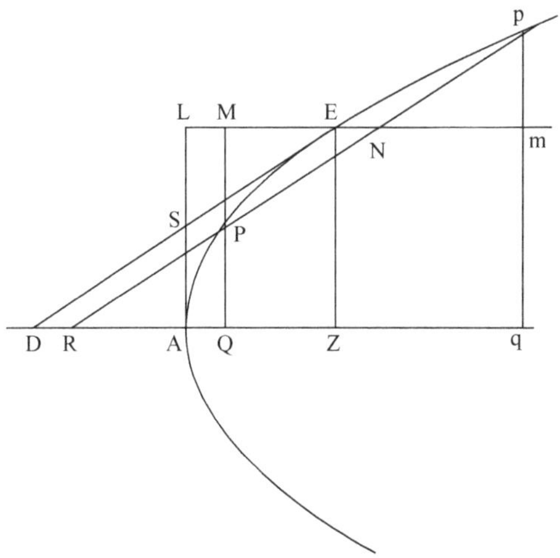

Fig. 7

expressed by a symptom of conic sections can be actually produced by cutting a cone with a plane.

The following is a detailed examination of propositions concerning the interchangeability of diameters. I begin with the case of the parabola. The results can be summarized as follows:

Let AE be a parabola with diameter AZ (Fig. 7). Any line EN parallel to AZ can be taken as a new diameter of the parabola; *i.e.* (1) there exists a group of parallel chords, each one of which (*e.g.* pP) is bisected by EN (*i.e.* pN = NP), (2) there exists a line l' (the *latus rectum* for the new diameter) such that $\mathbf{O}(l', \mathrm{EN}) = \mathbf{T}(\mathrm{pN})$ for any pP. Apollonius proves (1) in I 46 and (2) in I 49. These two propositions depend entirely on

> I 42: Let a parabola be given through the origin A with diameter AD, and draw, at an arbitrary point E of the parabola, the tangent line ED and the ordinate EZ. Through an arbitrary point P on the parabola, draw the lines PR and PQ, parallel to the tangent line ED and the ordinate EZ, to their intersections with the diameter AD. Complete the parallelogram AZEL. Then we have
>
> triangel PQR=parallelogram ALMQ.[14]

The proof is as follows:

Since DA = AZ (I 33, 35),
tri. EZD = par. ALEZ.
38 | And ALEZ: ALMQ = ZA: AQ
= $\mathbf{T}$(EZ):$\mathbf{T}$(PQ)(I 20)
= tri. EZD: tri. PQR.
Therefore tri. PQR. = par. ALMQ.

This proposition is often explained as a transformation of the symptom of the parabola. Indeed if we interpret the symptom (I 11) and this proposition algebraically, they are essentially the same. In this point, van der Waerden states:

> Thus we see that I 42 is nothing but a transformation of the equation of the parabola ... But the formulation is already of such a character, that it contains two diameters, AD and EL, and two tangent lines AL and ED. This prepares for interchanging the roles played by the points A and E.[15]

His remark is completely understandable in the context of modern mathematics, but remains misleading. While it is true that I 42 is derived from the symptom of the parabola, the discovery of this proposition is by no means a result of such a derivation. Its necessity is first due to another reason, and its truth is later reduced to the relation already known, *i.e.* the symptom. To make this point clear, let us look at I 46, and try to reconstruct its analysis (in the sense in Greek mathematics).

> I 46: In the same parabola as in I 42, any chord Pp parallel to the tangent ED is bisected by the line EN; *i.e.* PN = pN.

This proposition is clearly an indispensable part of the theorem of the interchangeability of the diameters. The proof goes as follows:

From I 42,
triangle PQR = parallelogram ALMQ
triangle pqR = parallelogram ALmq.

---

14 I have cited van der Waerden's paraphrase of this proposition. *SA*, pp. 252–253.
15 *SA*, p. 253.

| By substraction, 39

par. pqQP = par. qmMQ.

Take away the common part PQqmN, then

tri. pmN = tri. PMN.

Since the two triangles pmN and PMN are similar,

pN = PN.

From this proof, we can see that I 42 does not precede I 46. It is natural to assume that the effort to discover a proof for I 46 led to the recognition of I 42. The process of the analysis can be easily reconstructed. To prove that EN (one of the parallels to the original diameter AD) is a diameter, one must find a group of parallel chords which are bisected by EN. I 17 and 32 suggest that these chords are parallel to the tangent at the new vertex E. Then, a chord Pp, parallel to ED, is drawn as a candidate for a new ordinate. Next, it must be proved that PN = pN. This simple equality is "reduced" to the equality between the areas of triangles (PNM and pNm). This in turn is transformed into the equality of qmMQ and pqQP. The next step, in which the relation of I 42 is induced, seems hard to discover, although the consideration of a special case in which P coincides with A may give a hint to this transformation.

As a result, we arrive at I 42 through the analysis of I 46. Though this analysis is a mere reconstruction from the existing proof, it is likely that I 42 is first recognized through this analysis. At very least, it would be surprising if I 42 were derived from the symptom first, without any definite purpose, and its application only later found by chance. As a result, we should be wary of overemphasizing its nature as a transformation of the symptom.

The reconstructed analysis of I 46 reveals another character of Apollonius's argument. There, the equality of two lines is transformed to that of two triangles, then the latter equality into that of other figures through geometric process, *i.e.* adding or taking away the same area. This argument is puzzling at first sight, since we tend to assume that the equality between lines is simpler than that between areas. But in the analysis of I 46, the reduction of the former to the latter is a crucial step. This reduction enables Apollonius to transform the desired equality freely through the observation of the diagram (not through alleged algebraic operations). In other words, it is impossible for Apollonius to connect the symptom of the parabola and its tangent (I 11, 33, 35) to I 46 (PN = Np) without the aid of geometric intuition. The two auxiliary methods, the "geometric algebra" and the theory of proportions, are insufficient to solve this problem by themselves. In the *Conics*, many results are written in terms of "geometric algebra" and the theory of proportions. The theorems on the interchangeability of the diameters are also written in these terms, as we shall see later (propositions I 49–51). But these are not always convenient to the investigation of new results, for though Apollonius has some means of treating the relations in these terms, geometric intuition is indispensable for his purpose. As a result, he is required to transform the symptom (I 11) into the equality between "visible" areas (though the
symptom is already ex|pressed as an equality between square and rectangle, these 40
figures have been added afterwards and have no geometric relations with the conic section and the tangent lines, as noted by van der Waerden), in order to arrive at the desired result (I 46). Most of his argument is indeed algebraic in the sense that it can

be translated into modern algebraic operation, but crucial steps of his argument depend on geometric intuition, and some of the operations are used to transform the relations already known or desired into a form in which one can rely on geometric intuition.

Neither Zeuthen nor van der Waerden is willing to adimit this point, and they fail to pay due attention to I 46. Instead, they analyze I 42, explaining it as a derivation of the symptom through "algebraic" transformation, and thus emphasizing its character as another form of the symptom.[16] According to these scholars, it would appear that I 46 is merely a by-product of the algebraic operation. But as I noted earlier, it is not I 46 but I 42 that is a by-product of Apollonius's investigation. It strongly appears that scholars like Zeuthen were so convinced of the algebraic character of the *Conics* that they tended to overlook the significance of arguments which did not conform to the algebraic interpretation. Thus far I have pointed out one of the characteristics of the *Conics*, the intention of visualization. In the following argument it will become clearer that many other propositions in the *Conics* also confirm this characteristic.

Let us proceed to I 49 (Fig. 7). As PN = Np has been proved in I 46, it remains to prove that for some line l',

$$\mathbf{O}(l', \mathrm{EN}) = \mathbf{T}(\mathrm{pN}) \quad \cdots\cdots\cdots\cdots\cdots\cdots\cdots (※)$$

The proof is simple:

tri. SDA = tri. SLE, since AD = AZ = LE (I 33, 35)
Add to each figure ESAqm. Then,
EDqm = ALmq = tri. pqR (I 42)
Take away the common part NRqm, then
tri. pmN = DENR.

This is a "visualized" form of the desired symptom (※), just as I 42 is a visualized form of I 11, and the proof has been substantially accomplished; it remains merely to express the new *latus rectum* l' in terms of proportionality. The result is

SE : EL = l' : 2ED.

Here again, geometric transformation is the key of the proof. The process of expressing l', which may be called algebraic, does not involve an important step. It is a finishing touch to the result already found and confirmed by geometric intuition, and its aim is to bring the result into conformation with the standard style of statement as a proposition.

41 Let us examine the case of the hyperbola and the ellipse. I 43, 47 and 50 cor|respond respectively to I 42, 46 and 49 in the case of the parabola.

Neugebauer correctly pointed out that these propositions and their proofs of the case of the hyperbola were first found by analogy to those of the parabola.[17] Let us

---

[16] *Die Lehre*, pp. 102–103. *SA*, pp. 252–256.

[17] O. Neugebauer, "Apollonius Studien", *Quellen und Studien*, Bd. 2, pp. 215–254. Though Neugebauer refers only to the interrelation of I 42 and I 43, I extend this analogy to I 46 and I 47, since, as I have argued, I 42 should be regarded as a lemma for I 46.

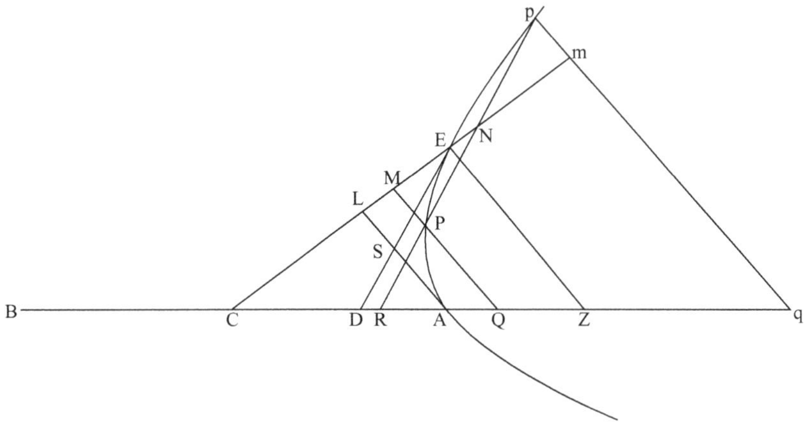

Fig. 8

recapitulate his argument. Fig. 7 is the diagram for I 42, 46 and 49, and Fig. 8 for I 43, 47 and 50.[18] I 43 states that

tri. PQR = ALMQ.

I 47, PN = Np.

Since these results correspond precisely to those in I 42 and 46, it is natural to assume that not only the result but also the proof for the theorem of interchangeability of diameters of the hyperbola has been modelled after that of the parabola. Indeed, one can see that the proof of I 47 corresponds literally to that in I 46.

Thus it is indisputable that Apollonius utilizes such analogies based on the observation of the diagrams. This fact is another support for Apollonius's dependence on geometric intuition. It also suggests that I 43, which is often interpreted as "symptom referred to the center", is nothing but a lemma for I 47. This proposition I 43 is an end result of the analysis of I 47, so that its truth is first expected as a sufficient condition for I 47. It can no more be justified to regard I 43 as if it were investigated for itself, calling it a "symptom referred to the center", than to regard I 42 as a mere transformation of the symptom of the parabola.

Let us examine Apollonius's argument in detail. As noted above, the proof of I 47 is completely conformable to that of I 46. On the other hand, the case for I 43 is far more complicated.[19] According to the extant text, this proposition is proved through two lemmata, namely, I 39 and 41. The former is derived from | the second result of 42

18 Note that in the case of the hyperbola and ellipse, the candidate for new diameter is the line which passes through the center of the curve.

19 Neugebauer gives an algebraic reconstruction for the proof of I 43, but it becomes more complicated than the original text. Moreover, as he admits, the reconstruction is not faithful to the text, and I find it unconvincing.

I 37 (which I have called I 37 (b)), and the latter is based on the symptom. But several parts of these propositions overlap with Pappus's lemmata for the *Conics*, leading to the conclusion that the original Apollonius's text probably lacked these parts, and the text which has come down to us is the result of an inclusion of Pappus's lemmata.[20] This makes it difficult to reconstruct Apollonius's line of thought. Moreover, Eutocius's commentary to the *Conics* complicates the situation even more, for he cites another proof of I 43, which he claims he found in other manuscripts.[21] This proof conforms better to that of I 42, the parallel proposition for the parabola.

Rather than seeking the original form of the proof of I 43, I would like instead to point out some crucial steps which must have been indispensable, whatever Apollonius's proof may have been.

In I 42, which is indisputably the model for I 43, the ratio ALMQ: ALEZ can be at once reduced to the ratio AQ: AZ, since LM is parallel to AQ. In I 43, where LM and AQ are not parallel, but intersect at C, one must find some other means to treat the area of ALMQ. The analysis based on the proof which Eutocius transmits us is as follows (Fig. 8): the expected equality

ALMQ = tri. PQR

is reduced to

ALEZ = tri. EZD,

because tri. PQR: tri. EZD
= **T**(PQ): **T**(EZ)
= **O**(AQ, QB): **O**(AZ, ZB) (I 21)
= **T**(CQ) − **T**(CA): **T**(CZ) − **T**(CA) (*Elem.* II 6)
= tri. CMQ − tri. CLA: tri. CEZ − tri. CLA
= ALMQ: ALEZ.

And the reduced equality (ALEZ = tri. EZD) is transformed into the form

tri. ESL = tri. ASD

then, into

tri. CED = tri. CAL,

which can be easily proved by

I 37 (a): **T**(CA) = **O**(DC, CZ).

It is clear thet *Elem.* II 6 plays a crucial role in connecting the expected equality to the symptom of the hyperbola. And in the proof which we see in the text of the *Conics*, the same proposition in the *Elem.* II is used in a similar way.

The examination of I 43 and 47 confirms my view that it is geometric intuition that plays a central role in the *Conics*. But this is not all. I have already argued that the

---

20 See J. L. Heiberg ed. *Apollonii Pergaei etc.*, Vol. 2, p. LIX.
21 *Ibid.*, Vol. 2, pp. 254–264.

process of visualization is a necessary element in the reliance on geometric intuition. The proof of I 43, examined above, suggests that the propositions in *Elem.* II play an important role in this process. Let me illustrate this point. The symptom of the hyperbola:

| **T**(PQ): **O**(AQ, QB) = *latus rectum* : *latus transversum* (Fig. 8) 43

is expressed in terms of "invisible" figures such as **T**(PQ) and **O**(AQ, QB). By *Elem.* II 6, **O**(AQ, QB) can be replaced by **T**(CQ) − **T**(CA), which in turn can be transformed into the difference of two similar triangles, *i.e.* a trapezium. Thus, by aid of the theory of proportions, *Elem.* II 6 makes an invisible rectangle **O**(AQ, QB) "visible". From this point of view, the significance of the two main auxiliary methods mentioned by Zeuthen is completely different. They are not methods of treating the lines and areas as general quantities in a way similar to modern algebra, but they are the means for transformation between "visible" and "invisible" forms of areas. The former is indispensable because it makes geometric intuition available, while the latter is adapted to the formal statement of results as propositions and is used in the expression of the symptoms.

Proposition I 43 of the *Conics* provides another important suggestion regarding the character of *Elem.* II. The equality

O(AQ, QB) = **T**(CQ) − **T**(CA) (Fig. 8)

is guaranteed by *Elem.* II 6 in the case of the hyperbola. On the other hand, in the case of the ellipse, the equality

O(AQ, QB) = **T**(CA) − **T**(CQ) (Fig. 9)

holds, because the point Q falls between A and B, and this equality is based on *Elem.* II 5. Here, we encounter a situation in which both of II 5 and 6 are naturally required. This is a notable result, for the significance of these two propositions has been polemical, this because the one seems to make the other unnecessary. This point will be discussed in the next chapter, where I argue that such alternate (mutually complementary) uses of propositions (according to the arrangement of the points)

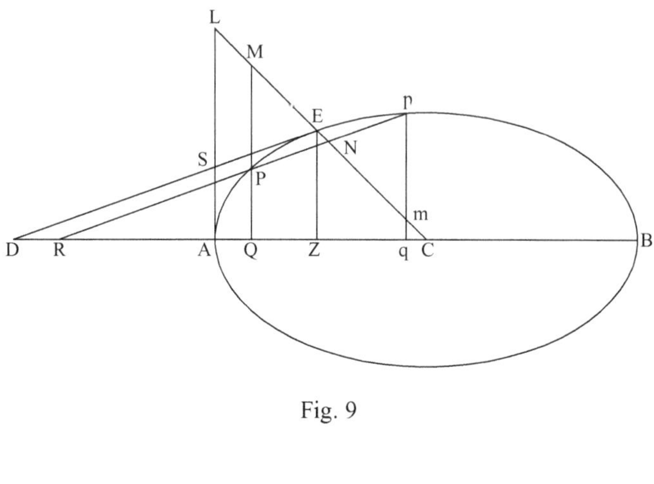

Fig. 9

have already been taken into account in the compilation of the *Elements*. For the moment, I confine myself to add that the alternate use of *Elem.* II 5. 6 is seen very often in the *Conics*.

44 As noted ealier, Proposition III 17 of the *Conics* is one of those used in the | solution of the four line locus, and the only one not involving the opposite sections. First, an examination of its preliminaries, namely, III 1–3.[22]

> III 1 (Fig. 10, 11): P, Q being any two points on a conic, if the tangent at P and the
> 45 diameter through Q meet in E, and the tangent at Q and the diameter | through P in T, and if the tangents intersect at O, then
>
> $$\text{tri. OPT} = \text{tri. OQE}$$
>
> III 2: (Keep the notation of III 1) If R be any other point on the conic, le tRU be drawn parallel to QT to meet the diameter through P in U, and let a parallel through R to the tangent at P meet QT and the diameters through Q, P in H, F, W respectively. Then
>
> $$\text{tri. HQF} = \text{HTUR}$$
>
> III 3: (Keep the notation of III 1,2) Take two points R′, R on the curve with points H′, F′, *etc.* corresponding to H, F, *etc.* And if RU, R′W′ intersect in I, and R′U′, RW in J, then
>
> $$\text{F}'\text{IRF} = \text{IUU}'\text{R}'$$
>
> $$\text{and}\ \ \text{FJR}'\text{F}' = \text{JU}'\text{UR}.$$

We note that III 2 is a result stemming directly from I 42, 43 and that III 3 is proved from III 2 through purely geometric operations, *i.e.* adding or removing the equal areas. As for III 3, Zeuthen is likely correct in asserting that the case of the hyperbola and the ellipse can be interpreted as a proposition to the effect that the quadrilateral CFRU is constant. It is probable that Apollonius understands III 3 in this way. But it is absurd and anachronistic to argue, as Zeuthen did, that the proposition represents a symptom of conics referred to the oblique axes,[23] for there is no evidence at all that Apollonius regards the quadrilateral CFRU in that way.

The significance of these preliminary propositions become fully clear in their applications in the power propositions. Let us examine III 17 (Fig. 10, 11):

> III 17: If OP, OQ be two tangents to any conic and Rr, R′r′ two chords parallel to them respectively and intersecting in J, then
>
> $$\mathbf{T}(\text{OP}) : \mathbf{T}(\text{OQ}) = \mathbf{O}(\text{RJ, Jr}) : \mathbf{O}(\text{R}'\text{J, Jr}').$$

The outline of the proof is as follows:

$$\mathbf{T}(\text{JW}) : \text{tri. JWU}' = \mathbf{T}(\text{WR}) : \text{tri. RWU},$$

$$\text{then} \quad \mathbf{T}(\text{RW}) - \mathbf{T}(\text{WJ}) : \text{RJU}'\text{U} = \mathbf{T}(\text{WR}) : \text{tri. RWU}$$

$$\therefore \quad \mathbf{O}(\text{RJ, Jr}) : \text{RJU}'\text{U} = \mathbf{T}(\text{WR}) : \text{tri. RWU} \ (\textit{Elem.}\ \text{II 5, 6})$$

$$= \mathbf{T}(\text{OP}) : \text{tri. OPT}.$$

---

22 The paraphrase below is based on Heath, *Apollonius*, pp. 84–87. I have omitted the diagram in the case of the ellipse.

23 *Die Lehre*, p. 98. Van der Waerden follows Zeuthen on this point. See *SA*, p. 256.

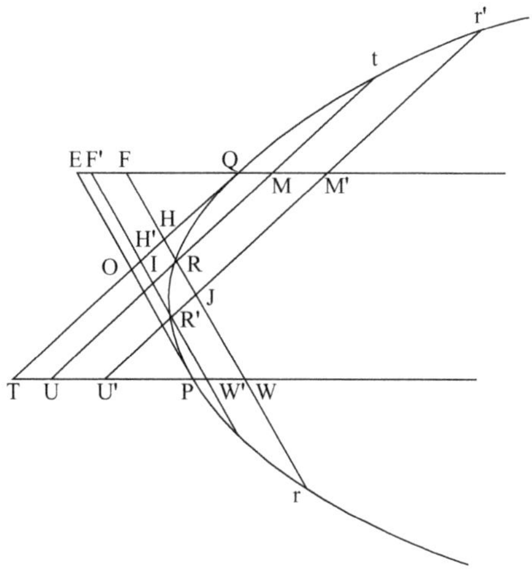

Fig. 10

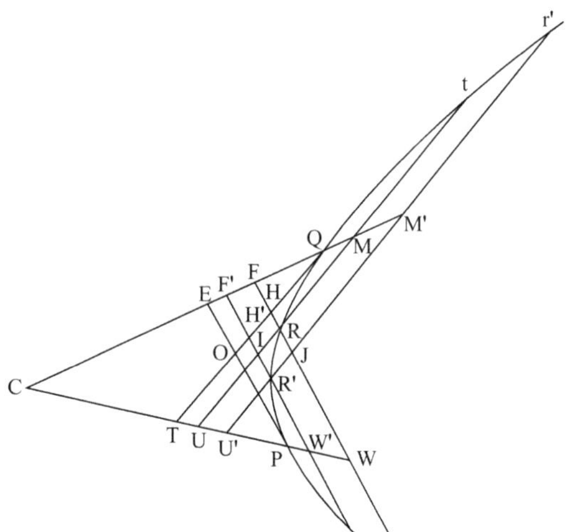

Fig. 11

We have already proved $RJU'U = R'JFF'$ (III 3)
and tri. $OPT =$ tri. OQE (III 1),
therefore $\mathbf{O}(RJ, Jr): R'JFF' = \mathbf{T}(OP)$: tri. OQE.
Similarly, $\mathbf{O}(R'J, Jr'): R'JFF' = \mathbf{T}(OQ)$: tri. OQE.

The remainder of the proof is simple and straightforward.

In this proof, the intention of visualization is manifest. **O**(RJ, Jr) is a rectangle contained by two line segments which lie in a line, making the rectangle invisible. The case is the same for **T**(OP). These invisible figures are transformed into visible ones: trapezium RJU′U and triangle OPT. This transformation is made possible by *Elem.* II
46 5, 6 and the theory of proportions. It is also noteworthy that both | *Elem.* II 5, 6 are necessary depending on whether the intersection J falls inside or outside of the curve.[24] Here, we have another example of the mutually complementary uses of these two propositions in *Elem.* II.

The above examination of the *Conics* has revealed an important characteristic in Apollonius's investigation, namely the essential dependence on geometric intuition. The alleged algebraic methods, *i.e.* "geometric algebra" and the theory of proportions, which have long been assumed to be the backbone of Apollonius's thought, bears only peripheral importance to the course of the investigations of new results. Their role can be summed up as follows:

(1) To transform known or desired relations into a form suitable for the use of geometric intuition;
(2) To prove the relation already anticipated by geometric intuition;
(3) To transform the results into a form suitable for formal statements as a proposition.

I 42 and the analysis in I 46 provide the example for (1). The proof of I 43 (together with I 37, 39, 41) is an example of (2). As an example of (3), we have seen in I 49, that the relation tri. pmN=DENR (Fig. 7) has been transformed into the form

$$\mathbf{O}(l', EN) = \mathbf{T}(pN),\ l' \text{ being such a line that } SE : EL = l' : 2ED$$

To make the situation clearer I have introduced the words "visible" and "invisible". Apollonius's investigation depends on geometric intuition which is valid only for "visible" figures, *i.e.* lines and areas as they are. But the results of the investigation thus obtained are not suited for statement as a proposition, and need to be transformed into a concise, general form, which is the expression in terms of "geometric algebra" and the theory of proportions. The main feature of this formal expression is the "invisibility" of the figures involved, such as "the square of the ordinate" and "the rectangle contained by two line segments which lie in a line". When the results are used again in later investigations, they are visualized to make the use of geometric intuition possible. This is the process I have summed up above as (3) and (1).

---

[24] Apollonius does not explicitly refer to the case in which the point J falls outside of the curve. In III 16, however, the limiting case of a chord coinciding with the tangent, J does fall outside the curve, and *Elem.* II 6 is required.

In the next chapter I will further examine the mutually complementary uses of *Elem.* II, showing that the object of the treatment in *Elem.* II was never "general quantities", but the lengths of lines conceived together with their positions and arrangements.

CHAPTER 2. ON THE INTERPRETATION OF THE *ELEMENTS*, BOOK II

The first ten propositions of the *Elements* II have been ususally interpreted as geometric expressions of algebraic theorems. As mentioned above, this interpretation
began with Zeuthen, who has called the book "geometric algebra". Follow|ing Mueller 47
I will refer to this explanation as the "algebraic interpretation."[25]

The core of the algebraic interpretation consists in the identification of the Euclidean propositions with algebraic equalities. But this simple identification involves difficulties, since two propositions may correspond to the same algebraic equality. For example, both II 5 and II 6 can be expressed by the equality:

$$(a + b)(a - b) = a^2 - b^2$$

and some other pairs of propositions which can likewise be represented by a single algebraic equality.[26]

II 5 and 6 are interpreted as solutions of the following sets of equations.

$$\begin{cases} x + y = p \\ xy = q \end{cases} \quad \text{(II 5)} \quad (1) \qquad \begin{cases} x - y = p \\ xy = q \end{cases} \quad \text{(II 6)} \quad (2)$$

It is well known that propositions of the *Data* (84, 85) support this explanation, and the application of areas (*Elem.* VI 28, 29), which is attributed to the Pythagoreans, can also be regarded as an extension of these problems.

This interpretation was first raised by Tannery,[27] who accepted the algebraic character of the propositions though he was unable to furnish evidence for the existence of the algebraic equations (1) (2). Half a century later, the discovery of Babylonian mathematical texts seemed to confirm this view, for these texts were thought to contain numerical solutions of these equations. Thus the "geometric algebra" was regarded as a geometrized version of Babylonian mathematics, and the algebraic interpretation became virtually an established assumption.[28]

This interpretation, however, remains controversial. It seems clear that II 5, 6 (as well as other propositions) are solutions to some other problems, since they make little sense when considered alone. But it remains questionable whether the problems involved are algebraic. The *Elements* does include cases in which II 5, 6 are utilized.

---

25 Mueller, *op. cit.* (see note 1), p. 42.

26 II 9 and 10 are another pair. The existence of these pairs are known as "double form" in Book II. I will refer to the propositions making up a pair "twin-propositions".

27 P. Tannery, "De la solution géométrique des problèmes du second degré; avant Euclide" (1882), in his *Mémoires scientifiques*, J. L. Heiberg and H. G. Zeuthen eds. (Toulouse and Paris, 1912), pp. 254–280.

28 I assume readers' knowledge of the course of events surrounding the interpretation of the "geometric algebra" after Zeuthen and Tannery. See *SA*, pp. 118–124.

For example, II 14 and II 11 use II 5, 6 respectively. On the other hand, although the algebraic interpretation assumes algebraic problems underlying the *Elem.* II, there is no direct evidence to confirm this assumption. Mueller has vigorously investigated this point. He has summarized the characteristics of the algebraic interpretation quoting Zeuthen and van der Waerden as follows:

1. The lines and areas of geometric algebra represent arbitrary quantities;
2. Geometric algebra is a translation of Babylonian algebraic methods;
3. The "line of thought" in much of Greek mathematics is "at bottom purely algebraic".[29]

48 | Mueller admits that the truth of any one of these would be sufficient to establish the algebraic interpretation excluding the geometrical one (that is, the interpretation of the propositions in book II as lemmata for geometric propositions). He examines these claims carefully and concludes that there is no conclusive evidence to establish any one of them. Mueller then asks whether geometric interpretation is possible. Attempting to find the propositions in the *Elements* where II 1–10 are used, he concludes after a minute and thoroughgoing examination, that II 5 and 6 can be interpreted as lemmata for II 14 and II 11 respectively.[30] According to Mueller, II 14 is a reworking of the proof of VI 13, and he argues convincingly that II 14 is a result of the effort to avoid the use of the theory of proportions in VI 13; the necessity of II 5 arises in this process of reworking.[31] For II 11, which is parallel to VI 30, Mueller provides a similar argument[32]. He reproduces the process in which the effort to construct the regular pentagon without the use of the theory of proportions led to the recognition of II 11 and II 6. Though Tannery has already pointed out the repetitions of the same propositions in the *Elements* II and VI (in the former in terms of "geometric algebra", in the latter in the language of the theory of proportions),[33] Mueller views this fact from completely different standpoint and proceeds to develop a successful argument.

But Mueller does not succeed in explaining all the propositions at issue, *i.e.*, II 1–10. He confesses that some propositions (II 1, 3, 8–10) are never or only implicitly used in the *Elements*, or only in places of questionable authenticity.[34] The existence of unexplainable propositions is more damaging to Mueller's than to the algebraic interpretation, for, according to the latter, these propositions can be viewed as examples to illustrate the method of the "geometric algebra".

In the following argument, I fundamentally follow Mueller in investigating the possiblity of the geometric interpretation.

My points are as follows.

1. For each of the propositions II 1–10, there exists evidence or, at least, probability that Euclid used them in other propositions, so that the difficulty Mueller has encountered disappears.

---

29 Mueller, *op. cit.*, p. 50.

30 Note that this claim has already been made by Arpád Szabó, in his *Anfänge der griechischen Mathematik* (1969).

31 Mueller, *op. cit.*, pp. 161–162.

32 *Ibid.*, pp. 168–170, 192–194.

33 Tannery, *op. cit.*, p. 274.

34 Mueller, *op. cit.*, pp. 300–302.

2. The double form in II 5, 6, II 9, 10 and II 4, 7 can be explained by the geometric interpretation. Their role is that of mutual complement in the geometric context, *i.e.*, they are used alternately depending upon the arrangement of points and lines.
3. In relation to 2 above, I claim that the object of the argument in the second book of the *Elements* is not quantities in general. The areas and length of lines are always considered and treated together with their positions.
4. | *Elem*. II is meant to prove the equalities between areas of "invisible" figures, by reducing them to "visible" ones, and to afford a set of propositions for the treatment of "invisible" figures. 49

In the previous chapter, I have shown some examples of the mutually complementary use of *Elem*. II 5, 6. These examples are found in the *Conics* I 43 (more precisely, in its preliminary lemmata) and III 17, and both propositions originate in Euclid's *Conic Elements*. So it is probable that Euclid himself is responsible for the use of *Elem*. II 5, 6 in this way. Moreover, this mutually complementary use of II 5, 6 can be found in the *Elements* itself. In III 35, 36, which is a special case (for the circle) of the *Conics* III 17, *Elem*. II 5, 6 are used exactly in the same way as in the *Conics*.

We can find similar examples for other twin-propositions in the *Elem*. II. As is well known, II 4 and II 7 are used in II 13 and II 12 respectively. The distinction of the cases in these two propositions is completely parallel to that of III 35, 36. For II 9, 10, although we cannot find their application in the *Elements*, the *Conics* III 27, 28 offer examples of the mutually complementary use of these propositions. In III 27, for example, one of *Elem*. II 9 and 10 is used according to whether the intersection O of the two chords Rr, R′r′ falls inside or outside the ellipse (Fig. 12), and the equality

$$\mathbf{T}(\mathrm{RO})+\mathbf{T}(\mathrm{Or})=2\mathbf{T}(\mathrm{Rw})+2\mathbf{T}(\mathrm{wO})$$

is proved in both cases. For the same *Elem*. II 9, 10, Pappus provides us with a simpler example. This is the proof of the theorem concerning the median line of a

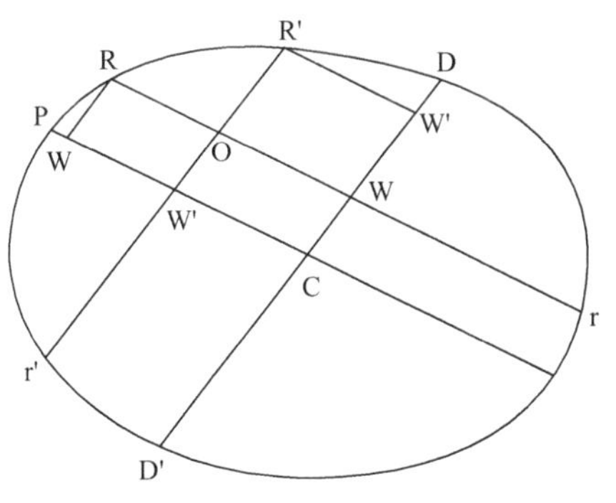

Fig. 12

triangle.[35] Though there is no evidence that Euclid did prove these propositions by *Elem.* II 9, 10, it is quite likely that Euclid knew some propositions for which II 9 and 10 are utilized in a mutually complementary way, and that Euclid inserted this pair of
50 propositions in the *Elements* for that reason. At very least, it is certain that | Euclid used the pairs II 5, 6 and II 4, 7 in this way, and he inserted another pair II 9, 10 in the same book of the *Elements*. If so, it would be rather unnatural to assume that Euclid could not find any case in which II 9, 10 are required in the same way as other pairs.

The case of *Elem.* II 2, 3 will make my argument more convincing. At first sight, these propositions appear to be nothing more than trivial special cases of II 1. Although these propositions have sometimes been explained as illustrations of the method of the geometric algebra (that is, some pedagogical intention is attributed to Euclid), it seems more reasonable to assume, with Heath, that their frequent necessity led Euclid to state them separately for sake of convenience.[36] But what Heath fails to mention is that II 2, 3 are a pair of twins the same as II 5, 6 *etc.* Pappus gives us an adequate example.[37] In one of the lemmata to the *Conics*, he proves

$\mathbf{O}(ZD, DE) = \mathbf{O}(AD, DB)$ (I 37 (b), Fig. 5, 6)

from $\mathbf{T}(ZA) = \mathbf{O}(DZ, ZE)$ (I 37 (a), Fig. 5, 6)

The proof is as follows.

For ellipse: take away $\mathbf{T}(ZD)$ from both sides of I 37 (a)

then, $\mathbf{O}(AD, DB) = \mathbf{O}(ZD, DE)$ (*Elem.* II 5 and 3 are used)

For hyperbola: take away the both sides of I 37 (a) from $\mathbf{T}(ZD)$

then, $\mathbf{O}(AD, DB) = \mathbf{O}(ED, DZ)$ (*Elem.* II 6 and 2)

It is clear that *Elem.* II 2, 3 play mutually complementary roles in this proof. Of course, it was Pappus, who thus used these propositions, and there is no evidence that Apollonius proved *Conics* I 37 in this way. We can argue still less about Euclid since we have no certainty that Euclid's *Conic Elements* was even so constructed as to include the same proposition. But Pappus's lemma shows at least a possibility of using II 2, 3 in a way parallel to II 5, 6 *etc.* It is natural to assume that Euclid was aware of this possibility and so decided to insert II 2 and 3, even though the latter II 3 is not utilized in other parts of the *Elements*.

This leaves two propositions (propositions 1 & 8) unexplained. Before proceeding to these, the general significance of the mutually complementary use of twin-propositions merits discussion. The double form would appear to furnish evidence that Euclid did not treat the length of lines as general quantities. This fact can be illustrated by reference to II 5, 6. In both of these propositions the rectangle contained by two line segments is at issue. If Euclid had regarded these line segments as representations of general quantities, thus neglecting their arrangement as insignificant, he would not have distinguished the two cases, for, in both propositions, the contents would have been the same.

---

35 Pappus, *La collection mathématique*, tr. by P. Ver Eecke (Paris, 1933, 1983), p. 662. T. L. Heath, *The Thirteen Books of the Elements*, Vol. 1, p. 401. (hereafter Heath, *Elements*)

36 Heath, *Elements*, Vol. 1, p. 377–78.

37 Pappus, *op. cit.*, pp. 723–24. In the following citation, I have put Pappus's proof into the context of the proposition in the *Conics*.

As Heath states, "The difference of the squares on two straight lines is equal to the rectangle contained by their sum and difference."[38] As a result, Euclid's motive in distinguishing the two cases must | have been the diversity of the arrangement of the 51
two line segments. In other words, he did not regard them as mere representations of quantities which can be placed arbitrarily, but as geometric existences the position and arrangement of which should also be considered. Euclid could not abstract the general quantity from line segments, or at least it was not convenient for him to perform such an abstraction. This leads to the conclusion that the first claim of the algebraic interpretation pointed out by Mueller cannot be supported.

Now for the two remaining propositions, *i.e.*, II 1 and II 8. We can find the application of II 8 in *Data* 86 which was cited by Tannery as an example of the solution of the equation by Euclid. Its content is as follows (Fig. 13):

> *Data* proposition 86.[39] If two straight lines contain a given area in a given angle and the square of the one is greater than the square of the other by a given area as in ratio, each of those lines will be given.
>
> Let the two straight lines AB and BG contain the given area AG in a given angle ABG and let **T**(GB) be greater than **T**(BA) by a given area as in ratio [*i.e.* the excess of **T**(GB) over a given area has a given ratio to **T**(BA)]. I say that each of AB and BG is also given.
>
> Since **T**(GB) is greater than **T**(BA) by a given area as in ratio, let the given area **O**(GB, BD) be taken away. Then the ratio of the remainder **O**(DG, GB): **T**(AB) is given.
>
> Since **O**(AB, BG) is given and **O**(GB, BD) is also given,
> the ratio **O**(AB, BG): **O**(GB, BD) is given.
>
> But **O**(AB, BG) : **O**(GB, BD) = AB : BD
> Hence the ratio AB : BD is given.
> Hence the ratio **T**(AB) : **T**(BD) is given.
>
> But **T**(AB) : **O**(BG, GD) is given.
> Therefore the ratio **O**(BG, GD) : **T**(DB) is also given.

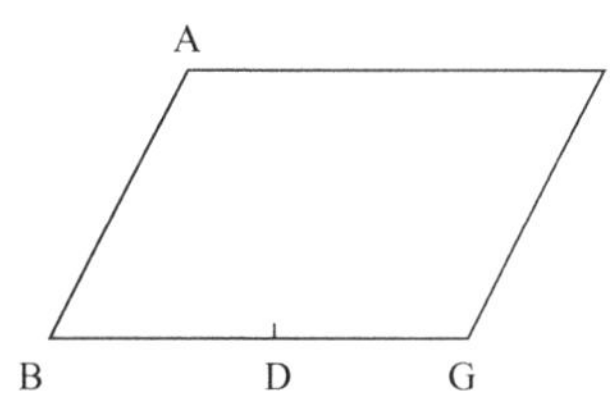

Fig. 13

[38] Heath, *Elements*, Vol. 1, p. 383.

[39] The translation of this proposition is based on S. Ito, *The Medieval Latin Translation of the Data of Euclid* (Tokyo, 1980). I have changed the notations in this translation, and corrected small deviations in the Latin translation according to the Greek text.

Hence the ratio 4**O**(BG, GD): **T**(BD) is also given.
Therefore the ratio 4**O**(BG, GD)+**T**(BD): **T**(BD) is given.

But 4**O**(BG, GD) + **T**(BD) = **T**(BG + GD) (*Elem*. II 8)
Therefore the ratio **T**(BG + GD) : **T**(BD) is also given.
52 | Hence the ratio (BG + GD) : BD is also given.

And *componendo*, the ratio 2GB : BD is given.
Hence the ratio GB : BD is given.

But GB : BD = **O**(GB, BD) : **T**(BD)
Therefore the ratio **O**(BG, BD) : **T**(BD) is given.

But **O**(GB, BD) is given.
Therefore **T**(BD) is also given. Therefore BD is given.
Hence BG is also given, for the ratio GB : BD is given and BD is given. And AG is given and the angle B is given. Therefore AB is also given. Therefore each of AB and BG is given.

Tannery interprets this proposition as the equation:

$$xy = A, \quad x^2 = my^2 + B$$

and explains its complicated solution, as the result of the avoidance of the introduction of the biquadratic term, $x^2y^2 = A^2$.[40]

But Tannery does not explain the reason why such a complicated equation appears in *Data*. *Data* contains three propositions (84–86) which can be construed as the solution of quadratic simultaneous equations. Propositions 84 and 85 are, as we have
53 noted, problem-versions of *Elem*. II 5, 6, respectively. Compared with | the simplicity of these two, the complexity of prop. 86 is striking. Assuming Euclid was developing techniques of "equations" in these propositions, the difference between 85 and 86 would represent a tremendous leap.

I believe that prop. 86 of *Data* was originally a problem concerning the intersection of two hyperbolas as follows (Fig. 14):

> Let there be given a hyperbola EP with the diameter BG and the secondary diameter BA. Let there be given another hyperbola PQ with asymptotes BA and BG. Then, the point of the intersection P is also given.

It is evident that *Data* 86 is mathematically identical to the above problem. Let the ordinate PG be drawn. As P is on the hyperbola PE, the ratio **T**(PG) : **O**(EG, GE′) is given.

As PG = AB and **O**(EG, GE′) = **T**(BG) − **T**(BE) (*Elem*. II 6)
the ratio **T**(BG) − **T**(BE) : **T**(AB) is given.

As **T**(BE) is given since the hyperbola PE is given, **T**(BG) is greater than **T**(AB) by the given area **T**(BE) as in ratio. On the other hand, as P is on the hyperbola PQ, **O**(AB, BG) is given.

Therefore it seems certain that *Data* 86 has its origin in the problem of finding the intersection of two hyperbolas. To assume this mode of interpretation is much more

---

40 Tannery, *op. cit*., pp. 262–63. See also *SA*, pp. 198–99.

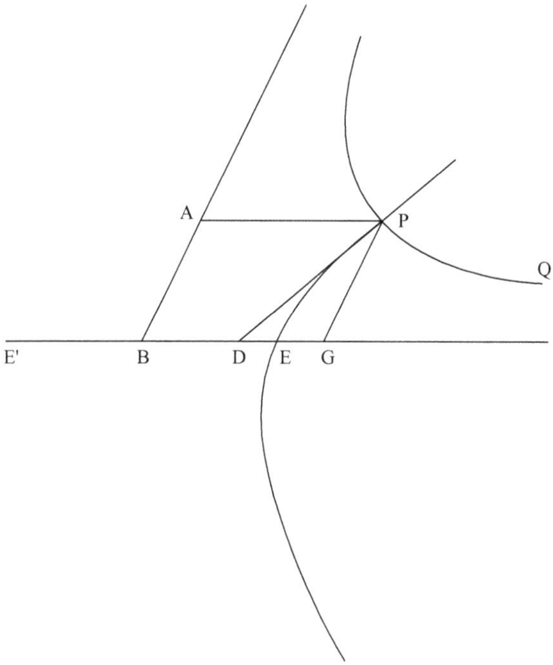

Fig. 14

natural than to regard the proposition as a simulataneous equation of obscure origin. There exists further confirmation for this interpretation. The propositions concerning the hyperbola gives hints to the solution of *Data* 86. The first step of the solution is to take away from **T**(BG) the given area by which it is greater than **T**(BA) as in ratio, and **O**(GB, BD) is made equal to this given area. In the hyperbola, the point D is the intersection of the tangent at P with the diameter because by *Conics* I 37 (a),

$$\mathbf{O}(BG, BD) = \mathbf{T}(BE)$$

and in the context of my interpretation, **T**(BE) is nothing but the given area by which **T**(BG) is greater than **T**(AE) as in ratio. In the next step, in *Data* 86, it is stated that the ratio of the remainder **O**(DG, GB) to **T**(AB) is given. According to my interpretation, this is nothing but

Conics I 37 (b): **T**(PG): **O**(DG, GB) = *latus rectum: latus transversum*.

This impressive coincidence is strong support for my interpretation. It is also noteworthy that this solution does not involve the use of any conic curve. Therefore *Data* 86 shows that a particular solid problem can be reduced to a plane problem. The remarkable nature of this result led Euclid to include it in the *Data*.

We have seen that the allegedly algebraic problem in the *Data* the solution of which requires *Elem*. II 8, is in fact a geometric problem. Further, it is quite likely that

Euclid's intention in inserting II 8 in the *Elements* was as part of the solution for geometric problems such as *Data* 86.

This leaves II 1. It seems most profitable to discuss this proposition from a different point of view, examining the proposition itself, since it appears that here is revealed Euclid's intention to elaborate those propositions which are now considered examples of the geometric algebra.

54 | *Elements* II 1 If there be two straight lines, and one of them be cut into any number of segments whatever, the rectangle contained by the two straight lines is equal to the rectangles contained by the uncut straight line and each of the segments.

> Let A, BC be two straight lines, and let BC be cut at random at the points D, E (Fig. 15);
>
> I say that the rectangle contained by A, BC is equal to the rectangle contained by A, BD, that contained by A, DE and that contained by A, EC.
>
> For let BF be drawn from B at right angles to BC;
> let BG be made equal to A,
> through G let GH be drawn parallel to BC,
> and through D, E, C let DK, EL, CH be drawn parallel to BG.
> Then BH is equal to BK, DL, EH.
>
> Now BH is the rectangle A, BC, for it is contained by GB, BC, and BG is equal to A;
> BK is the rectangle A, BD, for it is contained by GB, BD, and BG is equal to A;
> and DL is the rectangle A, DE, for DK, that is BG, is equal to A.
> Similarly also EH is the rectangle A, EC.
>
> Therefore the rectangle A, BC is equal to the rectangle A, BD, the rectangle A, DE and the rectangle A, EC. Therefore *etc.*[41]

This proposition appears quite strange, and one is puzzled about precisely what it is making claims. It seems to be a tautology, though it must have been a "proof" of

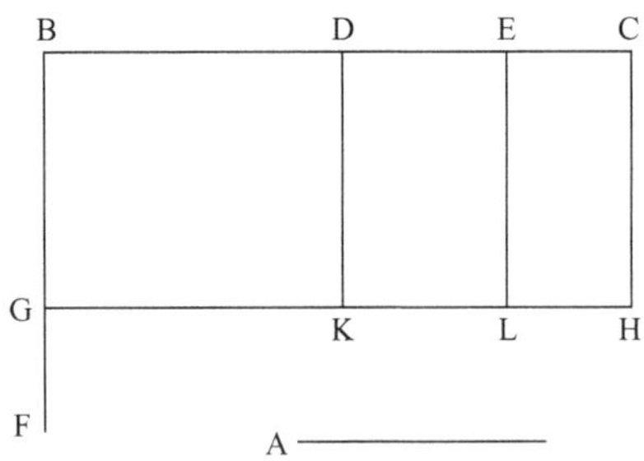

Fig. 15

[41] Heath, *Elements*, Vol. 1, p. 375.

something unknown on the basis of something known or admitted.[42] If we summarize Euclid's proof, he seems to admit

$$\mathrm{BCHG} = \mathrm{BDKG} + \mathrm{DELK} + \mathrm{ECHL} \quad \ldots\ldots\ldots\ldots\ldots (3)$$

and from this equality to prove

$$\mathbf{O}(\mathrm{A, BC}) = \mathbf{O}(\mathrm{A, BD}) + \mathbf{O}(\mathrm{A, DE}) + \mathbf{O}(\mathrm{A, EC}) \quad \ldots\ldots (4)$$

(4) must have been to Euclid less evident than (3). But what is the difference between
| (3) and (4)? Here the notion of "visible" and "invisible" figures seems to be useful. 55
(3) is a equality between "visible" figures and (4) is that between "invisible" ones. What Euclid has done is to reduce the equality between "invisible" figures to that of "visible" ones. The latter equality is evident by geometric intuition and it can be safely conjectured that Euclid thought it a sound basis for the proof of the former. As a result, I claim that II 1 is by no means trivial, for it extends the "visible" equality to the "invisible".

The further examination of some propositions in the *Elements* supports this view. In the proof of I 47 (the Pythagorean theorem), II 1 (more precisely, II 2) seems to be used, for Euclid states that (Fig. 16)

$$\mathrm{BL} + \mathrm{CL} = \mathrm{BDEC} \quad \ldots\ldots\ldots\ldots\ldots\ldots\ldots (5)$$

Does Euclid hereby commit *petitio principii?*[43] If we distinguish the two levels of figures, visible and invisible, we will see at once that Euclid's proof is correct. (5) is

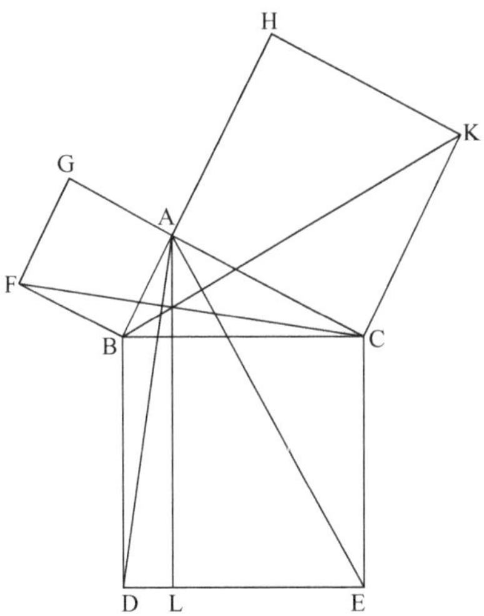

Fig. 16

[42] According to the algebraic interpretation, II 1 is a statement of the distributional law.

[43] W. R. Knorr claims it to be the evidence of diversity of the sources of Book I and II. See his *The Evolution of the Euclidean Elements* (Dordrecht, 1975), p. 179.

an equality between "visible" areas and equivalent to (3), the basis of the proof of II 1. As a result, there is no inconsistency in its use in I 47.

The case of XIII 10, the sole explicit example of the use of II 1–3 in the *Elements*, is in clear contrast to I 47, though the same equality appears in the proof. Euclid uses (Fig. 17)

$$\mathbf{O}(\mathrm{AB},\ \mathrm{BN}) + \mathbf{O}(\mathrm{BA},\ \mathrm{AN}) = \mathbf{T}(\mathrm{AB}) \quad \ldots\ldots\ldots\ldots\ldots (6)$$

This is the same as (5) if one draws a square on AB. But Euclid does not bother to draw it. Why? Because he depends on II 2. Thanks to this proposition, he is freed from
56 the necessity of proving (6) by reducing the "invisible" figures such as | **O**(AB, BN) to "visible" ones.

Now the significance of II 2 in this context is clear. It is an equality between "invisible" figures, and it simplifies the arguments on "invisible" figures by making it unnecessary to reduce the relations to those between "visible" figures. This interpretation is valid also for other propositions in *Elem*. II.

The distinction between "visible" and "invisible" figures which I have made in the examination of the propositions in the *Conics* has turned out to be useful in the interpretation of the second book of the *Elements*. Euclid has two different classes of geometric objects in his study, visible and invisible figures. And the aim of *Elem*. II seems to afford some typical equalities between invisible figures. A criticism might be raised that the invisible figures and their sides are substantially the same as quantities in general, and thus that my interpretation is at bottom algebraic. But it is a mistake to view the invisible figures in this way. They retain their geometric properties, since depending on the arrangement of the figures, one of the pair of twin-propositions is necessary.

Elem. II contains the propositions concerning the "invisible" figures for the solution of geometric problems, and these propositions are usually stated in pairs, the two propositions being used in mutually complementary way to solve a problem. And

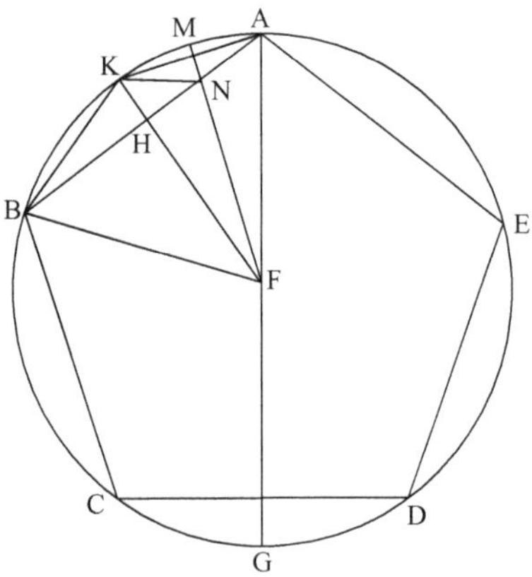

Fig. 17

this mutually complementary use of a pair of propositions is evidence that Euclid did not regard geometric magnitudes (areas and length of lines) as general quantities.

A word is on order regarding Zeuthen's remarks on twin-propositions. Pointing out that *Elem.* II 14 (he interprets this proposition as a problem of finding x for given a, b, such that $x^2 = ab$) could have been proved by II 6, as well as II 5, he states:

> Ob man den einen oder den anderen zu benutzen hat, beruht darauf, ob man—beim
> Beweise oder Herleitung, denn die Konstruktion ist dieselbe—damit be|ginnt eine 57
> der Strecken a und b entweder auf der anderen oder der Verlängerung der anderen
> abzutragen.[44]

This remark clarifies Zeuthen's views. His remark is, in a sense, correct. But his opinion fully relies on the assumption that II 14 is an algebraic problem concerning abstract general quantities, and that the lines which are the object of II 5, 6, 14 have been introduced afterwards to represent those quantities, so that their position and arrangement have nothing to do with the problem itself. In short, the lines can be "carried away" (abtragen) in Zeuthen's view because they were mere representations for quantities. But as noted earlier, the lines in II 5, 6 *etc.* cannot be "carried away". Evidence for the rectitude of this claim lies in the existence of the double form. Zeuthen's failure to recognize geometric significance in *Elem.* II was the result of his careless identificantion of the lines with quantities in algebra.

Finally, I discuss the following problem: those who argue for the algebraic interpretation of *Elem.* II tend to claim that propositions II 1–10 are illustration of the method of the geometric algebra by means of which other equalities are to be derived. Though the solution to this issue requires a thorough examination of a wide range of the Greek mathematical texts, it is the position of this paper that these propositions are meant to form a set of propositions necessary in Greek geometry and that no other equality is required. In support of this view, a refutation of Zeuthen's argument should be made.

Apollonius makes small skips in the course of his proofs in the *Conics*. Zeuthen argues that some of these skips involving the geometric algebra should be supplemented not by the propositions in *Elem.* II, but by a procedure illustrated in that book. Zeuthen presents the following example.[45] In the proof of *Conics* III 26, Apollonius assumes that (Fig. 18)

if $AB = CD$ then

$\mathbf{O}(EC, EB) = \mathbf{O}(AB, BD) + \mathbf{O}(ED, EA)$

| Zeuthen claims that this equality should be proved by drawing rectangles $B'C$ and 58
$A'D$, making EB' and EA' equal to EB and EA respectively (Fig. 19). Then the desired equality is apparent. On the other hand, Pappus affords a proof based on *Elem.* II 5, 6 for this equality. He takes the point Z in the midst of BC (Fig. 18). It is also the middle point of AD. By *Elem.* II, the following equalities hold:

$$\mathbf{O}(EC, EB) = \mathbf{T}(EZ) - \mathbf{T}(ZB) \quad \text{(II, 6)}$$
$$\mathbf{O}(AB, BD) = \mathbf{T}(AZ) - \mathbf{T}(ZB) \quad \text{(II, 5)}$$
$$\mathbf{O}(ED, EA) = \mathbf{T}(EZ) - \mathbf{T}(ZA) \quad \text{(II, 6)}$$

---

44 *Die Lehre*, p. 15.

45 *Die Lehre*, pp. 36–38.

Fig. 18

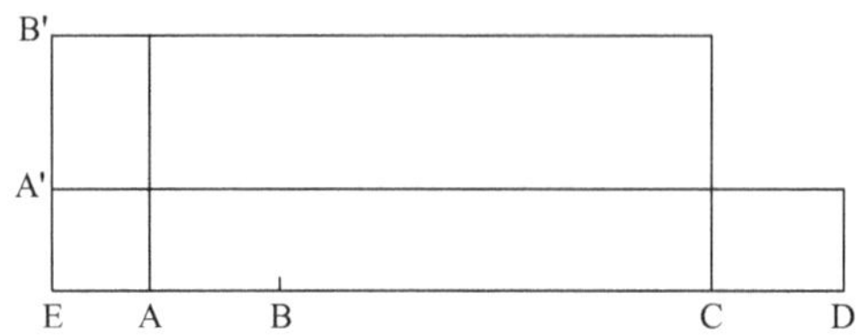

Fig. 19

The rest of the proof is relatively easy. Zeuthen criticizes Pappus's proof as "Pedantrie späterer Zeiten", and he approves Eutocius's commentray on which his own reconstruction is based.

Is Zeuthen right? Should we admit that Pappus did not understand the method of "geometric algebra" and stuck to the propositions themselves? On the contrary, I claim that Pappus's proof is most natural in this context, while Zeuthen's is akin to algebraic operations.

Let us examine the context in which this equality appears. In *Conics* III 26, the line ED is drawn to cut three branches of conjugate hyperbolas. The diameter OX, parallel to ED, is drawn. OY is the conjugate diameter to OX. Then OY bisects BC and AD by the propositions in Book II of the *Conics*. Though the point of intersection of OY and ED is not named in the text of the *Conics*, it must be at once clear for Apollonius that the rectangles such as **O**(EC, CD) can be transformed into the difference of two squares (Pappus later demonstrates) since Apollonius shows great mastery of *Elem.* II 5, 6 in the preceding arguments of the *Conics*. What Pappus needs is to name the midpoint of
59 BC, which has already appeared in the | diagram as the intersection of ED and OY.

For this reason it seems that Pappus's reconstruction is much more in conformation with the line of thought of Apollonius than the reconstruction of Eutocius and Zeuthen. Zeuthen's interpretation may have come from the following algebraic operations. Let EA $= x$, AB $=$ CD $= y$, BC $= z$. Then

$$\begin{aligned}\mathbf{O}(\mathrm{AB},\ \mathrm{BD}) + \mathbf{O}(\mathrm{ED},\ \mathrm{EA}) &= (y+z)y + (x+2y+z)x\\ &= (x+y+z)y + (x+y+z)x\\ &= (x+y+z)(x+y)\\ &= \mathbf{O}(\mathrm{EC},\ \mathrm{EB})\end{aligned}$$

This was likely the reason Zeuthen decided to approve this interpretation.

Thus Zeuthen's claim that Pappus's lemma contains improper use of the "geometric algebra" cannot be justified. In this example no equality is required except *Elem.* II 1–10. Though I cannot affirm that II 1–10 are sufficient for all the propositions, I have not, at least, seen any example for which II 1–10 are insufficient. Euclid's intention in

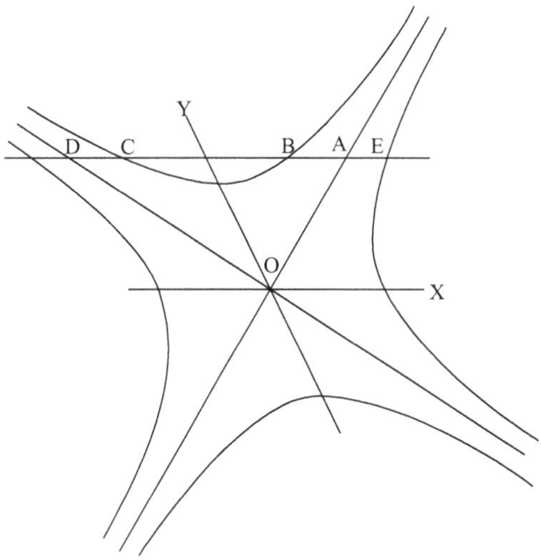

Fig. 20

the compilation of Book II was thus to afford a sufficient set of propositions concerning the "invisible" areas.

CONCLUSION

Book II of the *Elements* can be interpreted geometrically. The propositions II 1–10, those of the so-called geometric algebra, are lemmata for geometric arguments (the solution of geometric problems and the proof of geometric theorems). The necessity and significance of these propositions lies in the distinction of "visible" and "invisible" figures in Euclid. The propositions II 1–10 are those concerning invisible figures, and they must be proved by reducing invisible figures to visible ones, for one can apply to the latter the geometric intuition which is fundamental in Greek geometric arguments. Thus II 1–10 are a collection of the relations between "invisible" areas useful in works at the time of Euclid. We find some examples of their application in the *Elements*, but it was in the theory of the conic sections that these propositions were used repeatedly, and the significance of the distinction between visible and invisible figures was made clear. It is most likely that in the compilation of Book II of the *Elements*, Euclid intended to provide a set of propositions necessary to his *Conic Elements*. In connection with this claim, it is the assertion of this paper that interpretations attributing to Euclid some pedagogical intent in the compilation of *Elem.* II should be rejected. The view that some propositions in *Elem.* II are illustrations of the method of the "geometric algebra" is a contrivance to save the algebraic interpretation and has no positive evidence.

The double form *i.e.* the existence of twin-propositions, can be explained in the context of their application to the geometric arguments. They are used in mutually complementary ways according to the arrangements of points and lines in the problems and theorems to which they are applied. This also suggests that Euclid considered lines and areas not as representations of abstract quantities but as geometric entities, the arrangement of which is significant.

60 | Although I have avoided the argument regarding the origin of the "geometric algebra", I am inclined to think that Mueller's opinion which attributes its origin to the effort to avoid use of the theory of proportions, has much to commend it and I believe that the arguments presented here have given it even more substance by explaining the existence of propositions in *Elem.* II which Mueller left unexplained.

Since the four pairs of twin-propositions are used in mutually complementary ways in geometric context, any interpretation of the sources of these propositions which attributes different origins to the individual constituents of a pair, cannot be justified. Further, as Book II was compiled in order to afford a sufficient basis for geometric arguments involving invisible figures, we cannot assume that all the propositions II 1–10 can be traced back to the Pythagoreans. It might also be possible, for example, that some of the propositions such as II 5, 6, were recognized first, and others added later. At very least it would be meaningless to seek the origin of *each* proposition in some algebraic or arithmetic theory.

ADDITIONAL NOTES 2004

There is probably nobody who would not be tempted to add some after-thoughts if one's article is reprinted after almost twenty years. However, I have considered it better to use that time and energy to produce a new article, and I limit myself here to add some bibliographical notes. Page numbers are those in this volume.

—p. 157, note 35: Pappus's proposition is prop. 122 (Jones, 1986, p. 252; Hultsch, pp. 856-858).

—p. 158, note 37: prop. 170 (Jones, p. 300; Hultsch, p. 926).

—p. 159, note 39. Now [Taisbak 2003] provides English translation and commentary of the whole text of *Data*.

—p. 159ff.: The relation of *Data* 86 to hyperbola was already pointed out by Zeuthen in 1917. See [Taisbak 1996] and [Taisback 2003].

ADDITIONAL BIBLIOGRAPHY

Jones, A., 1986. *Pappus of Alexandria, Book 7 of the Collection.* 2 vols. New-York & Berlin: Springer.

Hultsch, ed. and trans., 1876-1878. *Pappi Alexandrini Collectionis quae supersunt.* 3 vols. Berlin.

Taisbak, Ch. M., 1996. 'Zeuthen and Euclid's Data 86: Algebra – or a Lemma about intersecting Hyperbolas?', *Centaurus* 38: 122-139.

Taisbak, Ch. M., 2003. *Euclid's Data or the Importance of Being Given.* Copenhagen: Museum Tusculanum Press.

G.E.R. LLOYD

THE MENO AND THE MYSTERIES OF MATHEMATICS

The principal object of the method of hypothesis introduced at *Meno* 86e ff is problem reduction. Vlastos,[1] p. 123, put it: 'the logical structure of the recommended method is entirely clear: when you are faced with a problematic proposition *p*, to "investigate it from a hypothesis," you hit on another proposition *h* (the "hypothesis"), such that *p* is true if and only if *h* is true, and then shift your search from *p* to *h*, and investigate the truth of *h*, undertaking to determine what would follow (quite apart from *p*) if *h* were true and, alternatively, if it were false.' That much is clear.[2] But almost everything else is obscure, thanks very largely to the obscurities in the mathematical example. This has generated an enormous secondary literature, and no interpretation can be said to be completely free from difficulty. The object of this paper is to attempt a new approach to that problem. I shall argue that the very obscurity of Plato's mathematical example is one of its *points*.

Let me first set out the text and a moderately straighforward rendering.

86e *λέγω δὲ τὸ ἐξ ὑποθέσεως ὧδε, ὥσπερ οἱ γεωμέ-*
τραι πολλάκις σκοποῦνται, ἐπειδάν τις ἔρηται αὐτούς, οἷον
περὶ χωρίου, εἰ οἷόν τε ἐς τόνδε τὸν κύκλον τόδε τὸ χωρίον
87 *τρίγωνον ἐνταθῆναι, εἴποι ἄν τις ὅτι "οὔπω οἶδα εἰ ἔστιν*
τοῦτο τοιοῦτον, ἀλλ' ὥσπερ μέν τινα ὑπόθεσιν προὔργου
οἶμαι ἔχειν πρὸς τὸ πρᾶγμα τοιάνδε· εἰ μέν ἐστιν τοῦτο τὸ
χωρίον τοιοῦτον οἷον παρὰ τὴν δοθεῖσαν αὐτοῦ γραμμὴν
παρατείναντα ἐλλείπειν τοιούτῳ χωρίῳ οἷον ἂν αὐτὸ τὸ
παρατεταμένον ᾖ, ἄλλο τι συμβαίνειν μοι δοκεῖ, καὶ ἄλλο
αὖ, εἰ ἀδύνατόν ἐστιν ταῦτα παθεῖν. ὑποθέμενος οὖν ἐθέλω
b *εἰπεῖν σοι τὸ συμβαῖνον περὶ τῆς ἐντάσεως αὐτοῦ εἰς τὸν*
κύκλον, εἴτε ἀδύνατον εἴτε μή."

| Knorr's version reads[3]: I say "from hypothesis" in the manner that the geometers often make inquiry, whenever someone has asked them, for instance about an area, whether this area here can be stretched out as a triangle in this circle here, one would say 'I don't yet know whether this is of such a sort, but I think that as a certain hypothesis the following will assist in the matter. If this area is such that the one who has 167

[1] Works cited by author alone are listed in the bibliography at the end of the article which offers a brief selection of the most important recent and earlier studies on the problem.
[2] One may compare already Aristotle's view, since he evidently has the *Meno*'s method of hypothesis in mind in his own account of *ἀπαγωγή* at *A Pr.* 69a20ff, 24ff.
[3] Knorr p. 71.

J. Christianidis (ed.), Classics in the History of Greek Mathematics, 169–183.

stretched (it) along its given line (makes it) fall short by an area such as is the stretched (area) itself, then it seems to me that a certain result follows, but a different result, if it is impossible that these things be done. Having hypothesized, then, I wish to say to you whether the result about its stretching in the circle is impossible or not.'

This passage contains the following major ambiguities or obscurities.[4]

(1) What is the χωρίον mentioned at 86e6, 87a3-4? Are we dealing with any area, a rectangle, a square, or even the square that had been used at 82c1 in the discussion with the slave-boy (though that was introduced as τετράγωνον χωρίον)?
(2) In the expression παρά τήν δοθεῖσαν αὐτοῦ γραμμήν how should δοθεῖσαν be taken? Which line is 'given'? Is this the diameter of the circle? Or a chord? Or some line associated not with the circle, but with the area/rectangle/square to be inscribed?
(3) In normal Greek αὐτοῦ in 87a4 should refer to χωρίον, not (as it is often taken) to the circle: but that takes us back to the second difficulty.[5]
(4) Does παρατείναντα in 87a5 refer to the so-called 'application of areas', or not? It should be noted that the normal Greek expression for the application of areas is παραβάλλειν and cognates.
(5) τοιούτῳ χωρίῳ οἷον, 87a5. How is 'by such as is' to be taken? Are the areas equal? Or similar (in the geometrical sense)? Or similar and similarly situated?[6]

Every interpretation proposed falls foul of one or more difficulties and all depend, to a greater or lesser extent, on allowing Plato the benefit of the doubt on one or more problems. All seem to have to help Plato along by disambiguating his ambiguous expressions. The approach I want to try out starts from questioning the validity of
168 doing so. By straining to make the | text yield a single determinate mathematical interpretation – which every reader should be able to understand and agree on – are we perhaps missing the point? If we can arrive at such an interpretation only after helping Plato along with a strenuous effort at disambiguation, we have to ask *why* we had to help him in that way, in other words why he was so obscure.[7] We cannot, of course, prejudge the issue of Plato's own mathematical competence, where indeed widely diverging views have been entertained, from Vlastos' suggestion that Plato is preening himself on his own expertise in geometry,[8] to the attempts made to condone some of the ambiguities on the grounds of the imprecision of mathematical terminology in his day.[9] We shall be returning to that issue in due course.

---

4 We may thus appropriately start our inquiry, as Myles Burnyeat points out to me, by raising, in effect, so many τί ἐστι questions.

5 As has recently been insisted upon by Karasmanis, ch. 4, p. 106, who argues that a circle's γραμμή is usually its circumference.

6 Myles Burnyeat has suggested to me a comparison with the vague expression ἀλλοῖον ἢ οἷον ἐπιστήμη in the ethical analogue at 87b7.

7 Several interpreters even assume as a matter of principle that the mathematical illustration has to be readily intelligible to the listening Meno and tend subsequently to rule out any reading deemed to have been beyond him. See, for example, Bluck, p. 449. However the search for some *simple* solution to the interpretation of this text has proved hopelessly elusive.

8 Vlastos, p. 123.

9 Compare, for example, Heath, p. 300 n. 1, Knorr, p. 71 (who suggests as a further possibility that the imprecisions may arise from certain dramatic and dialectical considerations, p. 73).

A comprehensive review of every existing interpretation would run to monograph length, but some of the more promising proposals may be briefly summarised to indicate their principal strengths and weaknesses. Thus Benecke[10] proposed that the square to be inscribed is the one already set out in the discussion with the slave-boy at 82c. The square when applied to the diameter of the circle falls short by an area that is not just similar, but equal, to the area you start with. In diagram 1, ABCD is the area that has to be inscribed in the circle. It is applied to the diameter ADL, and when so applied, falls short by the area DCML (equal to ABCD). The starting area, ABCD, may, then, be inscribed as a triangle, viz. ACL, in the given circle. | The advantages of this interpretation are that it 169
gives a determinate reference for the χωρίον mentioned at 86e6, and that it gives a strong sense to 'by such as is' at 87a5 (equal, not just similar). So it might be thought to score well on difficulties (1) and (5). However the major disadvantage is clear. If the inscription as so suggested is possible, all well and good. But if not, nothing follows: because even if the starting square is larger than a square on a radius of the circle, it might still be inscribable. The reason for this is simple. The largest possible triangle that can be inscribed in a circle is the equilateral,[11] not the right-angled isosceles whose hypotenuse is the diameter. So any square with area not greater than the equilateral triangle is inscribable, not just those with areas not greater than the right-angled isosceles.

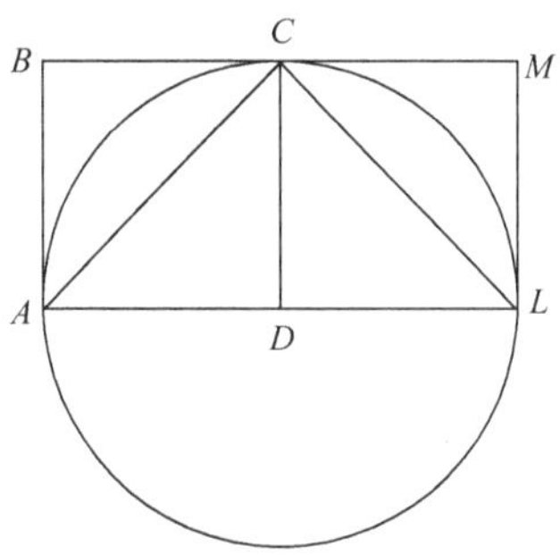

Diagram 1 (from Bluck, p. 447)

A second interpretation is that proposed by Butcher.[12] Here the starting-area to be inscribed is not a square, but a rectangle, applied (as in Benecke) along the diameter of the circle. If when so applied, it falls short by an area that is *similar* (not equal) to

---

[10] Cf. Heath, p. 302, Bluck, pp. 447f.

[11] This, as we shall see (pp. 173f.), is what would later have been called the *diorismos*. This sets the conditions on the given for a problem to be soluble. Knorr, n. 58 on p. 92, shows how an elementary proof can be given that the equilateral triangle is the greatest of all triangles inscribed in the same circle. Although the question of the date of the discovery of *diorismoi as such* is disputed (see below, n. 16), there is no reason to think that any feature of the proof of the *diorismos* needed for the Meno inscription problem would have posed any difficulty for Plato's contemporaries. Contrast the much more difficult question of whether the mathematics needed to *resolve the problem as reduced* in the Cook Wilson/Heath/Knorr interpretation would have been available in Plato's day (see below, n. 13).

[12] Cf. Heath, p. 300 n. 1, Bluck, pp. 442ff.

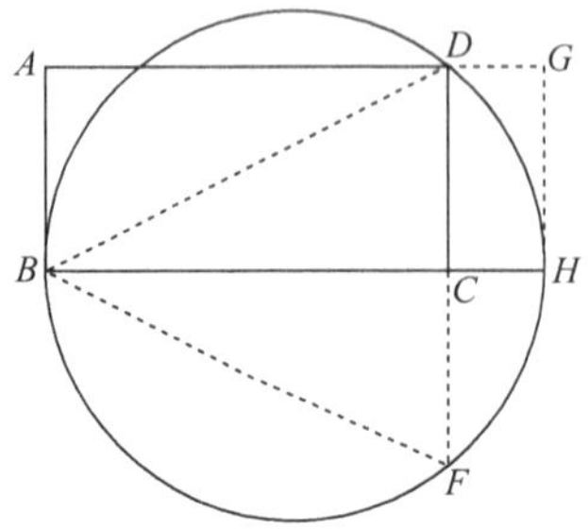

Diagram 2 (from Bluck, p. 443)

the rectangle itself, then, the inscription is possible. In diagram 2, ABCD is the starting-area, applied to the diameter of the circle, BCH. If it falls short by an area, DCHG, similar to itself, then the starting-area may be inscribed in the circle as the
170 triangle BDF. The | area by which it falls short, namely DCHG, will indeed be similar if the point D is on the circumference of the circle. By the properties of the circle (Euclid, *Elements* III 35) BC.CH = DC.CF. Since BH, the diameter, bisects DF, DC = CF and so BC.CH = DC^2. So BC:CD = CD:CH. Furthermore the triangle BDF equals ABCD, since Δ BCF = Δ BCD = $\frac{1}{2}$ the rectangle ABCD.

While the strength of this interpretation is that it gives a more general rendering to the area-inscription problem than Benecke's, it suffers – as Butcher himself appreciated – from the same major disadvantage, namely that if the condition suggested is *not* fulfilled, it may or may not be possible to inscribe the starting-area as a triangle.

I may note, in parenthesis, that a weaker interpretation of what the method of hypothesis itself intends has recently been proposed by Canto-Sperber in her edition. Her proposal is based on the distinction between the problem of the *construction* of the triangle, and that of settling the question of whether the starting-area is *inscribable* or not – where we may indeed agree that these are distinct problems, since knowing whether the area is inscribable or not does not, of itself, solve the problem of how to construct the triangle in question. Canto-Sperber's suggestion (p. 283) is that it is the construction of the triangle, not its inscription, that is subject to conditions that are at once necessary and sufficient, but that so far as the inscription itself goes, that is subject (merely) to the necessary condition that relates to falling short by a similar area. Butcher, on this view, gives the right interpretation of the example Plato has in mind, but the objection to it can be blocked if we allow that the stipulation of necessary and sufficient conditions relates only to the construction of the triangle.

But to that it should be said first that at the end of the mathematical illustration it is clearly stated (87bl-2) that we should be in a position to state whether or not the *inscription* is *im*possible. Moreover, secondly, in the ethical analogue necessary and sufficient conditions are envisaged, since it is clearly a matter of deciding on what basis virtue is teachable, and on what basis *not* so (87c8-9). Since neither on the Butcher nor on the Benecke interpretation are necessary and sufficient conditions for the inscription given, the objections against them still stand – unless we accept that Plato himself is either muddled or obfuscating. If Plato himself confused the necessary conditions for inscription with the necessary and sufficient conditions for the construction, then he is muddled. If he was aware of that distinction, but nevertheless allows the unguarded reader to think that the geometrical example yields necessary and sufficient conditions for the inscription, he is obfuscating.

| A third main line of interpretation stems from Cook Wilson and was adopted and 171
adapted by, among others, Heath (pp. 298ff) and Knorr (pp. 72ff). In this the starting-area is not already given as a rectangle. It is, rather, any rectilinear area (thus allowing the most general interpretation of *χωρίον* at 86e6 and avoiding the objections that can be mounted against taking that term to refer either to a rectangle or to a square). The task is, then, to inscribe this starting-area in the form of a rectangle, such that this rectangle falls short by an area similar to itself. In diagram 3, the starting-area is X: the rectangle ABCD (equal to X) is found such that, when applied to the diameter BCH, it falls short by the rectangle DCHG. For this to be the case, the point D must lie on the circumference of the circle. The crucial difference between Butcher and the Cook Wilson/Heath/Knorr interpretation is that for Butcher the rectangle ABCD is *given*, while on the Cook Wilson/Heath/Knorr account the point D has to be *found*. But if the point D is found on the circumference of the circle then by the same argument that has been mentioned already (Euclid *Elements* III 35) ABCD is similar to DCHG and the triangle BFD will solve the problem. Δ BFD = ABCD = the starting area X.

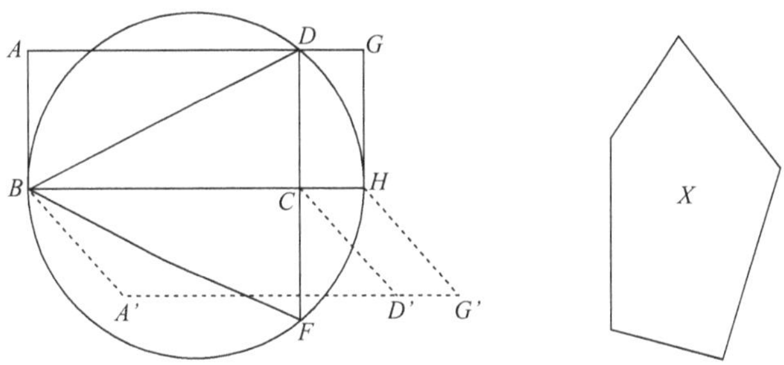

Diagram 3 (from Bluck, p. 445)

Moreover the second great advantage of this line of interpretation (apart from taking 'area' entirely generally) is that it provides both necessary *and sufficient* conditions for the inscription of the starting-area as a triangle in the circle. If the rectangle ABCD can be found, the problem can be solved, and if not, not. So one of the main objections to the interpretations of Benecke and Butcher does not hold against Cook Wilson, Heath and Knorr.

However it cannot be said that this third interpretation is free from problems, some of which are, to be sure, shared by other interpretations as well. The chief ones relate
to what on the Cook Wilson/Heath/Knorr view | Plato leaves unexplained in the text. 172
(1) The reduction of the starting-area X to an equivalent rectangular area is not explained nor justified. (2) There is no explicit reference, in the text, to the fact that the area as applied is a rectangle. (3) The 'given line' of 87a4 is assumed to be the diameter. (4) 'By such as' at 87a5 must be taken in the sense 'similar', not 'equal'.

Those indeterminacies are, perhaps, of minor importance. But we must add three other, more serious ones. (5) It is not explained how the problem as reduced is equivalent to the problem one started with. Of course *any* rectangle applied to the diameter and with the point D in diagram 3 on the circumference of the circle will fall short by another rectangle that is similar to it. This is by the above-mentioned properties of the circle. However the task is to find point D such that the rectangle ABCD is equal to the starting-area X: more on that in a moment. But the difficulty we must insist on is that we have no explanation, in the text, as to how the problem as so reduced solves the original problem. It does, to be sure, since once the rectangle ABCD is found, the triangle BDF can be constructed with the same area. Yet that move is not identified, let alone justified, though one should have thought it needed to be, especially if this is supposed to be a hypothesis that the mathematicians *can grant straightforwardly*.

(6) Nothing is said about how to go about the task of tackling the problem as reduced, that of finding the rectangle ABCD or more specifically the point D on the
173 circumference of the circle. The way the modern | commentators proceed can be explained with reference to diagram 4. To adapt Heath's discussion, the point required lies on the rectangular hyperbola that has as its asymptotes the diameter of the circle and the tangent drawn at its end-point. If the starting area is b^2 the equation of the hyperbola referred to its asymptotes as axes is $xy = b^2$. Thus the hyperbola passes through the point (Z) whose coordinates both equal b. If the hyperbola is tangent to the circle at that point, there is one solution – where the inscribed triangle will be equilateral. If the hyperbola cuts the circle at two points there will be two solutions.[13]

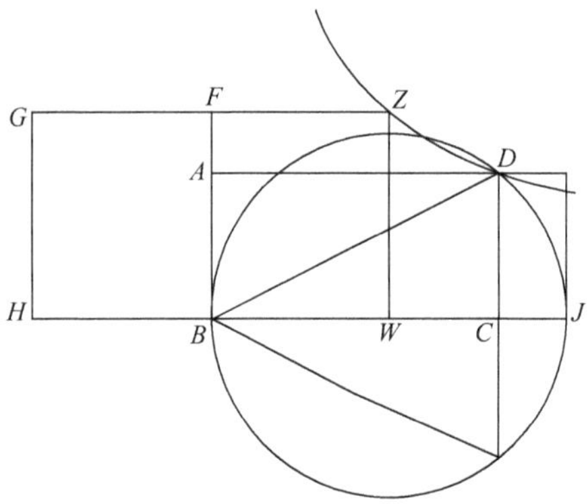

BC = x
AB = y
GH = FB = ZW = b
HB = BW = b
$xy = b^2$ = starting area

Diagram 4 (based on Knorr, p. 72)

In diagram 4, the square GFBH (b^2) is taken as equal to the starting area. Points Z and D are on the rectangular hyperbola that has as its asymptotes the diameter of the circle BCJ and the tangent drawn at its end point, BAF, where the product of the horizontal coordinate and the vertical one (i.e. in the case of D, x and y) is equal to the starting area.

---

[13] The question of whether the geometry required to tackle the problem as thus reduced in the Cook Wilson/Heath/Knorr interpretation was available in Plato's day has been much discussed. Certainly the normal method of obtaining the required curve would be by conics and would be hard to imagine in the period before Euclid. However Knorr, p. 73, has recently proposed that Menaechmus would have been able to obtain such a curve by point-wise constructions. It should, moreover, be stressed that even if the *problem as reduced* was beyond the reach of contemporary

(7) Finally nothing is said in the text concerning what later mathematicians would have called the διορισμός of the problem. Strictly the *diorismos* establishes the *conditions on the given* for a solution to a problem to be possible (and as such is quite distinct from problem reduction, cf ἀπαγωγή the finding of a simpler, more investigable problem that will enable one to solve a harder one). In the *Meno* inscription problem, the *diorismos* in the strict sense is clear. If the starting-area exceeds the equilateral triangle inscribed in the circle, no solution is possible. As just noted, under (6), if the starting-area equals the equilateral triangle, there is a single solution, and if less than that triangle, there are two. We must, however, remark that having a *diorismos* does not, of itself, solve the problem in any sense, even while it establishes the condition of its solubility. However, in the *Meno* no distinction is made between *diorismos* and an actual solution, and indeed no clear reference to a *diorismos* at all.

At this point it might appear that the Cook Wilson/Heath/Knorr interpretation cannot be on the right lines and must be abandoned. Yet equal or greater difficulties beset every other line of interpretation also. To mention just three: Heijboer takes the problem to be inscribing in any circle on one of the sides of a given rectangle a triangle with an equal area.[14] But | this involves taking ἐλλείπειν (87a5) not in the usual sense 174
of 'fall short of' (as in the application of areas) but in the quite strained and otherwise unattested sense of 'leave room above'. Moreover Heijboer has to assume that the starting-area is a rectangle that has one side (and specifically the longer side) as a *chord* of the circle. There is certainly no mention of that in the text.

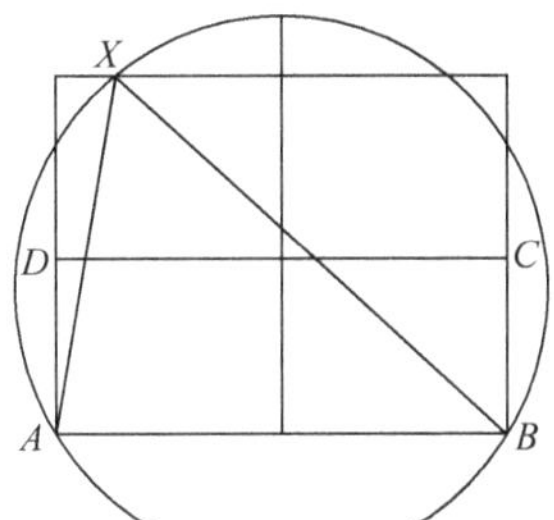

Diagram 5 (from Bluck, p. 453)

Farquharson adopts an appreciably more complex mathematical solution involving 'finding a chord of the circle such that a parallelogram equal to the given figure, described on half the chord, and placed so that one of its diagonals is also a chord, having one end at an extremity of the chord so found, is equal to a similar and similarly

mathematics in Plato's day, the *reduction* itself is a worthwhile step. We may compare the more famous case of the problem of the duplication of the cube. This was first reduced (by Hippocrates of Chios) to the problem of finding two mean proportionals *before* the solution to that problem was given by Archytas.

14 In diagram 5, from Bluck, p. 453, the given rectangle is ABCD, the inscribable triangle equal in area to it AXB. The given rectangle is assumed to have one of its two longer sides (AB) as a chord of the circle.

situated parallelogram described on the other half of the chord'.[15] Yet as Bluck pointed
175 out (p. 458) Far|quharson 'does not explain how the geometrician is going to find the
appropriate chord in any given case, or to determine whether such a chord can be found'.
Worse still, the hypothesis is not one that the geometrician applies or could apply.

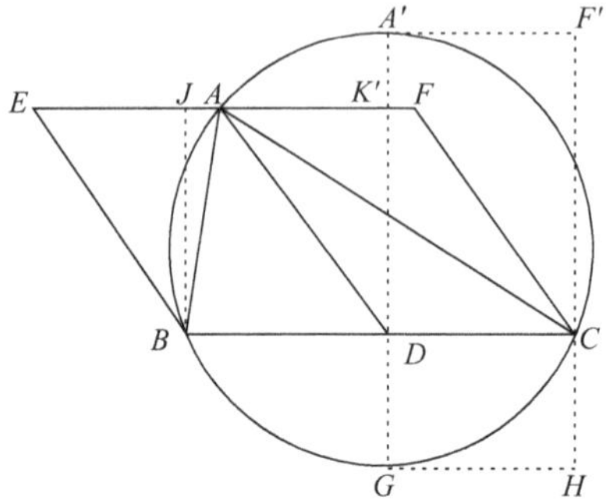

Diagram 6 (from Bluck, p. 457)

Bluck himself, however, adopted a modified Farquarson interpretation as 'freer than most from major difficulties' (p. 460). The modification consists in taking the chord as *given*. But that just seems opportunistic. After all the original problem is that of inscribing an area as a triangle in a circle. When reference is made to the 'given' line at 87a4, we can hardly allow that this is the line that *yields the solution*. What is given, in fact, has to be given in the *problem*, not as part of the solution.

Most of the proposed interpretations of the mathematical illustration reveal more about their authors' ingenuity than about any obvious sense that the text can be got straightforwardly to yield. This prompts me to ask why the illustration is so untransparent. Is it the case, as Vlastos (p. 123) suggested, that 'the geometrical example is ostentatiously technical' – where the reader would have needed considerable proficiency in this branch of mathematics to understand it? Or is it that some of the problems arise because mathematical terminology, in Plato's day, was still undeveloped and imprecise?[16] How far, in fact, to revert to the question I raised earlier, is Plato in control of this material himself?

---

15 In diagram 6, from Bluck, p. 457, ABC is the required triangle. The parallelograms on each half of the chord BDC are EADB and AFCD and both have diagonals (AB and AC) that are chords of the circle. EADB, AFCD and the triangle ABC itself are all equal to the required area. Each of the two parallelograms falls short on the line along which it is described (BDC) by a similar and similarly situated parallelogram (i.e. the other one of the pair). However how the chord BDC itself is to be found is not made clear.

16 Cf. above n. 9. It has been argued, in particular, that *diorismoi* were invented only after Plato. This is on the basis of a report in Proclus, *in Eucl.* I, 66.20ff, that attributed their introduction to Leon, the pupil of Neoclides (cf. e.g. Heath, p. 303). But that may have related to the introduction of the *term diorismos*. Heath, for one, took the *Meno* passage as evidence to show that the idea of a general *diorismos* defining the limits of possibility of solution was familiar even before Plato's time.

A digression on the earlier mathematical definition of σχῆμα at *Meno* 76a will serve to open up the main possible lines of argument. The definition offered by Socrates is στερεοῦ πέρας, and this is often hailed as an excellent model of what a definition should be. (Meno, it should be noted, does not so much accept it as take it for granted, turning immediately to ask about colour, 76a8, which Socrates in turn greets with the remark that he is ὑβριστής.) However the inadequacies of this definition of figure are considerable. We may first observe that the definiens, limit of solid, is used as a definition not of figure, but of *plane*, in Aristotle,[17] and of *surface*, in Euclid.[18] Moreover the definition in the *Meno* fails the test that it should be in terms that have already been defined.[19] There is a question, too, about its | universality. 176
While a plane figure may be treated as the limit of a solid, a solid figure (such as prism, pyramid, cone) may have several limits or boundaries. Finally it may be objected that it is not relevant to refer to *solid* in the definition of *figure* at all. Let us take triangle as an example of a plane figure and treat it as the limit of a solid. Which solid, we may ask? There are of course an infinite number of solids one of whose limits may be a triangle. The defence to this might be to insist that Plato's definition did not do any more than specify that a figure (triangle in this case) is the limit of *a* solid, *some* solid, that is. But that still miscues the point, which is that figure is (as Euclid had it *Elements* I Def. 14) contained by a boundary or boundaries. That fits both plane and solid figures. But while reference to solid is appropriate enough in the definition of surface, it is neither necessary nor desirable in a definition of figure.[20]

The definition of figure at 76, like the mathematical illustration at 86-87, raises questions both about the level of Plato's competence and about his motivations – and these two go together. Thus if we are prepared to cast doubts on Plato's competence (I shall call this interpretation alpha) we might take it that Plato himself considered the definition of figure perfectly adequate and indeed offered it as a model of proper definition. This would certainly give him a positive motive for introducing this definition: he aimed at this point to provide such a model. Yet it does not speak well for his competence, at least if we accept the arguments of the last paragraph concerning its shortcomings. Or maybe we can and should rescue Plato with the defence that this was the best definition available at the time. Then the charge of incompetence is transferred to the state of contemporary mathematics, and of course that may be where the problem lies.

Alternatively (on what I shall call interpretation beta) we might argue that Plato's control over the mathematical materials is beyond reproach, or at least perfectly adequate. But then questions concerning his motivations arise. Thus it may be that Plato himself realised that the proposed definition of figure is not watertight, but represented Meno, ὑβριστής as he is, accepting it all too impetuously. That would give Plato full marks for competence, or at least no bad marks on that score. But what then

---

17 *Topics* 141b22 (where Aristotle objects that this is to show the prior *via* the posterior).

18 *Elements* XI Def. 2.

19 *Meno* 79d referring back to 75d.

20 Vlastos, p. 122, claimed that Euclid's definition of figure in I Def. 14 is 'dictated by the architectonics of Euclid's treatise, which defers solid geometry to its latest books. So it is understandable that Euclid should want to make no reference to "solid" until then'. Plato *Meno* 76a4-7 is accordingly described by Vlastos as 'more expeditious' than Euclid's I Defs. 13 and 14. But that is opportunistic, since the reference to 'solid' in the definition of figure is not only not needed, it is undesirable.

177 are his | motives? We would have to take it that Plato is warning us, the readers, to be careful, not to be like Meno who perhaps assumes that he can tell a good definition and pass to the next subject without further ado. However if that is Plato's intention – to warn us to be wary of the definition on offer – I have to admit that the warning has gone very largely unheeded down the centuries by the commentators, most of whom (like Meno) rush to congratulate Plato on the splendid mathematical definition he offers us in this passage.

We may now return to the mathematical illustration of hypothesis, where questions of competence are far harder to resolve because of the indeterminacies in the passage as well as because of problems to do with how far contemporary mathematics had advanced by Plato's time. But that may just make the questions to do with motives the more urgent.

Of course on an alpha-style interpretation it is easy enough to condone Plato's motives. If he is not really in control of the mathematics of the inscription problems, there is not much more to be said. On this (alpha-style) view he has picked up *some* idea about mathematical methods and believes that may be handy to illustrate his method of hypothesis. But (on *this* view) the reasons for all the obscurities in the text are that Plato is none too sure of himself. Thus he may even have confused *diorismos* and problem reduction. He has some vague idea that mathematicians use problem shifts, and that will do as justification for his referring to them at the point when he introduces his own method of hypothesis for his own, specifically philosophical purposes. But to this it must be said that such a line of interpretation would be, if not a counsel of despair, at least moderately desperate.

But if then we try a beta-style interpretation of 86-87, what happens? If we assume that Plato is fully in control of whatever mathematics was available at the time, then the question that must be faced is this. Why did he write so obscurely at this point in the *Meno*? Why, first, did he choose such a difficult example, when plenty of simpler ones would have made his chief methodological point? Why, secondly, did he so underdescribe the mathematical illustration that was supposed to make clear how the method of hypothesis works? Thus if we take it that the Cook Wilson/Heath/Knorr interpretation is the, or among the, best on offer, the underdescription relates to the following points: first there is no reference to the *diorismos* (if the starting-area is greater than the inscribable equilateral triangle, then no solution is possible). Secondly, so far as the problem-reduction goes, none of the steps is explained and justified with the exception of the specification of the condition that there is some relationship ('by such as is') between the applied area and the area by which it falls short. Thirdly, there
178 are those | indeterminacies in the language of the description of the illustration that I set out at the outset.[21]

Vlastos who, as already noted, has Plato preening himself on his own expertise in geometry, added that he was 'warning his readers that if they have not already done a lot of work in that science they will have difficulty in following him, and this will be

[21] This is before we take into account the point that no advice is offered as to how to solve the problem as reduced. As we have noted (n. 13) how this might have been done before Euclid, and indeed whether it could be, are disputed. But that may not matter as much as is sometimes supposed, since the problem reduction itself is worthwhile even if the problem as reduced remained, for the time, beyond reach of a solution.

their loss, not his'. That, on a beta-style interpretation, is fine as far as it goes, but it does not go far enough. Nor does it address the chief difficulty, which is that the example is *unnecessarily* obscure. Certainly if Plato had wanted to encourage his readers to develop their mathematical skills, he could have done so by giving a *helpful* example, one that clearly showed the usefulness of *diorismos* as distinct from, and preliminary to, problem reduction, and then went on to explain the problem reduction step by step. But as it is, he does not – as everyone has to agree.

So there is a tension between the purported purpose of the mathematical example and its actual role. It purports to explain and make clear. It actually does nothing of the sort. An alpha-style interpretation has a simple explanation: Plato is not in control of these materials himself, or he merely mirrors the imprecisions of the mathematics of his day. But what can a beta-style interpretation offer on the problem – an interpretation that does not condone the obscurities of the example by an appeal to incompetence?

As a speculation I suggest the following. What we are given, at this point in the *Meno*, is an ultra-obscure mathematical example, and when we go through it, puzzling our way through the obscurities, the indeterminacies, the omissions, we find we have gone *through an initiation*.

Let me first do something to justify the talk of initiation and explain how I see initiation differing from *education*. In education the teacher does all he or she can to help the pupil, starting with elementary materials and proceeding gradually to more advanced ones, making sure at each stage that the pupil has grasped the simpler points before progressing to more complicated ones. The pupil does not usually start with *no* opinions on the subject, but an important feature of the process of learning will be distinguishing true from false opinions, and eventually progressing from true opinion to knowledge. In the *Meno* itself, the long exchange between Socrates and the | slave-boy, 179
82b-85b, offers as fine an example as anyone could wish of careful and painstaking *education*.

But by initiation I mean rather the process whereby the initiates themselves come to see some subject in a quite different light. They may even discover that what they had been taught, or what they thought they had learnt, at an early stage in the process, later turns out to be quite false. Thus in his study, *Ritual and Knowledge among the Baktaman of New Guinea* (Oslo and Yale, 1975), F. Barth showed how from childhood through to old age the Baktaman pass through a complex seven-stage process of initiation, where at each stage what had been presented as truth and wisdom is found at a later one to be falsehood, folly, madness.[22]

Now Greek initiations are not Baktaman ones. But that initiation in its Greek form is present as a theme in the *Meno* is clear. At 76e, after Socrates has offered his definition of colour as an effluence of figures, greeted by Meno as an 'excellent' answer, Socrates remarks that it is a τραγική one, and so it pleases Meno more than the answer

---

[22] Thus Barth, op. cit. p. 81, wrote: 'A recurrent theme of previous initiations has been that of *deception*. Sometimes, statements and promises have been made that were immediately exposed as lies by the next ritual act; sometimes information was made vaguely suspect by hints or evidence that it was rendered false or grossly incomplete by further secrets. Even the central revelation of one initiation – e.g. the showing of the bones to 2nd degree novices – was shown to be largely a hoax in a later initiation.'

that had been given to the definition of figure. When Meno has agreed to that, Socrates goes on to suggest that the latter is the better answer. 'And I think you too would think so, if you were not compelled, as you were saying yesterday, to go away before the mysteries and could stay and be initiated.'[23]

Meno, we may understand, is impatient: he could benefit from an initiation, indeed one into mathematics, but is too impatient to undergo that process. Moreover practice, training, study, on the mathematical examples, Socrates had earlier said (75a8), was useful for the investigation of the subject in hand, namely virtue. Socrates himself, of course, comes to introduce the theory of recollection at 81a ff as what he has heard from 'wise men and women', wise indeed in matters divine, priests and priestesses who have studied what they are concerned with so as to be able to give an account of it. Nor should we forget that at the very end of the dialogue (100a) someone who is
180 able to make a true statesman of another is com|pared, among the living; with what Homer said about Teiresias among the dead: he alone understands, the rest are flitting shades.[24]

The relevance of the theme of initiation to the mathematical example of *Meno* 86e ff might, then, speculatively, be represented as follows. Once you go through the example – and you will probably need a guide to help you – and you investigate different possibilities and finally see how it may work, once you have been through that process and look back at what Plato had originally given you in the text, you see *that* in a quite different light. Ah yes, you now say, we can take the 'given line' as the diameter of the circle. It is a 'given line' not *of* the circle, nor *of* the starting-area, but the given line *for* that starting-area.[25] What we need to do is to find the segment of the diameter such that the rectangle constructed on its base equals the starting-area. When the diagonal and the perpendicular of the rectangle meet at the circumference of the circle, the perpendicular will be a mean proportional between the two parts of the segments of the diameter.[26] Thus the rectangle will indeed fall short by a rectangle similar (but not equal) to itself. But *what was not in the text* was that we have to find a *certain point* giving a rectangle that corresponds to the starting-area. Nor indeed were we told how we go about finding such a point.

At the end of the investigation we can see that that was how we have to proceed, first to reduce one problem to another, and then to tackle the problem as so reduced. Only Plato did not exactly make any of that clear, merely indicating that there was some application of areas and providing the single clue that the area applied, and the area by which it falls short, are 'such as' one another. Indeed Plato could be taken as implying that all that is in question is settling the conditions of possibility for the solution of a problem (*diorismos* in the strict sense) – where that could have been done quite simply by reference to the inscribed equilateral triangle. While we can and must

---

23 *οἶμαι δὲ οὐδ' ἂν σοὶ δόξαι, εἰ μή, ὥσπερ χθὲς ἔλεγες, ἀναγκαῖόν σοι ἀπιέναι πρὸ τῶν μυστηρίων, ἀλλ' εἰ περιμείναις τε καὶ μυηθείης.*

24 Among further references to the topos of prophecy in the dialogue, there is 92c, where Anytus is compared to a *μάντις*, and 99c where ordinary soothsayers and diviners are said to speak the truth but know nothing of what they say.

25 This would be the resolution of the third difficulty mentioned at the outset, above p. 167, to do with the reference of *χωρίον*.

26 In diagram 3, as explained above, p. 170, BC:CD = CD:CH. In other words CD is the geometric mean between the two segments of the diameter, BC and CH.

distinguish between the preliminary *diorismos* (not itself a problem reduction) and the problem reduction itself, no such distinction is made explicit in the text – and an alpha-style interpretation might even countenance the possibility that Plato himself confused the two.

On the beta-style interpretation Plato's own mathematical competence is | not in 181
the dock: but the question *why* he wrote in the obscure way he did becomes the more urgent. The answer I suggest proceeds on the assumption that this is deliberate. I recognise that this may be too paradoxical for many, but those for whom that is the case will be forced back to or towards some version of an alpha-style interpretation, where for reasons of his own incomplete competence or the inadequacies of the mathematics of the day Plato is not in full control. I do not believe that such an interpretation can be *ruled out*. But it is certainly not the only, and it may not be the preferred, line of interpretation.

Plato's recommendation of hypothesis in philosophy has to get by and can get by perfectly well *on its own*. We do not need the mathematics to understand the method. In fact the way we do understand it is by reflecting on the philosophical, non-mathematical, materials we are given – certainly not by studying the complexities of the mathematical illustration. But the mathematical illustration, for all its complexities and obscurities, is *not pointless*: far from it. It is the very obscurities that provide its point, namely that we *stand in need of initiation* – not, of course, into the *method* but in the fields to which it is applied. The method, as I have argued, is clear enough from the philosophical example. But it is one thing to understand the method, another to use it to get the right results. Even if the procedures are clear enough from the philosophical example, that does not entitle anyone (Meno, or us) to feel confident that we can apply them correctly. There is nothing further needed by way of introduction into the method. But to be proper philosophers, proper mathematicians, that is another matter. The suggestion I offer is that Plato uses the mathematical example to make the point, by implication, about philosophy.

It is not that the mathematical example itself *actually* initiates us nor that Plato is interested in mathematical initiation for its own sake (not just in it, at least). It would be nice to think that Plato wanted positively to encourage his readers to become expert in mathematics. But in this text at least, it can hardly be said that his readers will be encouraged in the way they are in the slave-boy discussion. Rather in the mathematical illustration of hypothesis readers are – many of them – likely to be positively discouraged, even intimidated. First the obscurities are enough to put most people off – and Plato has certainly not done his best to make everything clear and simple, as all have to agree. But then once we persevere and work through the various possible geometrical interpretations, we discover just how opaque the original information in the text is.

Where the mathematical illustration serves Plato very well is in relation to the
warnings we derive from it about what to expect where the highest | philosophising 182
is concerned – about what the *Symposium* (210a) terms the τέλεα τε καὶ ἐποπτικά. Plato does not, in the *Meno*, say in so many words that the true understanding of virtue is only for initiates. To have said so directly would have transformed the nature of the exchanges between Socrates and Meno. With Meno as chief interlocutor, the dialogue ends in *aporia*, though with some dark hints about a Teiresias-figure. Meno himself, we are no doubt meant to understand, is not ready, not patient enough, for an

initiation. But the indirect lesson of the mathematical illustration of hypothesis may be that, like mathematics, the highest philosophy too is only for initiates.

As a further illustration of Meno's insouciance, that point may be acceptable enough. But there are, of course, negative features of the theme of the exclusiveness of philosophy that we may derive from this and other dialogues. In the mathematical illustration he has at the very least withheld some of the information needed for its interpretation. But it is not just a matter of what has been omitted, but at points also one of what has been included, where that contains material that is potentially misleading. Insofar as Plato is being *deliberately* obscure,[27] the view we gain of the expertise he thinks we need – to do mathematics or philosophy – is not at all encouraging. He was supposed to have put on the portals of the Academy the slogan μηδεὶς ἀγεωμέτρητος εἰσίτω. That is not exactly welcoming. But once we have struggled through the maze of the interpretation of 86e ff we might conclude that it would have been more appropriate to adopt Dante's inscription on the gates of hell: **Lasciate ogni speranza voi ch' entrate**.[28]

183 REFERENCES

August, E.F., *Zur Kenntniss der geometrischen Methode der Alten*, Berlin, 1843.
Benecke, A., *Ueber die geometrische Hypothesis in Platons Menon*, Elbing, 1867.
Blass, C. *De Platone mathematico*, Bonn, 1861.
Bluck, R.S., *Plato's Meno*, Cambridge, 1964.
Butcher, S.H. 'The Geometrical Problem of the *Meno* (p. 86E-87A)', *Journal of Philology* 17, 1888, 219-25.
Canto-Sperber, M., *Platon Ménon*. Paris, 1991.
Canto-Sperber, M., ed. *Les Paradoxes de la connaissance. Essais sur le Ménon de Platon*, Paris, 1991.
Cook Wilson, J., 'On the Geometrical Problem in Plato's Meno, 86E sqq.', *Journal of Philology* 28, 1903, 222-40.
Farquharson, A.S.L., 'Socrates' Diagram in the *Meno* of Plato, pp. 86E-87A', *Classical Quarterly*, 17, 1923, 21-6.
Gaiser, K., 'Platons *Menon* und die Akademie', *Archiv für Geschichte der Philosophie* 46, 1964, 241-92.
Gueroult, M., 'Note sur le "locus mathematicus" du "Ménon" 87a', *Revue philosophique de la France et de l'Etranger* 159, 1969, 126-46.
Heath, T.E., *A History of Greek Mathematics*, vol. 1, Oxford, 1921.
Heijboer, A, 'Plato "Meno" 86e-87a', *Mnemosyne* 4th ser. 8, 1955, 89-122.
Heitsch, E., 'Platons hypothetisches Verfahren im Menon', *Hermes* 105, 1977, 257-68.
Karasmanis, V., *The Hypothetical Method in Plato's Middle Dialogues*, unpubl. D. Phil. dissertation, Oxford, 1987.

[27] As Heath already noted, p. 302, Plato 'was fond of dark hints in things mathematical'. One of the most notorious examples is that of the nuptial number at *Republic* 546b-d. If the general point is taken, this may alter the approach to be adopted to the interpretation of the use of mathematics in such cases as the construction of the world-soul in *Timaeus* 35b ff.

[28] This paper is the outcome of a presentation to a seminar on the *Meno* held in Cambridge during the Lent and Easter terms 1991. I am most grateful to all those who participated in the discussion and who have corresponded with me on these issues subsequently, and my particular thanks go to Myles Burnyeat and Malcolm Schofield for characteristically forthright and constructive comments.

Knorr, W.R., *The Ancient Tradition of Geometric Problems*, Boston, 1986.
Meyers, J.I., 'Plato's Geometric Hypothesis: *Meno* 86e-87b', *Apeiron* 21, 1988, 173-80.
Sternfeld, R. and Zyskind, H., 'Plato's *Meno*: 86E-87A: Geometrical Illustration of the Argument by Hypothesis', *Phronesis* 22, 1977, 206-11.
Thomas, I., *Greek Mathematical Works, I Thales to Euclid*, London and Cambridge, Mass., 1939.
Vlastos, G., *Socrates*, Cambridge, 1991.

PART 3

STUDIES ON PROPORTION THEORY AND INCOMMENSURABILITY

Texts selected and introduced by Ken Saito

KEN SAITO

INTRODUCTION

The first two of the articles reprinted in this chapter represent the predominant approach to the pre-Euclidean mathematics up to the 1960s. The latter two are more recent reflexions and criticisms to this research style.

The assumption that the discovery of incommensurability had a strong impact on the style of Greek mathematics has been the predominant view until quite recently. Some even considered this discovery a real scandal. Now scholars reexamine this view and wonder whether such a historiography does not assume what should be proved.

The geometry of the *Elements* has a very particular style in its handling of geometric magnitudes. Numbers are never used to designate geometric magnitudes as lines, surfaces and solids, and a highly developed theory about ratio (defined as relation of two magnitudes) and proportion (sameness of ratio-relation according to some of the manuscripts of the *Elements*) is applied to argue the quantitative aspect of geometric entities.

The *Elements* present a number of theories. There is no preface nor any commentary that would have given a number of hints as to why Euclid decided to make use of the particular style of the *Elements*, and, furthermore, there are few, if any, documents about the shaping process of this style, especially concerning the pre-Eudoxean period. Proclus' excursus which probably derives from Eudemus is vague and insufficient; Plato and Aristotle offer abundant passages concerning mathematics, but these are of a different nature because their primary aim was not to supply a mathematical text, nor to describe the history of mathematics. Therefore, stories about the history of early Greek mathematics have been woven up from different kinds of such fragmentary testimonies. And the predominant issue in these stories has been the discovery of incommensurability.

In fact, the discovery of incommensurability plays important role in most of the historical accounts about early Greek mathematics given in the first half of the 20th century. Incommensurability *must have been* the scandal for Pythagoreans, since it *must have been* contradictory to the principle of their master: all is number. The so-called geometric algebra of Book II of the *Elements* was a transcription of Babylonian algebraic knowledge, because incommensurability forced the Greeks to represent the incommensurable square roots by means of geometric figures. Such theses are based on the assumption that the discovery of incommensurability – of which the precise date we don't know – gave an impact to mathematics, especially to the foundations of mathematics, and this assumption was in turn thought to be corroborated by such theses. We use the term incommensurability-historiography to encompass all these mutually-dependent and mutually-supporting theses about the impact of incommensurability.

Now, among the versions of the incommensurability-historiography, the strongest thesis can be seen in an article of Hasse and Scholz, who affirm that the discovery of incommensurability brought about the "foundational crisis" to Greek mathematics.[1]

J. Christianidis (ed.), Classics in the History of Greek Mathematics, 187–190.

Becker's argument may be said to belong to a milder version of incommensurability-historiography, for he assumes that incommensurability brought about certain difficulties to the naive theory of proportion. What made his article so influential was his claim that a specific remedy to the difficulty of proportion theory was soon available: it is a definition of proportionality (reconstructed by Becker) based on "anthyphairesis", i.e. the so-called Euclidean algorithm. Moreover, his claim was inspired by (if not strictly based on) two Aristotelian remarks, so that it enjoyed better documentary support.

Starting from this reconstructed definition of proportion, he tries to reconstruct a theory of proportion which *must have existed* between the discovery of incommensurability and the definition of proportion in Book V of the *Elements* ascribed to Eudoxus. Becker successfully demonstrated, from the anthyphairetic definition of proportion he had formulated, most of the propositions of Book V of the *Elements*. This reconstruction occupies large part of his article.

The influence of Becker's thesis was enormous, because it provided a mathematical procedure available to "reconstruct" mathematics about which we have little or no documentary evidence. The procedure of "anthyphairesis" could be tested or applied to any particular (real or imaginary) cases, thus stimulating a series of similar studies.

Becker's thesis is also complementary to the thesis of geometric algebra advocated by Tannery and Zeuthen half a century earlier. Both of these theses assume that incommensurability brought about some difficulty and that the efforts to remedy the situation are reflected in some idiosyncrasy of the style and content of Greek mathematics.

Thus the idea of "impact of incommensurability" has constituted an important part of a paradigm for the study of pre-Euclidean Greek mathematics, especially for the fifth century, where any serious investigation did not seem possible because of irrecoverable lack of documents. This paradigm has as its core-belief that incommensurability has caused difficulty in the foundation of proportion theory. Such a paradigm has been used for the mathematical reconstruction of ancient arguments such as anthyphairetic definition of proportion and on studies of pre-Eudoxean mathematics.

Such a historiography also seems to have encouraged the "archeological approach," another characteristic of the historiography of the twentieth century: if you find some idiosyncrasy in the *Elements* – roundabout arguments, avoidance of the use of proportion, etc. – you should assume it to be a reflexion of some earlier stage in the development of Greek mathematics. In short, the *Elements* were regarded as if it were an archeological site where one could excavate the remnants of earlier mathematics. Thus the *Elements* were used for the study, not of the mathematics at the moment of their compilation, but of the mathematics of a century before, or even earlier.[2]

Among the studies inspired by Becker's reconstruction approach, we here reprint von Fritz' article about the discovery of incommensurability. In this article, anthyphairesis is applied to the side and diagonal of a regular pentagon, and the resulting repetition *ad infinitum* of the same relation is conjectured to have been the very occasion of the discovery of incommensurability. Von Fritz attributes this hypothetical trial and discovery to Hippasus of Metapontum, following an infamous legend of scandalous disclosure of the secret knowledge of Pythagoreans about the dodecahedron. But it is extremely audacious to ascribe to somebody a very specific discovery with no documentary support. This speculative (though fascinating) paper has become so popular that the anthyphairesis of the side and diagonal of a pentagon was thought as if it were an established historical fact.

Thus von Fritz' article together with Becker's, represent the research trend in the first half of the 20th century, which we characterized by terms such as "incommensurability-historiography" and "reconstruction." Now let us turn to the studies criticizing this trend.

Freudenthal examines critically the foundation of the thesis that incommensurability must have brought a crisis in mathematics. Most of the documents, especially those which are more reliable, suggest nothing about the crisis or difficulty of mathematics caused by incommensurability. He argues persuasively that the documents suggest continuous development rather than rupture during the time incommensurability was probably discovered.

However, the theory of proportion in Book V of the *Elements* is no doubt aware of incommensurability, so that there seems to be no room to doubt that incommensurability stimulated new investigations about the theory proportion before and at the time of Eudoxus. Freudenthal himself accepts Becker's reconstruction while denying the "foundational crisis." Then what is wrong with Becker's "anthyphairetic proportion theory?" Knorr's paper, unpublished until very recently, though based on a talk in 1975, points out, as its title vividly shows, that the influence of 20th century mathematics and philosophy – i.e. the foundations of mathematics, and phenomenology – is what lies behind Becker's study. The research originating in Becker's article was casting the fragmentary documents and the results of mathematical reconstruction into the predetermined perspective supplied by phenomenological speculations.[3]

This approach was not altogether sterile, and sometimes such practices are useful when documents are desperately lacking. However, it is true that other significant approaches have been disregarded in the shadow of the enthusiasm for reconstructions, and Greek mathematics have always been investigated from more uncertain aspects of its "earlier developments" and "origins."

Recent historians have recognized the necessity and fertility of two other approaches. One is the concentration to the text: to examine the extant text itself, without easy translation to modern mathematical terms, and try to gather the interest and the technique of its author; it also implies not to look for traces of earlier phases in the text itself, but to understand the interests and the intentions of the original authors – though this is no easy task, and even impossible in a strict sense.

Another approach is concerned with the traditions of the texts, often making use of Arabic and Medieval Latin traditions, and make out what changes have occurred to the text we possess. This study is all the more necessary for a text like the *Elements* which has been continuously read, commented upon and edited. This is in part the study of the posterior periods late antiquity, Arabia, Medieval Latin West, but is also indispensable to get closer to the original texts.

Neither of these approaches would leave much room for reconstructions and speculations which fascinated the scholars and readers of the twentieth century. Reconstruction means, however, that we speak where documents keep silent. Then why can we be sure that we can speak correctly on behalf of the ancients? Now scholars are cautious, though this attitude might only mean that the pendulum has swung back too much into skepticism. Anyway, let us take advantage of this occasion and try to listen to the documents, which, though extremely taciturn about the fifth century, tell us many other things, if we tend an ear to them.

Osaka Prefecture University

NOTE

1 The impact of this paper is attested also by the fact that it is mentioned in all of the four articles reprinted here.

2 For example, the first four books of the *Elements* where the use of proportion is avoided and several propositions that could be proved more easily by the use of proportion theory are established without recourse to ratio and proportion. According to the archeological approach, these propositions are interpreted as the trace of an earlier period, when, after the discovery of incommensurability, the validity of proportion theory was put in doubt. In this case, a passage in Proclus' commentary is called to its support. However, this or other archeological interpretations are based on the assumption that Euclid did not change so much the materials he inherited from his predecessors.

3 Knorr, however, proposes another definition of proportion, based on an attentive analysis of Archimedes' argument, in his "Archimedes and the pre-Euclidean Proportion Theory," *Archives internationales d'histoire des sciences* 28(1978): 183–244, which is not reprinted here simply because it is too long. See also my "Phantom Theories of pre-Eudoxean Proportion," *Science in Context* 16(2003): 331–347.

OSKAR BECKER

EUDOXOS – STUDIEN I. EINE VOREUDOXISCHE PROPORTIONENLEHRE UND IHRE SPUREN BEI ARISTOTELES UND EUKLID.

1.

In der aristotelischen Topik (*Θ*. 3, p. 158 b 29–35) findet sich folgende Stelle:

ἔοικε δὲ καὶ ἐν ταῖς μαθήμασι ἔνια δι' ὁρισμοῦ ἔλλειψιν οὐ ῥᾳδίως γράφεσθαι, οἷον [καὶ] ὅτι ἡ παρὰ τὴν πλευρὰν τέμνουσα τὸ ἐπίπεδον ὁμοίως διαιρεῖ τήν τε γραμμὴν καὶ τὸ χωρίον.	Es scheint aber auch in der Mathematik Einiges wegen des Fehlens einer Definition nicht leicht zu beweisen zu sein, wie z. B. daß die ein Parallelogramm parallel zur Seite schneidende Gerade die Linie und die Fläche (d. i. die anliegenden Seiten und die Fläche des Parallelogramms) in demselben Verhältnis teilt.
τοῦ δὲ ὁρισμοῦ ῥηθέντος εὐθέως φανερὸν τὸ λεγόμενον. τὴν γὰρ αὐτὴν ἀνταναίρεσιν ἔχει τὰ χωρία καὶ αἱ γραμμαί. ἔστι δ' ὁρισμὸς τοῦ αὐτοῦ λόγου ⟨πρῶτος?⟩ οὗτος.	Wenn aber die Definition ausgesprochen ist, ist das Gesagte sogleich einleuchtend. Denn dieselbe „Antanairesis" haben die Flächen und die Linien. Es ist aber die ⟨erste?⟩ Definition desselben (d. i. des gleichen) Verhältnisses diese (d. h. die soeben angeführte).

Der Zusammenhang der Stelle ist belanglos; sie stellt nur eines der vielen Beispiele der Topik dar, die im allgemeinen noch zweifellos aus dem gemeinsamen Material der Akademie stammen. Der Text folgt der Ausgabe von *Strache-Wallies*, die auch das *καὶ* in l. 31 streicht. (Ich fügte l. 35 *πρῶτος* zur Erwägung bei, das sich in dem entsprechenden Zitat des Alexander-Kommentars zur Stelle in einer guten Handschrift (P) und in dem Erstdruck der Aldina findet.)

| Die Beziehung der Stelle auf den Satz, daß sich Parallelogramme von gleicher 312
Höhe wie ihre Grundlinien verhalten (Euklid Elem. VI, 1), der im Text noch in einer gewissen Spezialisierung, nämlich bezogen auf zwei einander gleichwinkelige Parallelogramme erscheint, ist evident und zum Überfluß durch den erwähnten Kommentar des Alexander von Aphrodisias gesichert, der sich auf S. 545, 1–21 (Wallies) mit unserer Stelle beschäftigt.

Es heißt da nämlich p. 545, 9–12 (nach der Walliesschen Textgestaltung): *ἔστι δὲ τὸ λεγόμενον, ὅτι ἐὰν ἐπίπεδον παραλληλόγραμμον ᾖ, ἀχθῇ δ' ἐν αὐτῷ ὁμοιᾷ τῶν πλευρῶν παράλληλος, ἡ παράλληλος ὁμοίως διαιρεῖ τὴν γραμμὴν καὶ τὸ πᾶν χωρίον, τουτέστιν ἐν τῇ αὐτῇ ἀναλογίᾳ.*

J. Christianidis (ed.), Classics in the History of Greek Mathematics, 191–209.

Damit ist aber noch nicht die Bedeutung des Wortes „Antanairesis“, das die Definition der Verhältnisgleichheit bestimmt, erklärt. Das sehr Merkwürdige ist nun, daß in den unmittelbar folgenden Zeilen des Kommentars diese Bedeutung – für den, der zu lesen versteht – in voller Klarheit zutage tritt, und damit sich ein überraschender Einblick in die Struktur einer voreuklidischen, ja wie ich meine, voreudoxischen Proportionenlehre eröffnet, die sich bereits auch auf irrationale Verhältnisse bezieht.

Der Kommentator fährt nämlich zunächst fort (p. 545, 12–17): *τοῦτο γὰρ ὁμοίως μὲν λεγόμενον οὐκ ἔστι γνώριμον· ῥηθέντος μέντοι τοῦ ὁρισμοῦ τοῦ ἀνάλογον γνώριμον γίνεται ὅτι ἀνάλογον ὑπὸ τῆς ἀχθείσης παραλλήλου τέμνεται ἥ τε γραμμὴ καὶ τὸ χωρίον.*

Und dann folgt die entscheidende Stelle:

ἔστι δὲ ὁρισμὸς τῶν ἀναλόγων, ᾧ οἱ ἀρχαῖοι ἐχρῶντο, οὗτος' ἀνάλογον ἔχει μεγέθη πρὸς ἄλληλα ὧν ἡ αὐτὴ ἀνθυφαίρεσις· αὐτὸς δὲ τὴνἀνθυφαίρεσιν ἀνταναίρεσιν εἴρηκε.

Es ist aber die Definition der proportionalen (Größen), deren sich die Alten bedienten, folgende: In Proportion stehen Größe zueinander, denen dieselbe „Anthyphairesis“ zukommt. Aristoteles aber nannte die Anthyphairesis „Antanairesis“[1]).

Zunächst freilich scheint durch die Aussage des Kommentators nicht viel gebessert. Wir erfahren zwar, daß es sich um eine fest formulierte Definition der Verhältnisgleichheit bei den „Alten“ handelt, aber ihr Inhalt bleibt dunkel. Er ist in der Tat bisher nicht verstanden worden. *Heiberg* (Mathematisches bei Aristoteles, S. 22) glaubt die „eudoxische“ Definition (Euklid El. V, def. 5) wiederzuerkennen, wozu der Text aber keinen begründeten Anlaß gibt. *Heath* (The thirteen books of Euclid's
313 | Elements, Vol. II, p. 120–21) lehnt dies mit Recht ab. Aber er selbst gibt keine positive Erklärung, hält die Ausdrücke „Antanairesis“ und „Anthyphairesis“ für „metaphysisch“ (im Sinne *Barrows*)[2]) und vag und versucht eine etymologische Deutung, die natürlich zu nichts Präzisem führt[2a]).

Trotzdem ist die Lösung sehr einfach. Alexander erklärt offenbar den für ihn veralteten und nicht mehr ohne weiteres verständlichen Ausdruck „Antanairesis“ durch den ihm geläufigen „Anthyphairesis“. Woher kann er den letzteren haben? Die nächstliegende Vermutung ist: aus Euklids Elementen, die er ja auch sonst öfter zitiert[3]) und als allgemein bekannt voraussetzt. In der Tat kommt nun zwar nicht das Verbalsubstantiv „Anthyphairesis“, wohl aber das zugehörige Verbum *ἀνθυφαιρεῖν* an

---

[1]) Diese Definition findet sich wörtlich wieder bei *Suidas* (Lexicon, sub verbo *ἀνάλογον*), wodurch ihr Wortlaut gesichert ist.

[2]) Lectiones mathematicae XVIII (The mathematical works, ed. Whewell, Cambridge 1860, p. 299).

[2a]) Nach Abschluß des Manuskripts dieser Arbeit erfuhr ich durch eine freundliche Mitteilung des Herrn Pater *Steele S. J.* (einem Mitglied des Bonner mathematisch-historischen Seminars), daß die richtige Deutung von *ἀνταναίρεσις* in unserer Stelle schon im Jahre 1917 von *H. G. Zeuthen* in seiner nur in dänischer Sprache veröffentlichten Abhandlung „Hvorledes Mathematiken i Tiden fra Platon til Euklid blev rational Videnskab“ (D. Kgl. Danske Vidensk. Selsk. Skrifter, naturvidensk. og mathem. Afd. 8. Raekke, 1. 5, Koebenhavn 1917) S. 108 (306) ff. gegeben worden ist. [Anm. bei der Korrektur.]

[3]) Z. B. im Kommentar zur aristot. Metaphysik, ad. p. 1051 a 27 (p. 568, 3; cf. p. 158, 1 Bonitz).

entscheidendster Stelle in den Elementen vor, nämlich in VII, 1, 2 und X, 2, 3[4]). Die letzte Stelle ist besonders wichtig. Dort bedeutet das Verb die Operation des sog. „Euklidischen Teilerverfahrens" („Algorithmus") zur Aufsuchung des (größten) gemeinsamen Maßes zweier Größen; im Falle seines Nichtabbrechens kennzeichnet es ihre Inkommensurabilität (X, 2). (In VII, 1, 2 ist entsprechend von Zahlen, ihrem (größten) gemeinsamen Teiler bzw. von ihrer Teilerfremdheit die Rede.)

Das Teilerverfahren zwischen zwei Größen a, b ist nun bekanntlich äquivalent mit der Kettenbruchentwicklung des Verhältnisses („Bruches") a:b. Das heißt aber: *Die „Definition der Alten*" besagte:

„*Zwei (rationale oder auch irrationale!) Verhältnisse sind gleich dann und nur dann, wenn sie dieselbe Kettenbruchentwicklung haben.*"

| Der historische und sachliche Sinn dieser Definition und die Möglichkeit, von ihr 314
aus eine *allgemeine* Proportionenlehre (sowohl den Verhältnissen als auch der Größenart nach) zu begründen, wird im folgenden untersucht werden.

Vorerst sei noch über die wörtliche Bedeutung von ἀνθυφαιρεῖν bemerkt, daß sie mit „gegenseitig wegnehmen" (z. B. Land im Kriege) wiedergegeben werden kann. Das Wort ἀνταναιρεῖν wird sogar schon speziell vom „sich gegenseitig Wegheben" von Posten in einer Rechnung (etwa wie Vermögen und Schulden) gebraucht. Das gegenseitige Abtragen der „Reste" im euklidischen Verfahren (das von den Griechen nicht unter dem Bilde einer Teilung, sondern eines Wegnehmens der Kleineren vom Größeren, so oft es geht, gesehen wurde) findet dort ein freilich viel primitiveres Vorbild, aus dem immerhin der Wortgebrauch ziemlich verständlich wird.

2.

Um in die historische Bedeutung der „anthyphairetischen" Definition des Verhältnisses einzudringen, stehen als Quellen – außer einer noch zu besprechenden Stelle aus der aristotelischen Zweiten Analytik und einigen kurzen Andeutungen bei Proklos – nur die Euklidischen Elemente zur Verfügung. In diesen Quellen wird die Anthyphairesis als Definition des Logos nicht erwähnt; wir sind daher auf indirekte Schlüsse angewiesen.

Daß die Euklidischen Elemente auf verschiedene Quellen zurückgehen, wird allgemein angenommen. Auf die vier ersten „elementaren" Bücher – gekennzeichnet durch das Fehlen des heute so genannten „archimedischen Axioms" und des allgemeinen Größenbegriffs (μέγεθος) – folgen die „höheren" Bücher V–XIII, die methodisch ganz von der Proportionenlehre beherrscht werden. Aus diesem Material heben sich auf Grund der Scholien, des Prokloskommentars zu I und des Papposkommentars zu X zwei Gruppen heraus, die bekannten voreuklidischen Mathematikern zugeschrieben werden dürfen: Buch VII–IX und der Anfang von X dem *Theätet*, Buch V und XII dem Eudoxos. (Buch XIII, das nach den Forschungen von *Eva Sachs*

---

[4]) Erstaunlicherweise hat das *Heath* übersehen, das Wort ἀνθυφαιρεῖν fehlt in seinem griechischen Wörterverzeichnis; es kommt allerdings nicht in der Form einer Definition vor. – Ich habe das Teilerverfahren selbst „Anthyphairesis" genannt (in meinem Aufsatz „Die diairetische Erzeugung der platonischen Idealzahlen", diese Zeitschr., Bd. I, S. 479) – ohne zu ahnen, daß dieses Wort in einem mathematisch relevanten Zusammenhang wirklich in einem erhaltenen griechischen Text vorkommt.

theätetische und eudoxische Bestandteile enthält, kann hier außer Betracht bleiben, die Stellung von VI und der „höheren“ Sätze von XI wird noch zu bestimmen sein.)

Es erhebt sich nun die Frage, ob auch die beiden Definitionen der Proportion (Verhältnisgleichheit), die euklidische (Elem. V, def. 5) und die anthyphairetische, sich den beiden bahnbrechenden Forschern zuschreiben lassen. Nämlich die „Definition der Alten“ dem Theätet und die vermutlich jüngere Definition Euklids dem Eudoxos.
315 Das letztere | ist ohne weiteres anzunehmen, denn das ganze V. Buch lebt sozusagen von seiner 5. (und 7.) Definition und wird dem Eudoxos ja allgemein gegeben. Das erstere erscheint ebenfalls sehr plausibel, wenn man bedenkt, daß am Anfang des VII. und X. Buchs, für Zahlen und stetige Größen, das Teilerverfahren entwickelt wird (VII, 1, 2 und X, 2, 3), während die 5. Definition des V. Buches in VII–X niemals explizit angewandt wird. Dazu kommt noch der im folgenden näher zu erörternde Umstand, daß gerade die anthyphairetische Definition aus dem elementaren Gleichheitsbegriff rationaler Verhältnisse (wie ihn VII, def. 20 gibt) durch einen beinahe stetigen Gedankengang entwickelt werden kann. Um die Gleichheit zweier schwer zu übersehenden rationalen Brüche festzustellen, muß man sie kürzen, d. h. den größten gemeinsamen Teiler von Zähler und Nenner beidemal durch den euklidischen Algorithmus bestimmen: man bemerkt dabei bald, daß die fraglichen Brüche dann und nur dann gleich sind, wenn der Algorithmus beide Male dieselben „Teilungszahlen“ der Reihe nach liefert, so daß man diese Eigenschaft unmittelbar als Gleichheitskriterium verwenden kann. Im rationalen Fall bricht das Verfahren ab, im irrationalen nicht. Aber auch dann hindert nichts, daß man die übereinstimmend verlaufende Anthyphairesis zum Kennzeichen der Verhältnisgleichheit nimmt. Damit hat man aber vom rationalen *μέρη*-Begriff aus den irrationalen *πηλικότης*-Begriff gewissermaßen stetig erreicht.

Aber gegen diese Auffassung wendet sich ein anscheinend entscheidender Einwand. Im X. Buch, dessen Anfangssätze Theätet unbedingt zuzuschreiben sind (ein Scholion zu X, 9 bezeichnet den Satz ausdrücklich als seine Entdeckung), wird zwar nicht unmittelbar von der „eudoxischen“ Definition, aber wohl vielfach von der allgemeinen Proportionenlehre, die in V ganz auf jener Definition aufgebaut ist, Gebrauch gemacht; das nach unserer Auffassung „ältere“ theätetische X. Buch wäre also in Wahrheit abhängig von dem „jüngeren“ eudoxischen V. Buch! Eine offenbare Absurdität.

Dieser Einwand kann nur durch ein Mittel widerlegt werden: Dadurch, daß gezeigt wird, daß eine „allgemeine“ Proportionenlehre, so wie sie als Fundament des X. Buchs gebraucht wird, auch auf Grund der anthyphairetischen Definition lückenlos und streng entwickelt werden kann. Dabei muß genau darauf geachtet werden, welche Weise und welcher Grad von Allgemeinheit im X. Buch benötigt wird, nämlich zwar Anwendbarkeit auf rationale wie irrationale Verhältnisse, aber nicht unbedingt für ganz beliebige Arten von Größen, sondern nur für Strecken und Rechtecke. Denn obwohl zu Anfang von X eine Reihe allgemeiner Größensätze formuliert wird, erfolgt ihre spätere Anwendung doch stets nur in dem erwähnten beschränkten Sinne.

316 | Es ist also unbedingt notwendig, eine auf der Anthyphairesis begründete Proportionenlehre im Umriß zu entwerfen, wie es im folgenden geschieht. Dabei wird zugleich ihre immanente Begrenztheit zutage treten und das Motiv verständlich werden, das Eudoxos antrieb, über sie hinauszugehen.

3.

Der folgende Rekonstruktionsversuch der voreudoxischen Proportionenlehre kann naturgemäß nicht beanspruchen, die historische Wahrheit unverfälscht darzustellen. Es kann sich nur darum handeln, eine *mögliche* anthyphairetische Theorie der Verhältnislehre zu geben, die irrationale Logoi mitumfaßt und zugleich klar erkennen läßt, wie weit sie eine Theorie für allgemeine Größen (μεγέϑη) sein kann. Ob und wann im Verlauf der geschichtlichen Entwicklung sie das wirklich gewesen ist, soll erst später untersucht werden.

Die folgende Betrachtung stellt also eine systematische mathematische Überlegung dar, die indessen immer versucht, mit den Denkmitteln der voreudoxischen und eudoxischen Zeit zu arbeiten und sich, wo angängig, an antikes Material anschließt.

Überblickt man das V. Buch der Elemente, so sieht man bald, daß die eudoxische Logosdefinition (enthalten in der 5. und 7. Definition) den ganzen Aufbau des Buchs weitgehend bedingt. Ihr Ersatz durch die alte anthyphairetische Definition bringt also einen ganz anderen Aufbau der Beweise mit sich, obwohl eine Reihe davon immerhin beibehalten werden kann. Vor allem die Gruppierung und Reihenfolge der Sätze ändert sich durchgreifend; einige werden sogar ganz überflüssig.

Um mit dem letzten Punkt zu beginnen: V, 14, 20, 21 dienen bei Euklid der „spezifisch eudoxischen" Vorbereitung von 16, 22, 23; eine ähnliche Rolle spielt 4. Diese Theoreme können also nicht in der alten Theorie gestanden haben; sie sind ihrer Begriffsbildung ganz fremd. Ferner enthalten die ganz elementaren Theoreme 1–3, 5, 6 überhaupt keine Proportionen; von ihnen können wir auch absehen.

Es bleiben also noch die Sätze 7–13, 15–19, 22–25 übrig. Diese gruppieren sich gemäß folgender Übersichtstafel:

Vorgruppe: Elementare Sätze: 1–3, 5–6.

Gruppe A: Für allgemeine Größen beweisbare Sätze:

I. Solche, die unmittelbar aus der anthyphairetischen Logos-definition folgen: 7, 11.
II. Solche mit *rein algorithmischen* Beweisen: 12 (15) und 19, 17 und 18.
| III. Solche, deren Beweis *Konvergenz*betrachtungen (d. h. das „archimed. Axiom") 317
braucht: 8–10, 13, (25).

Gruppe B: *Nicht* für allgemeine Größen beweisbare Sätze: 16, 22, 23, (24).
(*Wegfallend* oder *nachträglich* beweisbar:
A. für allg. Größen: 14. B. *nicht* für allg. Größen: 4, 20, 21.)

Der entscheidende Unterschied ist der zwischen den Gruppen A und B. Er ist, wie sich zeigen wird, dadurch bedingt, daß die Sätze der Gruppe B den Begriff der Multiplikation von Verhältnissen (Brüchen) enthalten – wie wir uns heute ausdrücken. Aber dieser Begriff läßt sich in der anthyphairetischen Theorie nicht für allgemeine Größen oder Verhältnisse (mit vielleicht irrationalen Werten) zwischen solchen definieren. Die Theoreme der Gruppe A benötigen diesen Multiplikationsbegriff dagegen noch nicht und sind gerade deshalb für allgemeine Größen beweisbar.

Wir gehen nun zu den einzelnen Beweisen über.

I. Die Definition der Gleichheit und Ungleichheit von Verhältnissen.

Definition A: Zwei Verhältnisse sind dann und nur dann *gleich*, wenn sämtliche Teilungszahlen ihrer Anthyphairesis übereinstimmen.

Definition B: Im Fall der Ungleichheit sind zwei Unterfälle zu unterscheiden: 1. Die erste differierende Teilungszahl steht im Algorithmus an *ungerader* Stelle, dann entspricht das größere Verhältnis der größeren Teilungszahl, das kleinere der kleineren.

2. Die erste differierende Teilungszahl steht an *gerader* Stelle, dann entspricht umgekehrt das größere Verhältnis der kleineren Teilungszahl, das kleinere aber der größeren.

In Zeichen: Ist $a:b=[x_0\, x_1\, x_2 \ldots]$; $c:d=[y_0\, y_1\, y_2 \ldots]$, so ist $a:b=c:d$ dann und nur dann, wenn $x_i=y_i$ für alle i; $a:b \neq c:d$ dann und nur dann, wenn es ein i gibt, für das $x_i \neq y_i$. Ist nun j das kleinste derartige i, so ist, falls $j+1$ ungerade, $a:b \gtrless c:d$, je nachdem $x_j \gtrless y_j$; falls aber $j+1$ gerade, ist umgekehrt $a:b \gtrless c:d$ je nachdem $x_j \lessgtr y_j$.

Logische Beziehungen zwischen Verhältnissen: Es geht aus den vorstehenden Definitionen sofort hervor, daß keine der Beziehungen größer, kleiner, gleich mit einer anderen zugleich bestehen kann. („Satz vom Widerspruch.“) *Dagegen folgt mit*
318 *finiten logischen Mitteln*[5]) | *nicht*, daß zwei Verhältnisse entweder gleich oder ungleich (größer oder kleiner) sind. (Wohl aber gilt, daß von zwei ungleichen Verhältnissen – was mehr besagt als von „zwei nicht gleichen Verhältnissen!“ – das eine entweder größer oder kleiner ist als das andere.)

D. h. es gilt ($\overline{}$ bedeutet „nicht“, $\rightarrow$ „impliziert“):

$$(1) \quad a:b=c:d \quad \rightarrow \quad \overline{a:b \gtrless c:d};$$

$$(2) \quad a:b \gtrless c:d \quad \rightarrow \quad \overline{a:b=c:d}.$$

Damit ist aber noch keineswegs gesagt, daß auch gilt:

$$(3) \quad \overline{a:b=c:d} \quad \rightarrow \quad a:b \gtrless c:d;$$

$$(4) \quad \overline{a:b \gtrless c:d} \quad \rightarrow \quad a:b=c:d.$$

In der Tat werden auch von Euklid im V. Buch derartige Argumentationen nicht benutzt.

Die Unverträglichkeit von Gleichheit und Ungleichheit von Logoi wird dagegen in den euklidischen Beweisen von V, 9, 10 stillschweigend vorausgesetzt. Sie ist auf Grund der eudoxischen Definitionen 5 und 7 nicht ohne weiteres einleuchtend, worauf schon *R. Simson* und *de Morgan* hingewiesen haben (vgl. *Heath*, Vol. II, p. 130/31, 156/57). Das kann daher rühren, daß noch die ältere Theorie nachwirkt.

II. Unmittelbare Folgesätze der Definition des Verhältnisses.

Satz 7. *Gleiche Größen haben zu derselben Größe dasselbe Verhältnis und dieselbe Größe hat zu gleichen Größen dasselbe Verhältnis.*

Der Satz ist in der anthyphairetischen Theorie nahezu selbstverständlich. Denn führt man für eine Größe eine gleiche in das Teilerverfahren ein, so ergeben sich die gleichen Multipla und dann nach dem Subtraktionsaxiom (Buch I, not. comm. 3) gleiche Reste. Das Verfahren verläuft also ganz unverändert, und ergibt somit die gleichen Teilungszahlen.

---

[5]) Im Sinne *D. Hilberts*, „intuitionistisch“ im Sinne *Brouwers*. Vgl. vorläufig A. *Heyting*, „Die Theorie der linearen Gleichungen in einer Zahlenspezies mit nichtkommutativer Multiplikation“ und „Zur intuitionistischen Begründung der projektiven Geometrie“, Math. Annal., Bd. 98, S. 465 ff., 491 ff. – Die Frage wird in einer zweiten Studie näher erörtert werden.

Auch das *zugehörige Porisma* ergibt sich sofort. Überhaupt ist in der alten Theorie ein Verhältnis stets mit dem größeren Glied als Vorderglied zu schreiben, weil das ja die Anthyphairesis verlangt. Die fast ausnahmslose griechische Gewohnheit, Verhältnisse mit einem größeren Vorderglied anzusetzen, geht vermutlich auf diesen Umstand zurück.

Satz 11. *Sind zwei Verhältnisse einem dritten gleich, so sind sie untereinander gleich.*

| Sind nämlich die Teilungszahlen der drei Verhältnisse x_i, y_i, z_i, so ist nach 319
Voraussetzung $x_i = z_i$ und $y_i = z_i$ für alle i, also nach I, not. comm. 1 $x_i = y_i$. Woraus sich nach Def. A die Behauptung ergibt.

Satz 13. *Ist ein Verhältnis gleich einem zweiten und das zweite größer als ein drittes, so ist das erste größer als das dritte.*

Hier sind nämlich die Teilungszahlen $x_i = y_i$ für alle i und es gibt eine erste y_j, die an ungerader Stelle j stehend, größer als z_j oder an gerader Stelle j stehend, kleiner als z_j ist. Dann ist aber auch in den entsprechenden Fällen x_j größer bzw. kleiner als z_j. Demnach ist nach Def. B das erste Verhältnis größer als das dritte.

III. Algorithmisch beweisbare Sätze der Gruppe A

Satz 12. *Wenn* $a:a' = b:b' = c:c' = \ldots = [x_0\, x_1\, x_2 \ldots]$, *so ist auch* $(a + b + c + \ldots):(a' + b' + c' + \ldots) = [x_0\, x_1\, x_2 \ldots] = a:a'$ *usw.*

Die Ausführung der Anthyphaireseis ergibt nämlich folgendes (die sich sukzessiv ergebenden Reste sollen mit $r_0\, r_1\, r_2 \ldots, s_0\, s_1\, s_2 \ldots, t_0\, t_1\, t_2 \ldots$ bezeichnet werden):

$$a = x_0\, a' + r_0, r_0 < a' \quad b = x_0\, b' + s_0, s_0 < b' \quad c = x_0\, c' + t_0, t_0 < c'$$
$$a' = x_1\, r_0 + r_1, r_1 < r_0 \quad b' = x_1\, s_0 + s_1, s_1 < s_0 \quad c' = x_1\, t_0 + t_1, t_1 < t_0$$
$$r_0 = x_2\, r_1 + r_2, r_2 < r_1 \quad s_0 = x_2\, s_1 + s_2, s_2 < s_1 \quad t_0 = x_2\, t_1 + t_2, t_2 < t_1$$
$$a + b + c + \ldots = x_0\, (a' + b' + c' + \ldots) + (r_0 + s_0 + t_0 + \ldots),$$
$$r_0 + s_0 + t_0 \ldots < a' + b' + c' + \ldots$$
$$a' + b' + c' + \ldots = x_1 (r_0 + s_0 + t_0 + \ldots) + (r_1 + s_1 + t_1 + \ldots),$$
$$r_1 + s_1 + t_1 \ldots < r_0 + s_0 + t_0 + \ldots$$
$$r_0 + s_0 + t_0 + \ldots = x_2 (r_1 + s_1 + t_1 + \ldots) + (r_2 + s_2 + t_2 + \ldots),$$
$$r_2 + s_2 + t_2 \ldots < r_1 + s_1 + t_1 + \ldots$$

Satz 15. $a:a' = ma:ma'$, *wo* m *eine natürliche Zahl.*

Das ist Spezialfall von (12) für $a = b = c = \ldots; a' = b' = c' = \ldots$

Satz 19. *Wenn* $a:a' = b:b'$, *dann ist* $(a - a'):(b - b') = a:a'$.

Der Beweis ist wie der von (12) für zwei Glieder, nur daß die Addition durch Subtraktion ersetzt ist.

Satz 17. $(a + a'):a' = (b + b'):b'$ *hat zur Folge* $a:a' = b:b'$.

Sei $(a + a'):a' = [x_0\; x_1\; x_2 \ldots]$, so ist $a:a' = [(x_0 - 1)\; x_1\; x_2 \ldots]$; da nun nach Voraussetzung ebenfalls $(b + b'):b' = [x_0\, x_1\, x_2 \ldots]$, ist analog $b:b' = [(x_0 - 1)\, x_1\, x_2 \ldots]$.
Also ist $a:a' = b:b'$. Falls $x_0 = 1$, wird | $x_0 - 1 = 0$, das bedeutet, daß $a < a'$ und also 320
durch Umkehrung (ἀνάπαλιν) des Verhältnisses $a:a'$ in $a':a$ die Anthyphairesis ermöglicht werden muß, ebenso natürlich bei $b:b'$.

Satz 18. *$a : a' = b : b'$ hat zur Folge $(a + a') : a' = (b + b') : b'$.*

Beweis analog dem von (17), nur mit Addition statt Subtraktion von 1 bei der ersten Teilungszahl der Anthyphairesis von $a{:}a'$.

IV. Die Sätze der Gruppe A, die das archimedische Axiom voraussetzen.

Satz 8. *Ist a' größer als a, so ist auch $a' : b$ größer als $a : b$ und $b : a$ größer als $b : a'$.*

Es sei $a' - a = D$. Bei dem ersten Schritt der Anthyphairesis von $a' : b$ erscheint dann statt des Restes r_0 der Anthyphairesis von $a : b$ der größere „Rest" $r_0 + D$. Ist nun schon $r_0 + D > b$, so wird die erste Teilungszahl z_0 des Algorithmus von $a : b$ im geänderten Algorithmus (von $a' : b$) bereits überschritten. Es ist alsdann $a' : b > a : b$. Im anderen Fall stimmen die beiden Algorithmen zunächst noch überein und an Stelle des zweiten Restes r_1 tritt der neue „Rest" $r_1' = r_0 - z_1 \mathrm{D} \leqslant r_0 - \mathrm{D}$. Man sieht, daß, solange keine neuen Teilungszahlen auftreten, die ungeraden Reste $r_0\, r_2\, r_4 \ldots$ um mindestens D vermehrt, die geraden $r_1\; r_3\; r_5 \ldots$ um mindestens D vermindert werden. Andererseits nehmen die Reste immer mehr ab und jeder Rest ist offenbar sogar kleiner als die Hälfte seines Vorvorgängers $r_{n+2} < \frac{1}{2}\, r_n$. Nach dem Satz X, 1, der aus dem archimedischen Axiom folgt, unterschreiten die Reste schließlich jede Größe, also auch D. Geschieht dies, so ändert sich aber die Teilungszahl des neuen Algorithmus spätestens beim nächsten Schritt und zwar, wie man sich leicht überlegt, stets in dem Sinne, daß $a' : b > a : b$ wird.

Gilt Satz X, 1 nicht, so wird der Beweis hinfällig. Der Satz 8 gilt in der Tat nicht notwendig, wenn die Differenz D gegenüber *a, b* im „nichtarchimedischen" Sinn „unendlich klein" ist. (Man denke, *a, b* seien geradlinige, D ein „hornförmiger" [Kontingenz-] Winkel. Dann ist ohne weitere Voraussetzungen gar nicht bestimmt, ob $(a + D) : b$ größer oder gleich $a : b$ ist.)

Satz 9 und 10. sind leichte Folgerungen aus 8, die wie bei Euklid bewiesen werden können. Bemerkenswert ist, daß keine „tranfiniten" Schlußweisen verwandt werden, sondern nur aus $a : b$ größer bzw. kleiner als $c : d$ gefolgert wird, daß $a : b$ nicht gleich $c : d$ ist. Dieser Schluß ist aber auch in der intuitionistischen („finiten") Logik zulässig.

Satz 14. läßt sich ebenfalls wie bei Euklid (aus 8–10, 13) ableiten und mit seiner Hilfe ist weiterhin **Satz 25** anthyphairetisch für allgemeine Größen beweisbar, worauf nicht näher eingegangen sei.

321 **V. Sätze der Gruppe B.**

Vorbemerkung. Die Sätze der Gruppe B sind im wesentlichen drei: 16, 22, 23. Alle drei hängen eng mit dem Begriff der Multiplikation von Größen mit Größen und von Verhältnissen mit Verhältnissen zusammen. Der tiefere Grund, warum diese Sätze nicht in der anthyphairetischen Theorie für allgemeine $\mu\varepsilon\gamma\acute{\varepsilon}\vartheta\eta$ beweisbar sind, liegt darin, daß die beiden genannten Multiplikationsaufgaben in dieser Theorie nicht unmittelbar lösbar sind. In unserer Sprache gesagt: Es gibt keine einfache Formel, die aus den Teilnennern zweier Kettenbrüche (den Teilungszahlen der entsprechenden Anthyphaireseis) die Teilnenner desjenigen Kettenbruchs zu berechnen gestattet, dessen Wert gleich dem Produkt der Werte der beiden ersten Kettenbrüche ist. Unsere übliche Multiplikationsformel für Verhältnisse $(a:b)\cdot(c:d)=ac:bd$ setzt voraus, daß die Produkte ac und bd definiert sind, auch wenn a, b, c, d allgemeine Größen sind. Das ist aber nicht der Fall, sondern definiert ist zunächst nur $a + b$ und $a+a=2a$, $a+a+a+3a$ usw., d. h. ma, wo m eine natürliche Zahl ist[6]).

Den Begriff der Multiplikation von Verhältnissen kennt indessen Euklid nicht eigentlich. Dagegen wird von ihm ein spezieller Fall davon, die „Komposition" der Verhältnisse $a:b$ und $b:c$ zu $a:c$ (d. h. $[a:b]\cdot[b:d]=a:c$), benutzt, wenn auch nicht definiert. Definiert wird nur der besondere Fall der duplicata und triplicata ratio oder proportio (διπλασίων und τριπλασίων λόγος). In VI, 23 (Beweis) heißt es: „ἀλλ' ὁ τῆς *K* πρὸς τὴν *M* λόγος συγκεῖται ἔκ τε τοῦ τῆς *K* πρὸς τὴν *Λ* λόγου καὶ τοῦ τῆς *Λ* πρός τὴν *M*." („Aber das Verhältnis von *K* zu *M* ist zusammengesetzt aus dem Verhältnis von *K* zu *Λ* und dem von *Λ* zu *M*.")

Betrachtet man nun die Sätze 16, 22, 23, so sieht man, daß sie abgeleitet werden können aus der Forderung der Eindeutigkeit und Kommutativität der oben genannten „Komposition".

Satz 22. *Wenn* $a:b=a':b'$ *und* $b:c=b':c'$, *so ist* $a:c=a':c'$.

Das bedeutet einfach die Eindeutigkeit der Komposition.

Satz 23. *Wenn* $a:b=b':c'$ *und* $b:c=a':b'$, *so ist* $a:c=a':c'$.

Ist die Operation $(a:b)\cdot(b:c)=a:c$ und $(b':c')\cdot(a':b')=(a':b')\cdot(b':c')=a':c'$ eindeutig, so gilt der Satz. Zu beachten die Vertauschung bei der zweiten Operation!

| **Satz 16.** *Wenn* $a:b=c:d$ *dann* $a:c=b:d$ 322

Es ist $(a:b)\cdot(b:c)=(b:c)\cdot(c:d)$ infolge der Eindeutigkeit und Kommutativität der Komposition. Also $a:c=b:d$.

Euklid ist nun nicht in der Lage, Eindeutigkeit und Kommutativität der Komposition als bewiesen vorauszusetzen. Immerhin gelingt es ihm im V. Buch, die Sätze 16, 22, 23 zu beweisen und damit die genannten Eigenschaften der Komposition sicher zu stellen. Aber nur durch einen eigentümlichen Schluß, der auf der spezifisch eudoxischen Definition der Verhältnisgleichheit beruht und also sicher selbst von Eudoxos stammt.

Am Beispiel von (16): Der Satz (14): „Wenn $a:b=c:d$, so ist, wenn a größer, kleiner, gleich c, auch entsprechend b größer, kleiner, gleich d", hat zur Folge $a:c=b:d$. Denn aus $a:b=c:d$ folgt $ma:mb=nc:nd$; also ist, wenn $ma >=< mc$, auch $nb >=< nd$, woraus nach Def. 5 (Eudoxos!) sich ergibt $a:c=b:d$.

Ganz analog sind die Beweise von (22), (23) gebaut.

Das Charakteristische an dieser Argumentation ist, die Einführung der Produkte ma, mb, nc, nd, wo m, n *willkürliche* natürliche Zahlen bedeuten. Damit ordnet Eudoxos die irrationalen Verhältnisse nach ihrer Größe in die Gesamtheit aller rationalen Verhältnisse $m:n$ ein. (Wir tun heute dasselbe bei *allen* unseren Theorien der reellen Zahlen.)

Es läßt sich unschwer zeigen, daß ohne die Einführung dieser Vergleichsverhältnisse, rein auf Grund der Sätze 7–11, 13–14, der Satz 16 ebensowenig zu

[6]) Man vergleiche etwa die Sachlage in der sog. Streckenrechnung (*Hilbert*, Grundlagen der Geometrie, § 15, *O. Hölder*, Streckenrechnung und projektive Geometrie, Berichte d. Sächs. Ges. d. Wiss. math.-phys. Kl. LXIII, S. 67ff.) und auch schon im gewöhnlichen Vektorkalkül. – Der Vergleich der antiken geometrischen Methoden mit der modernen axiomatischen Untersuchung ist eine zwar sehr interessante, aber auch umfangreiche, hier nicht angreifbare Aufgabe.

beweisen ist, wie mit Hilfe der „algorithmischen“ Sätze 12, 17–19 allein. Denn es gibt Funktionen zweier Variablen $f(x, y)$, die allen Sätzen 7–11, 13, 14 gehorchen, aber doch nicht der Bedingung 16.

Aus diesen Betrachtungen ergibt sich, daß von der anthyphairetischen Theorie aus keine direkte Beweismöglichkeit von V, 16 für allgemeine Größen besteht. Allerdings – das sei besonders betont – handelt es sich um keinen axiomatischen Sachverhalt. Beweisbar sind die Sätze 16, 22, 23 mittels des archimedischen Axioms und der üblichen anderen Voraussetzungen der Proportionenlehre. Andererseits sind zwar die „algorithmischen“ Proportionensätze ohne archimedisches Axiom beweisbar, aber nicht Satz 8 usw. Was durch Eudoxos hinzukommt, ist also nicht ein neues Axiom (wenn er V, def. 4 und X, 1 vielleicht auch zuerst formuliert hat, dann aber im Zusammenhang anderer Untersuchungen, nämlich der „Exhaustionsbeweise“ des XII. Buchs!), sondern eine neue Beweismethode und von dieser Beweismethode aus eine neue Definition der Verhältnisgleichheit.

Es muß ein Übergangsstadium (zum mindesten im Geiste des Eudoxos) zwischen beiden Theorien bestanden haben, in dem zwar noch die alte anthyphairetische Definition
323 in Geltung war, aber die spätere | eudoxische Definition als beweisbare charakteristische Eigenschaft zweier nach der alten Definition gleicher Verhältnisse fungierte. Es erwies sich dann als ökonomischer und durchsichtiger, die alte Definition ganz fallen zu lassen und die charakteristische Eigenschaft von vornherein als Definition zugrunde zu legen, womit allerdings ein ziemlich weitgehender Umbau der ganzen Theorie verbunden war.

Gewonnen wurde bei alledem schließlich eine ganz für beliebige Verhältnisse *allgemeiner Größen* gültige Theorie, während die alte in ihrem zweiten Teil auf solche Allgemeinheit zu verzichten gezwungen war.

Wir kehren nunmehr zu dieser alten Theorie und zwar zu eben diesem „zweiten Teil“ von ihr zurück.

Ohne die eudoxische Methode war man nicht im Besitz einer „Multiplikation“ von Verhältnissen allgemeiner Größen oder von diesen selbst. Dagegen war das „Produkt“ zweier Strecken der Sache nach seit alters her bekannt, nämlich das *Rechteck aus diesen Strecken.* Eine solche „Produktdarstellung“ spielt schon in der sogenannten „geometrischen Algebra“ des II. Buchs der Elemente eine ausschlaggebende Rolle, ferner im X., nach unserer Auffassung in seinen Anfängen „theätetischen“ Buch. (Zu vergleichen sind ferner die „ebenen Zahlen“ [ἐπίπεδοι ἀριθμοί] der ebenfalls „theätetischen Bücher“ VII–IX.) Diese Darstellung garantierte von vornherein die Eindeutigkeit und Kommutativität der „Multiplikation“[7]).

Für diese Art der Betrachtung ist Satz 16 der Schlüsselsatz des zweiten Teils der Theorie. Die Beweise gestalten sich wie folgt.

Satz 16. *Wenn* $a:b=c:d$, *dann* $a:c=b:d$.

Der Beweis kann entweder geführt werden nach der Methode von *Smith und Bryant*[8]) (1901) mit Hilfe von VI, 1 allein oder in engerem Anschluß an das historische Material in folgender Weise.

---

[7]) Vgl. zu der ganzen Frage der Multiplikation *H. Hankel*, Zur Geschichte der Mathematik im Altertum und Mittelalter, Leipzig 1874, S. 389; *H. G. Zeuthen*, Geschichte der Mathematik im Altertum und Mittelalter, Kopenhagen 1896, S. 44f., 145f.

[8]) *S. Heath*, The thirteen books of Euclid's elements, Cambridge 1908, Vol. II, p. 165f.

Der Beweis wird natürlich für alle Größenarten, für die er überhaupt möglich ist, gesondert geführt, und zwar zunächst für *Strecken*. Man muß den 1. und den 16. Satz des VI. Buchs heranziehen. Das letzte Theorem besagt: „Wenn vier Strecken proportioniert sind, ist das Rechteck aus den äußeren Strecken gleich dem Rechteck aus den inneren Strecken und umgekehrt." (Wenn $a:b=c:d$ ist $ad=bc$ und umgekehrt.) Er ist ein Spezialfall von VI, 14, der dasselbe mit Bezug auf gleich|winkelige Parallelo- 324
gramme sagt. Im Beweis von VI, 14 werden nur die Sätze V, 7, 9, 11 und VI, 1 (und zwar für Parallelogramme) benutzt. Die Sätze aus V gehören zu den im ersten Teil der alten Theorie bewiesenen; VI, 1 wird zwar von Euklid mit Hilfe von V, def. 5 bewiesen, ist aber gerade derjenige Satz, an dem Aristoteles in der Topikstelle, von der wir ausgingen, die „antanairetische" Definition der Verhältnisgleichheit entwickelt. Von ihm wissen wir also dokumentarisch, daß er durch parallele Antanairesis für γραμμαί und χωρία bewiesen wurde. Der alte Beweisweg von V, 16 läßt sich also so rekonstruieren: Beweis von V, 16 über V, 14 oder auch direkt aus VI, 1 und den genannten Proportionssätzen. Mit Hilfe von VI, 16 dann folgender Schluß: $a:b=c:d$, $ad=bc$ (VI, 16) oder $ad=cb$, $a:c=b:d$ (VI, 16). Damit ist man für Strecken am Ziel.

VI, 1 dient dann ebenfalls dazu, den Satz auf Parallelogramme zu übertragen; XI, 25 bewirkt das gleiche für Parallelepipeda, XII, 13 für Zylinder. (Die euklidischen Beweise sind sämtlich, wie bei VI, 1, anthyphairetisch durchzuführen.)

Bei allgemeinen krummlinigen Figuren, pyramidal und krummflächig begrenzten Körpern usw. stößt man aber auf Schwierigkeiten, die die Griechen höchstwahrscheinlich durch das Postulat der Existenz der vierten Proportionale in den betreffenden Fällen überwunden haben[9]).

Zu beweisen bleiben noch 22, 23; denn 24 ist aus ihnen eine leichte Folgerung, die wie bei Euklid bewiesen werden konnte. (14, 20, 21 waren, wie schon gesagt, in der alten Theorie nicht enthalten, obwohl sie natürlich sich leicht bzw. aus 16, 22, 23 rückwärts ergeben hätten.)

Satz 22. *Wenn* $a:b=a':b'$ *und* $b:c=b':c'$, *dann* $a:c=a':c'$.

Als Vorbild für den wahrscheinlich alten Beweis dieses Satzes kann der „theätetische" Beweis des analogen „ex aequali"-Satzes für Zahlen, VII, 14, dienen. Dieser verläuft so: Aus den beiden Voraussetzungen folgt durch ἐναλλάττειν $a:a'=b:b'$ und $b:b'=c:c'$. Daraus nach V, 11 $a:a'=c:c'$ und daraus (ἐναλλάξ) $a:c=a':c'$.

Heath (Vol. II, 314) erhebt gegen diesen Beweis den Einwand, daß durch ihn der Satz nur für vier Größen gleicher Art bewiesen sei, während er doch auch noch gelte, wenn a, b, c von anderer Art seien als a', b', c'. Indessen innerhalb des Gültigkeitsbereichs der anthyphairetischen Theorie macht das nichts aus. Denn man kann sich mit VI, 1 und den analogen Sätzen helfen[10]).

Satz 23. *Wenn* $a:b=b':c'$ *und* $b:c=a':b'$, *dann* $a:c=a':c'$.

Nach VI, 16 besagen die Voraussetzungen, daß $ac'=bb'$ und bb' | $=a'c$. Also ist 325
$ac'=a'c$ und, wieder nach VI, 16, $a:c=a':c'$. Auch dieser Beweis gilt zunächst nur für Strecken, kann aber leicht auf alle „erlaubten" Größenarten ausgedehnt werden.

---

9) S. die Ausführungen in der zweiten Studie.

10) Vgl. *Heath*, l. c. p. 166.

Es fehlt einzig noch **Satz 4 :** *Wenn* $a : b = c : d$, *dann* $ma : nb = mc : nd$. Der Beweis liegt auf der Hand : Aus der Voraussetzung ergibt sich durch V, 16 $a : c = b : d$, daraus durch V, 15 $ma : mc = nb : nd$, daraus wieder durch V, 16 die Behauptung. Ob der Satz allerdings in der alten Theorie vorkam, ist sehr zu bezweifeln.

Überhaupt ist bei aller Einfachheit des Aufbaus dieser rekonstruierten „alten Theorie" natürlich nicht gesagt, daß sie so in allen Stücken existiert hat. Es ist daher vielleicht nicht uninteressant, zu bemerken, daß Aristoteles scheinbar nur folgende Proportionensätze erwähnt : V, 12 (Eth. Nic. V, 7 [1131b 14]) und V, 11, 16, 24 (d. h. implizit 22), endlich den Satz: „$a : b = c : d$ impliziert, daß aus $a > b$ folgt $c > d$", der keineswegs, wie Heath meint, mit V, 14 äquivalent ist und ohne weiteres aus der antanairetischen Definition der Verhältnisgleichheit sich ergibt (diese sämtlich Meteorl. III, 5, 376a 11–26). Dabei ist merkwürdig, daß 23 fehlt und andererseits auch 8–10, 13. Überlegen wir, was in unserer Rekonstruktion sonst gebraucht wurde, so ist davon VI, 1, wie wir wissen, schon aristotelisch, dagegen ist über VI, 16 (bzw. 14) nichts bekannt. Nun brauchen wir VI, 16 nicht unbedingt, wenn wir V, 23 weglassen[11]), denn wir können ja V, 16 nach Smith' und Bryants Methode mittels VI, 1 allein beweisen und daraus V, 22, 24. Das ist deshalb interessant, weil der Beweis von VI, 16 den Satz V, 9 verwendet, den Aristoteles nicht erwähnt und der auf V, 8 zurückgeht. Das heißt wir können so alle Sätze der Gruppe A, die auf „Konvergenzbetrachtungen" beruhen, aus der Theorie herauslassen. Nun läßt sich nicht leugnen, daß die Konvergenzbetrachtung im Beweis von V, 8 bedeutend schwieriger ist als alles sonst in der Theorie ; es wäre also wohl möglich, daß sie gefehlt hat[11a]).

Wie dem auch sein mag, jedenfalls ist jetzt gezeigt worden, daß eine anthyphairetische Proportionentheorie einfach und in sich konsequent aufgebaut werden kann und zwar teilweise für allgemeine Größen, zum Teil aber auch nur für bestimmte einzelne Größenarten.

Die rekonstruierte Proportionenlehre genügt aber vollkommen, um die Theorie des X. Buchs der Elemente darauf zu gründen. Denn die dort verwendeten Größenarten beschränken sich auf Strecken und Rechtecke.

326 **4.**

Nachdem nunmehr der gegen unsere These von der theätetischen Proportionenlehre als Grundlage von Buch X erhobene Einwand als erledigt gelten kann, können die übrigen Euklid-Bücher auf ihr Verhältnis zum V. Buch hin geprüft werden.

Entscheidend wird dabei sein, wie weit gedankliche Grundlagen und sprachliche Ausdrucksweisen, die aus dem V. Buche stammen, außerhalb seines Bereichs verwendet werden. Es handelt sich dabei einmal um den Begriff und Namen der „allgemeinen Größe" (*μέγεθος*) und dann um die charakteristisch „eudoxischen" Definitionen 5 und 7.

---

11) Bemerkenswert ist, daß V, 23 in den Elementen nicht wieder benutzt wird, dagegen wohl bei Archimedes.

11a) V, 22 ist also für Strecken, nicht aber für allgemeine Größen, ohne archimedisches Axiom beweisbar.

Nun liegt die Sache, abgesehen von Buch X, ganz klar. Das Wort „μέγεθος“ kommt weder in Buch I–IV noch in VII–IX vor, dagegen vereinzelt in VI, XI, XII. Und zwar geschieht das letztere genau an den Stellen, an denen die 5. Definition des V. Buchs zitiert wird, nämlich in VI, 1, 33; XI, 25; XII, 13. In allen diesen Fällen handelt es sich um dieselbe feste Formel, die mit den Worten beginnt: „τεσσάρων δὴ ὄντων μεγεθῶν“, und dann die Anwendung der eudoxischen Definition der Verhältnisgleichheit ausspricht. Sonst kommt μέγεθος nicht in den genannten Büchern vor, mit einer Ausnahme, die aber die Regel durchaus bestätigt. Es ist dies das ausdrückliche wörtliche Zitat von X, 1 im Beweis von XII, 2[12]). In genauer Übereinstimmung damit wird V, def. 5 sonst nirgends verwendet und def. 7 erscheint überhaupt niemals außerhalb des V. Buchs.

Die genannten Fälle der Anwendung der eudoxischen Definition sind sämtlich genaue Analoga von VI, 1 und können ebenso wie dieser durch die aristotelische Topik belegte Fall ohne weiteres auch „antanairetisch“ behandelt werden. Mit anderen Worten: Es handelt sich bei diesem Vorkommen des allgemeinen Größenbegriffs und der eudoxischen Definition um eine äußerliche Retusche eines späteren Überarbeiters des theätetisch-eudoxischen Materials; es steht nichts im Wege, sie wieder rückgängig zu machen. Denn irgendwelche weiteren Eingriffe in die Beweise sind nicht vonnöten und zugleich verschwindet der den betreffenden Büchern sonst ganz fremde μέγεθος-Begriff.

Es ist wahrscheinlich, daß diese Überarbeitung erst von Euklid selbst herrührt. Die bekannten Worte des Proklos über ihn (in Eucl., p. 68, 7–9 Friedl.) gewinnen so eine ganz präzise Bedeutung:

„... Εὐκλείδης ὁ τὰ στοιχεῖα συναγαγὼν καὶ πολλὰ μὲν τῶν | Εὐδόξου συντάξας, πολλὰ δὲ τῶν Θεαιτήτου τελεωσάμενος ...“

„... Euklid, der die Elemente
zusammenbrachte und Vieles von 327
Eudo|xos (Herrührende) zusammenordnete, Vieles aber von Theätet (Stammende) vollendete.“

Das συντάττειν der eudoxischen Forschungen besteht in der Voranstellung des V. Buchs vor allem vor das XII. (das, wie wir später genauer zeigen werden, chronologisch der „allgemeinen Proportionenlehre“ eudoxischer Neuprägung vorzuordnen ist). Die Bücher VI, XI dienen als Vorbereitung von XII und schließlich werden noch VII–X dazwischen geschoben, vermutlich wegen X, 1. Das X. Buch (und vielleicht auch einiges in VII–IX) stellt dagegen die „Vollendung“ der theätetischen Untersuchungen dar (πολλὰ τῶν Θεαιτήτου τελεωσάμενος). Es ist daher auch verständlich, daß dort, wo die Mitarbeit Euklids intensiver war, der μέγεθος-Begriff eine größere Rolle spielt; er kommt dort in der Tat in Def. 1 und 2 und den Sätzen 1–8, 11–13, 15–16 vor.

Es bleibt jetzt allerdings noch eine sehr wichtige Frage zur Beantwortung übrig: Die Tragweite der anthyphairetischen Proportionenlehre erwies sich in ihrem zweiten Teil (Sätze der Gruppe B) als beschränkt auf Strecken und ihnen „verwandte“ Gebilde, auf die sich die für Strecken gefundene Sätze übertragen ließen, d. h. geradlinig begrenzte Figuren, Parallelepipede, Prismen, u. U. Zylinder. Reicht sie sachlich für die Beweise der Bücher VI, X–XII hin? Das ist zu bejahen für alle diese Bücher außer dem XII.

---

[12]) Das Zitat ist an sich seiner Form nach ganz ungewöhnlich und vielleicht eine spätere Hinzufügung, aber für unser Argument ist es ohnehin gleichgültig.

Im XII. Buch allerdings kommen Proportionen zwischen krummlinigen Gebilden vor, die streng genommen nicht mehr mit der voreudoxischen Lehre begründet werden können. Aber Euklid, wie überhaupt alle antiken Autoren, helfen sich mit dem Postulat der Existenz der vierten Proportionalen, das – stillschweigend – auch in den Fällen angewandt wird, in denen es nicht, wie für Strecken (vgl. VI, 12) konstruktiv bewiesen werden kann. Das Merkwürdige an der Sachlage ist nun, daß es tatsächlich möglich ist, das genannte Postulat im XII. Buch zu vermeiden und zwar auf Grund der neuen eudoxischen Verhältnis-definition. Die Frage ist allerdings verwickelter als es zunächst aussieht und wird erst an späterer Stelle behandelt werden. Die bisher vorliegende Lösung von *Hasse* und *Scholz*[13]) genügt nämlich noch nicht allen an sie zu stellenden Anforderungen; denn sie verwendet eine „transfinite“ Schlußweise (vollständige Disjunktion zwischen dem Gleich-, Größer- und Kleiner-Sein von *Verhältnissen*), die sonst nicht in der antiken Mathematik vorkommt. Aber wir werden zeigen, daß sich eine einwandfreie Lösung geben läßt. Nichtsdestoweniger macht
328 aber, wie schon | gesagt, weder Euklid noch sonst ein griechischer Mathematiker von dieser Möglichkeit Gebrauch.

Es besteht also der eigentümliche Tatbestand, daß die neue Proportionenlehre des Eudoxos in dem einzigen Punkte, in dem sie der alten anthyphairetischen in rein *mathematischer* Hinsicht überlegen war, nicht ausgenutzt wurde. Es gibt keinen einzigen Satz der griechischen Mathematik, der mit ihrer Hilfe zum erstenmal oder wenigstens strenger als früher bewiesen worden ist. Ihr Vorzug bleibt also auf rein „logischem“, gewissermaßen akademisch-philosophischem Gebiet. Im formalen *μέγεθος*-Begriff werden alle die einzelnen anschaulichen Fälle von Größe „in eine Idee zusammengeschaut“ (*εἰς μίαν ἰδέαν συνορῶντα* Phaedr. 273 e). Das ist ganz offenbar platonische Denkweise und in ihrer Betätigung wird man den Einfluß Platons selbst auf Eudoxos und seine unmittelbaren Nachfolger sehen müssen. Dabei sei hier dahingestellt, ob schon Eudoxos oder erst die Späteren (vielleicht Theudios[14]), Hermotimos oder gar erst Euklid selbst) den *Terminus* *μέγεθος* in die Aussagen der neuen Theorie einführten oder ob er selbst sich noch mit der unbestimmteren Ausdrucksweise des „absoluten“ Artikels „*τὸ*. . ., *τὰ* . . .“, wie sie Euklid noch in den Notiones Communes des I. und selbst manchen Theoremen des V. Buchs (wie z. B. 2–4) erhalten hat, begnügte.

Es spricht also nichts dafür, daß das XII. Buch dem V. nachfolgt.

Es fragt sich nun aber, ob den vorgebrachten negativen Gründen, nach denen für die Abhängigkeit der übrigen Euklid-Bücher von V. nichts angeführt werden kann, sich nicht positive Gründe für eine solche Unabhängigkeit zur Seite stellen lassen.

Eine Reihe solcher Gründe ist lange bekannt. In erster Linie die völlige Unabhängigkeit in der Beweisführung und zum größten Teil auch in den Definitionen der arithmetischen Bücher VII–IX. Hervorzuheben ist insbesondere die Definition der

---

13) *H. Hasse* und *H. Scholz,* Die Grundlagenkrise der griechischen Mathematik, Kantstudien, Bd. XXXIII, S. 4ff., insb. S. 26f.

14) Vgl. *Proclus* in Eucl., p. 67, 14–16 (Friedl.): *Θεύδιος . . . . καὶ γὰρ τὰ στοιχεῖα καλῶς συνέταξεν καὶ πολλὰ τῶν ὁρικῶν* (dies die überlieferte Lesart, die der von Solmsen empfohlenen *μερικῶν* m. E. vorzuziehen ist) *καθολικώτερα ἐποίησεν*. Theudios soll also viele Definitionen (*ὅροι*) allgemeiner gestaltet haben.

Verhältnisgleichheit selbst im kommensurablen Fall, VII, def. 20 (Heiberg), die sich durchaus *nicht* als Spezialfall von V, def. 5 darstellt. In zweiter Linie kommt die Verwendung eben dieser Definition VII, def. 20 innerhalb des X. Buchs in Frage. In den Beweisen von X, 5, 6 Euklid folgert das Bestehen einer Proportion $\Gamma : A = 1 : \Delta$ daraus, daß die Einheit die Zahl Δ ebenso oft mißt wie die Größe Γ die Größe A (*ἰσάκις ἄρα ἡ μονὰς τὸν Δ μετρεῖ ἀριθμὸν καὶ τὸ Γ μέγεθος τὸ Α· ἔστιν ἄρα ὡς τὸ Γ πρὸς τὸ Α, οὕτως ἡ μονὰς πρὸς | τὸν Δ*). Ähnlich in X, 6. Das Kriterium der Proportioniertheit 329
wird also (bei einer aus Zahlen und Größen gemischten Proportion!) nicht in der „allgemeinen“ Definition V, 5, sondern in der speziell für Zahlen gültigen VII, 20 gesucht. Das ist schon von *Robert Simson* bemerkt worden[15]). Ähnliche, wenn auch nicht so scharfe Einwände können gegen X, 9 erhoben werden, welcher Satz ja auch dem Theätet ausdrücklich vom Scholion zugeschrieben wird. Trotz der Verwendung des aus V stammenden *μέγεθος*-Begriffs wird also die tiefere und allgemeinere Betrachtungsart aus V nicht mitübernommen – ein Beleg für Euklids mitunter bis zur Inkonsequenz getriebene Ehrfurcht vor der Überlieferung.

Ein merkwürdigerweise wohl noch nicht in diesen Zusammenhang gebrachter weiterer Fall[16]) aus dem VI. Buch verdient Erwähnung: Im Beweis von Satz 9 wird aus der Proportioniertheit von vier Strecken geschlossen, daß, wenn die erste das Doppelte der zweiten ist, auch die dritte Strecke das Doppelte der vierten ist[17]). Das ist wiederum eine Anwendung nicht etwa der 5. Definition des V., sondern der 20. Definition des VII. Buches[18]), *das bei Euklid doch erst auf das VI. Buch folgt!* Also auch hier dieselbe Abhängigkeit von der alten „theätetischen“ Begriffsbildung.

5.

Das geschichtliche und textliche Material, das außer den „Elementen“ über die Frage der anthyphairetischen Proportionenlehre zur Verfügung steht, ist gering. Es muß daher jede Andeutung ausgenutzt werden. Da ist nun zu nennen eine schon oft behandelte Stelle in der aristotelischen Zweiten Analytik, I. Buch, Kapitel 5 (p. 74 a, 17–25).

καὶ τὸ ἀνάλογον ὅτι ἐναλλάξ, ᾗ ἀριθμοὶ καὶ ᾗ γραμμαὶ καὶ ᾗ στερεὰ καὶ ᾗ χρόνοι, ὥσπερ ἐδείκνυτό ποτε χωρίς, ἐνδεχόμενόν γε κατὰ πάντων μιᾷ ἀποδείξει δειχθῆναι· ἀλλὰ διὰ τὸ μὴ | εἶναι ὠνομασμένον τι πάντα ταῦτα ἕν, ἀριθμοὶ μήκη χρόνος στερεά, καὶ εἴδει

Und daß die Innenglieder einer
Proportion vertauschbar sind, – wie das
früher gesondert bewiesen wurde für
Zahlen, Strecken, Körper und Zeiten;
während es doch gezeigt werden konnte 330
bezüglich aller durch *einen* Beweis
|⟨gang⟩. Aber, weil Alles dieses ⟨als⟩

---

15) Euclidis Elementorum libri priores sex etc., Glaguae 1756, Liber quintus prop. C (p. 144), nota ad prop. C Lib. V (p. 360).

16) Doch erwähnt auch *R. Simson* (l. c. Lib. V. Prop. D et Nota, pp. 145, 361) den Satz VI, 9 als ungenügend bewiesen.

17) *ἀνάλογον ἄρα ἐστὶν ὡς ἡ ΓΔ πρὸς τὴν ΔΑ, οὕτως ἡ ΒΖ πρὸς τὴν ΖΑ. Διπλῆ δὲ ἡ ΓΔ τῆς ΔΑ· διπλῆ ἄρα καὶ ἡ ΒΖ τῆς ΔΑ.* (Aus dem Beweis von VI, 9.)

18) *Ἀριθμοὶ ἀνάλογόν εἰσιν, ὅταν ὁ πρῶτος τοῦ δευτέρου καὶ ὁ τρίτος τοῦ τετάρτου ἰσάκις ᾖ πολλαπλάσιος ἢ τὸ αὐτὸ μέρος ἢ τὰ αὐτὰ μέρη.* (VII, def. 20.) Nach diesem Vorbild muß die in VI, 9 zugrunde gelegte Definition gestaltet gewesen sein.

διαφέρει ἀλλήλων, χωρὶς ἐλαμβάνετο. νῦν δὲ καθόλου δείκνυται· οὐ γὰρ ᾗ γραμμαὶ ἢ ᾗ ἀριθμοὶ ὑπῆρχεν, ἀλλ' ᾗ τοδί, ὃ καθόλου ὑποτίθενται ὑπάρχειν.

Eines ⟨genommen⟩ nicht irgend etwas genannt war (irgendwie hieß), Zahlen, Längen, Zeit, Körper und dem Aussehen (der anschaulichen Gestalt) nach sich voneinander unterschied, wurde es gesondert erfaßt.

Jetzt aber wird die Eigenschaft im Ganzen bewiesen, denn sie bestand nicht, soweit es sich um Strecken oder Zahlen ⟨handelte⟩, sondern um das, was als „allgemein“ bestehend (subsistierend) vorausgesetzt wird.

An dieser Stelle ist in erster Linie höchst bemerkenswert die Erwähnung des *ἐναλλάξ*-Satzes V, 16. An und für sich könnte das ein beliebiges Beispiel sein. Bedenkt man aber, daß dieser Satz geradezu der Schlüsselsatz für die Proportionentheorie der Gruppe B ist, so wird die Annahme seiner bloß zufälligen Erwähnung sehr unwahrscheinlich. Ebenso wie Aristoteles in der Topikstelle VI, 1 einen fundamentalen Satz der anthyphairetischen Theorie nicht zufällig herangezogen haben kann. In beiden Fällen wurden vielmehr den Hörern seiner Pragmatie *bekannte* und wichtige Beispiele in Erinnerung gebracht, diese Anführungen sind zugleich Anspielungen auf berühmte wissenschaftliche Fragen der Zeit.

In der Tat war ja die Theoremgruppe B „einstmals“ (*ποτέ*) nicht „allgemein“ (*καθόλου*) beweisbar, sondern nur „gesondert“ (*χωρίς*). „Jetzt“ (*νῦν*) aber wird sie allgemein bewiesen, nämlich durch die neue eudoxische Theorie.

Indessen muß ein Zweites die Aufmerksamkeit erregen. Als Grund für die früher bestehende Unmöglichkeit, das *ἐναλλάξ* allgemein zu beweisen, wird nicht eine mathematische Schwierigkeit genannt, sondern eine rein „logische“, geradezu in dem wörtlichen Sinn von *λόγος*, nämlich *Rede.* Weil der Name fehlte für „dies Alles als Eines“, konnte der allgemeine Beweis nicht gegeben werden. Ist das nicht ein ganz anderer Gesichtspunkt als der, von dem aus wir die Sache ansehen? Man darf das Namengeben nicht in allzu engem, rein sprachlichem Sinn fassen: Aristoteles selbst gibt weder hier noch an der in vielem verwandten Stelle der Metaphysik (Buch *M* 2,
331 p. 1077a, 9–12)[19]) den Namen an, | sondern gebraucht Umschreibungen. Euklid sagt im V. Buch *μεγέθη*, aber gerade dieses Wort *μέγεθος* verwendet Aristoteles an der zweitgenannten Stelle zur Bezeichnung einer besonderen Größenart oder wenigstens mathematischen Gegenstandsart neben *ἀριθμός*, *στιγμή* und *χρόνος*, während ja die erste Stelle *ἀριθμοί, γραμμαί, στερεά, χρόνοι* bzw. *ἀριθμοί, μήκη, χρόνος, στερεά* hat. Gemeint ist also entweder *μῆκος* oder vielleicht auch, wie nach sonstigem

[19]) *ἔτι γράφεται ἔνια καθόλου ὑπὸ τῶν μαθηματικῶν παρὰ ταύτας τὰς οὐσίας. ἔσται οὖν καὶ αὕτη τις ἄλλη οὐσία μεταξὺ κεχωρισμένη τῶν τ' ἰδεῶν καὶ τῶν μεταξύ, ἣ οὔτε ἀριθμός ἐστιν οὔτε στιγμαὶ οὔτε μέγεθος οὔτε χρόνος.* Vgl. auch *Alexander* Aphrod., in Metaph. ad loc. (p. 705, 32–706, 25 Bonitz). – Dort wird übrigens p. 706, 1–2 auffallender- und nicht recht verständlicherweise VI, 16 neben I, not. comm. 3 als *καθόλου γραφόμενον* zitiert.

aristotelischen Sprachgebrauch nicht unwahrscheinlich[20]), die entweder ein- oder zwei- oder dreidimensionale Raumgröße[21]). Aristoteles erreicht also den reinen „formalen“ *μέγεθος*-Begriff, den wir Euklid zuzuschreiben pflegen, sprachlich noch nicht. Er meint aber den Begriff desjenigen Allgemeinen, das in irgendeiner Weise über den anschaulich erfaßbaren Gestaltarten (*εἴδη*) als mathematischer Gegenstand anzusetzen ist. (Den Charakter dieser *οὐσία* empfindet er freilich als sehr problematisch.[22]))

Und gerade die erste Erfassung dieses allgemeinen Begriffs, dieses „höheren Eidos“ über die anschaulichen Arten der bekannten mathe|matischen Gegenstände 332
hinaus ist es, was Eudoxos, dem Freunde Platons, zugeschrieben werden muß[23]). Gerade diese Erfassung und Durchsetzung des allgemeinen Logosbegriffs, des Logos zwischen beliebigen Größen, ist seine große Leistung als Platoniker!

Über die Art seines sprachlichen Ausdrucks dieses neuen Gedankens soll damit noch nichts gesagt sein. Auch das werde nicht entschieden, ob er als erster die Theoreme der Gruppe A, die sachlich allgemein gültig bereits in der anthyphairetischen Theorie beweisbar waren, zuerst allgemein formulierte oder ob das schon vor ihm geschah[24]). Auffallend ist jedenfalls, daß noch *Archimedes* das sog. „archimedische

---

[20]) Vgl. vorläufig den Bonitzschen Index Aristotelicus sub verbo *μέγεθος*, insbes. p. 449 a, 28–47.

[21]) Vgl. z. B. folgende Stellen: de coelo I, 1 (p. 268 a 7): *μεγέθους τὸ μὲν ἐφ’ ἓν γραμμή, τὸ δ’ ἐπὶ δύο ἐπίπεδον, τὸ δ’ ἐπὶ τρία σῶμα*; Met. Δ 13 (p. 1020 a 11): *μεγέθους τὸ μὲν ἐφ’ ἓν συνεχὲς μῆκος, τὸ δ’ ἐπὶ δύο πλάτος, τὸ δ’ ἐπὶ τρία βάθος*; vgl. auch de anim. II, 11 (p. 423 a 22) u. a.

[22]) Vgl. die Fortsetzung der zitierten Metaphysikstelle p. 1077 a 13–14: *εἰ δὲ τοῦτο ἀδύνατον, δῆλοί ὅτι κἀκεῖνα ἀδύνατον εἶαι κεχωρισμένα τῶν αἰσθητῶν* (cf. Alexander l. c.). – Weiterhin ist heranzuziehen Anal. Post. I, cap. 24 (p. 85 a 13 bis 86 a 30). Die Erörterung behandelt das Problem, ob der Beweis *καθόλου* oder *κατὰ μέρος* (= *χωρίς* an der Stelle p. 74 a 17–25) „besser“ (*βελτίων*) sei. Gegen den „allgemeinen“ Beweis wird geltend gemacht, daß er den Glauben erwecke, es gäbe etwas neben den einzelnen Dingen (*τι παρὰ τὰ καθ’ ἕκαστα*). So : ein allgemeines Dreieck, eine allgemeine Figur oder Zahl neben den einzelnen. Zur Erläuterung wird in Parenthese die *Proportionenlehre* angeführt, p. 85 a 37–b 1 : *προϊόντες γὰρ δεικνύουσιν, ὥσπερ περὶ τοῦ ἀνὰ λόγον, οἷον ὅτι ὃ* (fort. omittendum) *ἂν ᾖ τι τοιοῦτον, ἔσται ἀνὰ λόγον, ὃ οὔτε γραμμὴ οὔτ’ ἀριθμὸς οὔτε στερεὸν οὔτ’ ἐπίπεδον, ἀλλὰ παρὰ ταῦτά τι.* („In der Argumentation vorwärtsgehend zeigen sie [die Existenz der „allgemeinen Gegenstände“], wie bei den Proportionen, so z. B. : Wenn etwa etwas Derartiges [nämlich „Allgemeines“] sein sollte, wird es in Proportion stehen, [etwas] was weder Strecke noch Zahl noch Körper, noch Ebenen-[stück], sondern etwas ‚neben‘ diesen.“) – Später (p. 85 b 18–21) wird das ganze Argument zurückgewiesen : es sei nicht notwendig *τι παρὰ ταῦτα* anzunehmen, was ein Eines offenbar mache, *ὅτι ἓν δηλοῖ*, ebensowenig wie bei anderem, was kein „etwas“ (*τι*) bedeute, sondern eine Beschaffenheit (*ποιόν*) oder eine Beziehung (*πρός τι*) oder ein Tun (*ποιεῖν*). – Die Proportionen und Verhältnisse werden also als Beziehungen (*πρός τι* = *σχέσις* bei Euklid) aufgefaßt, die allgemein gelten können, ohne daß sie zwischen allgemeinen Gegenständen als Terminis stattfänden.

[23]) Vgl. Plato, Philebus 15A, Parmenides 128E–130A.

[24]) *Proclus*, in Euclid, p. 67, 3–5, sagt: *Εὔδοξος .... πρῶτος τῶν καθόλου καλουμένων θεωρημάτων τὸ πλῆθος ηὔξησεν*. Doch ist dies natürlich sehr vieldeutig. Zwar ist es richtig, daß die neuplatonischen Schriftsteller bei den *καθόλου θεωρήματα* vor allem an die Proportionenlehre denken, so *Jamblichus*, de communi mathematica scientia, cap. V, p. 18, 24 ss., 19, 2 ss. (ed. N. Festa) und *Proclus* selbst, in Eucl., p. 7, 11–8, 4 ; 9, 4 ss. ; 60, 24 ss. ; 391, 22 ss. Aber es werden auch andere Theoreme, z. B. der Satz von der Winkelsumme im allgemeinen Dreieck als *καθόλου* bezeichnet, so gerade Aristoteles, in unmittelbarer Fortsetzung der Bemerkung über das *ἐναλλάξ*, Anal. Post. I, 5, p. 74 a 25–b 4.

Axiom“ getrennt für die einzelnen anschaulichen Größenarten anführt[25]) und als *τάδε τὰ λήμματα* bezeichnet[26]). Da er für die mit dem Lemma beweisbaren Theoreme ausdrücklich Eudoxos zitiert, niemals aber Euklid, so kann in diesem gesonderten Anführen der Lemmata, wie *O. Toeplitz* bemerkt hat, sehr wohl ein Rückgang auf die Originalschriften des Eudoxos liegen[27]).

Endlich sei noch ein letzter Punkt, nicht ohne Zögern, erwähnt. *Zeuthen*[28]) hat die
333 These vertreten, die aus Platons „Theätet“ so be|rühmten Irrationalitätsbeweise für $\sqrt{3}$ bis $\sqrt{17}$, die Theodoros von Kyrene gegeben hat, seien auf Grund der Kettenbruchentwicklung (d. i. eben der Anthyphairesis) dieser irrationalen Verhältnisse $\sqrt{3}:1$ usw. geführt worden, durch den Nachweis ihrer Periodizität, also ihres Nichtabbrechens. *Toeplitz*[29]) hat dazu bemerkt, daß sich das Aufhören der Reihe dieser einzeln geführten Beweise bei $\sqrt{17}$ dadurch bedingt sein könne, daß die Periode der nächsten relevanten Wurzel, $\sqrt{19}$, wesentlich länger und der Beweis also wesentlich schwieriger sei, als bei den vorhergehenden.

Trifft diese Hypothese zu, so bedeutet das, daß bestimmte irrationale Verhältnisse vom Lehrer Theätets (!) durch ihre anthyphairetische Entwicklung als eindeutig gekennzeichnet angesehen worden wären. Und das ist, prinzipiell gesehen[30]), nichts anderes als die „antanairetische“ Definition der Verhältisgleichheit.

Indessen handelt es sich bei alledem nur um eine Hypothese, die keineswegs die einzig mögliche ist[31]).

* * *

---

[25]) Die Stellen sind: de sphaera et cylindro, postulatum 5 (Opp. ed. Heiberg, Vol. I, p. 8, 23–27), de lineis spiralibus, praefat. (Vol. II, p. 12, 6–11), quadratura parabolae, praefatio (Vol. II, p. 264, 8–22).

[26]) Die Wendung *τάδε τὰ λήμματα* steht l. c. Vol. II, p. 12, 7 allerdings – u. E. unbegründeterweise – nicht im Heibergschen Text (der *τόδε τὸ λῆμμα* liest), wohl aber durchweg in der Überlieferung, wie der Apparat ausweist.

[27]) Auch die Stelle aus quadrat. parab. (Vol. II, p. 264, 6–22) ist im Heibergschen Text entstellt. Zunächst ist, wie *Toeplitz* und *Solmsen* bemerkt haben, p. 264, 19 das Kolon nach dem Worte *ἴσον* in ein Komma zu verwandeln und wohl zweckmäßigerweise entsprechend in l. 16 das Komma nach *χρωμένοι* in ein Kolon. Dann ist m. E. l. 15 *αὐτῷ τῷ λήμματι* („von dem Lemma selbst“) nach der Überlieferung (Cod. A) gegen Heibergs Lesung *αὐτῷ ⟨τούτῳ⟩ τῷ λήμματι* wiederherzustellen, ebenso l. 22 *λήμματι* (oder vielleicht *λῆμμα [τι]*) gegen H.s Lesart *λῆμμα τι*. Die Beziehung auf Eucl. El. X, 1 ist irrig, mit *αὐτὸ τὸ λῆμμα* ist der Hilfssatz für Flächen (*χωρία*) gemeint, der diesem „ähnliche“ ist die entsprechende Fassung für Körper.

[28]) *H. G. Zeuthen*, Notes sur l'histoire des Mathématiques VIII (Sur la constitution des livres arithmétiques des Éléments d'Euclide et leur rapport à la question de l'irrationalité), Bulletin de l'Academie R. de Danemark, 1910, Nr. 5 (bes. S. 423ff.). – Sur l'origine historique de la connaissance des quantités irrationelles, ib. 1915, Nr. 3–4 (bes. S. 346ff.).

Vgl. *Heath*, l. c. Vol. III, p. 18–19, Vol. I, p. 398–401 („side“ and „diagonal“ numbers). – *A. E. Taylor*, Forms and Numbers: a study in Platonic metaphysics I. (Mind. N. S., Vol. XXXV, Nr. 140), insbes. p. 428ff.

[29]) Als Ergebnis aus einer gemeinsamen Kieler Seminarübung veröffentlicht bei *Hasse-Scholz*, a. a. O., S. 28–29.

[30]) Proclus sagt p. 66, 17–18: Leodamas, Archytas, Theätet, „*παρ' ὧν ἐπηυξήθη τὰ θεωρήματα καὶ προῆλθεν εἰς ἐπιστημονικωτέραν σύστασιν*“.

[31]) Vgl. die Aufsätze von *C. H. Müller* und *O. Toeplitz* in diesem Heft, in denen gewisse mit der Anthyphairesis möglicherweise konkurrierende (übrigens viel raschere) Näherungsverfahren aus platonischer Zeit belegt werden.

Damit sei diese erste Studie abgeschlossen. Die Aufgaben, mit denen sich die zweite beschäftigen soll, sind die bisher zurückgestellten drei Probleme (1) der logisch-sprachlichen Entwicklung des μέγεϑος-Begriffs bei den Mathematikern und Philosophen, (2) der scheinbar grundlosen Annahme der Existenz der vierten Proportionalen auch ohne konstruktiven Beweis und endlich (3) der Vermeidung der Anwendung. sogenannter ,,transfiniter“ Schlußweisen in der griechischen Mathematik nebst Untersuchung der Stellung der aristotelischen Philosophie (Logik und Physik bzw. Metaphysik) hierzu.

KURT von FRITZ

THE DISCOVERY OF INCOMMENSURABILITY BY HIPPASUS OF METAPONTUM*

The discovery of incommensurability is one of the most amazing and far-reaching accomplishments of early Greek mathematics. It is all the more amazing because, according to ancient tradition, the discovery was made at a time when Greek mathematical science was still in its infancy and apparently concerned with the most elementary, or, as many modern mathematicians are inclined to say, most trivial, problems, while at the same time, as recent discoveries have shown, the Egyptians and Babylonians had already elaborated very highly developed and complicated methods for the solution of mathematical problems of a higher order, and yet, as far as we can see, never even suspected the existence of the problem.

No wonder, therefore, that modern historians of mathematics have been inclined to disbelieve the ancient tradition which dates the discovery in the middle of the 5th century B.C.,[1] and that there has been a strong tendency to date the event much later, even as late as the first quarter of the 4th century.[2] But the question can hardly be decided on the basis of general considerations. It is the purpose of this paper to prove: 1) that the early Greek tradition which places the second stage of the development of the theory of incommensurability in the last quarter of the 5th century, and therefore implies that the first discovery itself was made still earlier is of such a nature that its authenticity can hardly be doubted, 2) that this tradition is strongly supported by indirect evidence, 3) that the discovery can have been made on the 'elementary' level
which, even | according to E. Frank and O. Neugebauer,[3] Greek mathematics had 243

* This article owes much to discussions of the early history of Greek mathematics which were carried on more than ten years ago between the author and Professor S. Bochner, now of Princeton University. This does not mean, of course, that Dr. Bochner has any part in whatever deficiencies the present article may have.

1 This tradition will be discussed below, pp. 244 ff (p. 213 in this volume).

2 The first to make an attempt to show that the discovery of incommensurability was 'late,' and certainly later than ancient tradition indicates, was Erich Frank in his book on *Platon und die sogenannten Pythagoreer* (Halle, Max Niemeyer, 1923). He does not commit himself to a definite date, but contends that the discovery cannot have been made before the last years of the 5th century (p. 228 ff.). O. Neugebauer, the most outstanding living authority on the earliest history of mathematics, goes even farther. In a letter to the author of the present paper he expressed the opinion that the discovery could not have been made before Archytas of Tarentum. Since Archytas was head of the government of Tarentum in 362 B.C., this seems to indicate that in his opinion the discovery was not made before the early 4th century at the earliest. It was also he who based his opinion on the 'trivial' character of 5th century Greek mathematics. In the present paper an attempt will be made to show that Greek mathematics in that period was in fact very elementary in many respects when compared with contemporary or earlier Babylonian and Egyptian mathematics, but by no means 'trivial.'

3 See the preceding note.

J. Christianidis (ed.), Classics in the History of Greek Mathematics, 211–231.

reached in the middle of the 5th century, 4) that the character of scientific investigation as developed in the early part of the 5th century makes it not only possible but very probable that the discovery was made at the time in which the late ancient tradition places it, and 5) that this late tradition itself contains some hints as to the way in which the discovery, in all likelihood, actually was made.

The earliest precise and definite tradition concerning a phase in the development of the theory of incommensurability is found in Plato's dialogue *Theaetetus*, p. 147 B. This dialogue was written in the year 368/67 B.C., shortly after the death of the mathematician Theaetetus after a battle in which he had been fatally wounded.[4] The fictive date of the dialogue is the year 399 B.C., that is, the year of the death of Socrates. In the first part of the dialogue the old mathematician Theodorus of Cyrene is represented as demonstrating to a group of young men, among them young Theaetetus, who is represented as a youngster of about seventeen, the irrationality of the square roots of 3, 5, 6, etc. up to 17. Though the dialogue itself is, of course, fictive, it seems hardly possible to assume that Plato, in a dialogue dedicated to the memory of a friend who has just died prematurely and who had had a very important part in the development of the theory of incommensurability and irrationality[5] would have attributed to someone else what was really his friend's own accomplishment. The inevitable conclusion, therefore, is that what Theodorus demonstrates in the introduction to the dialogue was actually known when Theaetetus was a boy of seventeen.[6]

Theodorus of Cyrene is represented as an old man in Plato's dialogue. According to an extract from Eudemus' history of mathematics[7] he was a contemporary of Hippocrates of Chios and belonged to the generation following that of Anaxagoras and preceding that of Plato. Since Anaxagoras was born in ca 500, and Plato in 428, this implies that Theodorus was born about 470 or 460, which agrees with Plato's
244 statement that he was an old man in 399. Plato | does not say that what Theodorus demonstrated to Theaetetus and the other youngsters in 399 was at that time an entirely new discovery, though the fact that he gave a proof for each one of the different cases separately shows that the theory had not yet reached a more advanced stage.[8] But even if we assume that Theodorus' demonstrations had been worked out for the first time not so very long before, Plato's dialogue would still indicate that the irrationality of the square root of 2, or the incommensurability of the side and diameter

---

4 This was proved by Eva Sachs in her dissertation *De Theaeteto mathematico* (Berlin, 1914). Her results in this respect seem absolutely certain and have been universally accepted.

5 For details see my article *Theaitetos* in Pauly-Wissowa, *Realencyclopädie*, vol. V A, p. 1351–72.

6 E. Frank (*op. cit.*, pp. 59, 228, and passim) and others have quoted a passage in Plato's *Laws* (p. 819c ff.) as a proof of their assumption that the discovery of incommensurability cannot have been made before the end of the fifth or the beginning of the fourth century. In this passage 'the old Athenian,' who is usually identified with Plato, says that he became acquainted with the discovery of incommensurability only late in his life and that it is a shame that 'all the Greeks' are still ignorant of the fact. It is quite clear that the latter statement is a rhetorical exaggeration since 'all the Greeks,' if taken literally, would include the Athenian himself, who by now obviously does know. The passage then proves nothing but that even striking mathematical discoveries in the fifth century did not become known to the general educated public. But this is also true of the fourth and third centuries.

7 In Proclus' commentary to Euclid's Elements, p. 66 Friedlein.

8 Concerning the probable steps from the first discovery to the theory of Theodoras, see infra pp. 254 ff.

of a square had been discovered by someone else. For it is difficult to see why he should have made Theodorus start with the square root of 3, unless he wished to give an historical hint that this was the point where Theodorus' own contribution to mathematical theory began. This in itself then would be quite sufficient to show that the discovery of incommensurability must have been made in the earlier part of the last quarter of the 5th century at the very latest, and since mathematical knowledge at that time traveled very slowly, may very well have been made earlier.[9]

What can be inferred from Plato's dialogue *Theaetetus* receives strong confirmation from indirect evidence which has been presented by H. Hasse and H. Scholz.[10] It is perhaps not necessary to accept their interpretation of the doctrines of Zenon of Elea in every detail. But there can hardly be any doubt that they have proved conclusively that there must have been a connection between some of Zeno's famous arguments against motion, and the discovery of incommensurability.[11] Since Zenon was born not later than 490 B.C., acceptance of the results of the treatise quoted would lead to the conclusion that the discovery of incommensurability must have been made not later than the middle of the 5th century, which is also the date indicated by ancient tradition.

In contrast to the tradition concerning the second phase of the development of the theory of incommensurability the tradition concerning the first discovery itself has been preserved only in the works of very late authors, and is frequently connected with stories of obviously legendary character.[12] But the tradition is | unanimous[13] in attributing 245
the discovery to a Pythagorean philosopher by the name of Hippasus of Metapontum.

Ancient tradition concerning the life and chronology of Hippasus is scanty. Iamblichus in his treatise *de communi mathematica scientia*[14] says that early Greek mathematical science made great progress through the work of Hippocrates of Chios and Theodorus of Cyrene, who followed upon Hippasus of Metapontum. Since Hippocrates and Theodorus are also mentioned together in the extract from the history of mathematics of Eudemus of Rhodes,[15] it seems likely that Iamblichus' note also goes back to the very reliable work of this disciple of Aristotle. According to this work Hippasus belonged to the generation preceding that of Theodorus (according to ancient usage this means an average difference of age of about 30–40 years), who in his turn was a contemporary of Hippocrates of Chios.

---

9 See note 6.

10 H. Hasse and H. Scholz, *Die Grundlagenkrisis der griechischen Mathematik*, Charlottenburg, Kurt Metzner, 1928, pp. 10 ff.

11 In contrast to this, E. Frank (*op. cit.*, pp. 219 ff.) has contended that the mathematical philosophy of the Pythagoreans which *preceded* the discovery of incommensurability presupposes the atomistic theory of Democritus and a fully developed theory of 'the subjectivity of sensual qualities.' The analysis of the early form of Pythagorean philosophy attempted below will, I hope, show that it has nothing whatever to do with Democritus' atomism, and is certainly no more dependent on a fully developed theory of the subjectivity of sensual qualities than the philosophy of Parmenides, who was born at least 60 years earlier than Democritus.

12 For instance, the story told by Iamblichus, that he was drowned in the sea, and that this was a divine punishment for his having made public the secret mathematical doctrines of the Pythagoreans.

13 The one seeming deviation from the unanimous tradition in Proclus, *op. cit.* (see note 7), p. 67, is obviously due to a corrupt reading (*ἀλόγων* for *ἀναλόγων* or *ἀναλογιῶν*) in some manuscripts.

14 Iamblichus, *De communi mathematica scientia*, 25, p. 77 Festa.

15 See note 7.

According to Iamblichus' *Life of Pythagoras*,[16] Hippasus had an important part in the political disturbances in which the Pythagorean order became involved in the second quarter of the 5th century, and which ended in the revolt of ca 445, which put an end to Pythagorean domination in southern Italy.[17] This agrees perfectly with the tradition which places him in the generation before Theodorus, who, as shown above, was born between 470 and 460. This confirmation is all the more valuable because the tradition of the political history of the Pythagoreans which was first collected by Aristoxenus of Tarentum and Timaeus of Tauromenium is, on the whole, quite independent from the ancient tradition of early Greek mathematics, which was first collected by Eudemus of Rhodes.

The mathematical achievements—apart from the discovery of incommensurability—ascribed to Hippasus by ancient tradition, are the following:

1. An anonymous scholion on Plato's *Phaedo*,[18] quoting a work on music by Aristotle's disciple Aristoxenus, says that Hippasus performed an experiment with metal discs. He had four metal discs of equal diameter made in such a way that the second disc was $1\frac{1}{3}$ times as thick, the third $1\frac{1}{2}$ times as thick, and the fourth twice as thick as the first one. He then showed that by striking any two of them the same harmony of sounds would be produced as by two strings whose lengths were in the same proportion as the thicknesses of the discs.
246 Theon | of Smyrna[19] attributes to him a similar experiment with four tumblers, the first of which was left empty, while the others were filled $\frac{1}{4}$, $\frac{1}{3}$, and $\frac{1}{2}$ with water.
2. Boethius[20] attributes to him a theory of the musical scale showing how the different musical harmonies can mathematically be derived from one another.
3. Iamblichus[21] says that Hippasus concerned himself with the theory of proportions and 'means' and was the first to change to 'harmonic mean' the name of what previously had been called the contrary, or, as some translate, the subcontrary, mean, the formula of which is $a/c = (a-b)/(b-c)$. But Nicomachus attributes this change in terminology to Philolaus.
4. According to Iamblichus,[22] Hippasus was also the first to draw or construct[23] the 'sphere consisting of 12 regular pentagons', or, as he says in another passage,[24] to inscribe the regular dodecahedron in a sphere and to make this

16 Iamblichus, *De Vita Pythagorae*, 257, p. 138 f. Deubner.

17 For the date see K. von Fritz, *Pythagorean Politics in Southern Italy* (Columbia University Press, 1940), pp. 77 ff.

18 Schol. in Plat. Phaed. 108d; see *Scholia Platonica*, ed. W. Chase Greene (Philol. Monographs publ. by Am. Philol. Ass., vol. VIII, 1938), p. 15. All the passages quoted in notes 18–24 are also collected, though sometimes in a slightly abbreviated form, in H. Diels, *Vorsokratiker*, Vol. 1.

19 Theo Smyrnaeus, *Expos. Rerum Mathem.*, p. 59 Hiller.

20 Boethius, *De Institutione musica*, 11, 10.

21 Iamblichus, *In Nicomachi arithmet. introd.*, p. 109 Pistelli.

22 Iamblichus, *De communi mathem. scientia* 25 (p. 77 Festa) and *Vita Pythag.* 18, 88 (p. 52 Deubner).

23 The Greek term *γράψασθαι* has both meanings.

24 *Vita Pyth.* 34,247 (p. 132 Deubner). The name of Hippasus is not mentioned in this passage, but since the same story is connected with the divulgation of the discovery as in the first passage, there can be no doubt that the reference is to Hippasus.

construction public, which was considered a criminal divulgation of Pythagorean secret knowledge.

Of these four statements the first and fourth are of special importance and must be carefully analyzed, while the second and the third are of a certain importance for our problem mainly in connection with the first one.

In regard to Hippasus' experiments it seems relevant to point out that in the period in which Hippasus lived other Greek philosophers also conducted scientific experiments, while after that time, with one possible exception,[25] we do not again hear of scientific experiments until the third century. In fact, the philosopher to whom most of these experiments are attributed, Empedocles (ca 490 to ca 430 B.C.), was a native of Sicily, lived for some time in southern Italy, and though not a Pythagorean himself, was undoubtedly influenced by Pythagorean thought.

The experiments attributed to Empedocles are the following: 1) an experiment to show that drinkable water could be extracted from the sea, in order to show that fish did not 'feed on' salt water, but on sweet water which could be extracted from it;[26] 2) an experiment with small open vessels filled with water and swung around on a cord, in order to prove the existence of what we would call a centrifugal force, which in his opinion prevented the celestial bodies from falling to the earth;[27] 3) an experiment with pulverized ore of various kinds and colors, | in order to show that the different ele- 247
ments when mixed in this way become inseparable, and their original qualities indistinguishable in the mixture;[28] 4) an experiment with a *clepsydra* or water-clock, in order to prove that seemingly completely empty vessels are actually filled with air.[29] This experiment and a similar one with leather bags is also attributed to Anaxagoras[30] (born in ca 500 B.C.).

The one possible exception to the statement that the known scientific experiments of the Greeks belong to the fifth and third (and later) centuries, but not to the fourth, is found in a passage from a work of Archytas, quoted literally by Nicomachus and Porphyrius.[31] In this fragment Archytas propounds the theory that sound is produced by a concussion of the air, that the pitch of the sound depends on and is proportional to the velocity of the motion producing it, and that if the velocities producing two sounds are in certain simple numerical ratios, well known musical harmonies result. The arguments by which these theories are supported are based on observations which *can* be made in everyday life, and without experimentation; but the way in which the observations are introduced strongly suggests that, though originally they may have been made incidentally, they were at least checked by being repeated in an experimental

---

25 See infra.

26 Empedocles, fragm. A 66 in H. Diels, *Die Fragmente der Vorsokratiker*, vol. 1.

27 Ibid., A 67.

28 Ibid., A 34.

29 Ibid., B 100. Here the description of the experiment is given in its original wording. Empedocles in fact does not describe it as an experiment made by himself, but as an illustrative analogy derived from the observation of a young girl playing with a water-clock. But this belongs to the poetical style, since Empedocles expounded all his philosophical and scientific theories in verse. The minute description of the process leaves no doubt whatever that Empedocles must have made the experiment himself.

30 Anaxagoras, fragm A 68/69 in H. Diels, *op. cit.*

31 Archytas, fragm. B 1 (Diels, *op. cit.*).

fashion. Archytas, however, does not claim to be the author of these theories and to have made personally the observations or experiments from which they are derived, but attributes them to mathematicians whose names he does not give. At the same time it is obvious that these theories and observations represent an advanced stage of scientific development as compared with the experiments of Hippasus and their results. For in the Archytas fragment Hippasus' demonstration of a way in which the same musical harmonies can be produced by any conceivable kind of sound-producing instrument is integrated with a general physical theory of sound. Since, on the other hand, both Hippasus and Archytas were Pythagoreans living in southern Italy, since Archytas, as shown above,[32] belonged to the second generation after Hippasus, and since, nevertheless, Hippasus and Archytas are sometimes mentioned together in ancient tradition[33] as having contributed to the development of a physical theory of sound, there really seems to be no reason to doubt that there actually existed a scientific tradition in one branch of the Pythagorean school through which a theory of sound was gradually developed. Since, finally, the authenticity of the fragment from
248 Archytas' *Har|monikos* can hardly be doubted, and as far as I can see never has been doubted, and since he clearly implies that the theory of sound had reached a rather advanced stage before he himself began to contribute to it, it is difficult to see how some scholars[34] could claim that ancient tradition projected into a much earlier time the accomplishments of a later period, when it attributed to Hippasus, a man belonging to the second generation before Archytas, the first beginnings of a theory which had reached a much more advanced stage before Archytas wrote his work.

Everything then seems to confirm the assumption that the experiments attributed to Hippasus by ancient tradition actually can have been made, and most probably were made, in Southern Italy in the middle of the fifth century, that is, when Hippasus is supposed to have lived in that region. To that extent, at least, the late tradition, which according to E. Frank and others, is of no value whatever, seems to be vindicated.

But what can Hippasus' experiments with discs and tumblers possibly have to do with the discovery of incommensurability? In order to show the interconnection, which is, of course, very indirect, it will be necessary to make a further analysis of the purpose and meaning of these experiments.

All the experiments ascribed to philosophers of the fifth century, as their description clearly shows, were obviously undertaken not so much in order to find out something new, but rather in order to support and verify an already existing theory, for instance, that the fish do not consume salt water as such, but extract sweet water from it, that the celestial bodies do not remain in the sky because they are lighter than air, etc. The same is true of the experiments attributed to Hippasus. That certain musical harmonies would be produced if the lengths of two strings of the same kind were in certain ratios had always been known. It had also been known in regard to flutes. From this double knowledge, then, the general assumption was derived that it would be so in all cases. What Hippasus did was, in a way, nothing but a verification of this assumption by means of various sound-producing bodies which were not ordinarily used as musical instruments. But two things are significant. Strings have, so to speak, only one dimension.

---

32 See supra p. 245 (p. 213/14 in this volume) and note 2.
33 For instance, Iamblichus, in *Nicom. arithm. intr.*, p. 109 Pistelli.
34 See E. Frank, *op. cit.*, p. 69 and passim.

In regard to flutes, too, especially if the different tones are produced on the same flute, one will not always think of the other two dimensions. When Hippasus used tumblers and discs, however, he had to point out that the discs, for instance, must be equal in two dimensions and differ only in the third if the musical harmonies are to be produced, but that it did not matter whether the third dimension was what usually was called length or thickness. In this way, then, the result can be most clearly formulated, namely, that the musical harmonies are completely independent of the material of which the sound-producing body consists, and of the special quality or color of the tones produced, and that the production of these harmonics depends exclusively on
simple one-dimensional numerical ratios. We hear then, further,[35] | that Hippasus was 249
not content with having proved this point but also investigated the mathematical relations between the ratios producing the most outstanding harmonies and tried to derive them mathematically from one another.

As long as Hippasus remained within the limits of the theory of music, all this, of course, could not lead to the discovery of incommensurability. But there are strong indications that he and his associates did not confine themselves to this special field.

Aristotle very frequently mentions the Pythagoreans or so-called Pythagoreans, and attributes to them the doctrine that 'all things are number.'[36] According to E. Frank these so-called Pythagoreans are not Pythagoreans at all,[37] but contemporaries of Plato who were deeply influenced by his philosophy.[38] If this were so it would be difficult to see why Aristotle, who should have known, never says a word about it, and always seems to imply that Plato's theory of numbers is later. It would also be possible to show that the comparatively very primitive Pythagorean theory cannot possibly be later than Plato's very complicated one. But this would require an analysis of considerable length, which fortunately is not necessary for the present purpose, since there is more direct evidence to show that there must have been Pythagoreans in the fifth century who had a doctrine similar to that ascribed to them by Aristotle.

Archytas in the long fragment quoted above[39] says that the same men who elaborated a theory of sound had also attained 'clear insight' into problems of astronomy, geometry, and arithmetic. Again, of course, he refers to what others had done before he wrote his work. Unfortunately, the passages in which he described the achievements of his

---

[35] See supra, p. 246, note 20 (p. 214 in this volume).

[36] The doctrine is expressed and explained in a great many different ways by Aristotle; for instance, that 'the elements of numbers are the elements of all things' (Metaph. 986a, 1 ff.), or that 'all things are composed of numbers' (ibid. 1080b, 16 f.), or that 'the things themselves are numbers' (ibid., 987b, 29 f.), or that 'number is the essence of everything' (ibid., 987a, 19). But the last expression uses specific Aristotelian terminology and is obviously an attempt to explain what appeared too odd in its original wording.

[37] *Op. cit.*, p. 68 ff.

[38] E. Frank lays great stress on the fact that Aristotle speaks often, though not in the majority of cases, of the 'so-called' Pythagoreans, and infers from this that he meant that they were not really Pythagoreans. In fact, there was an excellent reason for the use of the word 'so-called,' namely, that in Aristotle's time 'Pythagoreans' was the only name designating the adherents of a philosophical school or sect that was derived from the name of the founder; that is, it was an unusual expression. Confirmation of this can also be found in the fact that the only analogy to the name 'Pythagoreans' found in pre-Aristotelian literature (Herakleiteans in Plato's Theaet. 179e) is obviously used in fun.

[39] See supra p. 247 (p. 215 in this volume) and note 31.

predecessors in astronomy and geometry have not come down to us. But since he speaks of the clear insight which they had attained, it is not likely that it was only in music that they had arrived at a stage so advanced that it must have required a considerable time to attain it. Moreover, Archytas says that the sciences mentioned are intimately related to one another because all of them 'turn back' to 'the first (or fundamental) form of everything that is'. This seems a very advanced form of the doc-
250 trine which | Aristotle attributes to the 'so-called Pythagoreans'. Again, everything seems to indicate that the close connection between arithmetic, geometry, astronomy, and musical theory, as well as the somewhat crude theory that 'all things are numbers' must have been considerably older than Archytas, that is, at least as early as the middle of the fifth century.

In order to understand the origin and meaning of this latter doctrine, an analysis of the Greek terminology of the theory of proportions will be helpful. The Greek expression for proportion means literally 'the same ratio'. For our term 'ratio' the Greeks have two expressions: *diastema*, which means literally 'interval', and *logos*, which means literally 'word'. The first term clearly shows the connection of the early theory of proportions with musical theory.[40] But the second term is even more significant. The Greeks had two terms for 'word': *epos* and *logos*.[41] *Epos* means the spoken word, or the word which appeals to the imagination and evokes a picture of things or events. This is the reason why it is also specifically applied to epic poetry. *Logos* designates the word or combination of words in as much as they convey a meaning or insight into something.[42] It is this connotation of the term *logos* which made it possible for it in later times to acquire the meaning of an intrinsic law or the law governing the whole world.

If *logos*, then, is the term used for a mathematical ratio, this points to the idea that the ratio gives an insight into a thing or expresses its intrinsic nature. In the case of musical harmonies the harmony itself would be perceived by the ear, but it was the mathematical ratio which, in the mind of the Pythagoreans, seemed to reveal the nature of the harmony, because through it the harmony could be both defined and reproduced in different media.

It is easy to see how this general idea could be extended to astronomy, especially to the regular motions of the celestial bodies and the interrelations between their various cycles.[43] But it is the extension of the theory to geometry which is of special importance for our problem.

The mathematical theorem which is in tradition most closely connected with Pythagoras and the Pythagoreans, is the theorem that in a right-angled triangle the

---

40 This is also the case with the word *horos* designating the terms of a ratio or a proportion. See K. von Fritz, *Philosophie und sprachlicher Ausdruck bei Demokrit, Platon und Aristoteles* (New York, Stechert, 1938), p. 69.

41 As to the question of how early the term *logos* was used in the sense of ratio, see infra p. 261 f (p. 228 f in this volume).

42 This is also characteristic of the corresponding verb *legein*. In consequence, the Greeks can form the following sentence: N. N. says (there follows a literal quotation of his words) saying (there follows an interpretation of their meaning). It is clear that 'saying' in this sentence really means 'meaning.' The verb *eipein*, which corresponds to *epos* cannot be used in the latter sense. It is also significant that those stories which Herodotus, for instance, calls *logoi* are always stories with a moral, that is, with a meaning.

43 For details see my article on Oinopides of Chios in Pauly-Wissowa *Realencyclopaedie*, vol. 17, p. 2260–67.

sum of the squares on the sides including the right angle is equal to the square | on the 251
side subtending the right angle. Nobody who knows anything about the early history of Greek mathematics has ever doubted that the proof of this theorem given by Euclid in the first book of his Elements cannot have been found by Pythagoras or his early followers. This is also what the best ancient tradition says, since Proclus attributes this proof to Euclid himself.[44] Though at the time when, in the last quarter of the fifth century, Hippocrates of Chios elaborated his famous theory of the *lunulae*, the 'Pythagorean theorem' must have been considered valid for right-angled triangles whose sides are commensurable with one another and for triangles whose sides are incommensurable, and furthermore must have been extended to cover all similar figures erected on the sides of a right-angled triangle, it is not possible for us to find out exactly how the early Greek mathematicians proved or tried to prove the theorem in this general form, since there exists no tradition about it.[45] Fortunately, it is not necessary for our purpose to have this knowledge.

Again, the theory must have started from an observation which had been generally known long before the beginning of Greek philosophy, namely, that if one puts together three pieces of wood of the respective lengths of 3, 4, and 5, a right-angled triangle will result. In fact, this is an old form of a carpenter's square. Since the size of carpenter's squares was not standardized, it must also have been a matter of common knowledge that the absolute length of the sides of the triangle was irrelevant, and that all triangles whose sides were in that proportion were not only right-angled, but also 'similar' in shape. Finally, it seems to have been known of old that the sum of the squares of 3 and 4 was equal to the square of 5.

Even if we had no tradition about it we would have to conclude that the Pythagoreans must have been impressed by these facts as soon as they had begun to suspect that the nature of a good many things might be found in or expressed by numbers, especially since there is indirect evidence to show that even before Pythagoras the philosopher Thales (ca 620 to ca 540 B.C.) and his followers had concerned themselves with what we may call the ornamental shape of geometrical figures[46] and also seem to have connected this ornamental appearance especially with the angles. The fact, at least, that according to Proclus[47] they used the term 'similar angles' for what later was called 'equal' angles can hardly be explained otherwise.[48]

On the basis of this earlier development the Pythagoreans can hardly have | failed 252
to notice that any two triangles will be similar in shape if their sides are in proportion, though in actual fact in the earliest period this knowledge can have been an exact knowledge only in regard to triangles whose sides are commensurable with one another. Though this assumption is not supported by any direct tradition—probably

---

44 Proclus, *In primum Euclid. elem. librum Comment.*, p. 426 Friedlein.

45 For the various possibilities see the lucid exposition of Th. Heath in his commentated translation of Euclid's *Elements* (Cambridge 1926), vol. 1, pp. 352 ff.

46 For the evidence see Th. Heath, *A History of Greek Mathematics* (Oxford, 1921), vol. 1, pp. 130 ff.

47 *Op. cit.*, p. 250 Friedlein.

48 It is perhaps pertinent to observe that the historian Thucydides (I, 77) also uses the term 'similar' where he means equality of form (or in this case: procedure). For he uses the expression 'similar laws' where, as the context shows, he does not mean similar laws but what otherwise was called *isonomia* or equality before the law.

because it was too obvious to be especially mentioned—it follows not only from the general situation, but especially from the close analogy of the Pythagoreans' theory of music and their earliest theory of geometrical figures, which is attested everywhere. For just as they declared that the musical harmonies which are perceived by the ear 'are' really the numbers by which the proportionate lengths of the strings, etc., producing them are measured, so the geometrical figures, whose shape is perceived by the eye but cannot otherwise be either exactly determined or expressed in language, 'are' really the numbers or sets of numbers constituting the ratios of the lengths of their sides by which their shape is determined and can therefore be expressed.[49]

According to ancient tradition, the theory, before the discovery of incommensurability, was further extended in two directions. Proclus[50] credits Pythagoras with a formula which makes it possible to form any number of different rational right-angled triangles by finding pairs of numbers the sum of the squares of which is equal to a square number.[51] It is irrelevant for our purpose whether this formula is rightly attributed to Pythagoras personally, but one can safely assume that it belongs to the very oldest period of Pythagorean mathematics. For Proclus usually relies on the very excellent history of mathematics of Aristotle's disciple Eudemus of Rhodes; and in this case what he says seems all the more worthy of credit in that he does not claim too much and rather implies a criticism of the common tradition that Pythagoras 'proved' the 'Pythagorean theorem' in its general geometrical form.

Nevertheless, the formula marks a great advance. One has to interpret it in terms of Pythagorean philosophy in order to understand its importance in regard to our problem. In the theory discussed before, the shape of figures which are similar in the mathematical sense of the word is directly related to a definite set of integers. Two triangles, with the sides 3, 4, 5 and 8, 15, 17 respectively are not, on the other hand, *similar* in the sense of the (modern or Euclidean) mathematical term. But they are still 'similar' in regard to the ornamental element of one right angle; and this 'similarity' is not related to or expressed in one definite set of integers, but is related to the fact that
253 the two | sets of integers related to the two triangles enter into the same mathematical formula. What is important for our problem in this extension of the theory is merely that it shows how the Pythagoreans were not content with a simple theory but, with an extraordinarily inquisitive spirit, adapted this theory to ever more complicated problems.

The second extension of the Pythagorean theory which is important as a preparation for the discovery of incommensurability is the theory of polygonal numbers. This

49 The Pythagoreans were, of course, aware that triangles are the only rectilineal figures whose shape is definitely determined by the proportionate length of their sides. That they realized the importance of this fact for their theory seems proved by Theon's statement (*op. cit.*, pp. 40 ff.) that they divided all other rectilineal figures into triangles.

50 *Op. cit.* (see note 44), p. 428 Friedlein.

51 The formula, though expressed in a somewhat more complicated way amounts to the statement that if m be any odd number,

$$m^2 + \left(\frac{m^2-1}{2}\right)^2 = \left(\frac{m^2+1}{2}\right)^2.$$

theory, the beginnings of which ancient tradition, starting with Aristotle,[52] attributes also to the early Pythagoreans, was many centuries later developed, by Diophantus to what is now called indeterminate analysis. But for a long time it remained rather sterile from a purely mathematical point of view. This is probably the reason why Euclid disregarded it in the arithmetical section of his elements and why other high ranking mathematicians from the fourth century onwards have done likewise.

Just like the other geometrical theories of the Pythagoreans discussed so far, this theory is concerned with interrelations between numbers and geometrical figures. But in this case the figures are not drawn and formed by straight lines of certain proportionate measures, but are built up from dots. The theory then is concerned with the question from what numbers of dots arranged in a certain order the different polygons can be built.[53] It seems perfectly clear from the evidence presented so far that this theory is a natural product of the development of Pythagorean thought. It, therefore, certainly need not, as E. Frank contends,[54] be dependent on or, in its original form, even be influenced by the physical atomism of Democritus, which has an entirely different origin. Whatever chronological inferences E. Frank draws from this incidental affinity are, therefore, absolutely unwarranted.[55]

Though the 'atomism' of the theory of polygonal numbers seems most remote from the discovery of incommensurability it is here that we come nearest to our problem. All the Pythagorean doctrines discussed so far either are based on or result in a search for numbers, i.e., integers, from which geometrical figures with certain properties can be built up. In the course of these efforts the Pythagoreans can hardly have failed to wonder what numbers might be hidden in certain | well-known figures which 254
had not been built up in this way, for instance, the isosceles right-angled triangle, which was of special importance to the Pythagoreans because it was one-half of the square, the latter figure having become a mystical symbol in the Pythagorean community. In the case of the isosceles right-angled triangle, however, it is not possible to express the ratio between its sides in integers. It is perhaps not too far-fetched a speculation if one assumes that the early development of the theory of polygonal numbers was partly due to an attempt to overcome this difficulty by building up the polygons from dots rather than from straight lines. In fact, this seems all the more likely because here again the division of polygons and polygonal numbers into triangles and triangular numbers is one of the main points of the theory. Theon, for instance, points out[56] that an oblong number can be divided into two equal triangular numbers while a

---

52 The relevant passages have been collected by Heath, *History* (see note 46), 1, 76 ff.

53 Triangular numbers, for instance, are (1), 3, 6, 10, 15, like this:

etc.

54 *Op. cit.* (see note 2), p. 52 ff.

55 The passage in Aristotle, *De Anima*, 490a, 10 ff., where Aristotle quite correctly says that if one replaces Democritus' material atoms by immaterial dots the result is very similar to the quantitative theory of the Pythagoreans, need certainly not have chronological implications. But even if it had such implications, this would not prove anything, since Aristotle in this passage does not refer to the earliest form of the Pythagorean doctrine.

56 *Op. cit.*, p. 41 Hiller.

square number is made up of two unequal triangular numbers whose sides differ by one unit, namely,

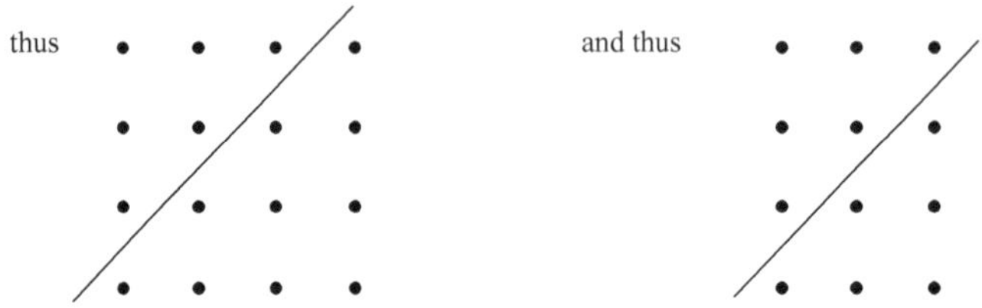

Fig. 1

But however this may be, men of the inquisitive spirit which characterized Hippasus and some of his Pythagorean contemporaries[57] can hardly have been satisfied with these arithmetical theorems as a substitute for the solution of the real problem, namely, the problem of the ratio between the sides of an isosceles right-angled triangle. This is again confirmed by ancient tradition; for what Plato says about Theodorus' demonstration of the irrationality of the square roots of 3, 5, 6, 7, etc. presupposes, as shown above,[58] that the irrationality of the square root of 2 had already been proved.

Fortunately, the original demonstration of the irrationality of the square root of 2 has been preserved in an appendix to the tenth book of Euclid's elements;[59] and that this demonstration is actually, at least in its general outline, the original one is attested
255 by Aristotle. One glance at this demonstration[60] | shows that it does not presuppose

---

57 See supra p. 245 ff. and p. 252 (p. 214 ff. and p. 220 in this volume).

58 See supra, p. 244 (p. 212/13 in this volume).

59 Euclid, *Elementa*, X, Append. 27, p. 408 ff. (This appendix is not included in Heath's translation of Euclid's Elements).

60 In literal translation this demonstration runs as follows: Let $ABCD$ be a square and AC its diameter. I say that AC will be incommensurable with AB in length.

For let us assume that it is commensurable. I say that it will follow that the same number is at the same time even and odd. It is clear that the square on AC is double the square on AB. Since then (according to our assumption) AC is commensurable with AB, AC will be to AB in the ratio of an integer to an integer. Let them have the ratio $DE{:}F$ and let DE and F be the smallest numbers which are in this proportion to one another. DE cannot then be the unit. For if DE was the unit and is to F in the same proportion as AC to AB, AC being greater than AB, DE, the unit, will be greater than the integer F, which is impossible. Hence DE is not the unit, but an integer (greater than the unit). Now since $AC:AB = DE:F$, it follows that also $AC^2:AB^2 = DE^2:F^2$. But $AC^2 = 2AB^2$ and hence $DE^2 = 2F^2$. Hence DE^2 is an even number and therefore DE must also be an even number. For if it was an odd number its square would also be an odd number. For if any number of odd numbers are added to one another so that the number of numbers added is an odd number the result is also an odd number. Hence DE will be an even number. Let then DE be divided into two equal numbers at the point G. Since DE and F are the smallest numbers which are in the same proportion they will be prime to one another. Therefore, since DE is an even number, F will be an odd number. For if it was an even number the number 2 would measure both DE and F, though they are prime to one another, which is impossible. Hence F is not even, but odd. Now since $ED = 2EG$ it follows that $ED^2 = 4EG^2$. But $ED^2 = 2F^2$, and hence $F^2 = 2EG^2$. Therefore F^2 must be an even number, and in consequence F also an even number. But it has also been demonstrated that F must be an odd number, which is impossible. It follows, therefore, that AC cannot be commensurable with AB, which was to be demonstrated.

any geometrical knowledge beyond the Pythagorean theorem in its special application to the isosceles right-angled triangle, which, as is well-known, can be 'proved' simply by drawing the figure in such a way that the truth of the theorem in that particular case is immediately visible.[61] Apart from this the demonstration remains in the purely arithmetical field; and since the early Pythagoreans speculated a good deal about odd and even numbers[62] the demonstration itself cannot have been beyond their reach.[63]

Yet if this demonstration of the irrationality of the square root of 2 was the only way in which incommensurability can have been discovered, one might still agree that there are good reasons for Frank's and Neugebauer's hesitation to attribute the discovery to the middle of the 5th century. The demonstration requires not only a good deal of abstract thinking, but also of strict logical reasoning. Apart from this, the labored language of the demonstration as given in the appendix in Euclid shows clearly with what difficulties the early Greek mathematicians had to struggle when elaborating a proof of this kind. In fact, this conclusion is all the more cogent because the demonstration, though somewhat more archaic in form than Euclid's own demonstrations, uses a form of presenting the argument in short concise sentences which has no parallel in Greek literature of the fifth century.[64] If, then, the proof as such, as the combined passages in Plato and Aristotle seem to indicate,[65] belongs to the fifth century, | it seems safe to assume that in its original form it was still more laborious. 256
Most significant, however, is the fact that the whole proof, as presented, uses the terms *commensurable* and *incommensurable*, just as Theodorus did in Plato's *Theaetetus*, as something already known. This seems to presuppose that incommensurability was already known when the demonstration was elaborated.

Since the form of the proof as it appears in the appendix to Euclid may not be the original one, the form of the proof in Euclid's appendix may not be sufficient to show with certainty that when the irrationality of the square root of 2 was demonstrated, the discovery of incommensurability as such had already been made, probably in a different mathematical object. But if one considers the further evidence presented above, the suspicion that such was the case becomes very strong. For it is difficult to believe that the early Greek mathematicians should have discovered the incommensurability of the diameter of a square with its side by a process of reasoning which was obviously so laborious for them if they had no previous suspicion that any such thing as incommensurability existed at all. If, on the other hand, they had already discovered the fact in a simpler way, it is perfectly in keeping with what we know of their methods

---

61 For examples, see Heath, *The Thirteen Books of Euclid's Elements*, vol. 1, p. 352.

62 See, for instance, Aristotle, *Physics*, 203a, 5 ff.; *Metaph.*, 986a, 22 ff.

63 Concerning the arithmetical premises of this demonstration and the probable deficiencies of its original form, see my article on Theodorus of Cyrene in Pauly-Wissowa, *RE*, vol. V A, p. 1817 and 1820 ff.

64 In order to illustrate this, one may compare the literal fragments of Zenon of Elea which show a very high degree of abstract thinking and also of close logical reasoning, but at the same time are written in a labored language with long and cumbrous sentences, while Aristotle (in the fourth century) and later writers who give an account of Zenon's theory, reproduce the same arguments in a sequence of very short sentences very similar to those found in the appendix to Euclid.

65 Plato, *Theaetetus* p. 147B ff. and Aristotle, *Analytica Priora*, 41a, 26–31 and 50a, 37. See also supra p. 244 and p. 251 (p. 212 f. and p. 219 in this volume).

to assume that they at once made every effort to find out whether there were other cases of incommensurability. The isosceles right-angled triangle in that case was the natural first object of their further investigations.

It is at this point that the tradition concerning Hippasus' interest in the dodecahedron, or 'the sphere out of 12 regular pentagons' has to be considered. There can be no doubt that Hippasus was not the author of the mathematical construction of the dodecahedron, as Iamblichus claims in one place.[66] Quite apart from other considerations, this is proved by the fact that the better tradition implies that this was an achievement of Theaetetus,[67] who belonged to the second generation after Hippasus. And in another passage, Iamblichus[68] himself claims merely that Hippasus 'drew' the regular dodecahedron, which is probably the original tradition.

That Hippasus was interested in the dodecahedron and in the dodecahedron as a 'sphere made of 12 regular pentagons' is very likely. For regular dodecahedra occurred in Italy as products of nature in the form of crystals of pyrite.[69] With the Pythagoreans' interest in geometrical forms these crystals must certainly have attracted their attention and evoked a desire to analyze their form mathematically. In addition, we know that the Pythagoreans used the pentagram, i.e., a regular pentagon with its sides prolonged to the point of intersection,[70] as a token of recognition. It is absolutely in the character of Hippasus as we now know him that he should have tried
257 to find out about the | numbers and ratios incorporated in the pentagram and regular pentagon. Could it then really be a mere coincidence that the same Hippasus is credited with the discovery of incommensurability and with an interest in the 'sphere consisting of 12 regular pentagons,' and that the regular pentagon is exactly the one geometrical figure in which incommensurability can be most easily proved?

How would the Pythagoreans have gone about it if they wanted to know the ratio between the lengths of two straight lines? Again, the method was an old one, known by craftsmen as a rule of thumb many centuries before the beginning of Greek philosophy and science, namely, the method of mutual subtraction,[71] by which one finds the greatest common measure. It is, of course, impossible to discover incommensurability by applying this method in the way in which craftsmen do it: with measuring sticks or measuring ropes. But if one looks at the pentagram or at a regular pentagon with all its diameters filled in—and we have seen that the Pythagoreans were interested in diameters—the fact that the process of mutual subtraction goes on infinitely, that therefore there is no greatest common measure, and that hence the ratio between diameter and side cannot be expressed in integers however great, is apparent almost at first sight. For one sees at once that the diameters of the pentagon form a new regular

---

66 See note 24.

67 For details see the article quoted in note 5, pp. 1364 ff.

68 See notes 22 and 23.

69 See F. Lindemann in *Sitz.-Berichte Akad. München, math.-phys. Klasse*, vol. 26, pp. 725 ff. Lindemann gives also evidence to show that dodecahedra were used as dice in Italy at a very early time, and that the regular dodecahedron seems to have had some religious importance in Etruria. Especially the latter fact, if known to the Pythagoreans, would naturally have increased their interest in the figure.

70 See Lucian, *De lapsu in salutando*, 5, and schol. Aristoph. *Nubes*, 609.

71 For evidence to show that the Pythagoreans used this method in mathematical theory, see infra p. 258 (p. 225 in this volume).

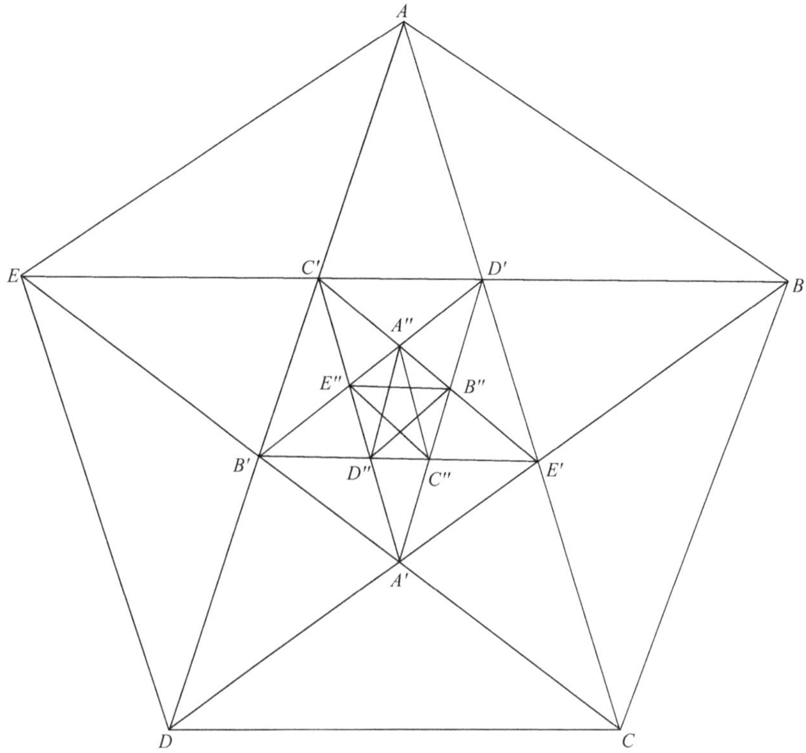

Fig. 2

pentagon in the centre, that the diameters of this smaller pentagon will again form a regular pentagon, and so on in an infinite process.

| It is then also very easy to see that in the pentagons produced in this way $AE = AB'$ 258
and $B'D = B'E'$ and therefore $AD - AE = B'E'$, and likewise $AE = ED' = EA'$ and $B'E' = B'D = B'E$ and therefore $AE - B'E' = B'A'$, and so forth ad infinitum, or, in other words, that the difference between the diameter and the side of the greater pentagon is equal to the diameter of the smaller pentagon, and the difference between the side of the greater pentagon and the diameter of the smaller pentagon is equal to the side of the smaller pentagon, and again the difference between the diameter of the smaller pentagon and its side is equal to the diameter of the next smaller pentagon and so forth in infinitum. Since ever new regular pentagons are produced by the diameters it is then evident that the process of mutual subtraction will go on forever, and that therefore no greatest common measure of the diameter and the side of the regular pentagon can be found.

One may, of course, still ask how the Pythagoreans could prove that $AE = AB'$ and $B'D' = B'E'$, etc. Now Proclus, probably getting his information from Eudemus of

Rhodes, states[72] that Thales was the author of the theorem that in an isosceles triangle the base angles are equal. In connection with this it is important to note that Aristotle[73] refers to an archaic proof of this proposition. He does not quote all the steps of this proof, but what he quotes shows that 'mixed angles,' i.e., angles formed by a straight line and the circumference of a circle, were used in the demonstration, and that in all likelihood the proof was based on a rather primitive method of superimposition.[74] It is clear that with this latter method the converse of the proposition could be proved without difficulty. It follows that the equality of AE with AB' and of $B'D$ with $B'E'$ could be derived from the equality of $\angle AEB'$ with $\angle AB'E$ and of $\angle B'DE'$ with $\angle B'E'D$, if these angles could be proved to be equal respectively.

As to this latter proof, the evidence is somewhat less definite. But Eudemus of Rhodes[75] attributes to the early Pythagoreans the proof that the sum of the internal angles in any triangle is equal to two right angles. From this theoream the general theorem that in any polygon the sum of the internal angles is equal to $2n-4$ right angles can very easily be derived, if one divides the polygon into triangles,[76] and we know[77] that the Pythagoreans constantly experimented with dividing polygons into triangles. The proposition furthermore that in any polygon the sum of the external angles is equal to four right angles is a mere corollary of the preceding proposition.[78] On the
259 basis of these propositions, finally, | the equality of the angles figuring in the demonstration suggested above can be very easily shown.

It follows that there is no reason whatever to disbelieve that Hippasus was able to demonstrate the incommensurability of the side with the diameter of a regular pentagon. For what is needed for the proof suggested is nothing but two fundamental geometrical propositions which concern the isosceles triangle and the sum of the angles in any triangle, and in addition the old time-honored method of finding the greatest common measure by mutual subtraction. All the rest is nothing but the simplest addition, subtraction and division. Of the two geometrical propositions, the first had undoubtedly been 'proved' in a very primitive way even before Pythagoras.[79] The second one was probably also proved in some such way, though we do not know exactly how.[80] But

---

72 *Op. cit.*, p. 250 f. Friedlein.

73 Aristotle, *Analyt. Pr.*, 41 b, 13 ff.

74 For details see Heath, *Elements* (see note 45), 1, 253.

75 Quoted by Proclus, *op. cit.*, 379 Friedlein.

76 The proof is quoted by Proclus, *op. cit.* After the polygon has been divided into triangles, the proposition about the sum of the angles of a triangle being known, the remainder of the proof is a simple addition.

77 See supra, p. 252, note 49 (p. 220 in this volume).

78 Aristotle refers to this proof as to something very well known in *Analyt. Post.* 99a, 19 ff and 85b, 38 ff.

79 It is an interesting fact that all the theorems which ancient tradition attributes to Thales are either directly concerned with problems of symmetry and 'provable' by super-imposition, or of such a kind that the first step of the proof was obviously based on a consideration of symmetry and the second step, which brings the proof to its conclusion, is a simple addition or subtraction. The much discussed Euclidean proof of the first theorem of congruence by superimposition seems, then, the last remnant of a method which once had been widely applied and with which Greek scientific geometry had started.

80 The proof attributed to the Pythagoreans by Eudemus seems to presuppose the famous fifth postulate of Euclid. But Aristotle (*An. Pr.*, 65a, 4) indicates that there existed an old mathematical

there can be no doubt whatever that its truth was known long before Hippasus. That the proofs of these theorems as existing in the middle of the fifth century did not come up to the Euclidean conception of a satisfactory proof is not to the point. For the question is not whether Hippasus could give a demonstration which in all its steps would have satisfied Euclid or Hilbert, but whether he was able to find a proof which at the level which mathematical theory had reached in his time was considered absolutely convincing, and as to this there can be no doubt. It is, perhaps, not unnecessary to point out specifically that the demonstration of incommensurability suggested above does not presuppose any geometrical construction in the strictly mathematical sense at all, as long as the Pythagoreans were able to draw a reasonably accurate regular pentagon in some way, and this can hardly be questioned, for a quite beautiful pentagram can be seen on a vase of Aristonophus which belongs to the seventh century B.C. This vase was found at Caere in Italy and is now in a museum in Rome. Neugebauer's argument, therefore, that the discovery of incommensurability could not have been made | by Hippasus since Oinopides, who belonged to the succeeding 260
generation, was still concerned with the most 'trivial'[81] mathematical constructions, has no validity.

There is, then, perhaps some justification for the claim that the analysis so far has proved what was promised in the introduction to this paper, namely, that the discovery of incommensurability can have been made in the middle of the fifth century, that the development of the Pythagorean doctrine of numbers as the essence of everything naturally led to this discovery, that ancient tradition contains strong hints as to the way in which the discovery actually was made, and last but not least, that Greek mathematics in that early period may have been very elementary,[82] but certainly was not trivial. It was not trivial because the Greeks had two peculiarities which the Egyptians and Babylonians obviously lacked. They were very prone to build up sweeping general theories on very scanty evidence. Of this the Pythagorean theory that 'all things are numbers' is a striking example. Yet at the same time they were not content with having such a theory, but made unremitting efforts to verify it in all directions. It was

demonstration about parallels and angles which involved a vicious circle. It seems, then, quite possible that the equality of alternate angles on parallels cut by a straight line was at first considered self-evident on the basis of considerations of symmetry, that then a faulty attempt to prove the proposition was made, and that finally Euclid tried to give the whole theory a sound foundation by his famous postulate. In this case the proof of the proposition concerning the sum of the angles of a triangle attributed to the early Pythagoreans by Eudemus may really be very old. But Geminus (in Eutocius' commentary on the Conica of Apollonius of Perge, vol. II, 170 of Heiberg's ed. of Apoll.) mentions a still older demonstration in which the proposition was proved first for the equilateral, then for the isosceles, and finally for the scalene triangle.

81 In my article on Oinopides (see note 43) I have tried to show that Oinopides' mathematical constructions were not 'trivial' either, if viewed in connection with the problems which he tried to solve. But the solution of the present problem is quite independent from the acceptance or rejection of this suggestion.

82 In the present article only so much mathematical knowledge has been attributed to the early Greek mathematicians as can be ascribed to them with the greatest approximation to certainty which a historical inquiry can attain. The attempt has then been made to show that *even if* their knowledge did not go beyond this, they nevertheless can have discovered incommensurability and by the nature of their theories and methods were naturally led to this discovery. But this does not imply that their knowledge must necessarily have been as limited and elementary as has been assumed in this paper.

on account of this second peculiarity that they discovered incommensurability in a very early period.

It is perhaps advisable to add a brief survey of the immediate consequences of the discovery of incommensurability for the further development of the theory of proportions. For this will confirm both the opinion concerning the general character of the early scientific investigations of the Greeks and some special suggestions which have been made in the course of the present inquiry.

The discovery of incommensurability must have made an enormous impression in Pythagorean circles because it destroyed with one stroke the belief that everything could be expressed in integers, on which the whole Pythagorean philosophy up to then had been based. This impression is clearly reflected in those legends which say that Hippasus was punished by the gods for having made public his terrible discovery.

But the consequences of the discovery were not confined to the field of philosophical speculation. *Logos* or ratio, as we have seen,[83] meant the expression of the essence of a thing by a set of integers. It had been assumed that the essence of anything could be
261 expressed in that way. Now it had been dis|covered that there were things which had no *logos*. When *we* speak of irrationality or incommensurability we mean merely a special quality of certain magnitudes in their relation to one another, and we speak even of a special class of irrational numbers. But when the Greeks used the term *alogos*, they meant originally, as the term clearly indicates, that there was no *logos* or ratio.

Yet this fact must have been very puzzling. It had been generally assumed that two triangles which were similar, i.e., which had the same ornamental appearance, though differing in size, had the same *logos*, i.e., that they could be expressed by the same set of integers. In fact, this is clearly the original meaning of the term *ho autos logos* (the same *logos*), which we translate by 'proportion.' But two isosceles right-angled triangles had still the same ornamental appearance, and therefore should have had the same *logos*. In fact, it seemed evident that their sides did have the same quantitative relation to one another. Yet they had no *logos*.

The way in which the Greeks amazingly soon after the stunning discovery of incommensurability began to deal with this problem is a much greater proof of their genius for and their tenacity in the pursuit of scientific theory than the discovery of incommensurability itself. For very soon[84] they began not only to extend the theory of proportion to incommensurables, but also established a criterion by which in certain cases it can be determined whether two pairs of incommensurables (which in the old sense have no *logos* at all) have the same *logos*. The terminological difficulty created by this seeming contradiction in terms is reflected by the fact that for some time the term *alogos* for irrational was replaced by the term *arrhetos* (inexpressable) which is merely another way of expressing what the term *alogos* originally meant. It is also interesting to see how the term *alogos* gradually came back. First the term *rhetos* (rational) is created in contrast to *arrhetos*. Then the term *arrhetos* disappears; and Theaetetus, who developed the theory of irrationality further, reintroduced the term

[83] See supra p. 249/50 (p. 217/18 in this volume).

[84] The famous demonstrations of Hippocrates of Chios, who belonged to the same generation as Theodorus of Cyrene, clearly presuppose that the theory of proportions at his time had already been adapted to incommensurables. See F. Rudio, *Der Bericht des Simplicius über die Quadraturen des Antiphon und des Hippocrates* (Leipzig, Teubner, 1907), and infra p. 262 (p. 229 in this volume).

alogos but used it only for 'higher' irrationalities, for instance of the form $\sqrt{a\sqrt{b}}$, while he called the simple irrationalities of the form $\sqrt{a}$ *dynamei monon rhetoi* (literally: rational only in the square). Finally, when *logos* had become a technical term and the incongruity of the statement that two pairs of *alogoi* have the same *logos* was no longer felt, the Greek mathematicians returned to the old terminology and called all irrationals *alogos*.[85] The fact that Theaetetus, who died in 369 B.C., had already begun to return to the old terminology is a very strong confirmation of the view that the discovery of incommensurability must have been made long before, and that the term *logos* for ratio, from which | *alogos* is derived, must certainly have been 262
used by the Pythagoreans before the middle of the fifth century.

The extension of the theory of proportion to incommensurables required an entirely new concept of ratio and proportion and a new criterion to determine whether two pairs of magnitudes which are incommensurable with one another have the same *logos*. The early solution of this problem is most ingenious. Instead of making the result of the process of mutual subtraction the criterion of proportionality, namely the two sets of integers determined by measuring two commensurable magnitudes with the greatest common measure found by mutual subtraction, they used the character of the process of mutual subtraction itself as the criterion of proportionality. They established this criterion by giving a new definition of proportionality which made it applicable to commensurables and incommensurables alike. In literal translation this definition says: *magnitudes have the same logos if they have the same mutual subtraction*.[86] It is interesting to see that in this definition the term *logos* has lost its original meaning. The sense of the definition is, then, that two sets of magnitudes are in proportion if in each case the process of mutual subtraction, even if going on in infinitum, nevertheless can be proved always to go in the same direction.

To show this is especially easy in the case of the diameters and sides of all regular pentagons, since in this case, the diameter being cut in the so-called golden section, it is evident that the process will always go exactly one step in each direction. But if the practical applicability of the new definition had been limited to this case it would have been of little use for the further development of mathematical theory. The most important case in which it is very easy to prove on the basis of the new definition that two pairs of magnitudes are in proportion is the proposition that rectangles and (since parallelograms can very easily be converted into rectangles of the same area) parallelograms of the same altitude are in proportion with their bases.

For it is easy to see that if b can be subtracted 5 times from a, B can also be subtracted 5 times from A, and if the remainder c can be subtracted 8 times from b, so can C from B, and so forth in infinitum.[87] This proposition is the foundation of the famous theorems of Hippocrates of Chios.

---

[85] For details see my article on Theaitetos (see note 5), p. 1361 f.

[86] See Aristotle, *Topica*, 158 b, 32 ff.

[87] In literal translation the passage in Aristotle runs like this: 'It seems that in mathematical theory some propositions cannot easily be proved on account of the lack of a definition (or: as long as the proper definition is lacking), as for instance the fact that a straight line cutting an area parallel to its side cuts the area and its base in the same proportion (literally: similarly). But as soon as the definition has been found (the truth of) the proposition is at once manifest. For the areas

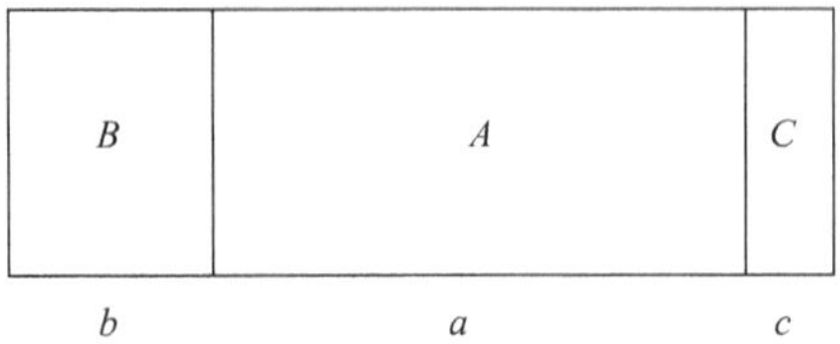

263 | Yet the usefulness of this new definition for the demonstration of geometrical propositions is still restricted to a rather limited field. The further expansion of the theory of proportions was made possible through the new and even more ingenious definition which was invented by Eudoxus of Cnidus and which runs as follows: *Magnitudes are said to be in the same logos, the first to the second and the third to the fourth, when, if any equimultiples whatever be taken of the first and the third and any equimultiples whatever of the second and the fourth, the former equimultiples alike exceed, are alike equal to, or alike fall short of, the latter equimultiples respectively taken in corresponding order.*[88]

If one compares the discovery of incommensurability (assuming that it was made in the manner suggested above) with these extensions of the theory of proportions, it seems evident that the discovery of incommensurability was by far the easiest step. For once the Pythagoreans became interested in the pentagram and the regular pentagon, anyone might be struck by the fact that the diameters will always form a new regular pentagon in the centre; and if, furthermore, the general Pythagorean doctrine required the determination of the 'logos' of diameters and sides, all the rest followed very easily. Of the two new definitions of proportion, that of Eudoxus is perhaps the most ingenious inasmuch as it required the greatest effort in abstraction. But the older

---

and their bases have the same mutual subtraction; and this is the definition of proportion (*ho autos logos*)! It seems strange that O. Becker in an article published in 1933 (Quellen und Studien zur Geschichte der Mathematik, Abteilung B, vol. 2, pp. 311 ff.) was the first to give the correct interpretation of the expression 'have the same mutual subtraction' in the passage quoted, while Heath, for instance, still called the definition 'metaphysical,' and said that it was difficult to see how any mathematical facts could be derived from the definition.

O. Becker in a most excellent analysis has also proved that the greater part of the 10th book of Euclid's Elements which contains a very elaborate theory of irrationals can be proved by means of this definition, while some of the propositions specifically ascribed to Eudoxus cannot be proved on the basis of this definition and presuppose the new definition Euclid V, def. 5. Since the most important propositions of the 10th book of Euclid are ascribed to Theaetetus, Becker drew the obvious conclusion that Theaetetus worked with the old definition quoted by Aristotle.

This is undoubtedly correct. But his interpretation of the rest of the passage in Aristotle seems to require a slight modification. Though Becker has seen that the 'areas' in Aristotle are in fact parallellograms, or rather, rectangles, he believes that the proposition about rectangles was from the beginning proved by an elaborate process of reasoning, which required that several other propositions had been proved first (*op. cit.*, p. 322). This is certainly not what Aristotle indicates, when he says that the truth of the proposition is manifest as soon as the definition is found. For this expression shows clearly that originally a direct application of the definition to the figure given above was considered sufficient proof of the proposition. This is an interesting parallel to the first demonstration of incommensurability in the pentagon as suggested above.

88 See Euclid, *Elements*, V, def. 5 and *Scholia in Euclid. Element.* V. 3 (Euclidis Opera. ed. I. L. Heiberg, vol. V, Leipzig, Teubner, 1889, p. 282.)

definition of proportion, by which the original concept of *logos* was replaced by a new one, which made it possible to apply the theory of proportion to incommensurables, was certainly by far the most important step in the development.

| The fact that the development from the discovery of incommensurability to Eudoxus took this course has also chronological implications. Eudoxus was born in 400 and died in 347 B.C.[89] His last work, which he left uncompleted, was a large geographical work in many volumes.[90] He was also the author of the method of exhaustion, of the theorem that the volume of a cone is one-third of the volume of a cylinder with the same base and altitude,[91] and undoubtedly of other stereometric theorems which must have been used in the proof of that proposition. All this would have been impossible without the new definition of proportion invented by Eudoxus. He therefore must have created this definition comparatively early in his life, hardly later than 370. It would, then, be little less than micraculous if the first discovery of incommensurability had been made 'in the time of Archytas' who, since he was head of the government of Tarentum in 362, can hardly have been born before 430. It is certainly much easier to believe that the discovery was made in the middle of the fifth century, as ancient tradition claims. 264

But the solution of the chronological problem is of importance mainly because it makes it possible to acquire a deeper insight into the way in which the Greeks laid the foundations of the science of mathematics and into the special qualities which enabled them within an amazingly short time to make a discovery which their Babylonian and Egyptian predecessors with all their highly developed and complicated methods had not made in many centuries of mathematical studies.

89 See K. von Fritz, 'Die Lebenszeit des Eudoxos von Knidos' in Philologus, 85 (1930). p. 478 ff.

90 See F. Gisinger, *Die Erdbeschreibung des Eudoxos von Knidos*, p. 5 ff.

91 See Archimedes, *Ep. ad Dositheum in De sphaera et Cylindro*, p. 4 Heiberg and *Ad Eratosth. Methodus*, p. 430 Heiberg.

H. FREUDENTHAL*

Y AVAIT-IL UNE CRISE DES FONDEMENTS DES MATHÉMATIQUES DANS L'ANTIQUITÉ?

1. Souvent dans l'histoire des mathématiques le problème de ses fondements s'est posé de nouveau et sous de nouveaux aspects. Cela ne veut pas dire que, à une époque donnée, il ait inquiété tous les mathématiciens. Les méthodes du calcul différentiel et intégral tel qu'il fut inventé par Newton et Leibniz ont été appliquées tranquillement quoiqu'on sût qu'il n'était pas bien fondé et malgré les paradoxes que ces méthodes impliquaient. Les paradoxes de l'infini connus depuis longtemps n'étaient jamais considérés comme des menaces sérieuses, mais plutôt comme des plaisanteries à la périphérie des mathématiques. La découverte des géométries non-euclidiennes, négligée d'abord, et thème de discussions ferventes plus tard, n'a jamais créé le climat d'une crise des fondements de la géométrie. En géométrie, on sent cette mentalité de crise plutôt après la découverte de l'insuffisance des fondements de la géométrie projective de von Staudt. En effet, ce problème de la continuité dans la géométrie a inquiété les géomètres jusqu'à la fin du 19e siècle. Les paradoxes de la théorie des ensembles auraient dû être sentis comme une menace, mais au moment où ils étaient formulés, la théorie des ensembles n'avait pas vraiment pénétré les mathématiques. Je ne sais pas qui a le premier parlé d'une crise des fondements des mathématiques, mais je suis sûr que ce terme a été inventé plus tard, à l'époque où l'on commençait à s'occuper sérieusement des fondements. Et je ne sais pas non plus qui a découvert une telle crise des fondements dans les mathématiques de l'Antiquité. Le petit livre
célèbre de Hasse et Scholz de 1928 est un | *terminus ante quem* pour l'usage de ce 44
terme, mais l'idée même est plus vieille, elle peut être retracée jusqu'à Tannery. A présent, la crise des fondements des mathématiques est considérée comme un fait historique établi sans doute possible, quoique les uns l'associent avec la découverte des grandeurs incommensurables et les autres avec la critique éléatique.

2. Le point où la découverte de l'incommensurabilité serait intervenue est la théorie de la proportionnalité et de la similitude des figures. On peut s'imaginer qu'une théorie exacte telle que celle que nous trouvons dans les Éléments d'Euclide, a été précédée par une théorie naïve où toutes les grandeurs comparables sont supposées commensurables. Que les aires de rectangles de même hauteur soient proportionnelles aux bases, que sur deux droites données un système de droites parallèles coupe des segments proportionnels, cela se déduit aisément des propriétés de congruence si les rapports donnés sont rationnels. En effet, si par exemple ce rapport est de 3 à 2, on divise l'une des bases (l'un des segments) en trois parties et l'autre en deux parties, et l'on réduit la comparaison des

* Conférence faite le 18 décembre 1965 pour la Société belge de Logique et de Philosophie des Sciences.

J. Christianidis (ed.), Classics in the History of Greek Mathematics, 233–241.

grandeurs à l'affaire de compter des aires et des segments congruents. Il est assez naturel de croire que chaque paire de grandeurs comparables possède une mesure commune. Si c'était vrai, cette démonstration serait correcte, et on peut s'imaginer qu'il y avait un temps où l'on pensait qu'elle était générale. A un certain moment on aurait découvert que le côté et la diagonale du carré n'ont pas de mesure commune. Cet exemple de grandeurs incommensurables invalide la théorie naïve des proportions et de la similitude. Tannery conclut que « la découverte de l'incommensurabilité... dut donc causer, en géométrie, un véritable scandale ». Ce qui est formulé comme une hypothèse par Tannery est considéré alors comme un fait établi pour lequel souvent l'autorité de Tannery est invoquée. On croit généralement que ce scandale logique a produit un choc, ou, comme disent Hasse et Scholz, une crise des fondements des mathématiques.

En effet, les arguments internes, mathématiques, en faveur de l'existence d'une telle crise sont accablants. En revanche on ne connaît guère d'arguments externes, historiques proprement dits. Le seul dont on s'est parfois servi est le récit antique suivant
45 lequel celui qui le premier a rendu public l'irrationnel aurait péri | dans un naufrage, parce que l'imprononçable et l'inimaginable auraient dû rester cachés—histoire qui s'attache au mot archaïque ἄρρητον (imprononçable) pour l'irrationnel. Pour cette fable il y a des parallèles où d'autres découvertes de l'école ésotérique des pythagoriciens sont rendues publiques et où celui qui a rompu le secret est puni par le destin. Ce serait le seul — et peu sérieux — témoignage externe du sentiment d'une crise, et on ne perd rien si l'on l'écarte.

D'autre part on sait très peu sur les mathématiques grecques d'avant la codification euclidienne. Il y en a quelque fragments, mais si je ne me trompe pas, aucun n'est considéré comme littéral. La source la plus féconde est toujours l'œuvre d'Euclide même — une compilation, dont on a isolé quelques parties et constaté des attributions. Les 5e, 6e et 12e livres des *Éléments* sont attribués à Eudoxe; aux pythagoriciens on attribue les 7e jusqu'au 9e, dont le dernier contient des parties archaïques tandis que le premier doit provenir du cercle d'Archytas; le 10e, qui est le plus volumineux, doit être l'œuvre de Théétète, ainsi que le 13e qui traite des corps réguliers.

3. Ce qui nous intéresse, c'est la théorie des grandeurs incommensurables. Telle qu'elle se trouve dans les *Éléments*, elle doit être l'œuvre d'Eudoxe qui vivait dans la première moitié du 4e siècle, un demi-siècle avant Euclide. Le point essentiel de cette théorie est une définition de l'égalité de deux rapports. On dit que

$$a : b \; = \; a' : b'$$

si les relations

$$ma \gtreqless nb \quad \text{et} \quad ma' \gtreqless nb'$$

impliquent l'une l'autre pour tous les nombres naturels m, n. Cette définition, où un rapport quelconque est localisé par rapport à la totalité des rapports rationnels, présage la localisation des nombres réels par rapport aux nombres rationnels au moyen des coupures de Dedekind.

Des indications chez Aristote (inconnues de Tannery)[1] montrent que cette théorie était précédée par une autre, celle de l'antanérèse, probablement l'œuvre de Théétète,

---

1 *Top.* VIII 3.158^{b}29-159^{a}1.

dont on trouve toujours des traces dans les *Éléments*. ἀνταναίρεσις veut dire | sous- 46
traction mutuelle et, en termes plus modernes, développement en fractions continues. Deux grandeurs a_0 et a_1 étant données, on ôte de la plus grande des multiples aussi grands que possible de la plus petite et l'on continue de cette manière indéfiniment,

$$\begin{array}{ll} a_0 = m_0a_1 + a_2, & a_2 < a_1, \\ a_1 = m_1a_2 + a_3, & a_3 < a_2, \\ \cdot & \cdot \\ \cdot & \cdot \\ \cdot & \cdot \end{array}$$

Deux rapports tels que $a_0 : a_1$ sont alors appelés égaux s'ils ont le même développement, la même antanérèse. Comparée à la méthode plutôt statique d'Eudoxe, celle de l'antanérèse est cinématique. Elle suggère, plus ou moins explicitement, des processus infinis accomplis dans un temps fini, mais bannis plus tard des mathématiques antiques.

4. Pour soutenir la thèse que l'auteur des 10e et 13e livres des *Éléments* était aussi l'inventeur, ou au moins un partisan de l'antanérèse, il est permis d'ajouter aux arguments connus deux autres que je crois nouveaux. Le 6e livre, autrement de structure homogène, se termine par un potpourri de trois propositions qui ne sont guère liées au reste du livre. Ce sont VI 31, une généralisation du théorème de Pythagore attribuée par Proclus à Euclide lui-même (ce qui ne veut pas dire qu'Euclide l'a inventée, mais plutôt que Proclus ne le trouvait pas dans les rédactions précédentes des *Éléments*), puis VI 32 qui est plutôt un lemme peu important dont la seule application se trouve dans XIII 17, et enfin VI 33 qui dit que dans des cercles égaux le rapport des angles égale celui des arcs correspondants. La démonstration de cette proposition consiste à répéter machinalement le cliché des raisonnements eudoxiens dans un cas où ils ne s'appliquent point; en effet, cela a peu de sens de comparer des angles et des arcs multipliés par des entiers arbitraires. Cette inconsistance a été bien remarquée et elle a donné lieu à l'hypothèse que VI 33 est une interpolation posteuclidienne. Cependant VI 33 se démontre aisément au moyen de l'antanérèse, où au lieu de multiplier on divise. Ce fait suggère la conclusion que la méthode antérieure du livre sixième était l'antanérèse. En remplaçant cette méthode par celle d'Eudoxe, Euclide (ou un de ses prédécesseurs) aurait dû réviser la démonstration de VI 36 qui ne se trouvait pas dans le traité d'Eudoxe, et dans cette affaire il n'aurait pas réussi de manière satisfaisante. Remarquons de plus que dans la récension de Théon, VI 36 est amplifié par l'assertion que les secteurs aussi se rapportent comme les angles. On dirait que ce n'est pas une
invention de Théon, mais qu'il l'a trouvé dans une des | rédactions préeuclidiennes. Vu l'im- 47
portance historique de cette addition pour la quadrature du cercle, on peut bien imaginer qu'elle est d'origine préeuclidienne. En général quelques autres passages dans les *Éléments* éliminés par la critique moderne comme des interpolations peuvent être plus anciens, omis par Euclide et rétablis plus tard.

Si VI 36 suggère l'antanérèse comme principe antérieur des 5e et 6e livres, c'est VI 35 qui n'est utilisé qu'en XIII, et qui par conséquent indique l'auteur de XIII comme celui qui a formulé V et VI au moyen de l'antanérèse. D'ailleurs, démontrer des lemmes dans un contexte systématique et bien avant de les appliquer, est un trait stylistique qui caractérise l'auteur de X et XIII.

L'autre argument pour accepter l'antanérèse comme base antérieure du 10e livre est X 9 qui par sa défectuosité logique contraste avec le reste de X. L'arrangeur a apparemment confondu la théorie eudoxienne des rapports avec la théorie arithmétique des livres VII-IX, tandis que dans l'original la théorie arithmétique doit être apparue comme cas spécial de l'antanérèse au moyen de X 2-3 où la commensurabilité est caractérisée par la finitude de l'antanérèse.

5. Évidemment ces théories de la proportionnalité et de la similitude étaient motivées par la découverte des grandeurs incommensurables. On ne sait pas quand cette découverte a été faite ou au moins quand elle a été publiée. Le début du 4e siècle serait un terme *ante quem*, et pour justifier l'hypothèse du choc donnant naissance à la théorie des incommensurables, on ne voudrait pas reporter ce moment trop en arrière.

D'autre part la tradition veut que la découverte de l'incommensurabilité fût l'œuvre des pythagoriciens ou de Pythagore lui-même. Pythagore mourut à la fin du 6e ou au début du 5e siècle, ce qui nous donnerait un intervalle de presque un siècle entre la découverte de l'incommensurabilité et la première théorie des grandeurs incommensurables. Mais n'oublions pas que, pour causer une crise, la découverte de l'incommensurabilité devait rencontrer une autre théorie qui devait sembler la contredire, une théorie de la proportionnalité et de la similitude des figures géométriques, une théorie assez systématique et non seulement un usage intuitif de similitudes.

Le seul document du 5e siècle qui contient des renseignements sur ce sujet est la célèbre quadrature des lunules d'Hippocrate. Plusieurs fois Hippocrate compare des segments de cercles semblables dont il sait que le rapport de leurs aires est le même que celui des carrés de leurs cordes. Mais il y a plus : d'après ce fragment de l'histoire
48 d'Eudème de Rhodos que nous devons à | Simplikios, Hippocrate aurait même *démontré* que des segments semblables ont le même rapport que les carrés de leurs cordes, et le fragment contient une esquisse, incomplète, de la démonstration. Il aurait d'abord démontré que des cercles ont le rapport des carrés de leurs diamètres et puis il aurait passé des cercles à leurs parties.

Hippocrate doit avoir vécu à peu près un demi-siècle avant Théétète et il doit avoir été le premier *στοιχειωτής*, le premier auteur d'*Éléments*, c'est-à-dire d'un système des mathématiques, qui aurait été suivi par une série d'autres dont le dernier fut celui d'Euclide. Presque tout ce que nous savons sur son œuvre se trouve dans ce fragment de l'histoire d'Eudème. D'après ce fragment, Hippocrate doit avoir eu des connaissances remarquables de la géométrie élémentaire, disons du contenu des quatre premiers livres des *Éléments* d'Euclide. Ce qui frappe le plus, c'est sa finesse, la technique de ses démonstrations et son niveau d'exactitude qui est fort élevé — par exemple il prend beaucoup de soin pour démontrer qu'un certain arc de cercle est plus grand qu'un demi-cercle quoique ce soit clair d'après la figure.

Ce fragment est évidemment ruineux pour l'hypothèse du scandale, du choc ou de la crise causés par la découverte de l'incommensurabilité. Qu'est-ce qui reste de cette crise évoquée par la découverte de l'incommensurabilité et vaincue par Eudoxe, si plus d'un demi-siècle avant Eudoxe on savait démontrer la proportionnalité des cercles et des carrés de leurs diamètres, qui est nécessairement fondée sur une théorie des rapports telle que celle d'Eudoxe? Plusieurs auteurs ont échappé à ce dilemme en déclarant nul et non avenu le témoignage d'Eudème. C'est bien étrange. Les fragments qu'on possède de l'histoire d'Eudème sont considérés avec le plus profond respect, on s'en sert comme de documents et d'arguments dignes de confiance, et le présent témoignage sur Hippocrate est le seul dont on rejette avec une unanimité sans précédent une partie, pour la seule raison qu'il contredit une hypothèse historique qui n'est soutenue que par des raisonnements internes, mais par aucun témoignage

historique. On nie la possibilité d'une démonstration exacte du théorème sur le rapport des aires des cercles précédant la théorie de la similitude d'Eudoxe de deux ou
trois générations. D'abord, on devrait remarquer que la théorie des proportions et | de 49
la similitude d'Eudoxe ne fut pas la première. Notons en passant qu'elle fut au moins précédée par celle de l'antanérèse, que Tannery n'avait pas encore remarquée et qui réduirait la lacune à deux générations au plus.

6. A mon avis l'histoire était beaucoup plus nuancée. Si l'on examine de plus près les remarques dans la littérature pré-euclidienne ou ayant leurs sources dans la littérature pré-euclidienne qui se rapportent à l'incommensurabilité, on rencontre le fait de l'existence de l'incommensurable non comme quelque chose de négatif ou de terrifiant, mais sous une forme plutôt positive. C'est toujours l'assertion qu'il y a des grandeurs commensurables *μήκει* ou *δυνάμει μόνον*, commensurables en longueur ou commensurables en puissance seulement. Je dois ajouter que c'est M. Szabó qui a le premier dûment mis l'accent sur ce fait historique absolument clair. Dans la littérature classique *δυνάμει* veut dire «en carré». Dans la littérature archaïque on le traduirait plutôt par «potentiellement»[1]. La diagonale et le côté du carré ne sont pas commensurables, mais ils le sont potentiellement, c'est-à-dire après avoir formé leurs carrés. Aussi longtemps que la notion de *δυνάμει* restait assez vague, on pouvait bien supposer que deux grandeurs quelconques étaient ou commensurables ou potentiellement commensurables. C'était Théétète qui avait raffiné cette classification dans le 10e livre des *Éléments* où il introduit une diversité de rapports autres que les commensurables et les commensurables en carré. Mais cette vieille attitude de substituer la commensurabilité potentielle à l'incommensurable se fait toujours sentir dans une dissertation platonicienne (probablement de Xénocrate) qui est conservée ensemble avec sa réfutation péripatéticienne «Sur les lignes indivisibles». Quoi qu'il en soit, aussi longtemps qu'on ne considérait que des rapports très spécialisés, on pouvait se passer d'une théorie générale des proportions et de la similitude, et je crois bien que cela caractérise la situation où Hippocrate se trouvait.

En effet, les rapports irrationnels qu'on rencontre dans les quadratures de lunules
d'Hippocrate sont par exemple $\sqrt{2}$, | $\sqrt{\frac{3}{2}}$, $\sqrt{6}$, c'est-à-dire des grandeurs commensu 50
rables en carré. Si les diamètres de deux cercles sont en raison $\sqrt{p}:\sqrt{q}$, où p et q sont des entiers, l'égalité des rapports de leurs aires et des rapports des carrés de leurs diamètres s'énonce avec les nombres entiers p et q, et aucune irrationnalité ou incommensurabilité n'entre dans le problème et sa solution. Des termes dont Hippocrate se sert, on peut conclure avec certitude qu'il ne comparait que des aires de surfaces dont le rapport était celui de nombres entiers. Dans ce cas, si l'on veut comparer les aires de polygones semblables, toutes les démonstrations se réduisent à des démonstrations d'égalité par rapport à l'aire de surfaces congruentes. Pour aborder les cercles on y inscrit ou circonscrit des polygones réguliers. La démonstration que les aires de ces

[1] Voir ARIST., *Metaph*. Δ12, 1019b33-34. Dans PLAT., *Théétète*, *δύναμις* veut dire racine carrée. Chez EUCL., *Élém.* il ne se trouve que dans X.

polygones ont même rapport que les aires des carrés de leurs diamètres est tout à fait élémentaire pour autant que ces rapports soient ceux de nombres entiers, mais c'est le seul cas dont Hippocrate devait s'être occupé. Reste le passage aux limites nécessaire pour passer des polygones aux cercles. Évidemment cela ne peut pas se faire d'une manière élémentaire. Mais dans le cas qui nous intéresse, celui de la commensurabilité en carré, c'est plus élémentaire que dans le cas général et cela ne suppose pas de théorie génerale des incommensurables. Si C, C′ sont les aires des deux cercles, Q, Q′ celles des carrés de leurs diamètres, P, P′ celles de polygones inscrits respectifs semblables, alors on sait que Q : Q′ $=p : p'$ où p, p' sont des nombres naturels, et par conéquent que P : P′ $=p : p'$. Cette proportion se traduit par une égalité p'P$=p$P′ où le multiple d'une aire est entendu comme l'aire d'un multiple concret de figures polygonales. Pour arriver à l'égalité p'C$=p$C′ qu'on veut démontrer, on doit seulement savoir que l'aire des polygones s'approche de celle des cercles, et cela pouvait se faire aisément par l'inclusion du cercle entre des polygones inscrits et circonscrits, même à l'époque d'Hippocrate. Évidemment ce passage à la limite présuppose le caractère archimédien du systéme de grandeurs, c'est-à-dire quelque analogue (formulé explicitement ou non) d'Euclide X 1. Mais ce qui est le plus essentiel : nulle part Hippocrate n'aurait besoin d'une théorie raffinée de la proportionnalité vu que tous
51 les rapports qu'il envisage sont rationnels et que | dans ce cas la similitude de figures polygonales est réduite à la congruence.

Quittons Hippocrate pour dire quelques mots sur Démocrite à qui on attribue la découverte du volume du cône (et probablement aussi de la pyramide). Dans le cas des volumes, les méthodes élémentaires conduisent moins loin que dans celui des surfaces planes. C'est déjà pour démontrer que deux pyramides à bases congruentes et de même hauteur sont égales qu'on doit faire appel à des processus infinis. Mais alors ce sont les mêmes processus que ceux dont Hippocrate devait se servir dans le cas des cercles. Si Démocrite se bornait à comparer des pyramides ou des cônes dont les bases étaient en rapport rationnel, il pouvait égaler Hippocrate pour ce qui concerne la précision de ses démonstrations.

7. Je viens d'affirmer que la proportionnalité de polygones semblables avec les carrés de leurs dimensions linéaires est élémentaire et ne nécessite aucune théorie élaborée de la proportionnalité pour autant que ce rapport soit rationnel. C'est en effet aisé, mais non-trivial, et en tout cas il faut se demander si cette démonstration était à la portée d'Hippocrate.

Il suffit de démontrer que, m étant rationnel, si l'on multiplie les côtés d'un rectangle par $\sqrt{m}$, l'aire sera multipliée par m. Pour fixer les idées, supposons $m=2$. En interprétant $\sqrt{2}a$ comme la diagonale du carré à côté a on démontre de façon élémentaire que $\sqrt{2}(a \pm b) = \sqrt{2}a \pm \sqrt{2}b$. En élevant cela au carré, on démontre alors par des méthodes de l'algèbre géométrique que $\sqrt{2}a \cdot \sqrt{2}b = 2ab$, ce qu'il fallait démontrer.

Remarquons que cette méthode est générale. Pour arriver à un corps non-archimédien il faut faire des adjonctions transcendantes au corps des rationnels. Aussi longtemps qu'on se borne à des adjonctions algébriques, on peut étendre le caractère archimédien du corps des rationnels aux corps élargis. Cela veut dire que la théorie des irrationnalités de Théétète (10e livre) ne dépend pas d'une théorie raffinée de la proportionnalité. Dans un certain sens, c'est implicite à la méthode de Théétète. On peut l'expliciter si, partout où Théétète traduit géométriquement l'extraction de la racine carrée d'une grandeur a par la construction de la moyenne proportionnelle de a et e, on se borne à une grandeur e fixe.

Il me semblait bon d'exposer pourquoi Théétète dans le 10e livre pouvait réussir avec des méthodes purement algébriques en principe. En fait, il pêchait contre ce principe et ce serait un anachronisme d'attribuer à Théétète le raisonnement moderne qui conduit à ce résultat. Quand même, plus j'étudie le 10e livre, plus mon respect augmente et plus j'ose attribuer d'intelligence plus profonde à son auteur.

8. Pour tirer une conclusion des analyses précédentes, je dirais que le problème d'une
théorie générale de la proportion|nalité et de la similitude ne se posait pas avant qu'on 52
se soit occupé de rapports surpassant en généralité ceux qui étaient au moins rationnels en carré. La première occasion où cela se faisait explicitement fut la classification des incommensurables par Théétète. Ce ne peut être par hasard que Théétète fut probablement aussi l'auteur de la première théorie générale de cette problématique dont une trace s'est conservée, la théorie de l'antanérèse. En tout cas, je ne vois pas de rupture qui pouvait justifier l'hypothèse d'une crise des fondements à cet égard.

9. Au début j'ai mentionné la critique éléatique comme l'autre cause de cette crise hypothétique des fondements des mathématiques. Ce problème-ci est encore plus difficile que celui de l'histoire de l'incommensurabilité. Zénon précédait Hippocrate d'une génération, absolument rien n'est connu sur l'état des mathématiques à son époque et il est bien hardi de supposer qu'à cette époque il y avait quelque chose comme des fondements des mathématiques. Mais, pire que cela, nous ne savons presque rien sur ce que Zénon enseignait. Toute la tradition à laquelle on peut attribuer quelque authenticité est contenue dans quelques phrases, disons dans 50 mots. Ce sont des réfutations de la pluralité de l'être formulées dans un style archaïque, très concis et exprimant une logique admirable. En général ce qui est plus populaire de Zénon, ce sont les quatre paradoxes contre le mouvement qui malheureusement ne sont connus que sous la forme dans laquelle Aristote les rapporte pour les réfuter. C'est un style tout à fait différent qui nous fait douter de la véridicité du rapport aristotélien. Dans l'interprétation de Tannery, l'intention de Zénon aurait été de réfuter l'atomisme, et cette interprétation est acceptée par presque tout le monde. Même la critique bien fondée de van der Waerden n'a pas pu changer cet état de choses.

En réalité, il n'y a aucune trace d'un atomisme d'avant Zénon; au contraire, il serait plus naturel de supposer que l'atomisme de la génération qui le suit dépend de sa critique—c'est un fait remarquable que la logique des pièces authentiques de Zénon peut conduire à un atomisme, et que l'atomisme d'espèce platonicienne représenté par Xénocrate est basé sur Zénon *expressis verbis*. Jamais dans l'antiquité on n'a interprété Zénon au sens que presque tous les modernes lui substituent. Par exemple, | là 53
où (d'après Porphyrios) Zénon dit que l'être, s'il était divisible, serait indéfiniment divisible, on omet la condition de divisibilité (qui serait à réfuter) et on dit que Zénon soutenait la divisibilité indéfinie. C'est la méthode usuelle de nos interprètes de Zénon : d'énoncer comme opinion de Zénon quelque résultat intermédiaire dans une démonstration *e contrario*. Ou on offre des interprétations franchement fantastiques comme chez Hasse et Scholtz : où Zénon dit « si l'être était une pluralité », on lit « s'il était permis de considérer la droite comme l'agrégat d'une infinité de segments infinitésimaux ».

10. Vu le très peu que nous savons sur Zénon, et l'absolument rien que les Anciens racontent sur des rapports de ses idées avec les mathématiques, tout ce qu'on pourrait dire sur l'influence qu'il aurait exercée dans l'histoire des mathématiques ou comme

catalyseur d'une crise de ses fondements est hypothèse pure. Il est bien possible que la méthode dialectique, dont l'invention est attribuée à Zénon, ait contribué à la formation du style mathématique grec, que Zénon soit l'inventeur du raisonnement par récurrence et que ses arguments atomistiques aient donné naissance à la méthode des indivisibles, enfant illégitime de la mathématique grecque, dont Archimède a rendu manifestes les vertus. Il est même sûr que le célèbre paradoxe d'Achille concourant avec la tortue a joué un rôle important dans l'analyse moderne du continu et dans la redécouverte de l'axiome dit d'Archimède. Mais rien n'indique que ce fut histoire répétée. Autant que nous le savons, les Anciens n'ont jamais entendu ces paradoxes au sens que nous sommes tentés d'y attacher. C'étaient des paradoxes contre le mouvement, et comme tels Aristote les analyse en s'abstenant d'arguments mathématiques qui autrement sont nombreux dans son œuvre et qu'il aurait rapportés s'il les avait connus dans un contexte mathématique. Ce qui, d'après le témoignage d'Aristote et d'autres, a le plus impressionné les Anciens dans ces paradoxes, c'était l'argument qu'un mouvement ne peut passer par une infinité d'étapes dans un temps fini — même le mouvement de la pensée. Il est bien possible que cet argument (et non ses conséquences paradoxales) a influencé le développement des mathématiques. Pour expliquer ce qu'est une limite, nous ne dédaignons pas l'image d'un processus infini
54 accompli dans une succession temporelle. C'est une | image très naturelle et qui doit avoir été essentielle dans la théorie de l'antanérèse, mais qui a disparu dans la théorie plutôt statique d'Eudoxe. Quoiqu'Aristote[1] nous ait expliqué la nécessité de technique de démonstration qui doit avoir conduit à la suppression de l'antanérèse en faveur de la méthode d'Eudoxe, il est bien possible que cette élimination des passages cinématiques à la limite et du sentiment cinématique de la continuité étaient en même temps une conséquence tardive de la critique de Zénon, peut-être reprise par l'école platonicienne. Toutefois, c'est une hypothèse, et affirmer que cette réforme a été précédée par une crise, c'est ajouter une hypothèse de plus.

11. La raison historique des théories raffinées de la proportionnalité était la découverte de l'incommensurabilité, et on ne voit pas où dans cet enchaînement des idées la critique éléatique a pu intervenir. C'est étonnant que les Grecs, loin d'accepter tacitement l'énoncé de l'axiome d'Archimède, l'aient formulé explicitement. Mais le langage dans lequel Archimède lui-même le formule montre que cet axiome doit son origine plutôt à la coexistence d'une mathématique des indivisibles (qu'il exclut) qu'à la critique éléatique.

12. La méthode d'Eudoxe est un remplaçant de notre système de nombres réels. Elle ressemble beaucoup à l'introduction de nombres réels par des coupures de Dedekind. Cependant les Grecs n'ont pas introduit de nombres réels. On parle de rapports de grandeurs géométriques qui peuvent être commensurables ou non, mais ce rapport n'est pas un nombre, elle est une paire de grandeurs ou plutôt une classe de paires à rapport égal. On pourrait dire que ce n'est qu'une différence de terminologie. Oui, c'est vrai, mais cette terminologie compliquée, ensemble avec d'autres complications superflues, a fini par étouffer les mathématiques grecques. Pour se développer de nouveau, la mathématique devait être libérée du corset que les Grecs lui avaient mis, et cela arriva quand leurs scrupules furent oubliés.

---

[1] *Anal. post.* I 5. 74^{a}19.

Mais on peut se demander pourquoi les Grecs n'ont pas introduit les nombres réels. Cette question est mal posée, et tout ce qu'on peut y répondre, c'est qu'ils n'ont pas introduit les | nombres rationnels non plus. Au moins dans les mathématiques. 55
Dans les comptes économiques, dans les calculs astronomiques et dans ceux d'Archimède où il détermine π, ils connaissaient bien les fractions, mais celles-ci étaient bannies des vraies mathématiques. Dans la théorie de la musique, qui fait partie des mathématiques, il y a des expressions qui rappellent une époque où les fractions n'étaient pas encore interdites. Par exemple, l'expression ἐπίτριτον pour le rapport 4 : 3 signifie verbalement « plus le troisième », c'est-à-dire quatre troisièmes.

Pourquoi les Grecs n'ont-ils pas admis les nombres rationnels en mathématiques? C'est en effet une question importante. Qu'est-ce qui a conduit à cette attitude étrange? Évidemment les nombres naturels doivent être le fondement de toute théorie des nombres. On comprend bien les raisons pour lesquelles les nombres négatifs n'ont été introduits que récemment. Mais les fractions étaient bien connues des Babyloniens, des Égyptiens, et des Grecs eux-mêmes. Pourquoi les a-t-on bannies des mathématiques?

C'est une attitude qui sent le dogmatisme, et il serait difficile de découvrir un autre motif. Kronecker a dit que les nombres naturels ont été créés par Dieu et que tout le reste est une œuvre humaine. Pour les Grecs, le reste était plutôt une invention des calculateurs, qui souillerait la science divine des mathématiques. Quelque part Platon blâme ceux qui brisent l'unité en fractions. Cette vénération de l'unité crue divine peut avoir deux sources historiques, éléatique ou pythagoricienne. L'être est un pour les Éléates, et l'unité est identifiée à Dieu par les pythagoriciens. Tout en reconnaissant la suprématie des nombres naturels, on aurait pu introduire les nombres rationnels par une définition axiomatique, comme Théétète et Eudoxe ont défini ce qui est égalité de deux rapports. On ne l'a pas fait, et ce refus fut funeste pour les mathématiques antiques. Y a-t-il eu une crise qui a conduit à cette étrange structure des mathématiques grecques? Nous n'en savons rien. Mais il est plus probable que c'étaient des dogmes philosophiques, d'origine pythagoricienne, éléatique ou platonicienne, qui l'avaient créée et que c'est le traditionalisme de la Grèce intellectuelle qui la conservait.

WILBUR R. KNORR

THE IMPACT OF MODERN MATHEMATICS ON ANCIENT MATHEMATICS

Edith Prentice Mendez found this lecture among Wilbur Knorr's papers after his death in March, 1997. Although Knorr probably never intended to publish it – and he surely would have attended to its occasional roughness – Ken Saito and I consider it an important methodological reflection on his just completed work on the early proportion theory,[1] but with much general interest. The three main examples he | discusses, 122
the theory of irrationals, the alleged foundations crisis in the fifth century and the problem of constructibility, remain important morality tales for contemporary researchers. Among specialists, the pendulum may have swung largely in the other direction, and for that reason, it is useful to quote a letter which warns against the opposed impediment to historical understanding. I thank Joseph Dauben for drawing it to my attention by sending me his transcription of it.

Wilbur Knorr to Joseph W. Dauben
Department of History and Philosophy of Science
Whipple Museum. University of Cambridge
March 27, 1975.

"[...] Now, research in the ancient materials is something of an art, and I know that many scholars are by temperament unsuited for it, as they themselves would agree. Basically, the Greek record is fragmentary; we possess a few mathematical treatises virtually complete, others in part, others in random snippets preserved by accident in derivative works, plus a small para-mathematical literature, the logical writings of Plato and Aristotle, for instance. In this circumstance, literalism would be disastrous. For instance, most of the complete treatises which have survived expound a highly formal type of advanced geometry. Does this mean the Greeks were weak in the traditional areas of practical geometry and arithmetic? It goes against reason to believe so. But some scholars ... would have us draw such a conclusion. Rather, at every step one must make allowance for the selective survival of documents. The formal geometry survived because it was also philosophically interesting (from the axiomatic viewpoint) and because it merited study by serious practitioners of geometry. But easily duplicatable computations were hardly worth preserving via manuscript traditions. What mathematician has ever preserved his rough figures, once the final draft of his study has been completed? Occasionally, papyri containing everyday arithmetic and geometric problems survive. These are invariably schoolboys' exercises, amazing for the modesty of their mathematical content. Interestingly, computation throughout

[1] Wilbur R. Knorr, *The Evolution of the Euclidean Elements*, Dordrecht: Reidel, 1975.

J. Christianidis (ed.), Classics in the History of Greek Mathematics, 243–253.

Greek antiquity – commercial arithmetic – was done by the Egyptian methods. But otherwise, we are left to surmise the nature of the whole from the upper most ten per cent. In this situation, a scholar with an imagination and a feeling for organizing incomplete evidence into rational frameworks can enjoy himself. But the end-products of such studies can never be much other than this or that degree of plausibility. I find this refreshing. But many find it appalling and seek the haven of documentary objectivity. I think that the student of mathematics from 1650 or so onward has the opposite problem of contending with more documentation than is manageable. Here, if ever one makes a general statement of fact, he must expect that in the materials he could not examine contrary patterns might emerge. But didn't Pascal develop this notion of the two types of reasoning? [...]"

We have provided all footnotes and hence are responsible for any failure to capture Knorr's allusions. I have also checked the quotations and adjusted some (including a slight clarification of the status of one quotation) and did some other minor editing. As to the alluring title, fans of the novelist David Lodge will no doubt recall the hapless Persse Mc Garrigle and his "The Influence of T.S. Eliot on Shakespeare" in *Small World* (1984).

Knorr left many other important papers, which I hope to bring out in due time.

Henry Mendell

TRANSCRIPT OF A LECTURE DELIVERED AT THE ANNUAL CONVENTION OF THE HISTORY OF SCIENCE SOCIETY, ATLANTA, DECEMBER 28, 1975. 123

Chairman: Prof. Joseph Dauben,
Lehman College, City University of New York[2]

When Joe suggested to me the possibility of speaking at this meeting, the topic then projected for the colloquium was nineteenth-century mathematics. I told him I was better prepared to speak on ancient mathematics than on 19th-century. But on thinking it over, I hit upon the idea of discussing the impact of modern mathematics on ancient mathematics.

Now, what ancient mathematics was and what ancient mathematicians did has not been influenced by more recent achievements, of course. But what we take ancient mathematics to have been is very strongly influenced by modern work, both in mathematics and in the philosophy of mathematics. It is this sensitivity of the historical criticism that I wish to examine, by way of a few examples from the study of pre-Euclidean geometry. – Afterwards, I will propose some general observations on historical method, based on these examples.

***

"Why didn't the Greeks construct the irrational numbers?" This question was the subject of an article by Heinrich Scholz in 1928.[3] Scholz was examining a polemical statement by Oswald Spengler, to the effect that the Greeks, overburdened by a concrete and plastic intellectual outlook, thereby missed the mathematical abstraction accessible to us now through our algebraic conceptions. Scholz rightly branded the observation nonsense. The Greeks were not blind to an extension of the number-
concept through some accidental failure of spirit. They rejected any such | extension 124
on scientific and philosophical grounds: the *arithmos* must be whole-number; even the rational numbers, a necessary preliminary to irrational numbers, were excluded from the classical number theory; the problem of irrationals was thus resolved by Eudoxus in a geometric manner instead.

---

2 The other paper in the symposium was by Winifred Wisan on "Galileo's Mathematical Method: A Reassessment." They had both just arrived at the ill-fated New School of Liberal Arts, an honors division of Brooklyn College, whose mission was immediately modified by an open admissions policy and which was to suffer under the budget crunches of New York City in the late seventies. As a result, the positions of each were terminated.

3 Heinrich Scholz, Warum haben die Griechen die Irrationalzahlen nicht aufgebaut, in Helmut Hasse und Heinrich Scholz, *Die Grundlagenkrisis der Griechischen Mathematik*, Berlin: Pan-Verlag, 1928, pp. 35–72.

Scholz' assessment is sound. But what we should at once notice is that such a debate could not have arisen before the successful resolution of the problem of irrational numbers by Weierstrass and Dedekind in the 19th century. Before that time, the Euclid-editors – Barrow and de Morgan, for instance – had to answer the charges of obscurity and verbosity levelled against Euclid in his definition of proportion in Book V.[4] But already in Dedekind's time a reversal was taking place: critics like Lipschitz[5] now questioned whether Dedekind had added anything to the Euclidean theory. Somewhat later, Thomas Heath (1921: 1926) judged that "the definition of equal ratios [by Eudoxus and Euclid] corresponds exactly to the modern theory of irrationals due to Dedekind.[6] ... It is word for word the same as Weierstrass' definition of equal numbers.[7] So far from agreeing in the usual view that the Greeks saw in their rational no *number* ... it is clear from Euclid V. that they possessed a notion of number in all its generality as clearly defined, nay almost identical with, Weierstrass' conception of it." This latter judgment, in which Heath follows the view of Max Simon,[8] is
125 undoubtedly overstated. Nevertheless, | we meet later writers like A.E. Taylor,[9] who insist on finding traces of the modern real-number concept in obscure passages from ancient authors. When Plato is reported to have described how "the One equalizes the Great-and-Small",[10] this is read as the definition of an irrational number as the limit of an alternating rational sequence. Again, a puzzling, and likely corrupt, passage from Aristotle, that "number is also predicated of that which is not commensurable",[11]

---

4 Cf. Thomas L. Heath, *The Thirteen Books of Euclid's* Elements, translation with introduction and commentary, 3 vols., 2nd ed., Cambridge: Cambridge University Press, 1926, vol. 2, pp. 121–122.

5 Knorr's source may be a letter from Richard Dedekind to Rudolf Lipschitz dated 6 July 1876, in which he quotes extensively from Lipschitz' previous letter to him. Lipschitz wonders if Dedekind's account of real number is merely Euclid, *Elements* V, def. 5, which he quotes in Latin, "rationem habere inter se magnitudines dicuntur, quae possunt multiplicatae sese mutuo superare [...]" (cf. Richard Dedekind, *Gesammelte mathematische Werke*, ed. by Robert Fricke, Emmy Noether, and Öystein Ore, vol. 3, Braunschweig, 1932, pp. 469–470. Lipschitz' letters are now published, *Briefwechsel mit Cantor, Dedekind, Helmholtz, Kronecker, Weierstrass und anderen*, ed. by Winfried Scharlau, Braunschweig: Vieweg, 1986. For this letter of 6 July 1876, see pp. 70–73.

6 Thomas L. Heath, *A History of Greek Mathematics*, Oxford: Clarendon Press, 1921, p. 327.

7 Heath, *op. cit.*, 1926, (see note 4), vol. 2, p. 124.

8 Maximilian Simon, *Euclid und die sechs planimetrischen Bücher*, Leipzig: Teubner, 1901, p. 110.

9 Alfred E. Taylor, *Plato: the Man and his Work*, 7th ed., London: Routledge, 1960, pp. 509–513.

10 Cf. Taylor, *ibid*., p. 512. Our primary source, Aristotle, *Metaphysics* M 8.1083^{b} 23–32, N 3.1091^{a} 23–5, attacks this view as part of Plato's position on number.

11 The text would appear to be *Met.* $D15.1021^{a}5$–6. Of the three principal manuscripts (labelled E, J, A^{b}) used by W.D. Ross in *Aristotle's Metaphysics* (Oxford: Oxford University Press, 1924), E and J and Alexander of Aphrodisias have: *kata mê summetron de arithmon legetai* (or *legontai*), and so it is printed in every text before Ross, and which he translates in his first Oxford translation (1908), "but this relation may involve a 'non-commensurate number'." A^{b} has instead: *kata mê summetron de arithmos ou legetai*, which Ross emends to: *kata mê summetrou de arithmos ou legetai* (number is not said of the non-commensurate). In general, where E, J, and Alexander agree against A^{b}, Ross sides with them against A^{b} (cf. introduction to his text, p. clxi), but not always (cf. $1008^{a}25$ and introduction p. clxii). All texts and most translations follow Ross (exceptions are translations by R. Hope and H. Apostle, who seem to translate unstated emendations along the lines of EJ). However, if E, J are in error, it still remains interesting that someone before Alexander (ca. 300 C.E.) wrote 'non-commensurable number', i.e. if they wrote it

has recently been used to affirm the conception of irrational *numbers* in the early 4th century B.C. In letting such evidence over-ride the unanimous restriction to whole numbers in the pre-Diophantine literature on number theory, these writers clearly betray a distortion of critical viewpoint owing to their awareness of the modern real-number concept.

Thus, the successful "arithmetization of the continuum" in the 19th century has had perceptible effects on the interpretation of the ancients. In a positive way, it has drawn new attention to certain areas, here the Eudoxean proportion theory, until then not fully understood or appreciated. But once such a problem in mathematics has received a modern solution, this solution tends to be given an absolute status and
| becomes a standard for judging prior work. Ancient work merits praise to the extent 126
it is *like* the modern (recall Heath's phrases: "corresponds exactly to", "is word for word the same as", "is almost identical with") – but the ancients are also *blamed* for failing to institute the full modern approach. The ancients solved the problem of irrationals, to be sure, but as *magnitudes*, not *numbers* – why didn't they succeed in constructing the irrational *numbers?* In other words, there is a sense that the concept of number necessarily and inevitably generalizes from the whole numbers to the rational and irrational numbers. In modern eyes, a mathematical concept, like number, once seen in a certain way, is now viewed as necessarily of this character.

This use of modern concepts as a standard for judging ancient work accounts not only for the negative critiques of such as Spengler, but also the implausible distortions of interpretation one reads in Taylor. Apparently, one cannot be satisfied that a fully competent mathematical system could be different in any important respect from the related modern work.

***

Scholz' examination of the Eudoxean study of irrationals was appended to a larger investigation into what he and Helmut Hasse called "the crisis of foundations in ancient mathematics".[12] Their joint article of 1928 was based on a set of courses on the work of Eudoxus, Weierstrass, Dedekind and Cantor. This notion of a "foundation crisis" had already appeared in Paul Tannery's study of Greek geometry in 1887.[13] Tannery *assumed* that the oldest Pythagorean geometry was built around the assumption of commensurability of all magnitudes. He concluded "the discovery of incommensurability by Pythagoras *must* thus have caused a veritable logical scandal, and to avoid it one *must* have tended to restrain the use of the principle of similarity as much as possible".[14]

---

intentionally. It is possible that Knorr refers to Julius Stenzel, Zur Theorie des Logos bei Aristoteles, *Quellen und Studien zur Geschichte der Mathematik, Astronomie und Physik*, Abt. B, Bd. I, 1929, pp. 34–66, in particular pp. 57–60 (reprinted in *Kleine Schriften zur griechischen Philosophie*, Darmstadt: Wissenschaftliche Buchgesellschaft, 1956, pp. 188–219, in particular pp. 210–212). However, the reference may well be to a more recent (and less sophisticated) interpretation of the passage.

12 See above note 3.

13 Paul Tannery, *La géométrie grecque*, Paris: Gauthier-Villars, 1887.

14 "La découverte de l'incommensurabilité par Pythagore dut donc causer, en Géométrie, un véritable scandale logique, et, pour y échapper, on dut tendre à restreindre autant que possible l'emploi du principe de similitude [...]" (*ibid.*, p. 98).

(Emphases mine.) The "scandal" was ultimately resolved through Eudoxus' theory of
127 proportion; Tannery remarks "it is easy to separate [the embarrassment of | its geometric form] from his theory – it sustains, without any disadvantage, comparison with modern expositions, so often defective".[15] The reference to "defective modern expositions" is especially interesting; for Tannery was well aware of the efforts in his own time to set the infinitesimal calculus on an adequate logical foundation.

Hasse and Scholz develop upon this interpretation. They remark of the irrationality of the square root of 2 that "the discovery of a case which cannot be comprehended in numbers, *must* naturally have shaken the idea of the 'arithmetica universalis' of the Pythagoreans."[16] (Emphasis mine.) Eudoxus' service in ending the crisis is explicitly likened to modern crises: "Just as in the past century and today, so also in the 2nd half of the 5th century, there was a severe foundations crisis."[17] The weak foundation of the limit-concept in infinitesimal calculus, they note, was remedied in the 19th century through the work of Abel, Cauchy, Weierstrass; the weak foundation of the set-concept was more recently remedied in the work of Hilbert, Brouwer; in antiquity it was the weak foundation of the ratio-concept, remedied by Eudoxus.

I cannot provide here a detailed study of the question of the ancient foundations crisis. For this, I defer to the discussion in my recent book on pre-Euclidean geometry[18] and to Hans Freudenthal's article of 1966.[19] But in labelling this "crisis" a "modern fiction", I ought to make one or two justifying remarks. First, there is no evidence of "restraint" in the use of proportions in geometry during the alleged crisis-period, say 450 to 350 B.C. The works of Hippocrates and Archytas, for instance, are indispensably based on such techniques. Second, on what grounds are we to believe that
128 the discovery of incommensurability was a *challenge* or | counter-example to naïve assumptions within the Pythagorean geometry? To be sure, this discovery was held to be significant: late writers suggest it was maintained as a secret of the school – but was it a challenge? Consider that the Pythagoreans based their natural philosophy on the conception of the world in terms of number and other mathematical categories, that is, in terms of certain abstract, rather than material principles. The discovery of incommensurability might well *support* this view: for this is a property of certain lines, for instance, the side and diameter of the square, which we can appreciate through a sequence of deductions to be necessarily true of these lines: yet no effort of practical measurement, no perception or procedure of an empirical character, could bring us to an awareness of this fact or of its certainty. Thus, the Pythagorean insistence on number as a fundamental principle could be *affirmed*: and we should note that the school never did relinquish its adherence to this principle.

---

[15] "Il serait facile de l'en dégager, et elle soutiendrait alors sans aucun désavantage la comparaison avec les expositions modernes, si souvent défectueuses." (*ibid.*)

[16] "Diese Entdeckung eines Falles, der nachweislich nicht mit Zahlen zu erfassen war, mußte naturgemäß die Idee der Arithmetica universalis aufs schwerste erschüttern." (Hasse und Scholz, Die Grundlagenkrisis in der griechischen Mathematik, in Hasse und Scholz as cited in note 3, pp. 4–34).

[17] "Genau wie im vergangenen Jahrhundert und heute, so lag auch damals, in der zweiten Hälfte des 5. Jahrhunderts, eine schwere Grundlagenkrisis der Mathematik vor." (*ibid.*, p. 12).

[18] As cited in note 1.

[19] Hans Freudenthal, Y avait-il une crise des fondements des mathématiques dans l'antiquité?, *Bulletin de la Société mathématique de Belgique*, 18 (1966), pp. 43–55.

When Tannery and Hasse and Scholz jump to the conclusion that the incommensurable was a counter-example to Pythagorean geometric method, they are thus already assuming the thesis of the foundations crisis. The logician and the philosopher, and following them, the historian might recognize that a certain result is paradoxical, and that it *ought* to provoke a crisis in the foundations of a given field of mathematics. But does the practising mathematician ever curtail his researches in accordance with such a challenge? In the 1820's Abel complained of the faulty state of the theory of infinite series – but his real complaint was that little was being done about it.[20] In the early 1920's Hermann Weyl who was occupied with constructing alternative models of the continuum in conformity with the intuitionist critique of logic and set-theory, wrote a paper on "the new foundations crisis in mathematics":[21] in it he criticized the mathematical profession for ignoring the implications of the Richard paradox in set-theory. Perhaps Hasse and Scholz, only a few years later studying the related ancient work, should have taken more seriously this | aspect of the 129
"crises". They may have recognized, for instance, that Plato's strong words against the faulty mathematical procedures of his time betokened a similar insensitivity by geometers to some crisis which *should have been underway*;[22] however, that Plato was uniquely positioned in the Academy to encourage such mathematicians as Eudoxus to address the problems of the reorganization of geometry on a satisfactory logical foundation.

***

Oskar Becker, more than any other modern scholar, has invigorated the study of ancient mathematics and logic. In a series of "Studies on Eudoxus" which appeared in the 1930's he investigated several logical problems in the Euclidean geometry.[23] His project was in large part inspired by the start made by Hasse and Scholz; with much greater detail, however, Becker sought to clarify the ancients' use of non-constructive assumptions, such as the principle of the excluded middle, the assumption of the existence of the fourth proportional, and others. But if Becker succeeded in showing that most of the Euclidean uses of the excluded middle can be brought into conformity with Brouwer's intuitionist criteria, for instance, we might well ask what is the significance

---

20 Niels Abel, Recherches sur la série $1 + m/1\, x + m(m-1)/(1 \cdot 2)x^2 + m(m-1)(m-2)/(1 \cdot 2 \cdot 3)x^3 +$?', *Journal für die reine und angewandte Mathematik*, vol. 1, Berlin 1826 (reprinted in *Œuvres Complètes*, 2 vols., Christiana: De Grondahl & Son. 1881), vol. 1, chap. xiv, pp. 219–250.

21 Hermann Weyl, Über die neue Grundlagenkrise der Mathematik, *Mathematische Zeitschrift*, 10 (1921), pp. 39–79 (reprinted in H. Weyl, *Gesammelte Abhandlungen*, ed. by K. Chandrasekharan, 4 vols., Berlin: Springer, 1968, vol. 2. pp. 143–180).

22 Plato, *Republic* vii 527 AB.

23 These are, from *Quellen und Studien zur Geschichte der Mathematik, Astronomie und Physik*, Abt. B: Eudoxos-Studien I: Eine voreudoxische Proportionenlehre und ihre Spuren bei Aristoteles und Euklid, 2 (1933), pp. 311–333; Eudoxos-Studien II: Warum haben die Griechen die Existenz der vierten Proportionale angenommen?, 2 (1933), pp. 369–387: Eudoxos-Studien III: Spuren eines Stetigkeitsaxioms in der Art des Dedekindschen zur Zeit des Eudoxos, 3 (1936), pp. 236–244; Eudoxos-Studien IV: Das Prinzip des ausgeschlossenen Dritten in der griechischen Mathematik, 3 (1936), pp. 370–388; Eudoxos-Studien V: Die eudoxische Lehre von den Ideen und den Farben, 3 (1936), pp. 389–410.

of that? If, again, he has shown that the ancients' use of the fourth proportional – reducible to weaker constructive assumptions, but not done so by them – indicates their implicit acceptance of Dedekind's axiom of continuity we might question whether Dedekind's axiom is indeed nothing more than the Platonic formula: "to that of which there is the greater and the smaller there is also the equal".[24] It would appear that Becker, like others mentioned earlier, is guilty of reading more formality into
130 the ancient work than was actually | there. But Becker's studies raise another more interesting problem about the objective of a historical analysis.

Becker was a proponent of the philosophy of Edmund Husserl. For several years in the 1920's he co-edited with Martin Heidegger and others the "Jahrbuch" of phenomenological research and philosophy. At this time Becker produced phenomenological analyses of subjects like mathematical existence and the logical foundations of geometry; his "Eudoxus-Studies" followed; and there after numerous books and articles on mathematical thought, both ancient and modern. Now, what did the commitment to the phenomenological outlook consist of? This philosophy was itself inspired by the mathematical work of the late 19th century. Husserl studied with Weierstrass and also with Kummer and Kronecker in Berlin in the 1880's. According to Becker, Husserl thus developed an ideal of mathematical exactitude as the appropriate standard for philosophy.[25] In an early work on the foundations of arithmetic (1887: 1891) Husserl[26] fused his mathematical training with the psychological principles of Brentano, with whom he studied at Vienna. This study of arithmetic, later renounced after Frege's withering attack on its "psychologism",[27] manifested important phenomenological features: the reduction of the meaning of such concepts as number and set to the activity of collecting and counting, for instance. Gradually, Husserl articulated the need for a precise description of the subjective processes necessarily involved in thought. Husserl hoped that, just as the new general theory of manifolds by virtue of its abstract character, could cover many different particular
131 theories, so also a general phenomenological theory could | provide the absolute basis for comprehending all thought. Ultimately, Husserl espoused a full transcendental philosophy, faced with many of the problems which had challenged such earlier rationalists as Plato and Descartes.

The phenomenologist thus seeks an absolute description of the thought-processes necessary for deductive thinking, as in mathematics and logic; he considers that these

---

[24] Plato, *Parmenides* 161D, cf. Becker, Eudoxos-Studien III (see previous note).

[25] Perhaps Knorr refers to the introduction to Oskar Becker, Beiträge zur phänomenologischen Begründung der Geometrie und ihrer physikalischen Anwendungen, *Jahrbuch für Philosophie und phänomenologische Forschung*, 6 (1923), pp. 385–560 (1–176), cf. pp. 385–388 (1–4), where Becker argues for the centrality of foundations for mathematics as central to Husserl's phenomenology. Cf. p. 386 (2): "Der Verfasser, dem wesentliche Teile jener Husserlschen Forschungen (in Vorlesungen, Übungen, Manuskripten, persölichen Unterredungen) zur Verfügung gestelt wurden, setzte sich die Aufgabe, jene Begründung und Aufklärung in ihren Grundzügen zu leisten und damit eine Brücke von Phänomenologie zur heutigen Mathematik und Physik zu schlagen."

[26] Edmund Husserl, *Über den Begriff der Zahl, psychologische Analysen*, Habilitationschrift, Universität Halle-Wittenberg, 1887, and his revised expansion, *Philosophie der Arithmetik: Psychologische und logische Untersuchungen*, Halle-Saale: Pfeffer, 1891.

[27] Gottlob Frege, review of Husserl's book (1891), in *Zeitschrift für Philosophie und philosophische Kritik*, 103 (1894), pp. 313–332.

processes, being absolute, do not depend for their understanding on a consideration of the particular individuals who happen to do the thinking or the particular circumstances, cultural or otherwise, within which the thinking occurs. What, then, are the objectives of *historical* analyses within this philosophy? Theodore Kisiel, writing of Husserl's investigation of the origins of geometry, answers thus: "the more basic considerations of the birth and becoming of science lie on the *a priori* level of meaning rather than the empirical level of facts. Hence as Husserl sees it, the 'essential history' . . . transcends the 'noisy events' of daily concerns . . . it is a history which can be traced even when the facts are no longer accessible [for instance, a study of the history of geometry is designed] to gain some insight into what the original but now submerged sense *must* have been when it *first* emerged."[28] (Emphases his.)

These remarks seem to clarify the purposes of such historical analyses as Becker's. For, his interest is in explicating the necessary logical relations among the several assumptions, some explicit, some implicit, in the ancient geometry; ostensible defects in Euclid may be removed by implicit axioms, "traces" of which may then be sought in other authors. These objectives seem consistent with the phenomenological search for how the historical development *must* have happened.

My uneasiness with this approach of Becker's is that it runs the risk of anachronism. The ancient mathematician was not working within the context of any such complete systems – either mathematical or philosophical – of the type presumed in Becker's philosophical analysis. Becker takes ideas and methods which he knows as necessary; he may thus read into past work what was not there, or at best only barely or intuitively perceived; conversely, he may miss the significance of other aspects which his | own 132
background leads him to view as accidental or superfluous.

An instance of the strengths and weaknesses of Becker's method may be seen in his reconstruction of a pre-Eudoxean proportion theory.[29] Although Eudoxus resolved the problem of proportions of incommensurable magnitudes, might not prior attempts have been made along alternative lines? A passage from Aristotle's *Topics* appears to indicate as much; "some things in geometry are difficult to prove, because of the lack of a definition" – a theorem is cited: "that when a rectangle is cut parallel to its side, the areas are in the same ratio as bases" – and the necessary definition is given: "magnitudes have the same ratio when their antanaireses are the same".[30] Becker was able to work out from this that all the theorems of the Euclidean proportion theory (Book V) *could* be proved via such a notion: using, instead of Eudoxus' definition, a test for equal ratio based on the Euclidean division algorithm (anthyphairesis = antanaireses). Further passages, as well as the internal structure of the *Elements*, seem to point to Plato's contemporary Theaetetus as the originator of this alternative theory.

Becker's thesis has several strengths and has thereby attracted numerous adherents. But notice that the starting-point of Becker's study coincides with his own philosophical objective of delineating the steps in the development of logic. For, on what grounds, other than mere assumption, are we to accept that Theaetetus, or any

---

28 Theodore Kisiel, Husserl on the History of Science, in Kockelmans (Joseph) and Kisiel (Theodore), eds., *Phenomenology and the Natural Sciences*, Evanston: Northwestern University Press, 1970, pp. 68–92, see pp. 69–70 for the quotation.

29 Cf. Oskar Becker, Eudoxos-Studien I (note 23).

30 *Topics* VIII.3 158b29–35.

predecessor of Eudoxus, took up the systematic study of the logical problems of proportion theory?

Suppose we reverse Becker's approach. What do we know about Theaetetus? Among other things, that he initiated the classification of irrational lines: the expanded theory of these survives as Euclid's Book X. Now, in Book X the division algorithm, anthyphairesis, is used to determine whether or not given lines possess a common measure, that is, whether or not they are commensurable. If we peruse the theorems in the book, we find that several results from proportion theory and the theory of similar rectangles are required, perhaps a dozen theorems in all. The theorem named by Aristotle, for instance, is one of them: it is required for all but the first 18
133 of the 115 theorems in the book. One | theorem whose anthyphairetic proof Becker found problematic is deceptively simple: "if $A : B = A : C$, then $B = C$"[31] – it is just the sort of mathematical fact which would pass notice until detected by a geometer specifically interested in foundations.

These facts suggest to me a pattern significantly different from that argued by Becker: that Theaetetus initiated the classification of irrationals and required a set of theorems on proportions to prove certain results about these irrationals – such as, that a line is designatable as a binomial irrational in one way only; that Theaetetus chose the division algorithm as the means of proving these theorems on proportions; and that later geometers, among them Eudoxus, in the course of extending this theory of irrationals and its use of proportions, detected those difficulties which then motivated the quest for a new definition of proportion and a revised theory based on it.

I will not here go into further details or seek to provide textual and technical supports beyond this (These are given in my book, mentioned earlier). For now, it should be clear that Becker's analysis might as well lead to the view of Theaetetus not as a predecessor of Eudoxus in the study of foundations, but rather as a mathematician interested in the geometric construction and description of irrational lines.

***

The basic issue, then, is this: what is a historical analysis intended to do? Here, I believe, each scholar is entitled to his own answer. But for me, a coherent philosophical synthesis, such as Becker and others seek, is only part of the objective. What I expect in addition is that a historical analysis conform to what it meant then and there to *be* a mathematician and to do mathematics. Of course, our ability to know these things is also restricted and thus becomes a task for historical analysis. But certainly, a purely logical or philosophical investigation will not suffice. We will find that wider considerations are necessary: certainly of the individual mathematician and the placement of his work in the setting of the labors of his contemporaries, but also frequently of other relevant cultural factors – which, depending on the culture, may include religious, social, educational or political elements.

134 | Without doubt, the philosophical reviews of ancient logic and mathematics done by Tannery, Hasse, Scholz, Becker and others, have enriched our appreciation with new hypotheses and new insights into the meaning of the concepts and methods

---

[31] *Elements* V-9; Becker, as cited in note 23, p. 320.

employed. But as I pointed out, a cultural dimension ought to be present in the analysis also – serving to subject the philosophical analysis to a wider documentary test, thereby to reduce the risk of anachronism, and also to keep alive the possibility of alternative views.

This last point is important. Whether or not you accept *my* view on the impact of the discovery of incommensurability on the Pythagorean philosophy or *my* view on Theaetetus' use of the division algorithm in proportion theory, you will accept, I believe, that my views are at least as tenable as the views they seek to modify. I believe the generalist in the history of mathematics especially should be made aware of this: he should recognize the degree of dependence which many historical treatments have on philosophical preconceptions. Above all, he should recognize the pervasiveness of what we could call a Platonic bias in the writing and research of the history of mathematics – that is, a conviction of the absoluteness and culture non-dependence of the concepts and truths of mathematics.

What are the *facts* about the history of ancient mathematics? In part, they are the research opinions of the very few specialists in this field – I have indicated how much these can be influenced by modern ideas and by philosophical assumptions. But the *facts* are also what is to be found in the survey histories – such as those by Heath, van der Waerden, Cantor, Hofmann,[32] and many others – and the dictionary and encyclopedia articles, and so on – the materials which the non-specialist is likely to consult – first and perhaps only.

In works of this general category it is a *fact*, for instance, that Greek mathematics from about 450 to 350 B.C. was in the throes of a paralyzing crisis of foundations. As
I and others have insisted, this is | at best a highly misleading – if not an entirely false 135
– description of pre-Euclidean geometry, strongly reflective of the concern over foundations among logicians in the early part of this century.

We do well to beware that many similar misconceptions derived from an insufficiently guarded application of modern mathematical conceptions may permeate what we commonly accept to be ancient mathematics.

In short, whenever we read about ancient mathematics, we ought to keep two warnings in mind: *cave modernum* and *cave Platonicum*.

---

32 Thomas L. Heath, *A History of Greek Mathematics*, Oxford: Clarendon Press, 1921 (*op. cit.* note 6); Bartel L. van der Waerden, *Ontwakende Wetenschap*, Groningen: Noordhoff, 1950, trans. by Arnold Dresden as *Science Awakening*, Groningen: Noordhoff, 1954: Moritz Cantor, *Vorlesungen über Geschichte der Mathematik*, 4 vols., 4th ed., Leipzig: Teubner, 1922: and Joseph E. Hofmann, *Geschichte der Mathematik*, 3 vols., Berlin: DeGruyter, 1963.

PART 4

STUDIES ON GREEK ALGEBRA

Texts selected and introduced by Jacques Sesiano

JACQUES SESIANO

INTRODUCTION

1. TRANSMISSION OF GREEK MATHEMATICS

The recovery of Greek mathematics in Europe began in the 12th century, when Latin translations of Greek works were made in Spain and southern Italy, mainly from the Arabic but sometimes from the Greek as well. New or improved versions appeared from Renaissance times onwards; sometimes Arabic manuscripts extant in European libraries were used as well. The last major original edition of a treatise in the Greek language was undoubtedly that of the palimpsest containing Archimedes' *Method* by Heiberg in the first years of the 20th century.

As is made clear in Toomer's article (p. 276), progress in our knowledge of Greek mathematics during the 20th century came almost exclusively from Arabic translations and, to a much lesser extent, from Greek papyri found in Egypt. Another of Toomer's statements (*ibid.*) cannot be emphasized strongly enough: we owe the preservation of Greek higher-level mathematical works, such as those of Archimedes, Apollonius and Diophantus, to very few or even one single copy extant in the early Middle Ages which fortunately happened to be copied by the Byzantines or translated by the Arabs. This helps us to understand why certain Greek texts, such as some of Diocles's treatises or Apollonius's *Cutting off of a ratio*, survive only in Arabic; why part of a text may be preserved in Greek whereas the Arabic version contains more, as is the case with Apollonius's *Conics*; why also the Arabic tradition may, unlike the Greek text we know, represent a commented version, as it does for Diophantus's *Arithmetica*.

Then considering how aleatory transmission was in the period from late antiquity to early Middle Ages, we realize how fortunate we are that Greek mathematical writings survived at all, but at the same time guess that a considerable number of works must have vanished for ever, although possibly recorded in ancient writings in the form of a summary, an excerpt, a quotation or sometimes by a single reference to their titles only. There are obvious reasons why some treatises were no longer copied: they were not comprehensive enough or too specialized, like Apollonius's *Unordered irrationals*; others were considered obsolete and superseded by textbooks in use during late antiquity, such as Euclid's *Conics*. In this way a whole part of Greek mathematics gradually disappeared during the first centuries of our era, sometimes whole topics as well. Thus we can only guess that astronomical models or mathematical techniques found in early Sanskrit texts go back to Greek science on the basis of small clues such as the vocabulary or underlying methods.

Neverthelesss, important new material may still come to light, as the recent discoveries mentioned by Toomer have proved. But this is unlikely to happen other than through the Arabic channel. Indeed, if ever there were a general assessment of the advances made in our knowledge of Greek mathematics during the 20th century, the

J. Christianidis (ed.), Classics in the History of Greek Mathematics, 257–263.

overall importance of Arabic sources would no doubt emerge, just as it will certainly characterize the research and progress in this century as well.

2. ELEMENTARY ALGEBRA

Not every aspect of Greek mathematics followed the royal road of being copied or translated, as some major works did. The transmission of mathematics of a more elementary level occurred through a channel which the Danish scholar J. Høyrup has aptly called 'sub-scientific tradition'. Thus, the main Greek *multiplication tables* for integers are completely preserved in Coptic and Armenian writings, whereas only fragments survive in Greek papyri; this is due to the fact that both Copts and Armenians kept the use of the Greek system of alphabetical numerals which had been transcribed in their own script. *Practical mathematics* is seen to survive even longer; thus, the Egyptian Abū Kāmil (ca. 900) complains in his treatise on mensuration about the rough formulae still being used by the official administration; they turn out to originate with Egyptian customs going back to the time of the Pharaohs, and must thus have remained in use throughout Greek and early Islamic times, despite the numerous reliable mensuration treatises composed meanwhile. *Recreational problems* are known from a few examples found in versified form in Book XIV of the *Anthologia græca*; there must have been considerably more, and some collections have managed to survive and reappear, occasionally with identical data, both in Arabic texts and in a Carolingian writing attributed to Alcuin, the *Propositiones ad acuendos iuvenes*; in both cases the link to antiquity, although evident, cannot be traced back. Finally, this kind of indirect or hidden transmission applies to *elementary algebra*, that is, the algebra taught in schools, which survives in only a few papyri but is seen to have maintained its influence in mediæval times, although no work on the subject was copied by Byzantines or translated into Arabic.

Speaking about algebra, we should mention a major change in our knowledge of its history, which occurred during the thirties. The deciphering of cuneiform mathematical tablets dating back to the years 1800 BC has led to the conclusion that linear systems, quadratic equations and some second-degree systems of equations were already being solved at that time, probably even before by Sumerian mathematicians. But the texts containing such problems are at first sight somewhat baffling: there are no theoretical hints, no justifications for the way followed, no description of an equation, but merely a sequence of computations ending with the determination of the solution, and sometimes the correctness of the result is proved. For single quadratic equations, there is clearly the application of a formula in a form somewhat similar to that we know today. For systems, the resolution seems intricate. Now Kurt Vogel pointed out that the treatment relied on the use of a few identities which were applied to the particular form of the problem. It looks as if the reader must have had in mind such identities, the aim of the resolution being to give him practice in choosing the appropriate identity, if need be by introducing a new unknown or transforming the system (a step which the text does not state explicitly). Vogel's results appeared in a journal of limited circulation, in an article which certainly deserved to be republished.

Vogel did not know then that such approaches are found in Greek papyri as well. As said above, the surviving documents are scarce.[1] Their similarity to

Mesopotamian documents is striking, as is their difference to what we commonly understand as Greek algebra, namely the higher-level one represented by Diophantus's *Arithmetica*.

Before turning to this other kind of algebra, we must at least put the question of a possible link between Mesopotamian and elementary Greek algebra. The use of Babylonian material is well attested for astronomy: their planetary records were being used around AD 150 by the major Greek astronomer Ptolemy, who says in his *Almagest* that from the reign of Nabonassar (ca. –720) "the ancient observations are, on the whole, preserved down to our time".[2] The situation of mathematics is not as clear: it does not show any real advance in Mesopotamia since the time of the early texts, even if there was a slight revival during the last centuries BC. An influence is thus conceivable. But an attractive supposition would be that treating systems by the application of basic identities is just a natural way of dealing with algebraic problems in the absence of a clear denomination of the unknowns and a suitable description of the equations.

3. HIGHER ALGEBRA

Greek algebra is identified with the work of Diophantus, of whom we know almost nothing. Even his lifetime is conjectural, although it is now generally accepted that it should probably be placed about AD 250, which can be considered as fairly late compared to the other known Greek mathematicians. Not surprisingly, he lived in Alexandria, the major centre of mathematical and, generally, scientific activity since the time of Euclid.

His historical importance is twofold. First, his *Arithmetica* represents the only preserved testimony of higher algebra in antiquity. Second, it is the study of his work that gave rise, through the remarks and commentaries of the Frenchman Fermat (1601-1665), to modern number theory. The Swiss Euler (1707-1783) still uses, but also considerably develops, Diophantus's algorithms in his widely-read *Algebra*. From then on the importance of Diophantus must be considered as being mainly historical.

As we gather from the *Arithmetica*'s introduction, the whole work originally consisted of thirteen "books", that is, parts. The first European studies on Diophantus, which started towards the end of the 16th century, were based on the Byzantine manuscripts reaching Italy at the latest after the fall of Constantinople. All these manuscripts contained six books only, and Diophantus's later influence and fame were entirely based on this part of the *Arithmetica*. The representation of his methods took a definitive form with the account of Heath, republished here, which provides a survey and synthesis of previous studies.[3]

All the more attention was given to the announcement that a part of the *Arithmetica*, in an Arabic translation attributed to Qusṭā ibn Lūqā (d. ca. 910), had been discovered by the Turkish orientalist Fuat Sezgin in Mashhad (Iran) in 1968.[4] Now it turned out that these books were different from those already known in Greek and, as a matter of fact, were the original Books IV to VII, so that the books numbered IV to VI in Byzantine times must have been later books, probably Books VIII to X. The form of the text is also different: algebraic symbolism has disappeared and the whole text has become purely rhetorical, which is, by the way, customary in Arabic algebraic treatises. A more important difference to the known Greek part is that the text is notably more prolix, for it gives in detail the computations, and then verifies that the solutions arrived at indeed fulfil the equations, what Diophantus himself did not bother to do. Internal evidence then showed this to be of Greek, and not Arabic, origin.

Two questions then arose. First, whether some more books of the *Arithmetica* could be obtained should another part of the Arabic translation come to light. Second, who the author of the commented version was. Although we do not possess the Arabic translation of the first three books, we know that it existed, for the Persian al-Karajī (*fl.* ca. 1000) reproduces practically all the problems of the first four books in his *Fakhrī*.[5] Furthermore, it seems that this part was also commented. But it now appears unlikely that the Arabs received more than the first seven books. As to the authorship of the commentary, we have no information from the Arabic sources. No Greek source refers to such a commentary either, but the Byzantines do, by crediting the daughter of the astronomer Theon of Alexandria, Hypatia (d. 415), with a commentary. It thus seems probable that the Arabs received the commented version and translated it, while the Greek text we know, apart from minor additions and interpolations, represents the original Diophantine composition.

As the number of known books has since increased from six to ten, a reassessment of Heath's account may be necessary. A survey of the ten extant books shows, however, that the change in our picture of the *Arithmetica* is more methodological than mathematical, so that Heath's account can still be considered as a satisfactory scientific evaluation of Diophantus.

Diophantus's General Introduction

In his introduction to the *Arithmetica*, Diophantus explains the basics of algebra for the beginner. First he presents the names and abbreviated forms, or symbols, of the powers of the unknown.[6] The two operations considered in Arabic times to be characteristic of algebraic reasoning are explained: when the equation has been established, we are to add on both sides the amount of the negative quantities so as to be left with an equation containing positive terms only (as was to remain traditional till the late 16th century); next, we are to remove from both sides like quantities.[7] Finally, Diophantus states that the problems presented should be reduced, "as much as possible", to the equality of one power of the unknown and a number. If the power is of the first degree, the solution will be rational, otherwise a condition of rationality must be met by the given quantities.

Book I

Book I itself can hardly be considered as representative of Diophantus's algebra. What we encounter there would have been familiar to a Greek schoolboy, or even to his Mesopotamian forerunners: mostly determinate problems, of the kind found in papyri. Diophantus thus starts on familiar ground, but, as a good teacher, he accustoms at the same time the reader to his algebraic symbolism and the technique for reaching a solution. In short, the surroundings are familiar but the road is new.

Books II-III

This slow training ends abruptly with Book II. For in Books II and III we encounter what is chacteristic of the *Arithmetica*: solving indeterminate problems ending with an algebraic expression in the unknown which must be made a square; that is, a (positive and rational) value of the unknown has to be found which will make the algebraic expression a numerical square. Generally Diophantus is content to give just one solution, although he knows that any number of further solutions can be derived from it: see p. 297 of Heath's account.[8]

On p. 287 of this same study, Heath repeats the opinion commonly held at that time on Diophantus: that general methods are not made evident in his work and so we have to find them out. This had, to a certain extent, been true, but must now be revised in the light of the contents and purpose of the Arabic books: indeed, we see there that Diophantus knows how to systematically apply the general methods taught somewhat casually in Book II.

Books IV-VII

In his introduction to the first of the newly-discovered books, Diophantus mentions that, as in the previous part, the problems will end with one term being equal to another. However, as was not the case in Books I-III, where the powers of the unknown remained limited to the first two degrees, higher powers will occur. Thus Diophantus introduces a third algebraic rule, which, like the other two, received in Islamic mathematics its own designation: when the terms in the final equation contain two powers of the unknown, we are to divide them both by that of lower degree to obtain, as before, a power equal to a number. Finally, Diophantus tells us that this part aims at providing "experience and skill", and this very same expression is repeated at the beginning of Book VII.

We are now in a position to say what is characteristic of the newly-discovered books: although we shall meet higher powers, the fundamental methods will remain the same and we shall just learn how to apply them to a wider range of problems. When we said before that the discovery of the four books in Arabic changed our picture of Diophantus from at least a methodological point of view, we meant this sort of respite Diophantus is giving the reader so that he may assimilate the principal methods taught before, mainly in Book II. Thus the commonly held idea that general methods must be just found out and are not really made evident is no longer tenable. For a general account of these methods, see Heath's presentation, pp. 296-8 and 301-4, to which must be added the method explained in problem II.19; all this is also summarized on pp. 6-7 of the Arabic Diophantus. Note too that in Arabic times already the general methods had been sorted out. This is done by al-Karajī, mentioned above, in another algebraical work of his, the *Badī'*.[9]

Books VIII-X (?)

That the last three Greek books belong to a later part of the work is now clear. Their level of difficulty is also higher, although the basic methods remain the same. The problems proposed often involve the resolution of an auxiliary problem if a first attempt has led to an impasse, either because the solution found was irrational or the equation impossible. This remains true for the last Greek book, which is devoted to the determination of right-angled triangles satisfying a condition on their sides (or areas).

Books XI-XIII (?)

As to the missing last three books, they must be considered as irretrievably lost since there is no indication that the Byzantines or the Arabs possessed them. In view of the discovery of the Arabic books and 20th-century research on Islamic mathematics, the question of their scope and contents has changed significantly. The problems of the known books end with the equality of two terms containing different powers of the unknown, a characteristic which, as we have seen, Diophantus pointed out himself. But, after stating in the introduction to the *Arithmetica* that this must be aimed at (for we thus have a rational solution or a simple condition of rationality), he then says that he will later show how a problem is solved when "two terms are left equal to one term". Heath took this (pp. 288-9) to mean the resolution of determinate quadratic equations. Since we now know that this was more or less common knowledge, there is hardly any doubt that it rather refers to indeterminate equations of the form $ax^2 + bx + c =$ a square, where the square will be taken either as m^2x^2 or as m^2, so that a positive and rational solution x depends upon a suitable determination of m.

An Arabic source may give a clue as to the kind of cases which were considered. In his *Algebra*, the Egyptian Abū Kāmil, already mentioned above, has a whole section on indeterminate equations, most of which are clearly like those Diophantus arrives at in the extant books. They are not taken from the *Arithmetica* but are clearly based on ancient methods akin to those of Diophantus, and it seems quite clear that they were drawn from or at least inspired by some other Alexandrian source. Particularly interesting are a few types in which we have precisely quadratic indeterminate equations involving three terms. Now it is very likely that such resolutions, which would fit into the Diophantine framework, were of the kind treated in the last three Greek books.[10] Since one characteristic of Diophantus is to create at will new problems with just a single method of solution, it is not unreasonable to suppose that a few such methods might have formed the basis for one or several books of his *Arithmetica*.

École Polytechnique Fédérale de Lausanne

NOTES

1 The characteristic ones of the second-degree are brought together in our study "An early form of Greek algebra", *Centaurus*, 40 (1998), pp. 276-302.

2 *Almagest* III.7, p. 166 of Toomer's translation, *Ptolemy's Almagest*, New York/Berlin 1984.

3 Unfortunately, Heath seems to be reluctant to mention other authors, except for criticism; thus he integrates a good deal of Paul Tannery's studies (now conveniently available in vol. I & II of his *Mémoires Scientifiques*, Paris-Toulouse 1912).

4 On an amusing later claim about this discovery, see the account by Kh. Jaouiche in *Annals of science*, 44 (1987), pp. 308-309; the pseudo-discoverer was in fact informed by Sezgin himself in 1973 about the existence of this manuscript.

[5] See the *Extrait du Fakhrî* by Fr. Woepcke, Paris 1853.

[6] Diophantus has just one symbol for his unknown. Thus, if one problem involves the resolution of another, intermediate one, the same symbol will have to be used for both unknowns; but, as the text clearly separates the two problems, no confusion arises.

[7] Diophantus did not use specific words for these two operations, whereas the Arabs did, and this has been indirectly recorded to the present day: for, in mediæval times, the term designating the first of these two operations, *al-jabr*, was both translated by *restauratio* and transcribed in Latin by *algebra*.

[8] On Heath's unjustified criticism of Diophantus, see p. 80, footnote, in the edition of the Arabic Diophantus.

[9] See our analysis of this part in *Archive for History of Exact Sciences*, 17 (1977), pp. 297-379.

[10] A summary of all the methods of Abū Kāmil is found in our study in *Centaurus*, 21 (1977), pp. 89-105; of those involving three terms on pp. 81-82 of the edition of the Arabic Diophantus.

KURT VOGEL

ZUR BERECHNUNG DER QUADRATISCHEN GLEICHUNGEN BEI DEN BABYLONIERN

Das bedeutendste Forschungsergebnis der letzten Jahre auf dem Gebiete der Geschichte der Mathematik ist die Entdeckung der quadratischen Gleichung bei den Babyloniern. Eine Reihe von Aufgaben, die zum Teil veröffentlicht sind[1], führt auf Gleichungssysteme mit zwei Unbekannten, wobei eine Gleichung linear ist, während die andere das Produkt oder die Quadratsumme der Unbekannten enthält. Manchmal ist nur die Lösung angegeben, in andern Fällen wird diese aber auch nach einer „Formel" vorgerechnet. Aus den teilweise recht verwickelten Problemstellungen, in denen sogar geometrische Gebilde verschiedener Dimension addiert werden[2], ergibt sich, daß die Aufgaben schon von dem geometrischen Hintergrund losgelöst sind, dem sie ursprünglich ihre Entstehung verdanken, so daß man sie vielfach als „Algebra" ansprechen muß. Besonders hochstehend sind gerade die ältesten Texte, vier Aufgaben auf einem vierseitigen Prisma im Louvre (AO 8862), aus der Zeit der 1. babylonischen Dynastie; die vorausgehenden Vorarbeiten sind also noch weit früher anzusetzen. Die wichtige Frage, ob man sich der Doppeldeutigkeit der Lösung bewußt war, wird von NEUGEBAUER, dem Herausgeber und Erklärer eines großen Teiles der in Frage stehenden Texte, bejaht. Er kommt auf Grund der dritten Aufgabe auf dem Louvre-Prisma zu dem „unabweislichen" Schluß[3], „daß man sich erstens darüber klar war, daß Gleichungen der hier behandelten Art zwei prinzipiell gleichwertige

---

1 H. S. SCHUSTER gibt in Quell(en und) Stud(ien zur Gesch. d. Math). B. 1, 1930, S. 194, Anm. 5 eine Zusammenstellung der quadratischen Gleichungen bei den Babyloniern, die noch durch die vier Aufgaben AO 8862 zu ergänzen ist. Die Aufgaben sind in folgenden Schriften zu finden:

a) TU = Tablettes d'Uruk (F. THUREAU-DANGIN, Textes Cunéiformes Tome VI, Tablettes d'Uruk, Paris 1922) aus 200—150 v. Chr.: 4 Aufgaben mit Auflösungsformel. S. hierzu H. S. SCHUSTER a. a. O. S. 194—200.

b) SKT = Straßburger Keilschrifttexte (C. FRANK, SKT, in: Schriften der Straßburger Wissensch. Ges. in Heidelberg, Neue Folge, 9. Heft, Berlin-Leipzig 1928) aus ca. 1900 v. Chr.: 4 Aufgaben mit Lösung, 9 Aufgaben nur teilweise mit Resultatangabe.
Zu Aufgabe: SKT 6 Rs. 6—8 siehe: QuellStud. B. 1, 1930, S. 124.
SKT 7 Vs. 1–12 siehe: QuellStud. B. 1, 1930, S. 124.
SKT 7 Vs. 13 bis Rs. 7 siehe: QuellStud. B. 1, 1930, 124—129.
SKT 7 RS 8 bis Rs 23 siehe: QuellStud. B. 1, 1930, 124—129.
S. ferner: QuellStud. B. 1, 1929, S. 78ff. und 1930, S. 199.

c) CT = Cuneiform Texts ... in the British Museum. Etwa 2000 v. Chr.: 2 Aufgaben. Hierzu Quell. Stud. B. 1, 1930, S. 80 und 1930, S. 124 Fußn. 7, 199.

d) Prismen Text AO 8862 im Louvre (F. THUREAU-DANGIN, Le prisme mathématique AO 8862, Revue d'Assyriologie 29, 1932, 1—10 und O. NEUGEBAUER in QuellStud. B. 2, 1932, 3—27). Vor 2000 v. Chr.; 4 Aufgaben, darunter 3 mit Auflösungsformel. Hierzu ferner: Revue d'Assyriologie 28, 1931, 196—198.

2 Dies war bisher erst für HERON bekannt. Opera IV (ed. HEIBERG) S. 380.

3 a. a. O. (Anm. 1 d) S. 22.

J. Christianidis (ed.), Classics in the History of Greek Mathematics, 265–273.

Ansätze" (einen für x bzw. für y) „zur Lösung gestatten und daß man zweitens auch die Doppeldeutigkeit der Lösung einer quadratischen Gleichung kannte (es sei denn, daß eine der Lösungen negativ wird)".

THUREAU-DANGIN schließt sich dieser Ansicht nicht an (a. a. O. [Anm. 1 d] S. 8); allerdings zieht er zur Beantwortung der Frage nur die beiden ersten Aufgaben heran, ohne auf die dritte einzugehen, auf die sich NEUGEBAUER hauptsächlich stützt. Ich möchte nun im folgenden ein Auflösungsverfahren für die eine Gruppe der quadratischen Gleichungen aufweisen, das vollkommen mit dem Text im Einklang steht und das gerade auf die jeweils vorliegenden Lösungen führt, ohne daß überhaupt eine Doppellösung in Frage kommt. Um es gleich vorauszunehmen: ich bin der Ansicht, daß der Rechner darauf ausging, zu dem gegebenen $x+y$ sich auch das $x-y$ (bzw. $(x-y)/2$) zu verschaffen, worauf sich dann leicht die beiden Unbekannten als $(x+y)/2 \pm (x-y)/2$ errechneten.

77 | Wenn wir die bisher veröffentlichten Aufgaben übersehen, so ergeben sich deutlich zwei Gruppen bzw. Gleichungssysteme. Die erste Gruppe hat folgende Form[4]: I) $x^2+y^2=A$; II) $y=Bx+C$. Statt der zweiten Gleichung findet sich auch die „Parameterdarstellung"

$$x = B'z + C',$$
$$y = B''z + C''.$$

Das Wesentliche ist dabei das Auftreten der Quadratsumme x^2+y^2 in der einen Gleichung. Eine dieser Gruppe zuzuzählende Aufgabe ist auch in der ägyptischen Mathematik des mittleren Reiches erhalten[5].

Die zweite Gruppe, für die ich das genannte Auflösungsverfahren unter Verwendung von $(x-y)/2$ nachweisen möchte, enthält nicht mehr die Quadrate, sondern das Produkt der beiden Unbekannten. Den einfachsten Fall zeigen die TU-Aufgaben, denen das Gleichungssystem

$$\text{I)}\ x + y = a,$$
$$\text{II)}\ x \cdot y = b,$$

zugrunde liegt. Hierbei hat b immer den Wert 1; daß aber die Lösung auch für andere b möglich war, zeigen die Beispiele aus AO 8862. Der Text der vier Aufgaben führt auf folgende Gleichungssysteme:

Nr. 1)
$$\text{I)}\ x \cdot y + x - y = 3{,}3\ [6]$$
$$\text{II)}\ x + y = 27;$$

Nr. 2)
$$\text{I)}\ xy + \frac{x}{2} + \frac{y}{3} = 15$$
$$\text{II)}\ x + y = 7;$$

---

4 Hierher gehören die Aufgaben in SKT 7.

5 Im Berliner Papyrus 6619; hierzu K. VOGEL im Archeion XII, 1930, S. 151 ff. und O. NEUGEBAUER in QuellStud. B. 1, 1930, S. 306 ff.

6 Wir schreiben 3,3 für $3 \cdot 60 + 3$, ferner 1,13,30 für $1 \cdot 60^2 + 13 \cdot 60 + 30$. Das „Sexagesimalkomma" wird durch einen Strichpunkt dargestellt. Also: $5;50 = 5 + 50/60$.

Nr. 3)
$$\text{I)}\ x \cdot y + (x + y) \cdot (x - y) = 1{,}13{,}20$$
$$\text{II)}\ x + y = 1{,}40\,;$$

Nr. 4)
$$\text{I)}\ x \cdot y = x + y$$
$$\text{II)}\ x + y + x \cdot y = 9.$$

Die letzte Aufgabe ist unvollständig. Die Lösung, die bei den andern drei Beispielen richtig vorgerechnet wird, fehlt hier. Bei der zweiten Gruppe ist noch einzureihen SKT 6 (s. S. 76 Anm. 1 d), eine Aufgabe mit Lösungsangabe ohne Ausrechnung. Sie lautet:

$$\text{I)}\ \frac{x \cdot y}{7} + \frac{x}{7} + \frac{y}{7} = 2$$
$$\text{II)}\ x + y = 5;\,50.$$

Alle diese Fälle enthalten als eine Gleichung $x + y = A$; in der andern kommt das Produkt xy vor, dessen Terminus (a-šà = Fläche) zeigt, wie auf Grund geometrischer Überlegungen die fortgesetzte Addition (z. B. von Flächenstreifen) zu einer selbständigen Rechenoperation, der Multiplikation führt. Die Voranstellung der Überschrift „Länge, Breite" vor die einzelnen Probleme als „Regiebemerkung" paßt gut zu der einen Gleichung $x + y = A$; sie weist darauf hin, daß es sich—wenigstens ursprünglich—darum handelte, Länge und Breite eines Rechteckes aus den Angaben ausfindig zu machen. Insbesondere galt es in dem einfachsten Fall, aus dem halben Umfang und der Rechteckfläche die beiden Seiten zu berechnen oder wie HERON in einem ähnlichen Fall sagt[7]: „sie (die Zahlen) auseinanderzulegen und jede Zahl zu finden".

Lösen wir das Gleichungssystem

$$\text{I)}\ x + y = a,$$
$$\text{II)}\ x \cdot y = b$$

auf, so erhalten wir mit unserer Formel $x = \frac{a}{2} \pm \sqrt{\left(\frac{a}{2}\right)^2 - b}$. Da x und y vertauschbar sind, führt die Auflösung nach y hier zu denselben Werten. Die Lösung ist eindeutig, da immer die Länge größer als die Breite vorauszusetzen ist. Im Text der TU-Aufgaben werden nun folgende Schritte vorgerechnet:

(1) $\frac{a}{2}$, (2) $\left(\frac{a}{2}\right)^2$, (3) $\left(\frac{a}{2}\right)^2 - b$,

(4) $\sqrt{\left(\frac{a}{2}\right)^2 - b}$, (5) $\frac{a}{2} + \sqrt{\left(\frac{a}{2}\right)^2 - b}$, (6) $\frac{a}{2} - \sqrt{\left(\frac{a}{2}\right)^2 - b}$.

---

[7] Heron (s. S. 76 Anm. 2) sagt: *Δοθέντων συναμφοτέρων τῶν ἀριθμῶν … ἐν ἀριθμῷ ἑνὶ διαστεῖλαι καὶ εὑρεῖν ἕκαστον ἀριθμόν.*

Die äußerliche Übereinstimmung mit unserer Endformel $x = \frac{a}{2} \pm \sqrt{\left(\frac{a}{2}\right)^2 - b}$
78 führte zu | der Ansicht, daß der Rechner „notwendigerweise den vollen Überblick über den ganzen Formelmechanismus der Aufgabe“ besaß[8]. Dies würde aber die Kenntnis der Doppeldeutigkeit von Schritt (4) voraussetzen, wofür keine Textbelege vorhanden sind.

Man kann aber auch auf folgende Weise zu den im Text niedergelegten Rechenschritten kommen: Schritt (1)—(3) wie oben:

$$(1)\ \frac{x+y}{2} = \frac{a}{2}, \quad (2)\ \left(\frac{x+y}{2}\right)^2 = \left(\frac{a}{2}\right)^2, \quad (3)\ \left(\frac{x+y}{2}\right)^2 - xy = \left(\frac{a}{2}\right)^2 - b.$$

Wegen der Identität $\left(\frac{x+y}{2}\right)^2 - xy = \left(\frac{x-y}{2}\right)^2$ bedeutet (3) auch: $\left(\frac{x-y}{2}\right)^2 = \left(\frac{a}{2}\right)^2 - b.$

Schritt (4) ist dann $\frac{x-y}{2} = \sqrt{\left(\frac{a}{2}\right)^2 - b}$. Schließlich wird mit

$$(5)\quad x = \frac{x+y}{2} + \frac{x-y}{2} = \frac{a}{2} + \sqrt{\left(\frac{a}{2}\right)^2 - b}$$

und mit $(6)\quad y = \frac{x+y}{2} - \frac{x-y}{2} = \frac{a}{2} - \sqrt{\left(\frac{a}{2}\right)^2 - b}$ gefunden.

Die Kenntnis der genannten Identität, die nur auf der Bekanntschaft mit den Formeln $(x \pm y)^2 = x^2 \pm 2\,xy + y^2$ beruht, wird man nach all dem, was man jetzt von der Mathematik der Babylonier weiß, ihnen unbedenklich einräumen. Sie ergab sich schon aus der einfachsten Beschäftigung mit rechnender Geometrie. Sie wurde auch von EUKLID, also vor der Zeit der TU-Texte, im Buch II, 5 der „Elemente“ geometrisch formuliert und bewiesen[9]. Der Satz, der die Lösung gerade des hier besprochenen Gleichungssystems I) $x + y = a$, II) $x \cdot y = b$[10] ermöglicht, heißt dort: *'Εὰν εὐθεῖα γραμμὴ τμηθῇ εἰς ἴσα καὶ ἄνισα, τὸ ὑπὸ τῶν ἀνίσων τῆς ὅλης τμημάτων περιεχόμενον ὀρθογώνιον μετὰ τοῦ ἀπὸ τῆς μεταξὺ τῶν τομῶν τετραγώνου ἴσον ἐστὶ τῷ ἀπὸ τῆς ἡμισείας τετραγώνῳ*. Daß dabei die Zwischenstrecke (*ἡ μεταξὺ τῶν τομῶν*)

[8] S. SCHUSTER a. a. O. S. 198.

[9] EUKLID, Opera I (ed. HEIBERG-MENGE) S. 128ff.

[10] Siehe zu der Behandlung der quadratischen Gleichung bei EUKLID: TH. L. HEATH, A History of Greek Mathematics, Oxford 1921, I, S. 151ff. und 379f. sowie J. TROPFKE, Geschichte der Elementarmathematik III, 2. Aufl., Berlin-Leipzig 1922, S. 37ff.

$\frac{a+b}{2} - b = \frac{a-b}{2}$ ist[11], ergibt sich auch ohne algebraische Umrechnung unmittelbar aus unterstehender Abbildung.

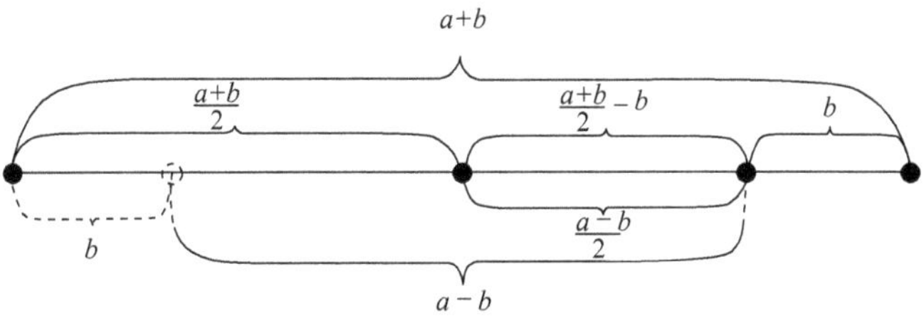

Wir haben sogar eine Beschreibung der Berechnung, und zwar bei DIOPHANT in dem zweiten Buch der Arithmetica[12]. Es heißt dort: εὑρεῖν δύο ἀριθμοὺς ὅπως ἡ σύνθεσις αὐτῶν καὶ ὁ πολλαπλασιασμὸς ποιῇ δοθέντας ἀριθμούς. Zuerst wird die unabweisbare Notwendigkeit betont (ἔστι δὲ τοῦτο πλασματικόν), daß $\left(\frac{a}{2}\right)^2 > b$ sein muß. DIOPHANT geht von der Differenz der beiden Größen $(x - y)$ aus, die als $2d$ (bei D. zwei Unbekannte = $ϛ\beta$) bezeichnet wird. Dann wird gezeigt, daß man die größere Zahl findet, wenn man $d = \frac{x-y}{2}$ zu der halben Summe $\frac{a}{2} = \frac{x+y}{2}$ addiert; es ist also $x = \frac{x+y}{2} + \frac{x-y}{2} = \frac{a}{2} + d$, während $\frac{a}{2} - d = \frac{x+y}{2} - \frac{x-y}{2}$ die kleinere Zahl y ergibt. Es kommt also nur darauf an, die halbe Differenz $\frac{x-y}{2}$ zu finden, genau das, was auch in den babylonischen TU-Texten im Schritt (1)—(4) geschieht. Bei den speziellen Werten $a = 20$ | und $b = 96$ bei DIOPHANT ist die halbe 79
Differenz $\sqrt{10^2 - 96} = 2$; x ist dann $\frac{a}{2} + 2, y = \frac{a}{2} - 2$. Die zweite Lösung $\sqrt{4} = -2$ kam für die damalige Zeit nicht in Betracht.

---

11 HEATH liest aus EUKLID II 5—wie HEIBERG—die „Formel“: $ab + \left(\frac{a+b}{2} - b\right)^2 = \left(\frac{a+b}{2}\right)^2$, während TROPFKE (Geschichte ..., II, 3. Aufl., 1932, S. 119) den Satz direkt als: $ab + \left(\frac{a-b}{2}\right)^2 = \left(\frac{a+b}{2}\right)^2$ deutet.

12 Opera I, 27, ed. TANNERY S. 60 ff. [*Note*: For the true meaning πλασματικόν, see pp. 192-3 in the edition of the Arabic Diophantus.]

Ein entsprechendes Lösungsverfahren scheint mir auch bei anderen Beispielen der zweiten Aufgabengruppe vorzuliegen. So lautet AO 8862 Nr. 1 in moderner Formulierung:

$$\text{I)}\ x \cdot y + x - y = 3{,}3 \qquad \text{II)}\ x + y = 27$$

oder allgemein[13]:

$$\text{I)}\ xy + x - y = \Sigma \qquad \text{II)}\ x + y = S.$$

Der Text enthält folgende neun Einzelschritte zur Berechnung der Unbekannten:

$$(1)\quad \Sigma + S, \qquad (2)\quad S + 2, \qquad (3)\quad \frac{S+2}{2},$$

$$(4)\quad \left(\frac{S+2}{2}\right)^2, \quad (5)\quad \left(\frac{S+2}{2}\right)^2 - (\Sigma + S), \quad (6)\quad \sqrt{\left(\frac{S+2}{2}\right)^2 - (\Sigma + S)},$$

$$(7)\quad \frac{S+2}{2} + \sqrt{} = x, \quad (8)\quad \frac{S+2}{2} - \sqrt{} = y', \quad (9)\quad y' - 2 = y.$$

NEUGEBAUER sieht in dieser Berechnung (Schritt 1—8) die Auflösung der für x umgeformten Gleichung $x^2 - x(2 + S) + S + \Sigma = 0$. Die Aufgabe sei „genau so durchgerechnet worden, wie man es heute auch machen würde“. Dies ergibt:

$$\left.\begin{matrix} x_1 \\ x_2 \end{matrix}\right\} = \frac{1}{2}(2 + S) \pm \sqrt{\frac{1}{4}(2 + S)^2 - (S + \Sigma)},$$

wobei die Ergebnisse 15 und 14 von (7) und (8) als x_1 und x_2 sich errechnen. Statt der jetzt notwendigen Doppellösung für y zeigt der Text aber nur $y_1 = x_2 - 2 = 12$ in (9).

Ich gebe den einzelnen Schritten eine andere Deutung. Die Addition des ziemlich unmotiviert auftretenden Summanden 2 in (2) schreibe ich der Absicht zu, durch die Substitution $y' = y + 2$ das komplizierte Problem auf die Normalaufgabe zu bringen. Es ist dann

$$(1)\quad \Sigma + S = 2x + xy = x(y + 2)$$

$$(2)\quad S + 2 = x + (y + 2) = x + y'.$$

Demnach ist das neue Problem:

$$\text{I)}\ xy' = 3{,}30$$

$$\text{II)}\ x + y' = 29.$$

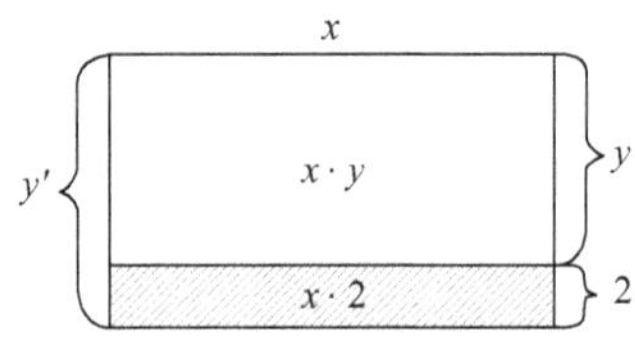

[13] Nach NEUGEBAUER a. a. O. (s. S. 76 Anm. 1 d) S. 12.

In den folgenden Schritten (3)—(8) wird dann alles in der oben geschilderten Weise durchgeführt. Die Größe y' ist dann nicht mehr „irrtümlich", sondern tatsächlich eine Breite. Geometrisch wirkt sich das Verfahren etwa in *obenstehender* Weise als ein Ansetzen eines Streifens an das alte Rechteck xy aus.

Ganz korrekt wird jetzt das in (9) berechnete $y = y' - 2$ als „ursprüngliche" bzw. „exakte" Breite bezeichnet und die Bemerkung „erstens" und „zweitens"[14] in (7) und (8) ist ganz am Platze. Eine Doppellösung tritt nicht auf, das Problem ist, ohne daß ein Wert übersehen wurde, eindeutig gelöst.

Bei AO 8862 Nr. 2 kommt eine weitere Schwierigkeit hinzu. Hier heißt das Gleichungssystem

$$\left.\begin{array}{r} \text{I)}\ xy + \dfrac{x}{2} + \dfrac{y}{3} = 15, \\ \text{II)}\ x + y = 7, \end{array}\right\} \text{ bzw. } \left\{\begin{array}{r} xy + \dfrac{x}{\alpha} + \dfrac{y}{\beta} = \Sigma, \\ x + y = S. \end{array}\right.$$

NEUGEBAUER nimmt die für y entwickelte quadratische Gleichung $y^2 - y\left(S - \dfrac{\beta - \alpha}{\alpha\beta}\right) + \Sigma - \dfrac{S}{\alpha} = 0$ an mit den Lösungen

$$\begin{Bmatrix} y_1 \\ y_2 \end{Bmatrix} = \frac{1}{2}(S - \gamma) \pm \sqrt{\frac{1}{4}(S - \gamma)^2 - \left(\Sigma - \frac{S}{\alpha}\right)}, \text{ wobei } \gamma = \frac{\beta - \alpha}{\alpha\beta} \text{ gesetzt ist.}$$

Wieder fehlt ein Wert, diesmal für x. Ich sehe in der Ausrechnung auch hier das Bestreben des Rechners, das Gleichungssystem in die Norm $xy = A$; $x + y = B$ zu | transformieren. Der erste Schritt hierzu ist 80

$$(1)\quad \frac{S}{\alpha} = \frac{x}{\alpha} + \frac{y}{\alpha}.$$

Der nächste Schritt

$$(2)\quad \Sigma - \frac{S}{\alpha}$$

mit darauffolgender Berechnung von $\gamma = \dfrac{\beta - \alpha}{\alpha\beta}$ bezweckt die Umwandlung $\Sigma - \dfrac{S}{\alpha} = xy + \dfrac{y}{\beta} - \dfrac{y}{\alpha} = y(x - \gamma)$. In (3) wird die zweite Gleichung $x + y = S$ übergeführt in $x + y - \gamma = S - \gamma$. Jetzt ist die Normalform hergestellt als

$$\text{I)}\quad x'y = \Sigma - \frac{S}{\alpha},$$

$$\text{II)}\quad x' + y = S - \gamma.$$

[14] Dem Fehlen dieser Bemerkungen in AO 8862 Nr. 3 möchte ich keine große Bedeutung beimessen. In TU fehlen sie ebenfalls. THUREAU-DANGIN übersetzt die betreffenden Stellen anders. Er meint: addiere zu dem ersten $\frac{a}{2}$ bzw. subtrahiere von dem zweiten $\frac{a}{2}$ den gefundenen Wert (nämlich die Wurzel), was geometrisch an der oben S. 78 gegebenen Abbildung klar wird.

Alles weitere entspricht dem Normalverfahren, das zu einer eindeutigen Lösung ohne Verlust eines Wertes führt. Dabei macht die „Bruchrechnung" $\frac{1}{\alpha} - \frac{1}{\beta} = \frac{\beta - \alpha}{\alpha\beta}$ dem arithmetischen Können der Babylonier im 3. Jahrtausend v. Chr. alle Ehre.

Auch die Aufgabe in SKT 6[15], bei der nur das Ergebnis ohne Berechnung genannt wird, sowie AO 8862 Nr. 4, von der nur die Angabe erhalten ist[16], lassen sich leicht, da es sich um einfache Fälle handelt, dem Normalverfahren der zweiten Gruppe einordnen.

Dagegen macht das Problem AO 8862 Nr. 3, das wegen des Auftretens der „Fläche" in der einen Gleichung zu der zweiten Gruppe gerechnet wurde, größere Schwierigkeiten. Es ist m. E. auf eine andere Weise gelöst worden. Die Gleichungen lauten:

$$\left.\begin{array}{ll} \text{I)} & xy + (x+y)(x-y) = 1{,}13{,}120 \\ \text{II)} & x + y = 1{,}40 \end{array}\right\} \quad \text{bzw.} \quad \begin{cases} xy + S(x-y) = \Sigma \\ x + y = S. \end{cases}$$

NEUGEBAUER nimmt eine Gleichung für y an: $y^2 + Sy + \Sigma - S^2 = 0$. Diese ergibt modern berechnet:

$$\left.\begin{array}{l} y_1 \\ y_2 \end{array}\right\} = -\frac{S}{2} \pm \sqrt{\left(\frac{S}{2}\right)^2 + (S^2 - \Sigma)} = \begin{cases} \frac{S}{2} - (S - \sqrt{}) = 40 = y \\ -2{,}20 \end{cases}$$

und

$$\left.\begin{array}{l} x_1 \\ x_2 \end{array}\right\} = S + \frac{S}{2} \mp \sqrt{} = \begin{cases} \frac{S}{2} + (S\sqrt{}) = 1{,}0 = x^{17} \\ 4{,}0. \end{cases}$$

Ich lege die einzelnen Schritte folgendermaßen aus:

$$(1)\quad S^2 = x^2 + 2xy + y^2, \qquad (2)\quad S^2 - \Sigma = 2y^2 + xy,$$

$$(3)\quad \frac{S}{2} = \frac{x+y}{2}, \qquad (4)\quad \left(\frac{S}{2}\right)^2 = \frac{x^2 + 2xy + y^2}{4},$$

---

15 S. S. 76 Anm. 1 b.

16 Nach THUREAU-DANGIN heißt das Gleichungssystem:

$$\text{I)}\quad x + y = xy, \quad \text{II)}\quad x + y + xy = 9.$$

NEUGEBAUER hält die Angaben für ungenügend. Seine Übersetzung „Wiederum Länge und Breite habe ich addiert und mit der Fläche entsprechend" (a. a. O. S. 23) stimmt aber doch gerade mit seiner Vermutung überein, daß der für „entsprechend" gebrauchte Terminus maḫâru = gegenüberstellen ein Fachwort für das Vorliegen einer Gleichung darstellt. THUREAU-DANGIN übersetzt: „En second lieu, j'ai additionné le flanc et le front: (la somme) est égale à la surface."

17 Die Übersicht soll nach NEUGEBAUER (a. a. O. S. 22) zeigen, wie der negative Wert für y_2 umgangen wurde. Bei meinem Lösungsvorschlag kommt ein negativer Wert überhaupt nicht in Frage.

$$(5)\quad (S^2 - \Sigma) + \left(\frac{S}{2}\right)^2 = \frac{x^2 + 6xy + 9y^2}{4} = \left(\frac{x + 3y}{2}\right)^2,$$

$$(6)\quad \sqrt{S^2 - \Sigma + \left(\frac{S}{2}\right)^2} = \frac{x + 3y}{2}.$$

Durch den nächsten Schritt soll wieder die halbe Differenz $\frac{x - y}{2}$ erreicht werden. Es ist weiter:

$$(7)\quad S - \sqrt{S^2 - \Sigma + \left(\frac{S}{2}\right)^2} = x + y - \frac{x + 3y}{2} = \frac{x - y}{2},$$

| $$(8)\quad \frac{S}{2} + \left[S - \sqrt{S^2 - \Sigma + \left(\frac{S}{2}\right)^2}\right] = \frac{x + y}{2} + \frac{x - y}{2} = x,$$ 81

$$(9)\quad \frac{S}{2} - \left[S - \sqrt{S^2 - \Sigma + \left(\frac{S}{2}\right)^2}\right] = \frac{x + y}{2} - \frac{x - y}{2} = y.$$

Bei dieser Erklärung, die mit dem überlieferten Text im Einklang steht, ist auch AO 8862 Nr. 3 eindeutig gelöst und es entfällt der weitgehende Schluß auf die Kenntnis der Doppeldeutigkeit der Lösungen[18]. Dagegen zeigt die Aufgabe eine virtuose Technik in der Schaffung eines vollständigen Quadrates, wie sie erst wieder bei DIOPHANT auftritt, weshalb gerade diese viertausendjährigen Probleme geeignet sind, ein klares Licht auf die Quellen zu werfen, die dem größten griechischen Logistiker noch zur Verfügung stehen mochten.

---

[18] Daß die Babylonier ein Gleichungssystem sowohl nach x als nach y umformen konnten, erscheint mir nach der Behandlung der Aufgaben in SKT 7 recht wahrscheinlich. Die Lösung dieser zur ersten Gruppe gehörenden Aufgaben ist noch nicht geklärt. NEUGEBAUER denkt sie sich nach unserer Methode berechnet. Sie stehen aber der Heronischen Lösung näher, bei der zur Schaffung eines quadratischen ersten Gliedes mit dem Koeffizienten von x multipliziert und nicht wie bei uns dividiert wurde. Hierzu TROPFKE a. a. O. (s. S. 78 Anm. 3) S. 40. Auffallend sind die der jeweiligen Angabe zugrundeliegenden Zahlenwerte (100, 10, 5, 20, 2225, 3125). Stammen sie aus dezimalen Gedankengängen?

G. J. TOOMER

LOST GREEK MATHEMATICAL WORKS IN ARABIC TRANSLATION

The works of the ancient Greek mathematicians which we possess represent only a few fragments from the wreck of the great treasure ship of Hellenistic mathematics. What has come down to us is little more than a reflection of the pedagogical interests of the schoolmen of late antiquity and Byzantine times, who caused to be copied only those works which were of some use in the curricula of their institutions of higher education at Alexandria, Antioch, Athens, Constantinople, and a very few other places. Their choice illustrates the impoverished intellectual climate of the Greek world in the millennium from the third to the thirteenth century A.D. Thus the compendium of elementary geometry which goes under the name of Euclid was transmitted through the schoolrooms, but none of the works on higher geometry which Euclid wrote (probably in the early third century B.C.) has been preserved; and only the first four books of Apollonius' *Conics*, which treat the elements of the theory, continued to be copied in Byzantine times: The last four books, which deal with more advanced topics, are lost in Greek.

Moreover, not only has the quantity of mathematical works which have come down to us been greatly diminished, but the kinds of mathematics which they represent give a very narrow and distorted picture of the actual range of the science in Hellenistic times, which was much wider than is popularly supposed. The conventional modern view of Greek mathematics is that it was essentially geometry, with some arithmetic, and this is indeed the impression that the surviving texts give, although the existence of a work like Archimedes' *Sandreckoner*, with its remarkable invention *ab ovo* of a place value system for expressing very large numbers, already suggests that this impression is too hasty. When we examine the evidence for lost mathematical works, we get a very different picture.

I will limit myself to just two examples. It appears that the famous astronomer Hipparchus wrote a work in which he derived the number of different combinations of 10 logical axioms. Were it not for a few references to this by the popular philosopher Plutarch, we should have no inkling that the topic of combinatorial arithmetic was broached by the Greeks at all.

Second, nowhere in extant Greek mathematical literature do we find the solution to the indeterminate equation (in integers)

$$by - ax = c$$

the simplest form of the "linear Diophantine equation," as it is called (somewhat misleadingly, since Diophantus never discussed equations of this particular type). Now the solution to this equation, using the "Euclidean algorithm," is found in Indian astronomical treatises from the fifth century A.D. on, and, although strict proof is lacking,

J. Christianidis (ed.), Classics in the History of Greek Mathematics, 275–284.

I have no doubt that, like most other procedures in these treatises, the solution was derived from Greek works of the earlier Hellenistic period which are otherwise lost to us. The moral is that one should be cautious in making confident pronouncements that "the Greeks did not know such-and-such a branch of mathematics."

Although most of the Greek mathematical works which have been lost can never be recovered (there is no hope, for instance, that we will ever see an original work of Hippocrates of Chios, Theaetetus, or one of the other pioneers from the earliest period of
33 Greek mathematics before 400 B.C.), there are two possible | sources of retrieval. One is through the discovery of papyri from Greco-Roman Egypt. Among the many thousands of Greek papyri dating from the third century B.C. to beyond the sixth century A.D. which have been excavated in the last 150 years are a number which are of a mathematical nature, but it must be admitted that, so far (unlike the astronomical papyri, which have revealed much new and surprising material), these have been disappointing, since they are either fragments of existing texts, such as those of Euclid, or collections of problems of a dismally low level. Much more fruitful has been the second source, Arabic translations of Greek mathematical treatises.

Serious interest in translating Greek scientific works into Arabic was part of the general intellectual ferment associated with establishment of the ʿAbbāsid caliphate, centered at Baghdad and ruling a vast territory in the Near East, Central Asia, and North Africa from 750 A.D. onward. Translations were first made in the late eighth century and the process continued for little more than 100 years, but during that period hundreds of works on all the sciences known to antiquity (including the pseudo-sciences of astrology and magic) were translated.

In the eighth and ninth centuries degeneration of the ancient Greek mathematical tradition was already well advanced, but the chances of finding unusual mathematical works were much better than seven centuries later, when western Europeans began actively to seek out Greek manuscripts. The situation in late antiquity is well documented by the *Mathematical Collection* of Pappus (taught at Alexandria in the fourth century), most of which survives in Greek. Although Pappus' own contributions are characterized by the sterile scholasticism of his time, he has preserved summaries of and excerpts from a large number of now lost Hellenistic mathematical works of the highest interest (which, through the medium of Commandino's Latin translation of Pappus, served as an inspiration for most of the best mathematicians of western Europe in the sixteenth and seventeenth centuries). It is clear that, for a significant number of these works, including some of the treatises of Apollonius, old manuscripts survived here and there down to the ninth century and were still available when the translators or their agents came west from Baghdad to the Byzantine Empire in search of Greek scientific works.

For a number of mathematical works, indeed, translators were apparently able to acquire only a single Greek manuscript (perhaps even the last manuscript of that work in existence). Thus the Arabic translation of Ptolemy's *Optics* was made from a manuscript which was defective at both beginning and end, so that the whole of Book 1 and the last part of Book 5 are lost irrevocably [1]. Moreover, although the brothers Banū Mūsā, who sponsored the translation of Apollonius' *Conics* in the ninth century, had access, like us, to several manuscripts of the first four books, they were evidently able to find only one manuscript which contained the later books, and that too was defective at the end, so that only Books 1 to 7 were translated and Book 8 remains lost, probably forever.

Some of the mathematical works which had been translated into Arabic and were later lost in Greek became known to western Europe during the Middle Ages through Latin translations from the Arabic, mostly made in Spain and Sicily during the twelfth century. Examples are the *Spherics* of Menelaus and the *Planispherium* of Ptolemy, both of which will be mentioned later. But it is probable that many of the most interesting treatises were simply not available in those Western outposts of Islam, which were in serious intellectual decline by the time the Latin translators began their search for Greek wisdom in Arabic dress.

The revival of interest in Greek mathematics in western Europe, which began at the end of the fifteenth century, reached its culmination in the seventeenth century, which also saw the beginnings of the collections of Arabic manuscripts in the great European libraries. The recovery of the first four books of Apollonius' *Conics* had been a seminal event in Renaissance mathematics (I mention only its use by Kepler in his *Astronomia Nova*), and mathematicians were eager to obtain the last four books as well. Although it was known quite early in the seventeenth century that there was hope of recovering these from the Arabic, there was no satisfactory presentation of them until 1710, when Edmund Halley provided a Latin translation from the Arabic of Books 5 to 7 in his great edition of the *Conics* [2].

Halley, who had learned Arabic expressly in order to edit the works of Apollonius, did a remarkable job for his time, and his translation of these three books has served as the basis for all later published translations of and commentaries on them. But his version has serious deficiencies from the point of view of the modern philologist. First, he treated the text as a living piece of mathematics (although his edition was at least 50 years too late for this; by 1710 its interest was largely historical), so he did not hesitate to "improve on" Apollonius' presentation, while retaining its mathematical essence. Second, he relied mainly on a single Arabic manuscript, Marsh 667 in the Bodleian Library (see Fig. 1), which, though an old and good one, is not without faults and omissions. Finally, he did not actually print the Arabic text, so that one could not check his translation (which is also not faultless). For the edition of the Arabic text which I am preparing I have the advantage of a wider manuscript base, in particular a manuscript in Istanbul written by the great mathematician ibn al-Haytham, best known in the West, under the name Alhazen, for his *Optics*; see Fig. 2.

| Halley had earlier published his translation from the Arabic of another lost work 34
of Apollonius, *On the Cutting-Off of a Ratio* [3], but after him there was no serious attempt to exploit the resources of European libraries to recover Greek scientific works from Arabic versions until the late nineteenth century. Among the treatises of minor interest which were recovered in this way one may mention the commentary of Pappus to Book 10 of Euclid (which is of more value for what it tells us about Apollonius than for Pappus' own comments) [4], and the treatise of Menelaus (late first century A.D.) on the mathematics of specific gravities (see Fig. 3) [5]. But the turning point came with the investigation of libraries of Arabic manuscripts in the lands of Islam.

By the twentieth century sizable collections of Arabic manuscripts had been assembled in Paris, London, Berlin, Leiden, and other libraries of Europe and North America. These were fairly well catalogued, and most of the contents known. But the great majority of Arabic manuscripts remained in Islamic lands (Turkey, the Near East, North Africa, Iran, and India). Although there too they were mostly in libraries, the situation was very different from that in Europe. For most of the libraries there was

Fig. 1. Bodleian Library, Marsh 667, f. 114 (the manuscript used by Halley for his translation of Apollonius' Conics *5 to 7). This page contains part of Book 5, Prop. 64 (on drawing minima to a parabola).*

no printed catalogue, and they were difficult of access, especially for non-Muslims.
Moreover, the sheer number of manuscripts was daunting. H. Ritter has estimated that
in the libraries of Istanbul alone there are 124,000 manuscripts [6]. This includes
Turkish and Persian as well as Arabic works and the majority are undoubtedly of a
35 religious | nature, but the task confronting the investigators of these treasures remains
a gigantic one.

In the area of mathematics, pioneering work was done by Max Krause, who published a list of the manuscripts containing mathematical works which he had inspected in the libraries of Istanbul, including a number of translations from the Greek [7]. (Krause went on to publish an edition of the Arabic text of the translation of Menelaus' *Spherics*, fundamental to the development of spherical trigonometry by the Greeks, before his death on the Russian front in World War II cut short a promising

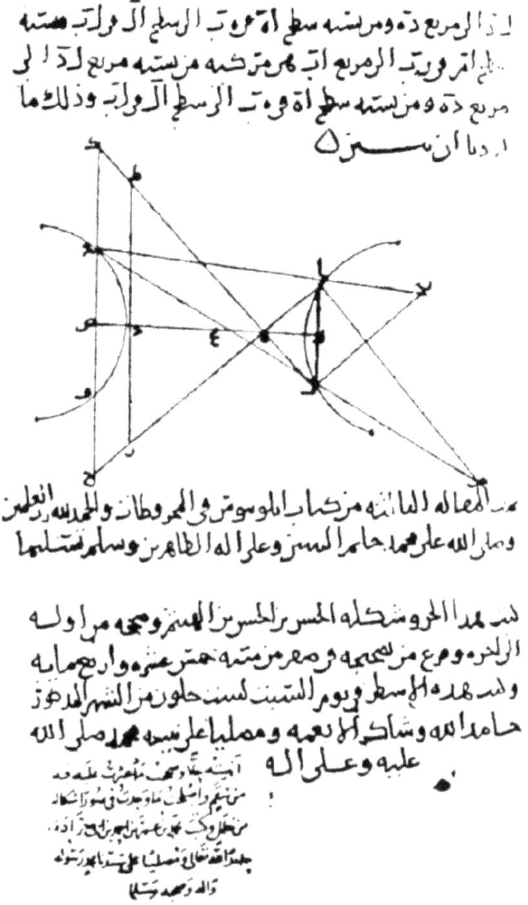

Fig. 2. Istanbul, Aya Sofya 2762, f. 135 (manuscript of Apollonius' Conics *in the hand of ibn al-Haytham). This page reads in part, "This part was written and provided with figures by al-Ḥasan ibn al-Ḥasan ibn al-Haytham, and he corrected it from the beginning to the end, and finished correcting it in [the month] Ṣafar of year 415 [of the Hijra]. And he wrote these lines on Saturday the sixth of the month [A.D. 1024, April 18th]."*

career.) But the great increase during the last 20 years in our information about the manuscript materials in all areas of Arabic studies is due principally to the work of Fuat Sezgin, whose investigations in the libraries of Islamic lands are now illuminating many different fields through his monumental history of Arabic literature [8]. Like

Fig. 3. Escorial, Arabic 960, f. 43. Title page of the unique manuscript of the treatise of Menelaus on specific gravities. It begins, "Menelaus to the Emperor Domitian on the clever method of determining the amount of a number of bodies mixed together," and goes on to tell the famous story of Archimedes and the crown of King Hieron.

many others, I am greatly in Prof. Dr. Sezgin's debt, not only for the storehouse of information in his publications, particularly on the Greek sources of Arabic works, but also for his personal help in locating and obtaining photographs of manuscripts of works which concerned me.

As a result of the efforts of the above scholars, and also of some progress in the publication of catalogues of Eastern libraries, a number of hitherto unknown Arabic translations of Greek mathematical works have come to light recently. Two of these are from the same source, the Shrine Library at Meshhed in Iran. The first | is *On Burning* 36
Mirrors by Diocles [9], a minor mathematician who lived about 200 B.C. and was thus a contemporary of Apollonius. Despite its title, the work is only partly concerned with burning mirrors and is in fact a short miscellany of topics in higher geometry; it is of unusual historical interest because it answers a number of questions (and raises new ones) about development of the theory of conic sections before Apollonius. Besides dealing with the focal property of the parabola (as one would expect), it also describes the first construction of the parabola from focus and directrix and presents two solutions to the classical problem of doubling the cube (finding two mean proportionals). The latter were already known, in part, from excerpts in the work of the late scholastic writer Eutocius, but the Arabic translation reveals how much Eutocius' presentation distorts the original.

A similar lesson about the untrustworthiness of the Greek mathematical tradition as it has come down to us through Byzantine hands can be learned from the second work preserved at Meshhed, part of the algebraic treatise of Diophantus (see Fig. 4) [10]. Diophantus' work, which deals with the solution of many types of indeterminate equations in rational (but not necessarily integer) numbers, was originally written in 13 books. Part of it is preserved in Greek and numbered as Books 1 to 6. The Arabic manuscript contains Books 4 to 7, and thus one would expect it to coincide in large part with the extant Greek text. In fact it is almost entirely different, and an examination of in|ternal consistency shows that the Arabic preserves the original Books 4 to 7 37
of Diophantus, which follow logically after Books 1 to 3 of the Greek, and that "Books 4 to 6" of the Greek must have been cobbled together mainly from the later Books 8 to 13 of the original work of Diophantus. The new material the Arabic translation brings is of great interest, and although it does not change our notions of Diophantus' mathematical methods, it considerably enlarges them.

Even when we possess the Greek original of an ancient work, the existence of an Arabic translation of it can often be illuminating. The reason is that the translations were mostly made in the ninth century, a date far earlier than that at which most of the Greek manuscripts on which we depend were written. A striking example is provided by the recent discovery of the Arabic translation of Book 8 of Pappus' *Mathematical Collection* [11] (apparently the only part of that work ever translated into Arabic). Not only does this material help to improve the corrupt text of the Greek archetype, but it also contains a whole section missing from our Greek version, a section of remarkable interest because it treats geometrical constructions in which the compass is restricted to a single opening. This topic has been discussed by mathematicians at various times, often in ignorance of similar work by their predecessors. Thanks to this discovery, we now know that Arabic work on the topic (that of Abū' l-Wafā' in the tenth century) was inspired by knowledge of the Greek treatment, and I am sure that the Greeks | fully 38
anticipated the results of Cardano and Ferrari in the sixteenth century; I would not

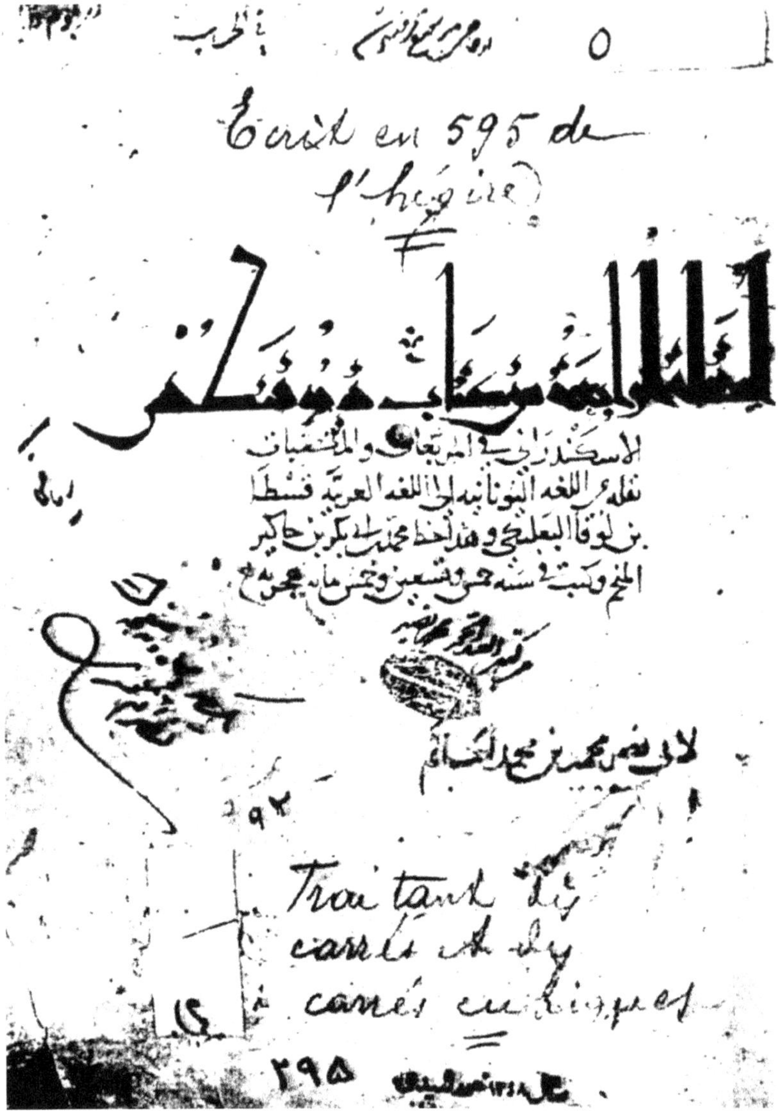

Figure 4. Meshhed, Shrine Library 295 (manuscript of Diophantus, Books 4 to 7, title page). Reads in part, "Fourth Book of the treatise of Diophantus of Alexandria on squares and cubes, which Qusṭā ibn Lūqā of Baalbek translated from the Greek language into the Arabic language."

be surprised if they also anticipated the work done independently by G. Mohr in the seventeenth century and Mascheroni in the eighteenth century [12].

The translation of Greek mathematical works into Arabic was an essential stimulus to the burgeoning of mathematics in the lands of Islam from the ninth century onward. The mathematicians who wrote in Arabic (and sometimes in Persian and Turkish) are in a true sense heirs of the Greeks, since they continued and enlarged the ancient tradition with original and sometimes profound contributions (in contrast to the Byzantine Empire, where the most that was done was to preserve some of the ancient legacy as a dead letter). Their works are worthy of study in their own right and also as a form of transmission of the ancient tradition. An extraordinary example which has only recently come to light is ibn al-Haytham's *Completion of the Conics of Apollonius*, in which the author supplies the theorems he supposes must have been in Book 8 of the *Conics* [13]. Although it is hardly successful as a "restoration" of the missing book, it is remarkable as an exercise in often very difficult problems in conics and displays the author's familiarity with and expertise in Apollonian methods.

The works of the mathematicians written in Arabic from the ninth to eleventh centuries also reveal that they had available translations of many other works of Greek mathematicians which are not known to be still extant. For example, they give references to or quotations from the *Analemma* of Diodorus (an early Hellenistic work on the mathematics of sundials), Apollonius' *Determinate Section* and *Tangencies*, and Archimedes' *Construction of the Regular Heptagon*. Even with our improved knowledge of the contents of Eastern libraries, there is a good chance that manuscripts of some of these or of other ancient works may yet turn up.

In the meantime there is plenty of work to do in making available material known to exist. We are still awaiting the long-promised edition of the Arabic text of Ptolemy's *Planispherium*, a work on stereographic projection, the mathematical basis of the astrolabe, of which there is a manuscript in Istanbul, but for which we still have to use the obscure and faulty medieval Latin translation (also made from the Arabic). We look forward to the publication of A. I. Sabra's edition and translation of the original text of ibn al-Haytham's *Optics* (a conscious attempt to supplement and improve on Ptolemy's optical work) [14], for which only a bad Renaissance printing of a medieval Latin translation is currently available. Finally, although much progress has been made recently in investigating and publishing the mathematical works written in Arabic during the great period following the Greek translations, what has actually been printed (to say nothing of translated) is only a small fraction of what is extant in manuscript.

REFERENCES

1. The Arabic translation has been lost in its turn, but for a bad Latin translation made from the Arabic in the twelfth century, see A. Lejeune, ed. (1956) *L'Optique de Claude Ptolémée*, Louvain
2. *Apollonii Pergaei Conicorum libri octo et Sereni Antissensis de sectione cylindri et coni libri duo* (1710) Oxford
3. *Apollonii Pergaei de sectione rationis libri duo* (1706) Oxford
4. William Thomson and Gustav Junge, eds. (1930) *The Commentary of Pappus on Book X of Euclid's Elements*, Cambridge, Mass.
5. J. Wurschmidt (1925) Die Schrift des Menelaus über die Bestimmung der Zusammensetzung von Legierungen, *Philologus* 80 : 377–409. Wurschmidt gives only a German translation. The Arabic text has never been printed

6. Hellmut Ritter (1953) Autographs in Turkish Libraries, *Oriens* 6: 63
7. Max Krause (1936) Stambuler Handschriften islamischer Mathematiker, *Quellen Stud. Gesch. Math. Astron. Phys. B3:* 437–532
8. Fuat Sezgin (1967–) *Geschichte des Arabischen Schrifttums*, Leiden (eight volumes published to date)
9. G. J. Toomer, ed. and trans. (1976) *Diocles on Burning Mirrors*, Sources in the History of Mathematics and Physical Sciences 1, Berlin-Heidelberg-New York
10. Jacques Sesiano, ed. and trans. (1982) *Book IV to VII of Diophantus' Arithmetica in the Arabic Translation Attributed to Qusṭā ibn Lūqā*, Sources in the History of Mathematics and Physical Sciences 3, New York-Heidelberg-Berlin
11*. The work will be published, with a translation, by D. E. P. Jackson in Sources in the History of Mathematics and Physical Sciences. See provisionally D. E. P. Jackson (1980) Towards a Resolution of the Problem of τὰ ἑνὶ διαστήματι γραφόμενα in Pappus' Collection Book VIII," *Classical Q.* 30: 523–533
12. On the history of the problem see W. M. Kutta (1897) Zur Geschichte der Geometrie mit constanter Zirkelöffnung, *Nova Acta Leopoldina* 81:71–101. Kutta, however, could not have known either of Mohr or of the Arabic Pappus
13*. J. Hogendijk, ed. and trans. (1983) (Doctoral thesis, Utrecht), to be published in Sources in the History of Mathematics and Physical Sciences
14*. Published so far: *The Optics of ibn al-Haytham* (1983) Books I–III, the Arabic text, Kuwait

**Note:* The forthcoming editions mentioned above which did appear are Hogendijk's *Ibn al-Haytham* [13] and Toomer's edition of the Arabic translation of Apollonius's *Conics*, Books V to VII (vol. 7, 1985, and 9, 1990 of the *Sources in the History of Mathematics and Physical Sciences*); the English of Sabra's *Optics of Ibn al-Haytham* [14] is now published and the rest is currently in press in Kuwait.

THOMAS L. HEATH

DIOPHANTUS' METHODS OF SOLUTION

BEFORE I give an account in detail of the different methods which Diophantus employs
for the solution of his problems, so far as they can be classified, it is worth while to
quote some remarks which Hankel has made in his account of Diophantus[1]. Hankel,
writing with his usual brilliancy, says in the place referred to, "The reader will now be
desirous to become acquainted with the classes of indeterminate problems which
Diophantus treats of, and with his methods of solution. As regards the first point, we
must observe that included in the 130 (or so) indeterminate problems, of which
Diophantus treats in his great work, there are over 50 different classes of problems,
strung together on no recognisable principle of grouping, except that the solution of
the earlier problems facilitates that of the later. The first Book is confined to determi-
nate algebraic equations; Books II. to V. contain for the most part indeterminate prob-
lems, in which expressions involving in the first or second degree two or more
variables are to be made squares or cubes. Lastly, Book VI. is concerned with right-
angled triangles regarded purely arithmetically, in which some linear or quadratic
function of the sides is to be made a square or a cube. That is all that we can pronounce
about this varied series of problems without exhibiting singly each of the fifty classes.
Almost more different in kind than the problems are their solutions, and we are com-
pletely unable to give an even tolerably exhaustive review of the different turns which
his procedure takes. Of more general comprehensive methods there is in our author no
trace discoverable: every question requires a quite special method, which often will
not serve even for the most closely allied problems. It is on that | account difficult for 55
a modern mathematician even after studying 100 Diophantine solutions to solve the
101st problem; and if we have made the attempt, and after some vain endeavours read
Diophantus' own solution, we shall be astonished to see how suddenly he leaves the
broad high-road, dashes into a side-path and with a quick turn reaches the goal, often
enough a goal with reaching which we should not be content; we expected to have to
climb a toilsome path, but to be rewarded at the end by an extensive view; instead of
which our guide leads by narrow, strange, but smooth ways to a small eminence; he
has finished! He lacks the calm and concentrated energy for a deep plunge into a sin-
gle important problem; and in this way the reader also hurries with inward unrest from
problem to problem, as in a game of riddles, without being able to enjoy the individ-
ual one. Diophantus dazzles more than he delights. He is in a wonderful measure
shrewd, clever, quick-sighted, indefatigable, but does not penetrate thoroughly or
deeply into the root of the matter. As his problems seem framed in obedience to no
obvious scientific necessity, but often only for the sake of the solution, the solution
itself also lacks completeness and deeper signification. He is a brilliant performer in

---

[1] *Zur Geschichte der Mathematik in Alterthum und Mittelalter*, Leipzig, 1874, pp. 164–5.

J. Christianidis (ed.), Classics in the History of Greek Mathematics, 285–327.

the *art* of indeterminate analysis invented by him, but the *science* has nevertheless been indebted, at least directly, to this brilliant genius for few methods, because he was deficient in the speculative thought which sees in the True more than the Correct. That is the general impression which I have derived from a thorough and repeated study of Diophantus' arithmetic."

It might be inferred from these remarks of Hankel that Diophantus' object was less to teach methods than to obtain a multitude of mere results. On the other hand Nesselmann observes[1] that Diophantus, while using (as he must) specific numbers for numbers which are "given" or have to be arbitrarily assumed, always makes it clear how by varying our initial assumptions we can obtain any number of particular solutions of the problem, showing "that his whole attention is directed to the explanation of the *method*, to which end numerical examples only serve as means"; this is proved by his frequently stopping short, when the method has been made sufficiently clear, and the remainder of the work is mere straightforward calculation. The truth seems to
56 be that there is as much in the shape of general | methods to be found in Diophantus as his notation and the nature of the subject admitted of. On this point I can quote no better authority than Euler, who says:[2] "Diophantus himself, it is true, gives only the most special solutions of all the questions which he treats, and he is generally content with indicating numbers which furnish one single solution. But it must not be supposed that his method was restricted to these very special solutions. In his time the use of letters to denote undetermined numbers was not yet established, and consequently the more general solutions which we are now enabled to give by means of such notation could not be expected from him. Nevertheless, the actual methods which he uses for solving any of his problems are as general as those which are in use today; nay, we are obliged to admit that there is hardly any method yet invented in this kind of analysis of which there are not sufficiently distinct traces to be discovered in Diophantus."

In his 8th chapter, entitled "Diophantus' treatment of equations,"[3] Nesselmann gives an account of Diophantus' solutions of (1) Determinate, (2) Indeterminate equations, classified according to their kind. In chapter 9, entitled "Diophantus' methods of solution,"[4] he classifies these "methods" as follows:[5] (1) "The adroit assumption of unknowns," (2) "Method of reckoning backwards and auxiliary questions," (3) "Use of the symbol for the unknown in different significations," (4) "Method of Limits," (5) "Solution by mere reflection," (6) "Solution in general expressions," (7) "Arbitrary determinations and assumptions," (8) "Use of the right-angled triangle."

At the end of chapter 8 Nesselmann observes that it is not his solutions of equations that we have to wonder at, but the art, amounting to virtuosity, which enabled Diophantus to avoid such equations as he could not technically solve. We look (says

---

1 *Algebra der Griechen*, pp. 308–9.

2 *Novi Commentarii Academiae Petropolitanae*, 1756–7, Vol. VI. (1761), p. 155 = *Commentationes arithmeticae collectae* (ed. Fuss), 1849, I. p. 193.

3 "Diophant's Behandlung der Gleichungen."

4 "Diophant's Auflösungsmethoden."

5 (1) "Die geschickte Annahme der Unbekannten," (2) "Methode der Zurückrechnung und Nebenaufgabe," (3) "Gebrauch des Symbols für die Unbekannte in verschiedenen Bedeutungen," (4) "Methode der Grenzen," (5) "Auflösung durch blosse Reflexion," (6) "Auflösung in allgemeinen Ausdrücken," (7) "Willkürliche Bestimmungen und Annahmen," (8) "Gebrauch des rechtwinkligen Dreiecks."

Nesselmann) with astonishment at his operations, when he reduces the most difficult problems by some surprising turn to a quite simple | equation. Then, when in the 9th 57
chapter Nesselmann passes to the "methods," he prefaces it by saying: "To give a complete picture of Diophantus' methods in all their variety would mean nothing else than copying his book outright. The individual characteristics of almost every problem give him occasion to try upon it a peculiar procedure or found upon it an artifice which cannot be applied to any other problem. ... Meanwhile, though it may be impossible to exhibit all his methods in any short space, yet I will try to describe some operations which occur more often or are particularly remarkable for their elegance, and (where possible) to bring out the underlying scientific principle by a general exposition and by a suitable grouping of similar cases under common aspects or characters." Now the possibility of giving a satisfactory account of the methods of Diophantus must depend largely upon the meaning we attach to the word "method." Nesselmann's arrangement seems to me to be faulty inasmuch as (1) he has treated Diophantus' solutions of equations, which certainly proceeded on fixed rules, and therefore by "*method*," separately from what he calls "methods of solution," thereby making it appear as though he did not look upon the "treatment of equations" as "methods"; (2) the classification of the "Methods of solution" seems unsatisfactory. Some of the latter can hardly be said to be *methods* of solution at all; thus the third, "Use of the symbol for the unknown in different significations," might be more justly described as a "hindrance to the solution"; it is an *inconvenience* to which Diophantus was subjected owing to the want of notation. Indeed, on the assumption of the eight "methods," as Nesselmann describes them, it is really not surprising that no complete account of them could be given without copying the whole book. To take the first, "the adroit assumption of unknowns." Supposing that a number of essentially different problems are proposed, the differences make a different choice of an unknown in each case absolutely necessary. That being so, how could a rule be given for all cases? The best that can be done is to give a number of typical instances. Precisely the same remark applies to "methods" (2), (5), (6), (7). The case of (4), "Method of Limits," is different; here we have a "method" in the true sense of the term, *i.e.* in the sense of an *instrument* for solution. And accordingly in this case the method can be exhibited, as I hope to show later on; (8) also deserves to some extent the name of a "method."

| In one particular case, Diophantus formally states a method or rule; this is his rule 58
for solving what he calls a "double-equation," and will be found in II. 11, where such an equation appears for the first time. Apart from this, we do not find in Diophantus' work statements of method put generally as book-work to be applied to examples. Thus we do not find the separate rules and limitations for the solution of different kinds of equations systematically arranged, but we have to seek them out laboriously from the whole of his work, gathering scattered indications here and there, and to formulate them in the best way that we can.

I shall now attempt to give a short account of those methods running through Diophantus which admit of general statement. For the reasons which I have stated, my arrangement will be different from that of Nesselmann; I shall omit some of the heads in his classification of "methods of solution"; and, in accordance with his remark that these "methods" can only be adequately described by a transcription of the entire work, I shall leave them to be gathered from a perusal of my reproduction of Diophantus' book.

I shall begin my account with

I. DIOPHANTUS' TREATMENT OF EQUATIONS.

This subject falls naturally into two divisions: (A) Determinate equations of different degrees, (B) Indeterminate equations.

(A) Determinate Equations.

Diophantus was able without difficulty to solve determinate equations of the first and second degrees; of a cubic equation we find in his *Arithmetica* only one example, and that is a very special case. The solution of simple equations we may pass over; we have then to consider Diophantus' methods of solution in the case of (1) Pure equations, (2) Adfected, or mixed, *quadratics*.

(1) *Pure Determinate Equations*

By *pure* equations I mean those equations which contain only one power of the unknown, whatever the degree. The solution is effected in the same way whatever the
59 exponent of the term in the | unknown; and Diophantus treats pure equations of any degree as if they were simple equations of the first degree.

He gives a general rule for this case without regard to the degree[1]: "If a problem leads to an equation in which any terms are equal to the same terms but have different coefficients, we must take like from like on both sides, until we get one term equal to one term. But, if there are on one side or on both sides any negative terms, the deficient terms must be added on both sides until all the terms on both sides are positive. Then we must take like from like until one term is left on each side." After these operations have been performed, the equation is reduced to the form $Ax^m = B$ and is considered solved. The cases which occur in Diophantus are cases in which the value of x is found to be a rational number, integral or fractional. Diophantus only recognises one value of x which satisfies this equation; thus, if m is even, he gives only the positive value, excluding a negative value as "impossible." In the same way, when an equation can be reduced in degree by dividing throughout by any power of x, the possible values, $x = 0$, thus arising are not taken into account. Thus an equation of the form $x^2 = ax$, which is of common occurrence in the earlier part of the book, is taken to be merely equivalent to the simple equation $x = a$.

It may be observed that the greater proportion of the problems in Book I. are such that more than one unknown quantity is sought. Now, when there are two unknowns and two conditions, both unknowns can easily be expressed in terms of one symbol. But, when there are three or four quantities to be found, this reduction is much more difficult, and Diophantus shows great adroitness in effecting it: the ultimate result being that it is only necessary to solve a simple equation with one unknown quantity.

(2) *Mixed Quadratic Equations*

After the remarks in Def. 11 upon the reduction of equations until we have one term equal to another term, Diophantus adds:[2] "But we will show you afterwards how, in

---

1 Def. 11.

2 ὕστερον δέ σοι δείξομεν καὶ πῶς δύο εἰδῶν ἴσων ἑνὶ καταλειφθέντων τὸ τοιοῦτον λύεται.

the case also when two terms are left equal to a single term, such an equation can be solved." That is to say, he promises to explain the solution of a mixed quadratic equation. In the *Arithmetica*, as we possess the book, this promise is not fulfilled. The first
| indications we have on the subject are a number of cases in which the equation is 60
given, and the solution written down, or stated to be rational, without any work being shown. Thus

$$\text{(IV. 22) } x^2 = 4x - 4, \text{ therefore } x = 2;$$

$$\text{(IV. 31) } 325x^2 = 3x + 18, \text{ therefore } x = \tfrac{78}{325} \text{ or } \tfrac{6}{25};$$

$$\text{(VI. 6) } 84x^2 + 7x = 7, \text{ whence } x = \tfrac{1}{4};$$

$$\text{(VI. 7) } 84x^2 - 7x = 7, \text{ hence } x = \tfrac{1}{3};$$

$$\text{(VI. 9) } 630x^2 - 73x = 6, \text{ therefore } x = \tfrac{6}{35};$$

$$\text{and (VI. 8) } 630x^2 + 73x = 6, \text{ and } x \text{ is rational.}$$

These examples, though proving that Diophantus had somehow arrived at the result, are not in themselves sufficient to show that he was necessarily acquainted with a regular method for the solution of quadratics; these solutions might (though their variety makes it somewhat unlikely) have been obtained by mere *trial*. That, however, Diophantus' solutions of mixed quadratics were not merely empirical is shown by instances in V. 30. In this problem he shows that he could approximate to the root in cases where it is not "rational." As this is an important point, I give the substance of the passage in question: "This is not generally possible unless we contrive to make $x > \frac{1}{8}(x^2 - 60)$ and $< \frac{1}{5}(x^2 - 60)$. Let then $x^2 - 60$ be $> 5x$, but $x^2 - 60 < 8x$. Since then $x^2 - 60 > 5x$, let 60 be added to both sides, so that $x^2 > 5x + 60$, or $x^2 = 5x$ + some number > 60; therefore x must not be less than 11." In like manner Diophantus concludes that "$x^2 = 8x$ + some number less than 60; therefore x must be found to be not greater than 12."

Now, solving for the positive roots of these two equations, we have

$$x > \tfrac{1}{2}(5 + \sqrt{265}) \text{ and } x < 4 + \sqrt{76},$$

or

$$x > 10.6394\ldots \text{ and } x < 12.7177\ldots$$

It is clear that x may be < 11 or > 12, and therefore Diophantus' limits are not strictly accurate. As however it was doubtless his object to find *integral* limits, the limits 11 and 12 are those which are obviously adapted for his purpose, and are *a fortiori* safe.

In the above equations the other roots obtained by prefixing the negative sign to the radical are negative and therefore would be of no use to Diophantus. In other cases
of the kind occurring | in Book V. the equations have both roots positive, and we have 61
to consider why Diophantus took no account of the smaller roots in those cases.

We will take first the equations in V. 10 where the inequalities to be satisfied are

$$72x > 17x^2 + 17 \qquad (1).$$

$$72x < 19x^2 + 19 \qquad (2).$$

Now, if α, β be the roots of the equation

$$x^2 - px + q = 0 \qquad (p, q \text{ both positive}),$$

and if $\alpha > \beta$, then
(*a*) in order that $x^2 - px + q$ may be > 0

we must have $x > \alpha$ or $< \beta$,

and (*b*) in order that $x^2 - px + q$ may be < 0

we must have $x < \alpha$ and $> \beta$.

(1) The roots of the equation

$$17x^2 - 72x + 17 = 0$$

are $$\frac{36 \pm \sqrt{1007}}{17}\text{; that is, } \frac{67.73\ldots}{17} \text{ and } \frac{4.26\ldots}{17};$$

and, in order that $17x^2 - 72x + 17$ may be < 0, we must have

$$x < \frac{67.73\ldots}{17} \text{ but } > \frac{4.26\ldots}{17}.$$

(2) The roots of the equation

$$19x^2 - 72x + 19 = 0$$

are $$\frac{36 \pm \sqrt{935}}{19}\text{; that is, } \frac{66.577\ldots}{19} \text{ and } \frac{5.422\ldots}{19};$$

and, in order that $19x^2 - 72x + 19$ may be > 0, we must have

$$x > \frac{66.577\ldots}{19} \text{ or } < \frac{5.422\ldots}{19}.$$

Diophantus says that x must not be greater than $\frac{67}{17}$ or less than $\frac{66}{19}$. These are again doubtless intended to be *a fortiori* limits; but $\frac{66}{19}$ should have been $\frac{67}{19}$, and the more correct way of stating the case would be to say that, if x is not greater than $\frac{67}{17}$ and not less than $\frac{67}{19}$, the given conditions are *a fortiori* satisfied.

Now consider what alternative (if any) could be obtained, on Diophantus' prin-
62 ciples, if we used the lesser positive roots of the | equations. If, like Diophantus, we were to take *a fortiori* limits, we should have to say

$$x < \tfrac{5}{19} \text{ but } > \tfrac{5}{17},$$

which is of course an impossibility. Therefore the smaller roots are here useless from his point of view.

This is, however, not so in the case of another pair of inequalities, used later in V. 30 for finding an auxiliary x, namely

$$x^2 + 60 > 22x,$$
$$x^2 + 60 < 24x.$$

The roots of the equation

$$x^2 - 22x + 60 = 0$$

$$\text{are } 11 \pm \sqrt{61}\text{; that is, } 18.81\ldots \text{ and } 3.18\ldots;$$

and the roots of the equation $x^2 - 24x + 60 = 0$

$$\text{are } 12 \pm \sqrt{84}\text{; that is, } 21.16\ldots \text{ and } 2.83\ldots.$$

In order therefore to satisfy the above inequalities we must have

$$x > 18.81\ldots \text{ or } < 3.18\ldots,$$

and

$$x < 21.16\ldots \text{ but } > 2.83\ldots.$$

Diophantus, taking *a fortiori* integral limits furnished by the greater roots, says that x must not be less than 19 but must be less than 21. But he could also have obtained from the smaller roots an integral value of x satisfying the necessary conditions, namely the value $x = 3$; and this would have had the advantage of giving a smaller value for the auxiliary x than that actually taken, namely 20.[1] Accordingly the question has been raised[2] whether we have not here, perhaps, a valid reason for believing that Diophantus only knew of the existence of roots obtained by using the positive sign with the radical, and was unaware of the solution obtained by using the negative sign. But in truth we can derive no certain knowledge on this point from Diophantus' treatment of the particular equations in question. Thus, *e.g.*, if he chose to use the first of the two equations

$$72x > 17x^2 + 17,$$
$$72x < 19x^2 + 19,$$

for the purpose of obtaining an upper limit *only*, and the second | for the purpose of 63
obtaining a lower limit *only*, he could *only* use the values obtained by using the positive sign. Similarly, if, with the equations

$$x^2 + 60 > 22x,$$
$$x^2 + 60 < 24x,$$

---

1 This is remarked by Loria (*Le scienze esatte dell' antica Grecia*, V. p. 128). But in fact, whether we take 20 or 3 as the value of the auxiliary unknown, we get the same value for the original x of the problem. For the original x has to be found from $x^2 - 60 = (x - m)^2$ where m is the auxiliary x; and we obtain $x = 11\frac{1}{2}$ whether we put $x^2 - 60 = (x - 20)^2$ or $x^2 - 60 = (x - 3)^2$.

2 Loria, *op. cit.* p. 129.

he chose to use the first in order to find a lower limit *only*, and the second in order to find an upper limit *only*, it was not open to him to use the values corresponding to the negative sign[1].

For my part, I find it difficult or impossible to believe that Diophantus was unaware of the existence of two real roots in such cases. The numerical solution of quadratic equations by the Greeks immediately followed, if it did not precede, their geometrical solution. We find the geometrical equivalent of the solution of a quadratic assumed as early as the fifth century B.C., namely by Hippocrates of Chios in his *Quadrature of lunes*[2], the algebraic form of the particular equation being $x^2 + \sqrt{\frac{3}{2}} \cdot ax = a^2$. The complete geometrical solution was given by Euclid in VI. 27–29: and the construction of VI. 28 corresponds in fact to the *negative* sign before the radical in the case of the particular equation there solved, while a quite obvious and slight variation of the construction would give the solution corresponding to the *positive* sign. In VI. 29 the solution corresponds to the positive sign before the radical; in the case of the equation there dealt with the other sign would not give a "real" solution[3]. It is true that we do not find the negative sign taken in Heron any more than in Diophantus, though we find Heron[4] stating an approximate solution of the equation

$$x(14 - x) = 6720/144,$$

without showing how he arrived at it; x is, he says, approximately equal to $8\frac{1}{2}$. It is clear however that Heron already possessed a scientific method of solution. Again, the author of the so-called *Geometry* of Heron[5] practically states the solution of the equation

$$\tfrac{11}{14}x^2 + \tfrac{29}{7}x = 212$$

in the form
$$x = \frac{\sqrt{(154 \times 212 + 841)} - 29}{11},$$

64 | showing pretty clearly the rule followed after the equation had been written in the form

$$121x^2 + 638x = 212 \times 154.$$

We cannot credit Diophantus with less than a similar uniform method; and, if he did not trouble to give two roots where both were real, this seems quite explicable when it is remembered that he did not write a text-book of algebra, and that his object throughout is to obtain a single solution of his problems, not to multiply solutions or to show how many can be found in each case.

In solving such an equation as

$$ax^2 - bx + c = 0,$$

---

1 Eneström in *Bibliotheca Mathematica* IX_3, 1908–9, pp. 71–2.

2 Simplicius, *Comment. in Aristot. Phys.*, ed. Diels, p. 64, 18; Rudio, *Der Bericht des Simplicius über die Quadraturen des Antiphon und des Hippokrates*, 1907, p. 58, 8–11.

3 *The Thirteen Books of Euclid's Elements*, Cambridge, 1908, Vol. II. pp. 257–267.

4 Heron, *Metrica*, ed. Schöne, pp. 148–151. The text has 8 as the approximate solution, but the correction is easy, as the inference immediately drawn is that $14 - x = 5\frac{1}{2}$.

5 Heron, ed. Hultsch, p. 133, 10–23. See M. Cantor, *Geschichte der Math.* I_3, p. 405.

it is our modern practice to divide out by a in order to make the first term a square. It does not appear that Diophantus divided out by a; rather he multiplied by a so as to bring the equation into the form

$$a^2x^2 - abx + ac = 0;$$

then, solving, he found

$$ax = \tfrac{1}{2}b \pm \sqrt{(\tfrac{1}{4}b^2 - ac)},$$

and wrote the solution in the form

$$x = \frac{\tfrac{1}{2}b \pm \sqrt{(\tfrac{1}{4}b^2 - ac)}}{a},$$

wherein his method corresponds to that of the Pseudo-Heron above referred to.

From the rule given in Def. 11 for removing by means of addition any negative terms on either side of an equation and taking equals from equals (operations called by the Arabians *aljabr* and *almukābala*) it is clear that, as a preliminary to solution, Diophantus so arranged his equation that all the terms were positive. Thus, from his point of view, there are three cases of mixed quadratic equations.

Case 1. From $mx^2 + px = q$; the root is

$$\frac{-\tfrac{1}{2}p + \sqrt{(\tfrac{1}{4}p^2 + mq)}}{m},$$

according to Diophantus. An instance is afforded by VI. 6. Diophantus namely arrives at the equation $6x^2 + 3x = 7$, which, if it is to be of any service to his solution, should give a rational value of x; whereupon he says "the square of half the coefficient[1] of
x | together with the product of the absolute term and the coefficient of x^2 must be 65
a square number; but it is not," *i.e.* $\tfrac{1}{4}p^2 + mq$, or in this case $(\tfrac{3}{2})^2 + 42$, must be a square in order that the root may be rational, which in this case it is not.

Case 2. Form $mx^2 = px + q$. Diophantus takes

$$x = \frac{\tfrac{1}{2}p + \sqrt{(\tfrac{1}{4}p^2 + mq)}}{m}.$$

An example is IV. 39, where $2x^2 > 6x + 18$. Diophantus says: "To solve this take the square of half the coefficient of x, *i.e.* 9, and the product of the absolute term and the coefficient of x^2, *i.e.* 36. Adding, we have 45, the square root[2] of which is not[3] < 7.

---

1 For "coefficient" Diophantus simply uses πλῆθος, multitude or number; thus "number of ἀριθμοί" = coeff. of x. The absolute term is described as the "units."

2 The "square root" is with Diophantus πλευρά, or "side."

3 7, though not accurate, is clearly the nearest integral limit which will serve the purpose.

Add half the coefficient of x, [and divide by the coefficient of x^2]; whence x is not <5." Here the form of the root is given completely; and the whole operation of finding it is revealed. Cf. IV. 31, where Diophantus remarks that the equation $5x^2 = 3x + 18$ "is not rational. But 5, the coefficient of x^2, is a square plus 1, and it is necessary that this coefficient multiplied by the 18 units and then added to the square of half the coefficient of x, namely 3, that is to say $2\frac{1}{4}$, shall make a square."

Case 3. Form $mx^2 + q = px$. Diophantus' root is

$$\frac{\frac{1}{2}p + \sqrt{(\frac{1}{4}p^2 - mq)}}{m}.$$

Cf. in V. 10 the equation already mentioned, $17x^2 + 17 < 72x$. Diophantus says: "Multiply half the coefficient of x into itself and we have 1296; subtract the product of the coefficient of x^2 and the absolute term, or 289. The remainder is 1007, the square root of which is not[1] >31. Add half the coefficient of x, and the result is not >67. Divide by the coefficient of x^2, and x is not $>67/17$." Here again we have the complete solution given. Cf. VI. 22, where, having arrived at the equation $172x = 336x^2 + 24$, Diophantus remarks that "this is not always possible, unless half the coefficient of x multiplied into itself, *minus* the product of the coefficient of x^2 and the units, makes a square."

For the purpose of comparison with Diophantus' solutions of quadratic equations we may refer to a few of his solutions of

66 | (3) *Simultaneous Equations Involving Quadratics*

Under this heading come the pairs of equations

$$\left.\begin{aligned} \xi + \eta &= 2a \\ \xi\eta &= B \end{aligned}\right\}, \qquad \text{(I. 27.)}$$

$$\left.\begin{aligned} \xi + \eta &= 2a \\ \xi^2 + \eta^2 &= B \end{aligned}\right\}, \qquad \text{(I. 28.)}$$

$$\left.\begin{aligned} \xi - \eta &= 2a \\ \xi\eta &= B \end{aligned}\right\}. \qquad \text{(I. 30.)}$$

I use Greek letters to distinguish the numbers which the problem requires us to find from the one unknown which Diophantus uses and which I shall call x.

In the first two of the above problems, he chooses his x thus. Let, he says,

$$\xi - \eta = 2x \qquad (\xi > \eta).$$

Then it follows, by addition and subtraction, that

$$\xi = a + x, \quad \eta = a - x.$$

Consequently, in I. 27,

[1] As before, the nearest *integral* limit.

$$\xi\eta = (a + x)(a - x) = a^2 - x^2 = B,$$

and x is found from this "pure" quadratic equation.

If we eliminate ξ from the original equations, we have

$$\eta^2 - 2a\eta + B = 0,$$

which we should solve by completing the square $(a - \eta)^2$, whence

$$(a - \eta)^2 = a^2 - B,$$

which is Diophantus' ultimate equation with $a - \eta$ for x.

Thus Diophantus' method corresponds here again to the ordinary method of solving a mixed quadratic, by which we make it into a pure quadratic with a different x.

In I. 30 Diophantus puts $\xi + \eta = 2x$, and the solution proceeds in the same way as in I. 27.

In I. 28 the resulting equation in x is

$$\xi^2 + \eta^2 = (a + x)^2 + (a - x)^2 = 2(a^2 + x^2) = B.$$

(4) *Cubic Equation*

There is no ground for supposing that Diophantus was acquainted with the algebraical solution of a cubic equation. It is true that there is one cubic equation to be found in the *Arithmetica*, but it is only a very particular case. In VI. 17 the problem leads to the equation

$$x^2 + 2x + 3 = x^3 + 3x - 3x^2 - 1,$$

| and Diophantus says simply "whence x is found to be 4." All that can be said of this 67
is that, if we write the equation in true Diophantine fashion, so that all the terms are positive,

$$x^3 + x = 4x^2 + 4.$$

This equation being clearly equivalent to

$$x(x^2 + 1) = 4(x^2 + 1),$$

Diophantus no doubt detected the presence of the common factor on both sides of the equation. The result of dividing by it is $x = 4$, which is Diophantus' solution. Of the other two roots $x = \pm\sqrt{(-1)}$ no account is taken, for the reason stated above.

It is not possible to judge from this example how far Diophantus was acquainted with the solution of equations of a degree higher than the second.

I pass now to the second general division of equations.

(B) Indeterminate Equations.

As I have already stated, Diophantus does not, in his *Arithmetica* as we have it, treat of indeterminate equations of the first degree. Those examples in Book I. which would lead to such equations are, by the arbitrary assumption of a specific value for one of the required numbers, converted into determinate equations. Nor is it likely that indeterminate equations of the first degree were treated of in the lost Books. For, as

Nesselmann observes, while with indeterminate quadratic equations the object is to obtain a *rational* result, the whole point in solving indeterminate simple equations is to obtain a result in *integral* numbers. But Diophantus does not exclude fractional solutions, and he has therefore only to see that his results are *positive*, which is of course easy. Indeterminate equations of the first degree would therefore, from Diophantus' point of view, have no particular significance. We take therefore, as our first division, indeterminate equations of the second degree.

(I) *Indeterminate Equations of the Second Degree*

The form in which these equations occur in Diophantus is invariably this: one or two (but never more) functions of the unknown quantity of the form $Ax^2 + Bx + C$ or simpler forms are to be made rational square numbers by finding a suitable value for x. Thus the most general case is that of solving one or two equations of the form $Ax^2 + Bx + C = y^2$.

68 | *(a) Single equation.* The single equation takes special forms when one or more of the coefficients vanish or satisfy certain conditions. It will be well to give in order the different forms as they can be identified in Diophantus, premising that for "$=y^2$" Diophantus simply uses the formula *ἴσον τετραγώνῳ*, "is equal to a square," or *ποιεῖ τετράγωνον*, "makes a square."

1. Equations which can always be solved rationally. This is the case when A or C or both vanish.

Form $Bx = y^2$. Diophantus puts for y^2 any arbitrary square number, say m^2. Then $x = m^2/B$.

Ex. III. 5: $2x = y^2$, y^2 is assumed to be 16, and $x = 8$.

Form $Bx + C = y^2$. Diophantus puts for y^2 any square m^2, and $x = (m^2 - C)/B$. He admits fractional values of x, only taking care that they are "rational," *i.e.* rational and positive.

Ex. III. 6: $6x + 1 = y^2 = 121$, say, and $x = 20$.

Form $Ax^2 + Bx = y^2$. Diophantus substitutes for y any multiple of x, as $\frac{m}{n}x$; whence $Ax + B = \frac{m^2}{n^2}x$, the factor x disappearing and the root $x = 0$ being neglected as usual. Thus $x = \frac{Bn^2}{m^2 - An^2}$.

Exx. II. 21: $4x^2 + 3x = y^2 = (3x)^2$, say, and $x = \frac{3}{5}$.

II. 33: $16x^2 + 7x = y^2 = (5x)^2$, say, and $x = \frac{7}{9}$.

2. Equations which can only be rationally solved if certain conditions are fulfilled. The cases occurring in Diophantus are the following.

Form $Ax^2 + C = y^2$. This can be rationally solved according to Diophantus

(α) When A is positive and a square, say a^2.

Thus $a^2x^2 + C = y^2$. In this case y^2 is put $= (ax \pm m)^2$;

therefore $$a^2x^2 + C = (ax \pm m)^2,$$

and $$x = \pm \frac{C - m^2}{2ma},$$

(m and the doubtful sign being always assumed so as to give x a positive value).

(β) When C is positive and a square number, say c^2.

Thus $Ax^2 + c^2 = y^2$. Here Diophantus puts $y = (mx \pm c)$;

| therefore $$Ax^2 + c^2 = (mx \pm c)^2,$$ 69

and $$x = \pm \frac{2mc}{A - m^2}.$$

(γ) When one solution is known, any number of other solutions can be found. This is enunciated in the Lemma to VI. 15 thus, though only for the case in which C is negative: "Given two numbers, if, when one is multiplied by some square and the other is subtracted from the product, the result is a square, then another square also can be found, greater than the aforesaid square which has the same property." It is curious that Diophantus does not give a general enunciation of this proposition, inasmuch as not only is it applicable to the cases $\pm Ax^2 \pm C = y^2$, but also to the general form $Ax^2 + Bx + C = y^2$.

Diophantus' method of finding other greater values of x satisfying the equation $Ax^2 - C = y^2$ when one such value is known is as follows.

Suppose that x_0 is the value already known and that q is the corresponding value of y.

Put $x = x_0 + \xi$ in the original expression, and equate it to $(q - k\xi)^2$, where k is some integer.

Since $$A(x_0 + \xi)^2 - C = (q - k\xi)^2,$$

it follows (because by hypothesis $Ax_0^2 - C = q^2$) that

$$2\xi(Ax_0 + kq) = \xi^2(k^2 - A),$$

whence $$\xi = \frac{2(Ax_0 + kq)}{k^2 - A},$$

and $$x = x_0 + \frac{2(Ax_0 + kq)}{k^2 - A}.$$

In the second Lemma to VI. 12 Diophantus does prove that the equation $Ax^2 + C = y^2$ has an infinite number of solutions when $A + C$ is a square, *i.e.* in the particular case where the value $x = 1$ satisfies the equation. But he does not always bear this in mind; for in III. 10 the equation $52x^2 + 12 = y^2$ is regarded as impossible of solution although $52 + 12 = 64$, a square, and a rational solution is therefore possible. Again in III. 11 the equation $266x^2 - 10 = y^2$ is regarded as impossible though $x = 1$ satisfies it.

The method used by Diophantus in the second Lemma to VI. 12 is like that of the Lemma to VI. 15.

Suppose that $A + C = q^2$.

Put $1 + \xi$ for x in the original expression $Ax^2 + C$, and equate it to $(q - k\xi)^2$, where k is some integer.

| Thus $$A(1 + \xi)^2 + C = (q - k\xi)^2,$$ 70

and it follows that $$2\xi(A + kq) = \xi^2(k^2 - A),$$

so that $$\xi = \frac{2(A + kq)}{k^2 - A},$$

and $$x = 1 + \frac{2(A + kq)}{k^2 - A}.$$

It is of course necessary to choose k^2 such that $k^2 > A$.

It is clear that, if $x = 0$ satisfies the equation, C is a square, and therefore this case (γ) includes the previous case (β).

It is to be observed that in VI. 14 Diophantus says that a rational solution of the equation

$$Ax^2 - c^2 = y^2$$

is impossible *unless A is the sum of two squares.*

[In fact, if $x = p/q$ satisfies the equation, and $Ax^2 - c^2 = k^2$,

we have $$Ap^2 = c^2q^2 + k^2q^2,$$

or $$A = \left(\frac{cq}{p}\right)^2 + \left(\frac{kq}{p}\right)^2.]$$

Lastly, we have to consider

$$\text{Form } Ax^2 + Bx + C = y^2.$$

This equation can be reduced by means of a change of variable to the preceding form wanting the second term. For, if we put $x = z - \frac{B}{2A}$, the transformation gives

$$Az^2 + \frac{4AC - B^2}{4A} = y^2.$$

Diophantus, however, treats this form of the equation quite separately from the other and less fully. According to him the rational solution is only possible in the following cases.

(α) When A is positive and a square, or the equation is

$$a^2x^2 + Bx + C = y^2.$$

Diophantus then puts $y^2 = (ax - m)^2$, whence

$$x = \frac{m^2 - C}{2am + B}. \qquad \text{(Exx. II. 20, 22 etc.)}$$

(β) When C is positive and a square, or the equation is

$$Ax^2 + Bx + c^2 = y^2.$$

Diophantus puts $y^2 = (c - mx)^2$, whence

$$x = \frac{2mc + B}{m^2 - A}. \qquad \text{(Exx. IV. 8, 9 etc.)}$$

71 | (γ) When $\frac{1}{4}B^2 - AC$ is positive and a square number. Diophantus never expressly enunciates the possibility of this case; but it occurs, as it were unawares, in IV. 31. In that problem

$$3x + 18 - x^2$$

is to be made a square. To solve this Diophantus assumes

$$3x + 18 - x^2 = 4x^2,$$

which leads to the quadratic $3x + 18 - 5x^2 = 0$; but "the equation is not rational." Accordingly the assumption $4x^2$ will not do; "and we must find a square [to replace 4] such that 18 times (this square + 1) + $(\frac{3}{2})^2$ may be a square." Diophantus then solves the auxiliary equation $18(m^2 + 1) + \frac{9}{4} = y^2$, finding $m = 18$. He then assumes $3x + 18 - x^2 = (18)^2x^2$, which gives $325x^2 - 3x - 18 = 0$, "and $x = 78/325$, that is 6/25."[1]

---

[1] With this solution should be compared the much simpler solution of this case given by Euler (*Algebra*, tr. Hewlett, 1840, Part II. Arts. 50–53), depending on the separation of the quadratic expression into factors. (Curiously enough Diophantus does not separate quadratic expressions into their factors except in one case, VI. 19, where however his purpose is quite different: he has made the sum of three sides of a right-angled triangle $4x^2 + 6x + 2$, which has to be a cube, and, in order to simplify, he divides throughout by $x + 1$, which leaves $4x + 2$ to be made a cube.)

Since $\frac{1}{4}B^2 - AC$ is a square, the roots of the quadratic $Ax^2 + Bx + C = 0$ are real, and the expression has two real linear factors. Take the particular case now in question, where Diophantus actually arrives at $3x + 18 - x^2$ as the result of multiplying $6 - x$ and $3 + x$, but makes no use of the factors.

We have $$3x + 18 - x^2 = (6 - x)(3 + x).$$

Assume then $$(6 - x)(3 + x) = \frac{p^2}{q^2}(6 - x)^2,$$

and we have $$p^2(6 - x) = q^2(3 + x),$$

$$x = \frac{6p^2 - 3q^2}{p^2 + q^2},$$

where p, q may be any numbers subject to the condition that $2p^2 > q^2$. If $p^2 = 9$, $q^2 = 16$, we have Diophantus' solution $x = \frac{6}{25}$.

In general, if $$Ax^2 + Bx + C = (f + gx)(h + kx),$$

we can put $$(f + gx)(h + kx) = \frac{p^2}{q^2}(f + gx)^2,$$

whence $$q^2(h + kx) = p^2(f + gx),$$

and $$x = \frac{fp^2 - hq^2}{kq^2 - gp^2}.$$

This case, says Euler, leads to a fourth case in which $Ax^2 + Bx + C = y^2$ can be solved, though neither A nor C is a square, and though $B^2 - 4AC$ is not a square either. The fourth case is that in

72 | It is worth observing that from this example of Diophantus we can deduce the reduction of this general case to the form

$$Ax^2 + C = y^2$$

wanting the middle term.

Assume, with Diophantus, that $Ax^2 + Bx + C = m^2x^2$: therefore by solution we have

$$x = \frac{-\frac{1}{2}B \pm \sqrt{\frac{1}{4}B^2 - AC + Cm^2}}{A - m^2},$$

---

which $Ax^2 + Bx + C$ is the sum of two parts, one of which is a square and the other is the product of two factors linear in x. For suppose

$$Ax^2 + Bx + C = Z^2 + XY,$$

where $$Z = dx + e, \quad X = fx + g, \quad Y = hx + k.$$

We can then put $$Z^2 + XY = \left(Z + \frac{p}{q}X\right)^2,$$

whence $$Y = 2\frac{p}{q}Z + \frac{p^2}{q^2}X,$$

or $$hx + k = 2\frac{p}{q}(dx + e) + \frac{p^2}{q^2}(fx + g),$$

that is, $$x(p^2f + 2pqd - q^2h) = kq^2 - 2pqe - p^2g.$$

Ex. 1. Equation $2x^2 - 1 = y^2$.

Put $$2x^2 - 1 = x^2 + (x+1)(x-1) = \left\{x + \frac{p}{q}(x+1)\right\}^2.$$

Therefore $$x - 1 = 2\frac{p}{q}x + \frac{p^2}{q^2}(x+1),$$

and $$x(p^2 + 2pq - q^2) = -(p^2 + q^2).$$

As x^2 is alone found in our equation, we can take either the positive or negative sign and we may put

$$x = \frac{p^2 + q^2}{p^2 + 2pq - q^2}.$$

Ex. 2. Equation $2x^2 + 2 = y^2$.

Here we put $$2x^2 + 2 = 4 + 2(x+1)(x-1).$$

Equating this to $$\left\{2 + \frac{p}{q}(x+1)\right\}^2,$$

we have $$2(x-1) = 4\frac{p}{q} + \frac{p^2}{q^2}(x+1),$$

or $$x(p^2 - 2q^2) = -(2q^2 + 4pq + p^2),$$

and x is rational provided that $\frac{1}{4}B^2 - AC + Cm^2$ is a square. This condition can be fulfilled if $\frac{1}{4}B^2 - AC$ is a square, by the pre|ceding case. If $\frac{1}{4}B^2 - AC$ is not a square, we 73
have to solve (putting, for brevity, D for $\frac{1}{4}B^2 - AC$) the equation

$$D + Cm^2 = y^2,$$

and the reduction is effected.

(b) Double-equation. By the name "double-equation" Diophantus denotes the problem of finding one value of the unknown quantity which will make two different functions of it simultaneously rational square numbers. The Greek term for the "double-equation" occurs variously as διπλοϊσότης, διπλῆ ἰσότης or διπλῆ ἴσωσις. We have then to solve the equations

$$\left.\begin{aligned} mx^2 + \alpha x + a &= u^2 \\ nx^2 + \beta x + b &= w^2 \end{aligned}\right\}$$

in rational numbers. The necessary preliminary condition is that each of the two expressions can severally be made squares. This is always possible when the first term (in x^2) is wanting. This is the simplest case, and we shall accordingly take it first.

1. *Double-equation of the first degree.*

Diophantus has one general method of solving the equations

$$\left.\begin{aligned} \alpha x + a &= u^2 \\ \beta x + b &= w^2 \end{aligned}\right\},$$

taking slightly different forms according to the nature of the coefficients.

(α) First method of solution of

$$\left.\begin{aligned} \alpha x + a &= u^2 \\ \beta x + b &= w^2 \end{aligned}\right\}.$$

---

and $$x = \frac{p^2 + 4pq + 2q^2}{2q^2 - p^2}.$$

It is to be observed that this method enables us to solve the equation

$$Ax^2 - c^2 = y^2$$

whenever it can be solved rationally, *i.e.* whenever A is the sum of two squares ($d^2 + e^2$, say). For then

$$Ax^2 - c^2 = d^2x^2 + (ex - c)(ex + c).$$

In cases not covered by any of the above rules our only plan is to try to discover *one* solution empirically. If one solution is thus found, we can find any number of others; if we cannot discover such a solution by trial (even after reducing the equation to the simplest form $A'x'^2 + C = y'^2$), recourse must be had to the method of continued fractions elaborated by Lagrange (cf. *Oeuvres*, II. pp. 377–535 and pp. 655–726; additions to Euler's *Algebra*).

This method depends upon the identity

$$\{\tfrac{1}{2}(p+q)\}^2 - \{\tfrac{1}{2}(p-q)\}^2 = pq.$$

If the difference between the two expressions in x can be separated into two factors p, q, the expressions themselves are equated to $\{\frac{1}{2}(p \pm q)\}^2$ respectively. Diophantus himself (II. 11) states his rule thus.

"Observing the difference [between the two expressions], seek two numbers such that their product is equal to this difference; then equate either the square of half the difference of the two factors to the lesser of the expressions or the square of half the sum to the greater."

74 | We will take the general case and investigate to what particular classes of cases the method is applicable, from Diophantus' point of view, remembering that his cases are such that the final quadratic equation in x always reduces to a simple equation.

Take the equations

$$\left.\begin{aligned} \alpha x + a &= u^2 \\ \beta x + b &= w^2 \end{aligned}\right\}.$$

Subtracting, we have

$$(\alpha - \beta)x + (a - b) = u^2 - w^2.$$

We have then to separate $(\alpha - \beta)x + (a - b)$ into two factors; let these be p, $\{(\alpha - \beta)x + (a - b)\}/p$.

We write accordingly

$$u \pm w = \frac{(\alpha - \beta)x + a - b}{p},$$

$$u \mp w = p.$$

Thus $$u^2 = \alpha x + a = \frac{1}{4}\left\{\frac{(\alpha - \beta)x + a - b}{p} + p\right\}^2;$$

therefore $$\{(\alpha - \beta)x + a - b + p^2\}^2 = 4p^2(\alpha x + a),$$

or

$$(\alpha - \beta)^2 x^2 + 2x\{(\alpha - \beta)(a - b + p^2) - 2p^2\alpha\} + (a - b + p^2)^2 - 4ap^2 = 0,$$

that is,

$$(\alpha - \beta)^2 x^2 + 2x\{(\alpha - \beta)(a - b) - p^2(\alpha + \beta)\} + (a - b)^2 - 2p^2(a + b) + p^4 = 0.$$

Now, in order that this equation may reduce to a simple equation, either

(1) The coefficient of x^2 must vanish, so that

$$\alpha = \beta,$$

or (2) The absolute term must vanish, that is,

$$p^4 - 2p^2(a + b) + (a - b)^2 = 0,$$

or
$$\{p^2 - (a + b)\}^2 = 4ab,$$
so that ab must be a square number.

Therefore either a and b are both squares, in which case we may substitute c^2 and d^2 for them respectively, p being then equal to $c \pm d$, or the ratio of a to b is the ratio of a square to a square.

With respect to (1) we observe that on one condition it is not necessary that $\alpha - \beta$ should vanish, *i.e.* provided we can, before solving the equations, make the coefficients of x the same in both expressions by multiplying either equation or both equations by some square number, an operation which does not affect the problem, since a square multiplied by a square is still a square. | In other words, it is only necessary that 75
the ratio of α to β should be the ratio of a square to a square[1].

Thus, if $\alpha/\beta = m^2/n^2$ or $\alpha n^2 = \beta m^2$, the equations can be solved by multiplying them respectively by n^2 and m^2; we can in fact solve the equations

$$\left.\begin{aligned} \alpha m^2 x + a &= u^2 \\ \alpha n^2 x + b &= w^2 \end{aligned}\right\},$$

like the equations

$$\left.\begin{aligned} \alpha x + a &= u'^2 \\ \alpha x + b &= w'^2 \end{aligned}\right\},$$

in an infinite number of ways.

Again, the equations under (2)

$$\left.\begin{aligned} \alpha x + c^2 &= u^2 \\ \beta x + d^2 &= w^2 \end{aligned}\right\}$$

---

[1] Diophantus actually states this condition in the solution of IV. 32 where, on arriving at the equations

$$\left.\begin{aligned} 8 - x &= u^2 \\ 8 - 3x &= w^2 \end{aligned}\right\},$$

he says: "And this is not rational because the coefficients of x have not to one another the ratio which a square number has to a square number."

Similarly in the second solution of III. 15 he states the same condition along with an alternative condition, namely that a has to b the ratio of a square to a square, which is the second condition arrived at under (2) above. On obtaining the equations

$$\left.\begin{aligned} 4x + 3 &= u^2 \\ 6\tfrac{1}{2}x + 5\tfrac{1}{2} &= w^2 \end{aligned}\right\},$$

Diophantus observes "But, since the coefficients in one expression are respectively greater than those in the other, neither have they (in either case) the ratio which a square number has to a square number, the hypothesis which we took is useless."

Cf. also IV. 39 where he says that the equations

$$\left.\begin{aligned} 8x + 4 &= u^2 \\ 6x + 4 &= w^2 \end{aligned}\right\},$$

are possible of solution because there is a square number of units in each expression.

can be solved in two different ways according as we write them in this form or in the form

$$\left.\begin{aligned} \alpha d^2x + c^2d^2 &= u'^2 \\ \beta c^2x + c^2d^2 &= w'^2 \end{aligned}\right\},$$

obtained by multiplying them respectively by d^2, c^2, in order that the absolute terms may be equal.

I shall now give those of the possible cases which we find solved in Diophantus' own work. These are equations

(1) of the form

$$\left.\begin{aligned} \alpha m^2x + a &= u^2 \\ \alpha n^2x + b &= w^2 \end{aligned}\right\},$$

76 | a case which includes the more common one where the coefficients of x in both are *equal*;

(2) of the form

$$\left.\begin{aligned} \alpha x + c^2 &= u^2 \\ \beta x + d^2 &= w^2 \end{aligned}\right\},$$

solved in two different ways according as they are written in this form or in the alternative form

$$\left.\begin{aligned} \alpha d^2x + c^2d^2 &= u'^2 \\ \beta c^2x + c^2d^2 &= w'^2 \end{aligned}\right\}.$$

General solution of Form (1) or

$$\left.\begin{aligned} \alpha m^2x + a &= u^2 \\ \alpha n^2x + b &= w^2 \end{aligned}\right\}.$$

Multiply by n^2, m^2 respectively, and we have to solve the equations

$$\left.\begin{aligned} \alpha m^2n^2x + an^2 &= u'^2 \\ \alpha m^2n^2x + bm^2 &= w'^2 \end{aligned}\right\}.$$

The difference is $an^2 - bm^2$; suppose this separated into two factors p, q.

Let
$$u' \pm w' = p,$$
$$u' \mp w' = q;$$

therefore
$$u'^2 = \tfrac{1}{4}(p + q)^2, \quad w'^2 = \tfrac{1}{4}(p - q)^2,$$

and
$$\alpha m^2n^2x + an^2 = \tfrac{1}{4}(p + q)^2,$$

or $$\alpha m^2 n^2 x + bm^2 = \tfrac{1}{4}(p - q)^2.$$

Either equation gives the same value of x, and

$$x = \frac{\tfrac{1}{4}(p^2 + q^2) - \tfrac{1}{2}(an^2 + bm^2)}{\alpha m^2 n^2},$$

since $pq = an^2 - bm^2$.

Any factors p, q may be chosen provided that the resulting value of x is *positive*.

Ex. from Diophantus:

$$\left.\begin{aligned} 65 - 6x &= u^2 \\ 65 - 24x &= w^2 \end{aligned}\right\};$$ (IV. 32.)

therefore $$\left.\begin{aligned} 260 - 24x &= u'^2 \\ 65 - 24x &= w^2 \end{aligned}\right\}.$$

The difference $= 195 = 15 \cdot 13$, say;

therefore $\tfrac{1}{4}(15 - 13)^2 = 65 - 24x$; that is, $24x = 64$, and $x = \tfrac{8}{3}$. 77

| General solution (first method) of Form (2), or

$$\left.\begin{aligned} \alpha x + c^2 &= u'^2 \\ \beta x + d^2 &= w'^2 \end{aligned}\right\}.$$

In order to solve by this method, we multiply by d^2, c^2 respectively and write

$$\left.\begin{aligned} \alpha d^2 x + c^2 d^2 &= u^2 \\ \beta c^2 x + c^2 d^2 &= w^2 \end{aligned}\right\},$$

u being supposed to be the greater. The difference $= (\alpha d^2 - \beta c^2)x$. Let the factors of this be px, q. Therefore

$$u^2 = \tfrac{1}{4}(px + q)^2,$$
$$w^2 = \tfrac{1}{4}(px - q)^2.$$

Thus x is found from the equation

$$\alpha d^2 x + c^2 d^2 = \tfrac{1}{4}(px + q)^2.$$

This equation gives

$$p^2 x^2 + 2x(pq - 2\alpha d^2) + q^2 - 4c^2 d^2 = 0,$$

or, since $pq = (\alpha d^2 - \beta c^2)$,

$$p^2 x^2 - 2x(\alpha d^2 + \beta c^2) + q^2 - 4c^2 d^2 = 0.$$

In order that this may reduce to a simple equation, as Diophantus requires, the absolute term must vanish, or $$q^2 = 4c^2d^2,$$

and $$q = 2cd.$$

Thus our method in this case furnishes us with only *one* solution of the double-equation, q being restricted to the value $2cd$, and the solution is

$$x = \frac{2(\alpha d^2 + \beta c^2)}{p^2} = \frac{8c^2d^2(\alpha d^2 + \beta c^2)}{(\alpha d^2 - \beta c^2)^2}.$$

Ex. from Diophantus. This method is only used in one particular case (IV. 39), where $c^2 = d^2$ as the equations originally stand, the equations being

$$\left.\begin{aligned} 8x + 4 &= u^2 \\ 6x + 4 &= w^2 \end{aligned}\right\}.$$

The difference is $2x$, and q is necessarily taken to be $2\sqrt{4}$, or 4; the factors are therefore $\frac{1}{2}x$, 4.

Therefore $$8x + 4 = \tfrac{1}{4}\left(\tfrac{1}{2}x + 4\right)^2,$$

and $$x = 112.$$

General solution (second method) of Form (2) or

$$\left.\begin{aligned} \alpha x + c^2 &= u^2 \\ \beta x + d^2 &= w^2 \end{aligned}\right\}.$$

78 | The difference $= (\alpha - \beta)\,x + (c^2 - d^2)$.
Let the factors of this be p, $\{(\alpha - \beta)\,x + c^2 - d^2\}/p$.
Then, as before proved (p. 74)*, p must be equal to $(c \pm d)$.
Therefore the factors are

$$\frac{\alpha - \beta}{c \pm d}x + c \mp d, \quad c \pm d,$$

and we have finally

$$\alpha x + c^2 = \frac{1}{4}\left(\frac{\alpha - \beta}{c \pm d}x + c \mp d + c \pm d\right)^2$$

$$= \frac{1}{4}\left(\frac{\alpha - \beta}{c \pm d}x + 2c\right)^2.$$

Therefore $$\left(\frac{\alpha - \beta}{c \pm d}\right)^2 x^2 + 4x\left\{\frac{c(\alpha - \beta)}{c \pm d} - \alpha\right\} = 0,$$

which equation gives two possible values for x. Thus in this case we can find by our method *two* values of x, since one of the factors p may be either $(c + d)$ or $(c - d)$.

*See p. 303 in this volume.

Ex. from Diophantus. To solve the equations

$$\left.\begin{aligned} 10x + 9 &= u^2 \\ 5x + 4 &= w^2 \end{aligned}\right\}. \qquad \text{(III. 15.)}$$

The difference is here $5x + 5$, and Diophantus chooses as the factors $5, x + 1$. This case therefore corresponds to the value $c + d$ of p. The solution is given by

$$\left(\tfrac{1}{2}x + 3\right)^2 = 10x + 9, \text{ whence } x = 28.$$

The other value, $c - d$ of p is in this case excluded, because it would lead to a negative value of x.

The possibility of deriving any number of solutions of a double-equation when one solution is known does not seem to have been noticed by Diophantus, though he uses the principle in certain cases of the single equation (see above, pp. 69, 70)*. Fermat was the first, apparently, to discover that this might always be done, if one value a of x were known, by substituting $x + a$ for x in the equations. By this means it is possible to find a positive solution, even if a is negative, by successive applications of the principle.

But, nevertheless, Diophantus had certain peculiar artifices by which he could arrive at a second value. One of these artifices (which is made necessary in one case by the unsuitableness of the value of x found by the ordinary method) gives a different way of solving a double-equation from that which has been explained, and is used only in one special case (IV. 39).

| (β) Second method of solving a double-equation of the first degree. 79

Consider only the special case

$$hx + n^2 = u^2,$$
$$(h + f)x + n^2 = w^2.$$

Take these expressions, and n^2, and write them in order of magnitude, denoting them for convenience by A, B, C.

$$A = (h + f)x + n^2, \quad B = hx + n^2, \quad C = n^2.$$

Therefore $$\frac{A - B}{B - C} = \frac{f}{h}, \text{ and } \left.\begin{aligned} A - B &= fx \\ B - C &= hx \end{aligned}\right\}.$$

Suppose now that $$hx + n^2 = (y + n)^2;$$

therefore $$hx = y^2 + 2ny,$$

and $$A - B = \frac{f}{h}(y^2 + 2ny),$$

or $$A = (y + n)^2 + \frac{f}{h}(y^2 + 2ny);$$

---

*See pp. 297, 298 in this volume.

thus it is only necessary to make this expression a square.

Assume therefore that

$$\left(1+\frac{f}{h}\right)y^2+2n\left(\frac{f}{h}+1\right)y+n^2=(py-n)^2;$$

and any number of values for y, and therefore for x, can be found, by varying p.

Ex. from Diophantus (the only one), IV. 39.

In this case there is the additional condition of a limit to the value of x. The double-equation

$$\left.\begin{aligned}8x+4&=u^2\\6x+4&=w^2\end{aligned}\right\}$$

has to be solved in such a manner that $x<2$.

Here $(A-B)/(B-C)=\frac{1}{3}$, and B is taken[1] to be $(y+2)^2$.

Therefore $$A-B=\tfrac{1}{3}(y^2+4y);$$

therefore $$\begin{aligned}A&=\tfrac{1}{3}(y^2+4y)+y^2+4y+4\\&=\tfrac{4}{3}y^2+\tfrac{16}{3}y+4,\end{aligned}$$

which must be made a square.

80 | If we multiply by $\frac{9}{4}$, we must make

$$3y^2+12y+9=\text{a square},$$

where y must be <2.

Diophantus assumes

$$3y^2+12y+9=(3-my)^2,$$

whence $$y=\frac{6m+12}{m^2-3},$$

and the value of m is then taken such as to make $y<2$.

It is in a note on this problem that Bachet shows that the double-equation

$$\left.\begin{aligned}\alpha x+a&=u^2\\\beta x+b&=w^2\end{aligned}\right\}$$

can be rationally solved by a similar method provided that the coefficients satisfy either of two conditions, although none of the coefficients are squares and neither of

1 Of course Diophantus uses the same variable x where I have for clearness used y. Then, to express what I have called m later, he says: "I form a square from 3 minus *some number* of x's, and x becomes *some number* multiplied by 6 and then added to 12, divided by the difference by which the square of *the number* exceeds 3."

the ratios $\alpha : \beta$ and $a : b$ is equal to the ratio of a square to a square. Bachet's conditions are:

(1) That, when the difference between the two expressions is multiplied or divided by a suitably-chosen number, and the expression thus obtained is subtracted from the smaller of the original expressions, the result is a square number, or
(2) That, when the difference between the two expressions is multiplied or divided by a suitably-chosen number, and the smaller of the two original expressions is subtracted from the expression obtained by the said multiplication or division, the result is a number bearing to the multiplier or divisor the ratio of a square to a square.[1]

---

[1] Bachet of course does not solve equations in general expressions (his notation does not admit of this), but illustrates his conditions by equations in which the coefficients are specific numbers. I will give one of his illustrations of each condition, and then set the conditions out more generally.

Case (1). Equations
$$\left.\begin{aligned} 3x + 13 &= u^2 \\ x + 7 &= w^2 \end{aligned}\right\};$$
difference $2x + 6$

The suitably-chosen number (to *divide* by in this case) is 2;
$$\tfrac{1}{2}\,(\text{difference}) = x + 3,$$
and (lesser expression) $-\tfrac{1}{2}$ (difference) $= x + 7 - (x + 3) = 4$, that is, a square.

We have then to find two squares such that
$$\text{their difference} = 2\ (\text{difference between lesser and } 4).$$
Assume that the lesser $= (y + 2)^2$, 2 being the square root of 4.

Therefore
$$\begin{aligned} (\text{greater square}) &= 3\ (\text{lesser}) - 8 \\ &= 3y^2 + 12y + 4. \end{aligned}$$
To make $3y^2 + 12y + 4$ a square we put
$$3y^2 + 12y + 4 = (2 - py)^2,$$
where p must lie between certain limits which have next to be found. The equation gives
$$y = \frac{12 + 4p}{p^2 - 3}.$$
In order that y may be positive, p^2 must be > 3; and in order that the second of the original expressions, assumed equal to $(y + 2)^2$, may be greater than 7 (it is in fact $x + 7$), we must have $(y + 2) > 2\tfrac{3}{4}$ (an *a fortiori* limit, since $2\tfrac{3}{4} > \sqrt{7}$), or $y > \tfrac{3}{4}$.

Therefore
$$4p + 12 > \tfrac{3}{4}\,(p^2 - 3),$$
or
$$16p + 57 > 3p^2.$$
Suppose that $3p^2 = 16p + 53\tfrac{2}{3}$, which gives $p = 7\tfrac{2}{3}$.

Therefore $p < 7\tfrac{2}{3}$, but $p^2 > 3$.

Put $p = 3$ in the above equation; therefore
$$3y^2 + 12y + 4 = (2 - 3y)^2,$$

81 | 2. *Double-equation of the second degree.*

or the general form

$$\left.\begin{array}{r} Ax^2 + Bx + C = u^2 \\ A'x^2 + B'x + C' = w^2 \end{array}\right\}.$$

---

and $$y = 4.$$

Therefore $$x = (y + 2)^2 - 7 = 29.$$

Case (2). Equations $$\left.\begin{array}{r} 6x + 25 = u^2 \\ 2x + 3 = w^2 \end{array}\right\};$$

difference $$4x + 22$$

The suitable-number (again to *divide* by) in this case is 2;

$$\tfrac{1}{2}\,(\text{difference}) = 2x + 11,$$

and $$\tfrac{1}{2}\,(\text{difference}) - (\text{lesser expression}) = 8 = 2 \cdot 4,$$

where 2 is the divisor used and 4 is the ratio of a square to a square.

Hence two squares have to be found such that

$$(\text{their difference}) = 2\,(\text{sum of lesser and } 8).$$

If the lesser is y^2, the greater is $3y^2 + 16 = (4 - py)^2$, say.

Bachet gives, as limits for p,

$$p \not> 4\tfrac{1}{2}, \quad p^2 > 3,$$

and puts $p = 3$. This gives $y = 4$, so that $x = 6\tfrac{1}{2}$.

Let us now state Bachet's conditions generally.

Suppose the equations to be

$$\left.\begin{array}{r} \alpha x + a = u^2 \\ \beta x + b = w^2 \end{array}\right\}.$$

The difference is $(\alpha - \beta)\,x + (a - b)$.

This has to be multiplied by $\frac{\beta}{\alpha - \beta}$ which is the "suitable" factor in this case, and, if we subtract the product from $\beta x + b$, we obtain

$$b - \frac{\beta}{\alpha - \beta}(a - b), \quad \text{or} \quad \frac{\alpha b - a\beta}{\alpha - \beta}.$$

(1) The first of Bachet's conditions is that

$$\frac{\alpha b - a\beta}{\alpha - \beta} = a \ \text{ square} = p^2/q^2, \text{ say}.$$

(2) The second condition is that

$$\frac{a\beta - \alpha b}{\alpha - \beta} = \frac{p^2}{q^2} \cdot \frac{\beta}{\alpha - \beta},$$

These equations are much less thoroughly treated in Diophantus than those of the first degree. Only such special instances | occur as can be easily solved by the methods which we have described for equations of the first degree. 82

---

or $$\frac{a\beta - \alpha b}{\beta} = \text{a ratio of a square to a square.}$$

It is to be observed that the first of these conditions can be obtained by considering the equation

$$(\alpha - \beta)^2x^2 + 2x\{(\alpha - \beta)(a - b) - p^2(\alpha + \beta)\} + (a - b)^2 - 2p^2(a + b) + p^4 = 0,$$

obtained on page 74 above*.

Diophantus only considers the cases in which this equation reduces to a simple equation; but the solution of it as a *mixed* quadratic gives a rational value of x provided that

$$\{(\alpha - \beta)(a - b) - p^2(\alpha + \beta)\}^2 - (\alpha - \beta)^2\{(a - b)^2 - 2p^2(a + b) + p^4\} \text{ is a square,}$$

that is, if

$$p^4\{(\alpha + \beta)^2 - (\alpha - \beta)^2\} + 2p^2\{(a + b)(\alpha - \beta)^2 - (\alpha^2 - \beta^2)(a - b)\} \text{ is a square,}$$

which reduces to $$\alpha\beta p^2 + (\alpha - \beta)(\alpha b - a\beta) = \text{a square} \qquad \text{(A).}$$

This can be solved (cf. p. 68 above**), if

$$\frac{\alpha b - a\beta}{\alpha - \beta} \text{ is a square.} \qquad \text{(Bachet's first condition.)}$$

Again take Bachet's second condition

$$\frac{a\beta - \alpha b}{\beta} = \text{a square} = \frac{r^2}{s^2} \text{ say,}$$

and substitute $\beta r^2/s^2$ for $a\beta - \alpha b$ in the equation (A) above.

Therefore $$\alpha\beta p^2 - (\alpha - \beta)\beta\frac{r^2}{s^2} = \text{a square,}$$

or $$\alpha\beta p'^2 - (\alpha - \beta)\beta = \text{a square.}$$

This is satisfied by $p' = 1$; therefore (p. 69)*** any number of other solutions can be found.

The second condition can also be obtained directly by eliminating x from the equations

$$\left.\begin{aligned}\alpha x + a &= u^2\\ \beta x + b &= w^2\end{aligned}\right\};$$

for the result is $$\frac{\alpha}{\beta}w^2 + \frac{a\beta - \alpha b}{\beta} = u^2,$$

which can be rationally solved if

$$\frac{a\beta - \alpha b}{\beta} = \text{a square.}$$

---

* See p. 302 in this volume.
** See p. 297 in this volume.
*** See p. 297 in this volume.

The following types are found.

(1) $$\left.\begin{aligned}\rho^2x^2 + \alpha x + a &= u^2\\ \rho^2x^2 + \beta x + b &= w^2\end{aligned}\right\}.$$

The difference is $(\alpha - \beta)x + (a - b)$, and, following the usual course, we may, *e.g.*, resolve this into the factors

83 $$(a - b)\left\{\frac{\alpha - \beta}{a - b}x + 1\right\};$$

as usual, we put

$$\rho^2x^2 + \alpha x + a = \frac{1}{4}\left(\frac{\alpha - \beta}{a - b}x + 1 + a - b\right)^2,$$

or $$\rho^2x^2 + \beta x + b = \frac{1}{4}\left(\frac{\alpha - \beta}{a - b}x + 1 - a + b\right)^2.$$

In order that x may be rational a condition is necessary; thus x is rational if

$$\rho = \frac{1}{2}\frac{\alpha - \beta}{a - b}.$$

This is the case in the only instance of the type where a is not equal to b, namely (III. 13)

$$\left.\begin{aligned}4x^2 + 15x &= u^2\\ 4x^2 - x - 4 &= w^2\end{aligned}\right\};$$

the difference is $16x + 4$, and the resolution of this into the factors 4, $4x + 1$ solves the problem.

In the other cases of the type $a = b$; the difference is then $(\alpha - \beta)x$, which is resolved into the factors

$$\frac{\alpha - \beta}{2\rho}, \quad 2\rho x;$$

and we put $$\rho^2x^2 + \alpha x + a = \frac{1}{4}\left(\frac{\alpha - \beta}{2\rho} + 2\rho x\right)^2,$$

or $$\rho^2x^2 + \beta x + a = \frac{1}{4}\left(\frac{\alpha - \beta}{2\rho} - 2\rho x\right)^2,$$

whence $$\frac{\alpha + \beta}{2}x = \left(\frac{\alpha - \beta}{4\rho}\right)^2 - a.$$

Exx. from Diophantus:

$$\left.\begin{aligned} x^2 + x - 1 &= u^2 \\ x^2 - 1 &= w^2 \end{aligned}\right\} \qquad \text{(IV. 23.)}$$

and

$$\left.\begin{aligned} x^2 + 14x + 1 &= u^2 \\ x^2 + 1 &= w^2 \end{aligned}\right\}. \qquad \text{(VI. 8.)}$$

(2) The second type found in Diophantus[1] is

$$\left.\begin{aligned} x^2 + \alpha x + a &= u^2 \\ \beta x + a &= w^2 \end{aligned}\right\},$$

where one equation has no term in x^2, and $\rho = 1$, $a = b$.

| The difference $x^2 + (\alpha - \beta)x$ is resolved into the factors 84

$$x(x + \alpha - \beta);$$

---

[1] It is perhaps worth noting that the method of the "double-equation" has a distinct advantage in this type of case. The alternative is to solve by the method of Euler (who does not use the "double-equation"), *i.e.* to put the linear expression equal to p^2 and then, substituting the value of x (in terms of p) in the quadratic expression, to solve the resulting equation in p. But the difficulties would generally be great. Take the case of VI. 6 where

$$\left.\begin{aligned} x^2 + 1 \\ 14x + 1 \end{aligned}\right\} \text{have to be made squares.}$$

If $14x + 1 = p^2$, $x = (p^2 - 1)/14$;

therefore $$x^2 + 1 = \frac{(p^2 - 1)^2}{14^2} + 1 \text{ has to be made a square,}$$

or $$p^4 - 2p^2 + 197 = \text{a square.}$$

This does not admit of solution unless we could somehow discover empirically *one* value of p which would satisfy the requirement, and this would be very difficult. [*Note*: $p = 7$ is a solution.]

Let us take an easier case for solution by this method,

$$\left.\begin{aligned} x^2 + 1 &= u^2 \\ x + 1 &= w^2 \end{aligned}\right\},$$

which is solved by Euler (*Algebra*, Part II. Art. 222), and let us compare the working of the two methods in this case.

I. *Euler's method*. Assuming $x + 1 = p^2$ and substituting $p^2 - 1$ for x in the quadratic expression, we have

$$p^4 - 2p^2 + 2 = \text{a square.}$$

This can only be solved generally if we can discover one possible value of p by trial; this however is not difficult in the particular case, for $p = 1$ is an obvious solution.

To find others we put $1 + q$ instead of p in the expression to be made a square; this gives

$$1 + 4q^2 + 4q^3 + q^4 = \text{a square.}$$

and we put
$$\beta x + a = \tfrac{1}{4}(\alpha - \beta)^2,$$
which gives x.

85 | Exx. from Diophantus:

$$\left.\begin{aligned} x^2 - 12 &= u^2 \\ 6\tfrac{1}{2}x - 12 &= w^2 \end{aligned}\right\}, \qquad \text{(V. 1.)}$$

This can be solved in several ways.

1. Suppose $$1 + 4q^2 + 4q^3 + q^4 = (1 + q^2)^2;$$

thus $$4q^2 + 4q^3 = 2q^2, \text{ whence } q = -\frac{1}{2}, \quad p = \frac{1}{2} \text{ and } x = -\frac{3}{4}.$$

2. Suppose $$1 + 4q^2 + 4q^3 + q^4 = (1 - q^2)^2;$$

thus $$4q^2 + 4q^3 = -2q^2, \text{ and } q = -\frac{3}{2}, \quad p = -\frac{1}{2} \text{ and } x = -\frac{3}{4}.$$

3. Suppose $1 + 4q^2 + 4q^3 + q^4 = (1 \pm 2q \pm q^2)^2$;

and we find, in either case, that $q = -2$, so that $p = -1$, $x = 0$.

4. Suppose $1 + 4q^2 + 4q^3 + q^4 = (1 + 2q^2)^2$;

and we have $4q^3 + q^4 = 4q^4$, whence $q = \frac{4}{3}$, $p = \frac{7}{3}$ and $x = (\frac{7}{3})^2 - 1 = \frac{40}{9}$.

This value of x satisfies the conditions, for
$$x + 1 = \left(\frac{7}{3}\right)^2, \quad x^2 + 1 = \frac{1681}{81} = \left(\frac{41}{9}\right)^2.$$

The above five suppositions therefore give only two serviceable solutions
$$x = -\frac{3}{4}, \quad x = \frac{40}{9}.$$

To find another solution we take one of the values of q already found, say $q = -\frac{1}{2}$, and substitute $r - \frac{1}{2}$ for q. This gives $p = 1 + q = r + \frac{1}{2}$, and we substitute this value for p in the expression $p^4 - 2p^2 + 2$.

We have then $\dfrac{25}{16} - \dfrac{3}{2}r - \dfrac{1}{2}r^2 + 2r^3 + r^4$ to be made a square, or
$$25 - 24r - 8r^2 + 32r^3 + 16r^4 = \text{a square}.$$

1. We take $5 + fr \pm 4r^2$ for the root, so that the absolute term and the term in r^4 may disappear. We can make the term in r disappear also by putting $10f = -24$ or $f = -\frac{12}{5}$ We then have
$$-8r^2 + 32r^3 = r^2(f^2 \pm 40) \pm 8fr^3.$$

(a) The upper sign gives
$$-8 + 32r = 40 + f^2 + 8fr,$$

$$\left.\begin{aligned} x^2 + 1 &= u^2 \\ 14x + 1 &= w^2 \end{aligned}\right\}, \qquad \text{(VI. 6.)}$$

$$\left.\begin{aligned} x^2 - 6144x + 1048576 &= u^2 \\ x + 64 &= w^2 \end{aligned}\right\}. \qquad \text{(VI. 22).}$$

---

and $$r = (f^2 + 48)/(32 - 8f) = \frac{21}{20};$$

thus $$p = \frac{31}{20}, \text{ and } x = p^2 - 1 = \frac{561}{400}.$$

(b) The lower sign gives

$$-8 + 32r = -40 + f^2 - 8fr,$$

and $$r = (f^2 - 32)/(32 + 8f) = -\frac{41}{20};$$

thus $$\text{p} = -\frac{31}{20}, \text{ and } x = \frac{561}{400} \text{ as before.}$$

We have therefore $$x + 1 = \left(\frac{31}{20}\right)^2, \text{ and } x^2 + 1 = \left(\frac{689}{400}\right)^2.$$

2. Another solution is found by assuming the root to be $5 + fr + gr^2$ and determining f and g so that the absolute term and the terms in r, r^2 may vanish; the result is

$$f = -\frac{12}{5}, \quad g = -\frac{172}{125}, \quad r = \frac{2fg - 32}{16 - g^2} = -\frac{1550}{861},$$

whence $$p = -\frac{2239}{1722}, \quad x = \frac{2047837}{2965284}, \quad x + 1 = \left(\frac{2239}{1722}\right)^2$$

and $$x^2 + 1 = \left(\frac{3603685}{2965284}\right)^2.$$

II. *Method of "double-equation."*

$$x^2 + 1 = u^2,$$
$$x + 1 = w^2.$$

The difference $= x^2 - x$.

(1) If we take as factors x, $x - 1$ and, as usual, equate the square of half their difference, or $\frac{1}{4}$, to $x + 1$, we have

$$x + 1 = \frac{1}{4},$$

or $$x = -\frac{3}{4}.$$

(2) If we take $\frac{1}{2}x$, $2x - 2$, as factors, half the sum of which is $\frac{5}{4}x - 1$, so that the absolute terms may disappear in the resulting equation, we have

86 | The absolute terms in the last case are made equal by multiplying the second equation by $(128)^2$ or 16384.

(3) One separate case must be mentioned which cannot be solved, from Diophantus' standpoint, by the foregoing method, but which sometimes occurs and is solved by a special artifice.

The form of double-equation is

$$\left.\begin{aligned}\alpha x^2 + ax &= u^2\\ \beta x^2 + bx &= w^2\end{aligned}\right\} \qquad \begin{aligned}&(1),\\ &(2).\end{aligned}$$

Diophantus assumes $$u^2 = m^2x^2,$$

whence, by (1), $$x = a/(m^2 - \alpha),$$

$$\frac{25}{16}x^2 - \frac{5}{2}x = x^2,$$

or $$\frac{9}{16}x = \frac{5}{2},$$

and $$x = \frac{40}{9}.$$

(3) To find another value by means of the first, namely $x = -\frac{3}{4}$ we substitute $y = -\frac{3}{4}$ for x in the original expressions. We then have to solve

$$y^2 - \frac{3}{2}y + \frac{25}{16} = u^2,$$
$$y + \frac{1}{4} = w^2.$$

Multiply the latter by $\frac{25}{4}$ so as to make the absolute terms the same, and we must have

$$\frac{25}{4}y + \frac{25}{16} = w'^2.$$

Subtract from the first expression, and the difference is $y^2 - \frac{31}{4}y = y(y - \frac{31}{4})$; then, equating the square of half the difference of the factors to the smaller expression, we have

$$\frac{1}{4}\left(\frac{31}{4}\right)^2 = \frac{25}{4}y + \frac{25}{16},$$

so that $$961 = 400y + 100.$$

Therefore

$$y = \frac{861}{400}, \text{ and } x = y - \frac{3}{4} = \frac{561}{400}, \quad x + 1 = \left(\frac{31}{20}\right)^2, \quad x^2 + 1 = \left(\frac{689}{400}\right)^2.$$

(4) If we start from the known value $\frac{40}{9}$ and put $y + \frac{40}{9}$ for x in the equations, we obtain Euler's fourth value of x, namely $\frac{2047837}{2965284}$.

Thus all the four values obtained by Euler are more easily obtained by the method of the "double-equation."

| and, by substitution in (2), we derive that 87

$$\beta\left(\frac{a}{m^2-\alpha}\right)^2+\frac{ba}{m^2-\alpha}\text{ must be a square,}$$

or

$$\frac{a^2\beta+ba(m^2-\alpha)}{(m^2-\alpha)^2}=\text{a square.}$$

We have therefore to solve the equation

$$abm^2+a(a\beta-\alpha b)=y^2,$$

and this form can or cannot be solved by the methods already given according to the nature of the coefficients[1]. Thus it can be solved if $(a\beta-\alpha b)/a$ is a square or if a/b is a square.

Exx. from Diophantus:

$$\left.\begin{aligned}6x^2+4x&=u^2\\6x^2+3x&=w^2\end{aligned}\right\},\qquad\text{(VI. 12.)}$$

$$\left.\begin{aligned}6x^2-5x&=u^2\\6x^2-3x&=w^2\end{aligned}\right\}.\qquad\text{(VI. 14.)}$$

(II) *Indeterminate Equations of a Degree Higher than the Second*

(a) Single equations. These are properly divided by Nesselmann into two classes; the first comprises those problems in which it is required to make a function of x, of a degree higher than the second, a square; the second comprises those in which a rational value of x has to be found which will make any function of x, not a square, but a higher power of some number. The first class of problems requires the solution in rational numbers of

$$Ax^n+Bx^{n-1}+\cdots+Kx+L=y^2,$$

the second the solution of

$$Ax^n+Bx^{n-1}+\cdots+Kx+L=y^3,$$

---

[1] Diophantus apparently did not observe that the above form of double-equation can be reduced to one of the first degree by dividing by x^2 and substituting y for $1/x$, when it becomes

$$\alpha+ay=u'^2,$$
$$\beta+by=w'^2.$$

Adapting Bachet's second condition, we see that the equations can be rationally solved if $(\beta a-\alpha b)/a$ is a square, which is of course the same as one of the conditions under which the above equation $abm^2+a\,(a\beta-\alpha b)$ can be solved.

for Diophantus does not go beyond making a certain function of x a cube. In no instance, however, of the first class does the index n exceed 6, while in the second class n does not (except in a special case or two) exceed 3.

88 | *First Class.* Equation

$$Ax^n + Bx^{n-1} + \cdots + Kx + L = y^2.$$

The forms found in Diophantus are as follows:

1. Equation $\qquad Ax^3 + Bx^2 + Cx + d^2 = y^2,$

Here, as the absolute term is a square, we might put for y the expression $mx + d$, and determine m so that the coefficient of x in the resulting equation vanishes. In that case

$$2md = C \quad \text{and} \quad m = C/2d;$$

and we obtain, in Diophantus' manner, a simple equation for x, giving

$$x = \frac{C^2 - 4d^2B}{4d^2A}.$$

Or we might put for y the expression $m^2x^2 + nx + d$, and determine m, n so that the coefficients of x, x^2 in the resulting equation both vanish, in which case we should again have a simple equation for x. Diophantus, in the only example of this form of equation which occurs (VI. 18), makes the first supposition. The equation is

$$x^3 - 3x^2 + 3x + 1 = y^2,$$

and Diophantus assumes $y = \frac{3}{2}x + 1$, whence $x = \frac{21}{4}$.

2. Equation $\qquad Ax^4 + Bx^3 + Cx^2 + Dx + E = y^2.$

In order that this equation may be solved by Diophantus' method, either A or E must be a square. If A is a square and equal to a^2, we may assume $y = ax^2 + \frac{B}{2a}x + n$, determining n so that the term in x^2 vanishes. If E is a square $(= e^2)$, we may write $y = mx^2 + \frac{D}{2e}x + e$, determining m so that the term in x^2 in the resulting equation may vanish. We shall then, in either case, obtain a simple equation in x.

The examples of this form in Diophantus are of the kind

$$a^2x^4 + Bx^3 + Cx^2 + Dx + e^2 = y^2,$$

where we can assume $y = \pm ax^2 + kx \pm e$, determining k so that in the resulting equation the coefficient of x^3 or of x may vanish; when we again have a simple equation.

Ex. from Diophantus (IV. 28):

$$9x^4 - 4x^3 + 6x^2 - 12x + 1 = y^2.$$

Diophantus assumes $y = 3x^2 - 6x + 1$, and the equation reduces to

$$32x^3 - 36x^2 = 0, \text{ whence } x = \tfrac{9}{8}.$$

| Diophantus is guided in his choice of signs in the expression $\pm\, ax^2 + kx \pm e$ by 89
the necessity for obtaining a "rational" result.

Far more difficult to solve are those equations in which, the left-hand expression being bi-quadratic, the odd powers of x are wanting, *i.e.* the equations $Ax^4 + Cx^2 + E = y^2$ and $Ax^4 + E = y^2$, even when A or E is a square, or both are so. These cases Diophantus treats more imperfectly.

3. Equation $\qquad Ax^4 + Cx^2 + E = y^2,$

Only very special cases of this form occur. The type is

$$a^2x^4 - c^2x^2 + e^2 = y^2,$$

which is written

$$a^2x^2 - c^2 + \frac{e^2}{x^2} = y'^2.$$

Here y' is assumed to be ax or e/x, and in either case we have a rational value for x.

Exx. from Diophantus:

$$25x^2 - 9 + \frac{25}{4x^2} = y^2. \qquad \text{(V. 27.)}$$

This is assumed to be equal to $25x^2$.

$$\frac{25}{4}x^2 - 25 + \frac{25}{4x^2} = y^2, \qquad \text{(V. 28.)}$$

where y^2 is assumed to be equal to $25/4x^2$.

4. Equation $\qquad Ax^4 + E = y^2.$

The case occurring in Diophantus is $x^4 + 97 = y^2$ (V. 29). Diophantus tries one assumption, $y = x^2 - 10$, and finds that this gives $x^2 = \frac{3}{20}$, which leads to no rational result. Instead, however, of investigating in what cases this equation can be solved, he simply drops the equation $x^4 + 97 = y^2$ and seeks, by altering his original assumptions, to obtain, in place of it, another equation of the same type which can be solved in rational numbers. In this case, by altering his assumptions, he is able to replace the refractory equation by a new one, $x^4 + 337 = y^2$, and at the same time to find a suitable substitution for y, namely $y = x^2 - 25$, which brings out a rational result, namely $x = \frac{12}{5}$. This is a good example of his characteristic artifice of "Back-reckoning,"[1] as Nesselmann calls it.

---

1 "Methode der Zurückrechnung und Nebenaufgabe."

5. Equation of sixth degree in the special form

$$x^6 - Ax^3 + Bx + c^2 = y^2.$$

90 | It is only necessary to put $y = x^3 + c$, and we have

$$-Ax^2 + B = 2cx^2,$$

or

$$x^2 = \frac{B}{A + 2c},$$

which gives a rational solution if $B/(A + 2c)$ is a square.

6. If, however, this last condition does not hold, as in the case occurring IV. 18, $x^6 - 16x^3 + x + 64 = y^2$, Diophantus employs his usual artifice of "back-reckoning," which enables him to replace the equation by another, $x^6 - 128x^3 + x + 4096 = y^2$, where the condition is satisfied, and, by assuming $y = x^3 + 64$, x is found to be $\frac{1}{16}$.

Second Class. Equation of the form

$$Ax^n + Bx^{n-1} + \cdots + Kx + L = y^3.$$

Except for such simple cases as $Ax^2 = y^3$, $Ax^4 = y^3$, where it is only necessary to assume $y = mx$, the only cases occurring in Diophantus are of the forms

$$Ax^2 + Bx + C = y^3,$$

$$Ax^3 + Bx^2 + Cx + D = y^3.$$

1. Equation $$Ax^2 + Bx + C = y^3.$$

There are only two examples of this form. First, in VI. 1 the expression $x^2 - 4x + 4$ is to be made a cube, being already a square. Diophantus naturally assumes $x - 2 =$ a cube number, say 8, and $x = 10$.

Secondly, in VI. 17 a peculiar case occurs. A cube is to be found which exceeds a square by 2. Diophantus assumes $(x - 1)^3$ for the cube and $(x + 1)^2$ for the square, and thus obtains the equation

$$x^3 - 3x^2 + 3x - 1 = x^2 + 2x + 3,$$

or

$$x^3 + x = 4x^2 + 4,$$

previously mentioned (pp. 66–7)* which is satisified by $x = 4$. The question arises whether it was accidentally or not that this cubic took so simple a form. Were $x - 1$, $x + 1$ assumed with knowledge and intention? Since 27 and 25 are, as Fermat observes,[1] the only integral numbers which satisfy the conditions, it would seem that Diophantus so chose his assumptions as to lead back to a known result, while apparently making them arbitrarily with no particular reference to the end desired. Had this
91 not | been so, we should probably have found him, here as elsewhere in the work, first

* See pp. 295–296 in this volume.

1 Note on VI. 17, *Oeuvres*, I. pp. 333–4, II. p. 434. The fact was proved by Euler (*Algebra*, Part II. Arts. 188, 193). See note on VI. 17 *infra* for the proof.

leading us on a false tack and then showing us how we can correct our assumptions. The fact that he here makes the right assumptions to begin with makes us suspect that the solution is not based on a general principle but is empirical merely.

2. Equation $$Ax^3 + Bx^2 + Cx + D = y^3.$$

If A or D is a cube number, this equation is easy of solution. For, first, if $A = a^3$, we have only to write $y = ax + \frac{B}{3a^2}$, and we obtain a simple equation in x.

Secondly, if $D = d^3$, we put $y = \frac{C}{3d^2}x + d$.

If the equation is $a^3x^3 + Bx^2 + Cx + d^3 = y^3$, we can use either assumption, or we may put $y = ax + d$, obtaining a simple equation as before.

Apparently Diophantus used the last assumption only; for in IV. 27 he rejects as impossible the equation

$$8x^3 - x^2 + 8x - 1 = y^3,$$

because $y = 2x - 1$ gives a negative value $x = -\frac{2}{11}$, whereas either of the other assumptions gives a rational value.[1]

(b) Double-equations. There are a few examples in which, of two functions of x, one is to be made a square, and the other a cube, by one and the same rational value of x. The cases are for the most part very simple; *e.g.* in VI. 19 we have to solve

$$\left.\begin{aligned} 4x + 2 &= y^3 \\ 2x + 1 &= z^2 \end{aligned}\right\};$$

thus $y^3 = 2z^2$, and $z = 2$.

A rather more complicated case is VI. 21, where we have the double-equation

$$\left.\begin{aligned} 2x^2 + 2x &= y^2 \\ x^3 + 2x^2 + x &= z^3 \end{aligned}\right\}.$$

Diophantus assumes $y = mx$, whence $x = 2/(m^2 - 2)$, and we have

$$\left(\frac{2}{m^2 - 2}\right)^3 + 2\left(\frac{2}{m^2 - 2}\right)^2 + \frac{2}{m^2 - 2} = z^3,$$

or $$\frac{2m^4}{(m^2 - 2)^3} = z^3.$$

[1] There is a special case in which C and D vanish, $Ax^3 + Bx^2 = y^3$. Here y is put equal to mx, and $x = B/(m^3 - A)$. Cf. IV. 6, 28(2).

92 | To make $2m^4$ a cube, we need only make $2m$ a cube or put $m = 4$. This gives $x = \frac{1}{7}$. The general case

$$\left.\begin{aligned} Ax^3 + Bx^2 + Cx &= z^3 \\ bx^2 + cx &= y^2 \end{aligned}\right\}$$

would, of course, be much more difficult; for, putting $y = mx$, we have

$$x = c/(m^2 - b),$$

and we have to solve

$$A\left(\frac{c}{m^2 - b}\right)^3 + B\left(\frac{c}{m^2 - b}\right)^2 + C\left(\frac{c}{m^2 - b}\right) = z^3,$$

or

$$Ccm^4 + c(Bc - 2bC)m^2 + bc(bC - Bc) + Ac^3 = u^3,$$

of which equation the above corresponding equation is a very particular case.

Summary of the Preceding Investigation.

1. Diophantus solves completely equations of the first degree, but takes pains to secure beforehand that the solution shall be positive. He shows remarkable address in reducing a number of simultaneous equations of the first degree to a single equation in *one* variable.
2. For determinate equations of the second degree he has a general method or rule of solution. He takes, however, in the *Arithmetica*, no account of more than one root, even where both roots are positive rational numbers. But, his object being simply to obtain *some* solution in rational numbers, we need not be surprised at his ignoring one of two roots, even though he knew of its existence.
3. No equations of a degree higher than the second are solved in the book except a particular case of a cubic.
4. Indeterminate equations of the first degree are not treated of in the work. Of indeterminate equations of the second degree, as $Ax^2 + Bx + C = y^2$, only those cases are fully dealt with in which A or C vanishes, while the methods employed only enable us to solve equations of the more general forms

   $$Ax^2 + C = y^2 \quad \text{and} \quad Ax^2 + Bx + C = y^2$$

   when A, or C, or $\frac{1}{4}B^2 - AC$ is positive and a square number, or (in the case of $Ax^2 \pm C = y^2$) when one solution is already known.

93 | 5. For double-equations of the second degree Diophantus has a definite method when the coefficient of x^2 in both expressions vanishes; the applicability of this method is, however, subject to conditions, and it has to be supplemented in one or two cases by another artifice. Of more complicated cases we find only a few examples under conditions favourable for solution by his method.

6. Diophantus' treatment of indeterminate equations of degrees higher than the second depends upon the particular conditions of the problems, and his methods lack generality.

7. More wonderful than his actual treatment of equations are the clever artifices by which he contrives to avoid such equations as he cannot theoretically solve, *e.g.* by his device of "back-reckoning," instances of which would have been out of place in this chapter and can only be studied in the problems themselves.

I shall not attempt to class as "methods" certain headings in Nesselmann's classification of the problems, such as (a) "Solution by mere reflection," (b) "Solution in general expressions," of which there are few instances definitely so described by Diophantus, or (c) "Arbitrary determinations and assumptions." The most that can be done by way of describing these "methods" is to quote a few characteristic instances. This is what Nesselmann has done, and he regrets at the end of his chapter on "Methods of Solution" that it must of necessity be so incomplete. To understand and appreciate the various artifices of Diophantus it is in fact necessary to read the problems themselves in their entirety.

With regard to the "Use of the right-angled triangle," all that can be said of a general character is that only "rational" right-angled triangles (those namely in which the three sides can all be represented by rational numbers) are used in Diophantus, and accordingly the introduction of the "right-angled triangle" is merely a convenient way of indicating the problem of finding two square numbers, the sum of which is also a square number. The general form used by Diophantus (except in one case, VI. 19, *q.v.*) for the sides of a right-angled triangle is $(a^2 + b^2)$, $(a^2 - b^2)$, $2ab$, which expressions clearly satisfy the condition

$$(a^2 + b^2)^2 = (a^2 - b^2)^2 + (2ab)^2.$$

The expression of the sides of a right-angled triangle in this form Diophantus calls "forming a right-angled triangle from the numbers a and b." His right-angled triangles are of course formed from *particular* numbers. "Forming a right-angled | triangle 94
from 7, 2" means taking a right-angled triangle with sides $(7^2 + 2^2)$, $(7^2 - 2^2)$, $2 \cdot 7 \cdot 2$, or 53, 45, 28.

II. METHOD OF LIMITS.

As Diophantus often has to find a series of numbers in ascending or descending order of magnitude, and as he does not admit negative solutions, it is often necessary for him to reject a solution which he has found by a straightforward method because it does not satisfy the necessary condition; he is then very frequently obliged to find solutions which lie *within certain limits* in place of those rejected.

1. A very simple case is the following: Required to find a value of x such that some power of it, x^n, shall lie between two given numbers. Let the given numbers be a, b. Then Diophantus' method is to multiply both a and b by 2^n, 3^n, and so on, successively, until some nth power is seen which lies between the two products. Thus suppose that c^n lies between ap^n and bp^n; then we can put $x = c/p$, in which case the condition is satisfied, for $(c/p)^n$ lies between a and b.

Exx. In IV. 31 (2) Diophantus has to find a square between $1\frac{1}{4}$ and 2. He multiplies both by a square, 64; this gives 80 and 128, and 100 is clearly a square which lies between them; therefore $(\frac{10}{8})^2$ or $\frac{25}{16}$ satisfies the prescribed condition.

Here, of course, Diophantus might have multiplied by any other square, as 16. In that case the limits would have become 20 and 32, between these lies the square 25, which gives the same square $\frac{25}{16}$ as that before found.

In VI. 21 a sixth power ("cube-cube") is sought which shall lie between 8 and 16. The sixth powers of the first four natural numbers are 1, 64, 729, 4096. Multiply 8 and 16 by 2^6 or 64, and we have as limits 512 and 1024, between which 729 lies. Therefore $\frac{729}{64}$ is a sixth power satisfying the given condition. To multiply by 729 in this case would not give us a solution.

2. Sometimes a value of x has to be found which will give some function of x a value intermediate between the values of two other functions of x.

Ex. 1. In IV. 25 it is necessary to find a value of x such that $8/(x^2+x)$ shall lie between x and $x+1$.

The first condition gives $8 > x^3 + x^2$.

95 | Diophantus accordingly assumes that

$$8 = (x + \tfrac{1}{3})^3 = x^3 + x^2 + \tfrac{1}{3}x + \tfrac{1}{27},$$

which is greater than $x^3 + x^2$.

Thus $x = \frac{5}{3}$ satisfies one condition. It is also seen to satisfy the second condition, or $\frac{8}{x^2+x} < x + 1$. Diophantus, however, says nothing about the second condition being satisfied; his method is, therefore, here imperfect.

Ex. 2. In V. 30 a value of x has to be found which shall make

$$x > \tfrac{1}{8}(x^2 - 60) \text{ but } < \tfrac{1}{5}(x^2 - 60),$$

that is,

$$\left.\begin{matrix} x^2 - 60 > 5x \\ x^2 - 60 < 8x \end{matrix}\right\}.$$

Hence, says Diophantus, x is not less than 11 and not greater than 12. We have already spoken (p. 60 sqq.)* of his treatment of such cases.

Next, the problem in question requires that $x^2 - 60$ shall be a square. Assume then that

$$x^2 - 60 = (x - m)^2,$$

and we have

$$x = (m^2 + 60)/2m.$$

Since, says Diophantus, x is greater than 11 and less than 12, it follows that

$$m^2 + 60 > 22m \text{ but } < 24m;$$

and m must therefore lie between 19 and 21 (cf. p. 62 above)**. He puts $m = 20$, and so finds $x = 11\frac{1}{2}$.

* See p. 289 in this volume.
** See p. 291 in this volume.

III. METHOD OF APPROXIMATION TO LIMITS.

We come, lastly, to a very distinctive method called by Diophantus παρισότης or παρισότητος ἀγωγή. The object of this is to solve such problems as that of finding two, or three, square numbers the sum of which is a given number, while each of them approximates as closely as possible to one and the same number.

This method can be best exhibited by giving Diophantus' two instances, in the first of which *two* such squares, and in the second *three*, are required. In cases like this the principles cannot be so well indicated with general symbols as with concrete numbers,
which have the advantage that their properties are immediately | obvious, and the 96
separate expression of conditions is rendered unnecessary.

Ex. 1. Divide 13 into two squares each of which > 6 (V. 9.)

Take half of 13, or $6\frac{1}{2}$, and find what small fraction $1/x^2$ added to it will make it a square: thus

$$6\tfrac{1}{2} + \frac{1}{x^2}, \text{ or } 26 + \frac{1}{y^2}, \text{ must be a square.}$$

Diophantus assumes

$$26 + \frac{1}{y^2} = \left(5 + \frac{1}{y}\right)^2, \text{ or } 26y^2 + 1 = (5y+1)^2,$$

whence $y = 10$ and $1/y^2 = \frac{1}{100}$, or $1/x^2 = \frac{1}{400}$; and

$$6\tfrac{1}{2} + \tfrac{1}{400} = \text{a square, } (\tfrac{51}{20})^2.$$

[The assumption of $(5y+1)^2$ is not arbitrary, for assume $26y^2 + 1 = (py+1)^2$, and y is then $2p/(26 - p^2)$; since $1/y$ should be a *small* proper fraction, 5 is the most suitable and the smallest possible value for p, inasmuch as $26 - p^2 < 2p$ or $p^2 + 2p + 1 > 27$.]

It is now necessary, says Diophantus, to divide 13 into two squares the sides of which are both as near as possible to $\frac{51}{20}$.

Now the sides of the two squares into which 13 is naturally decomposed are 3 and 2, and

$$3 \text{ is } > \tfrac{51}{20} \text{ by } \tfrac{9}{20},$$

$$2 \text{ is } < \tfrac{51}{20} \text{ by } \tfrac{11}{20}.$$

But, if $3 - \frac{9}{20}$, $2 + \frac{11}{20}$ were taken as the sides of two squares, the sum of the squares would be

$$2(\tfrac{51}{20})^2 = \frac{2 \cdot 2601}{400},$$

which is > 13.

Accordingly Diophantus puts $3 - 9x$, $2 + 11x$ for the sides of the required squares, where therefore x is not exactly $\frac{1}{20}$ but near it.

Thus $$(3 - 9x)^2 + (2 + 11x)^2 = 13,$$

and Diophantus obtains $\quad x = \frac{5}{101}$.

The sides of the required squares are $\frac{257}{101}, \frac{258}{101}$.

[It is of course a necessary condition that the original number, here 13, shall be a number capable of being expressed as the sum of two squares.]

97 | Ex. 2. Divide 10 into three squares such that each of them is >3 (V. 11).

[The original number, here 10, must of course be expressible as the sum of three squares.]

Take one-third of 10, or $3\frac{1}{3}$, and find what small fraction of the form $1/x^2$ added to $3\frac{1}{3}$ will make a square; *i.e.* we have to make $30 + \frac{9}{x^2}$ a square, or $30y^2 + 1$ a square, where $3/x = 1/y$.

Diophantus assumes

$$30y^2 + 1 = (5y + 1)^2,$$

whence $y = 2$ and therefore $1/x^2 = \frac{1}{36}$; and $3\frac{1}{3} + \frac{1}{36} = \frac{121}{36}$, a square.

[As before, if we assume $30y^2 = (py+1)^2$, $y = 2p/(30 - p^2)$; and, since $1/y$ must be a small proper fraction, $30 - p^2$ should be $< 2p$, or $p^2 + 2p + 1 > 31$. Accordingly Diophantus chooses 5 for p as being the smallest possible integral value.]

We have now, says Diophantus, to make each of the sides of our required squares as near as may be to $\frac{11}{6}$.

Now $$10 = 9 + 1 = 3^2 + (\tfrac{3}{5})^2 + (\tfrac{4}{5})^2,$$

and $3, \frac{3}{5}, \frac{4}{5}$ are the sides of three squares the sum of which is 10.

Bringing $(3, \frac{3}{5}, \frac{4}{5})$ and $\frac{11}{6}$ to a common denominator, we have

$$\left(\frac{90}{30}, \frac{18}{30}, \frac{24}{30}\right) \text{ and } \frac{55}{30}.$$

And $$3 > \frac{55}{30} \text{ by } \frac{35}{30},$$

$$\frac{3}{5} < \frac{55}{30} \text{ by } \frac{37}{30},$$
$$\frac{4}{5} < \frac{55}{30} \text{ by } \frac{31}{30}.$$

If now we took $3 - \frac{35}{30}, \frac{3}{5} + \frac{37}{30}, \frac{4}{5} + \frac{31}{30}$ for sides of squares, the sum of the squares would be $3(\frac{11}{6})^2$ or $\frac{363}{36}$, which is > 10.

Accordingly Diophantus assumes as the sides of the three required squares

$$3 - 35x,\ \tfrac{3}{5} + 37x,\ \tfrac{4}{5} + 31x,$$

where x must therefore be not exactly $\frac{1}{30}$ but near it.

Solving $$(3 - 35x)^2 + (\tfrac{3}{5} + 37x)^2 + (\tfrac{4}{5} + 31x)^2 = 10,$$

or $$10 - 116x + 3555x^2 = 10,$$

we find $$x = \frac{116}{3555};$$
the required sides are therefore

$$\frac{1321}{711}, \frac{1285}{711}, \frac{1288}{711},$$

and the required squares

$$\frac{1745041}{505521}, \frac{1651225}{505521}, \frac{1658944}{505521}.$$

| Other instances of the application of the method will be found in V. 10, 12, 13, 14, where, however, the squares are not required to be nearly equal, but each of them is subject to limits which may be the same or different; *e.g.* sometimes each square is 98
merely required to be less than a given number (10, say), sometimes the squares lie respectively between different pairs of numbers, sometimes they are respectively greater than different numbers, while they are always subject to the condition that their sum is a given number.

As it only lies within the scope of this survey to explain what we actually find in Diophantus, I cannot do more than give a reference to such investigations as those of Poselger in his "Beiträge zur unbestimmten Analysis" published in the *Abhandlungen der Königlichen Akademie der Wissenschaften zu Berlin aus dem Jahre 1832*, Berlin, 1834. One section of this paper Poselger entitles "Annäherungs-methoden nach Diophantus," and obtains in it, on Diophantus' principles, a method of approximation to the value of a surd which will furnish the same results as the method of continued fractions, with the difference that the "Diophantine method" is actually quicker than the method of continued fractions, so that it may serve to expedite the latter.

PART 5

DID THE GREEKS HAVE THE NOTION OF COMMON FRACTION? DID THEY USE IT?

Texts selected and introduced by Jean Christianidis

JEAN CHRISTIANIDIS

INTRODUCTION

The subject matter of this chapter should be placed normally among the issues treated in the previous chapter, dealing with Greek Algebra (and Logistic). Its separate treatment is due to the fact that the history of fractions has recently become a topic of independent study among historians of ancient mathematics, though a comprehensive book on the topic does not yet exist, as Jim Ritter points out in his introduction of the collective work *Histoire des fractions, fraction d'histoire*.[1] This bibliographic lacuna is particularly notable for the period from the end of the Middle Ages until the utilization and popularization of techniques involving decimal fractions in Western Europe at the end of the sixteenth century. With respect to Antiquity the gap is less significant, though the disagreement among historians as to the appropriate criteria of what constitutes a fraction leaves room for simplistic views, which, in particular, occurred frequently in the works of the older historiography.

Recently a number of important studies, concerning the status of fractions and the techniques of their treatment, appeared in the historiography of Greek mathematics. We must mention, in particular, the article of Wilbur R. Knorr "Techniques of fractions in ancient Egypt and Greece",[2] as well as David H. Fowler's recently published studies,[3] which are also partly included in his book *The Mathematics of Plato's Academy*.[4] The conclusion drawn by the studies of these historians is that the Greek logistic tradition is really rich in manipulations of unit fractions (also called "parts" or "*moria*"); on the other hand, the situation concerning common fractions is less clear. Knorr believed that we could recognize manipulations of common fractions in Akhmîm's mathematical papyrus as well as in the Heronian tradition; but Fowler has shown in a rather convincing way that "what are presented there as manipulations of common fractions m/n are clearly manipulations of the descriptions 'the nth of m' which are conceived throughout the texts under discussion in terms of unit fractions".[5] The papers that are reprinted in this chapter represent the approaches of these two historians.

Fowler's research is by no means limited to the two sources mentioned above. On the contrary, it covers a vast portion of the textual tradition of ancient computational mathematics. After having examined thoroughly a significant number of papyri and of medieval manuscripts of ancient texts, he is in a position to answer whether Greek mathematicians used the manipulation of common fractions such as $p/q + r/s = (ps + qr)/qs$ and $p/q \times r/s = pr/qs$.

I would like to quote, by way of example, a number of passages where Fowler summarizes his conclusions:

- "Nowhere do I find any convincing evidence for the proposal that the 'Greeks' used anything like our notations for common fractions and our ways of performing fractional arithmetic".[6]

J. Christianidis (ed.), Classics in the History of Greek Mathematics, 331–336.

- "Did the Greek mathematicians of the fifth and fourth centuries BC have at their disposal the manipulations of common fractions such as $p/q + r/s = (ps + qr)/qs$, $p/q \times r/s = pr/qs$? I believe that we have no good evidence on which to argue that they did".[7]
- "I shall argue in this chapter that we have no evidence for any conception of common fractions p/q and their manipulations such as, for example, $p/q \times r/s = pr/qs$ and $p/q + r/s = (ps + qr)/qs$, in Greek mathematical, scientific, financial, or pedagogical texts before the time of Heron and Diophantus; and even the fractional notations and manipulations found in the Byzantine manuscripts of these late authors may have been revised and introduced during the medieval modernization of their minuscule script".[8]
- "The very few instances there that can be cited as illustrating notions of common fractions appear, on closer scrutiny, more probably to be abbreviations of unresolved descriptions of divisions that are still conceived as sums of unit fractions, and all can be more naturally explained as relaxations of stylistic conventions about how these divisions should be evaluated and expressed. Possibly the abbreviations did then evolve into our conceptions of common fractions, and certainly the practice and popularity of common fractions developed, particularly among Italian mathematicians, from the ninth or tenth centuries onwards. These new fractional notations and conceptions may then have been adopted by the scribes and readers of the medieval manuscripts, and so infiltrated and corrupted the evidence to be found there; on this more work needs to be done".[9]

These quotations from Fowler's works should not be interpreted as a disagreement on my part with his conclusions. On the contrary, I am in substantial agreement with them. The most important reason for which I chose to quote these particular passages is to stress that both the studies of D.H. Fowler as well as those of W.R. Knorr do not actually refer to the works of Diophantus, suggesting once more that Diophantus still remains an isolated case in the historiography of Greek mathematics. Thus, the purpose of this text is to try to overcome this particular lacuna. But first a brief remark ought to be made concerning the third passage quoted above, where Fowler mentions once the name of Diophantus. In my opinion, the investigation of the concept of common fraction in the field of ancient mathematics should focus on the operational rather than the stylistic and notational level. For this reason, I think that the cautiousness exhibited by Bernard Vitrac, when he says that "je ne puis me prononcer sur le traitement des fractions dans le traité de Diophante, *car je n'ai pas pu examiner les différents manuscripts*"[10], is of minor importance as regards the analysis of this concept in ancient mathematical thought. Fowler, on the other hand, seems to acknowledge that we should not exaggerate the symbolic part and he accepts that the core of the problem is situated at the operational level. In fact, he notes: "Just one example of some operation such as the addition, subtraction, multiplication, or division of two fractional quantities, expressed directly as something like 'the *n*th of *m* multiplied by the *q*th of *p* gives *nq*th of *mp*' and *clearly unrelated, by context, to any conception in terms of simple and compound parts*, could be fatal to my thesis that we have no good evidence for the Greek use or conception of common fractions. I know of no such example".[11]

In fact, we are going to see that such an example exists in Diophantus' *Arithmetica*. This is, in a sense, not surprising, given that Diophantus' work belongs to Indeterminate

(Diophantine) Analysis, if we may use this modern term, which is practiced in the field **Q** of *rational numbers*. This particular analysis, called by Jean Itard "analyse diophantienne ancienne", must be distinguished clearly from the Indeterminate Analysis which is practiced in the ring **Z** of integer numbers, that is, in the sense in which it was going, later, to be understood by C.-G. Bachet sieur de Mériziac, P. Fermat and the mathematicians of the 17th century. Itard calls the latter "analyse diophantienne moderne". The difference between the two is not merely limited to the fact that the numbers accepted by the ancient Diophantine Analysis as given or unknown are rational numbers, whereas the modern Diophantine Analysis only accepts integers. It is rather that the methods of solution are different. Let me quote what Robert D. Carmichael, a specialist in the field of Indeterminate Analysis, wrote about this particular point: "Dans certains cas, la recherche des solutions rationnelles et celle des solutions entières sont deux problèmes essentiellement équivalents. Ceci est évidemment vrai dans le cas de l'équation $x^2+y^2=z^2$. Dans certains autres cas, les deux problèmes sont essentiellement différents ainsi que l'on peut le voir immédiatement d'après une équation telle que $x^2+y^2=1$. Le nombre des solutions entières est évidemment fini; de plus ces solutions sont dépourvues d'intérêt. Par contre, le nombre des solutions rationnelles est infini, et elles ne sont pas toutes sans intérêt ...".[12]

The numbers treated by Diophantus in his *Arithmetica* are always non-negative rational numbers. Diophantus does not formulate, for any problem, any kind of restriction that would ensure a whole-number solution. The solving methods he uses are not convenient for such a solution.[13] The problems that would be equivalent to a first-degree indeterminate equation, such as the "problem of remainders" or the so-called "problem of the 100 fowls", though they are found in almost every ancient and medieval mathematical tradition,[14] are totally absent from his work. In fact, problems of this kind require a whole-number solution and not a rational one. Diophantus' *Arithmetica* belongs, as I have already stated, to rational indeterminate analysis, and for this reason it is inconceivable, unless one presupposes that its author was indeed familiar with common fractions and their manipulation. My purpose in the rest of this introduction is to confirm this statement.

In his *Arithmetica* Diophantus expands essentially the numerical field in the following sense. It is true that he gives, at the beginning of the introduction which precedes the first book, the traditional definition of number as "made up of some multitude of units". This definition completely conforms to the Greek concept of *arithmos* which means "a definite number of definite things", and it reminds us, *prima facie*, the corresponding definition in the seventh book of Euclid's *Elements*. Diophantus, however, differs from Euclid because he also accepts as "units" the fractional parts of the unity.[15]

Thus:

- He forms entities such as "781543 units in part of 9699920" in the problem V.8 or "5358 units in part of 10201" in the problem V.9.
- He calls these entities "numbers" (*arithmoi*).
- He accepts these entities as solutions to problems.
- And, most important, he manipulates these entities as numbers, as we are going to show in what follows.

The claim that he considers the above-mentioned expressions as numbers and that he accepts those numbers as solutions to problems, becomes clear simply by the fact that in the problem IV.31 he asks, "to divide unity into two parts such that, if given numbers are added to them respectively, the product of the two sums gives a square". If we take into account that the given numbers are 3 and 5 then the conditions of the problem, stated in algebraic language, are: $\alpha + \beta = 1$, $(\alpha + 3)(\beta + 5) = \square$. The solution is $\alpha = 6/25$, $\beta = 19/25$.

In the same vein, in the problem V.11 he asks, "to divide unity into three parts such that, if we add the same number to each of the parts, the results are all squares". The sought numbers in this case are 228478 units in part of 505521, 142381 units in part of 505521, and 134662 units in part of 505521, and indeed their sum is 1. In the next problem (V.12) he asks, "to divide unity into three parts such that, if three different given numbers be added to the parts respectively, the results are all squares". In this case the sought numbers are 140447 units in part of 707281, 502557 units in part of 707281, and 64277 units in part of 707281, and their sum is 1. All the above expressions are considered by Diophantus as numbers.

We could cite a lot of similar cases in which Diophantus asks to find one or more "numbers" so that one or more conditions are satisfied by them and that, at the end of the solving process, they are expressed with the formula "*m* units in part of *n*". This shows that Diophantus conceives such statements as numbers.

Now we have to consider also whether Diophantus treats them as numbers too, that is, whether he performs arithmetical manipulations using those expressions. In my opinion, the answer to this question is a definite yes; Diophantus treats the expressions "*m* units in part of *n*" as he treats the numbers.[16] In order to support this assertion I just have to mention a single example, the problem IV.36. Let us proceed to the examination of this particular problem.

PROBLEM IV.36[17]

Trouver trois nombres tels que le produit de deux quelconques d'entre eux ait un rapport donné avec leur somme.

Proposons donc que le produit des premier et second nombres soit le triple de leur somme; que le produit des seconde et troisième nombres soit le quadruple de leur somme, et que le produit des premier et troisième nombres soit quintuple de leur somme.[18] *Posons que le second*[19] *nombre est 1 arithme; donc, en vertu du lemme,*[20] *le premier nombre sera 3 arithmes en partie de 1 arithme moins 3 unités, et pareillement le troisième nombre sera 4 arithmes en partie de 1 arithme moins 4 unités.*[21] *Il faut enfin que le produit des premier et troisième nombres soit le quintuple de leur somme. Mais le produit des premier et troisième nombres est 12 carrés d'arithme en partie de 1 carré d'arithme plus 12 unités moins 7 arithmes.*[22] *D'autre part, la somme des premier et troisième nombres est 7 carrés d'arithme moins 24 arithmes en partie de 1 carré d'arithme plus 12 unités moins 7 arithmes;*[23] *ce que l'on obtient de la manière suivante. Lorsqu'il faut additionner des fractions (moria) telles que 3 arithmes en partie de 1 arithme moins 3 unités, et 4 arithmes en partie de 1 arithme moins 4 unités, on multiplie inversement les arithmes de la fraction par les dénominateurs, notamment 3 arithmes par le dénominateur de l'autre fraction, c'est-à-dire par 1 arithme moins 4 unités, et, d'autre part, 4 arithmes par le dénominateur de l'autre fraction, c'est-à-dire par 1 arithme moins 3 unités. On obtient ainsi comme somme 7 carrés d'arithme moins 24 arithmes en partie de produit des dénominateurs, c'est-à-dire par 1 carré d'arithme plus 12 unités moins 7 arithmes.*[24]

In our terms, Diophantus performs here the multiplication as well as the addition of the fractions $3x/(x-3)$ et $4x/(x-4)$. He gives the results

$$\frac{3x}{x-3} \times \frac{4x}{x-4} = \frac{12x^2}{x^2+12-7x},$$

$$\frac{3x}{x-3} + \frac{4x}{x-4} = \frac{7x^2-24x}{x^2+12-7x},$$

and he explains that the last one is taken out of the rule

$$\frac{3x}{x-3} + \frac{4x}{x-4} = \frac{3x \cdot (x-4) + 4x \cdot (x-3)}{(x-3) \cdot (x-4)}.$$

It follows that, according to the last passage quoted from Fowler's *The Mathematics of Plato's Academy*, we have here an example of multiplication and addition of two fractional quantities expressed as "*m* things in part of *n* multiplied by *p* things in part of *q* gives *mp* things in part of *nq*" and "*m* things in part of *n* added to *p* things in part of *q* gives *mq* plus *np* in part of *nq*". This provides good evidence that Diophantus had knowledge of the concept of common fraction as well as of the elementary rules of its arithmetical treatment.

University of Athens

NOTES

1 P. Benoit, K. Chemla, J. Ritter (eds.), *Histoire des fractions, fraction d'histoire*, Basel/Boston/Berlin, 1992, p. x.

2 W.R. Knorr, "Techniques of fractions in ancient Egypt and Greece", *Historia Mathematica*, 9 (1982), pp. 133-171.

3 D.H. Fowler, "Logistic and fractions in early Greek mathematics: a new interpretation", in P. Benoit, K. Chemla, J. Ritter (eds.), *Histoire des fractions, fraction d'histoire*, pp. 133-147; "Ratio and proportion in Early Greek Mathematics", in A. Bowen (ed.), *Science and Philosophy in Classical Greece*, New York, 1991; "Hibeh papyrus I 27: an early example of Greek arithmetical notation" *Historia Mathematica*, 10 (1983), pp. 344-359 (joint paper with E.G. Turner).

4 D.H. Fowler, *The Mathematics of Plato's Academy. A New Reconstruction*, second edition, Oxford, 1999.

5 D.H. Fowler, *The Mathematics of Plato's Academy. A New Reconstruction*, 1st ed., 1990, p. 265, note 91.

6 D.H. Fowler, "Logistic and fractions in early Greek mathematics: a new interpretation", p. 134.

7 D.H. Fowler, *The Mathematics of Plato's Academy*, p. 195.

8 D.H. Fowler, *The Mathematics of Plato's Academy*, p. 226.

9 D.H. Fowler, *The Mathematics of Plato's Academy*, p. 264.

10 B. Vitrac, "Logistique et fractions dans le monde hellénistique", in *Histoire des fractions, fraction d'histoire*, p. 162, note 38 (the emphasis is mine).

11 D.H. Fowler, *The Mathematics of Plato's Academy*, pp. 264-5.

12 R.D. Carmichael, *Analyse indéterminée*, traduit de l'Anglais par A. Sallin, Paris, 1929, p. 3.

13 See J. Christianidis, "Les interprétations de la méthode de Diophante", *Neusis*, 3 (1995), pp. 109-132 (in Greek); of the same author, "Une interprétation byzantine de Diophante", *Historia Mathematica*, 25 (1998), pp. 22-28.

14 J. Christianidis, "On the History of Indeterminate problems of the first degree in Greek Mathematics", in K. Gavroglu, J. Christianidis, E. Nicolaidis, (eds.), *Trends in the Historiography of Science*, Dordrecht/Boston/London, 1994, pp. 237-247.

15 According to Jacob Klein "by a fraction Diophantus means nothing but *a number of fractional parts*" (*Greek Mathematical Thought and the Origin of Algebra*, translated by E. Brann, New York, 1992, p. 137).

16 I would like to thank Fabio Acerbi for bringing to my attention that W.R. Knorr has arrived at the same conclusion in his paper "What Euclid Meant: On the Use of Evidence in Studying Ancient Mathematics". See A. Bowen (ed.), *Science and Philosophy in Classical Greece*, pp. 119-163.

17 We use the French translation of *Arithmetica* by Paul ver Eecke because Heath's English translation is not literal. The only modification to the French translation is that I have substituted "en partie de" instead of "fractionnés par", the expression by which Ver Eecke renders the Greek "*en moriô*".

18 Algebraic transcription: $XY = 3 \cdot (X + Y)$, $YZ = 4 \cdot (Y + Z)$, $XZ = 5 \cdot (X + Z)$.

19 P. ver Eecke has "premier".

20 It refers to the lemma that precedes immediately this proposition: "To find two numbers indeterminately such that their product has to their sum a given ratio".

21 Algebraic transcription: Let $Y = x$; therefore $X = 3x$ in part of $x - 3$, and $Z = 4x$ in part of $x - 4$.

22 Algebraic transcription: $XZ = (3x$ in part of $x - 3) \times (4x$ in part of $x - 4) = 12x^2$ in part of $x^2 + 12 - 7x$.

23 Algebraic transcription: $X + Z = (3x$ in part of $x - 3) + (4x$ part of $x - 4) = 7x^2 - 24x$ in part of $x^2 + 12 - 7x$.

24 Algebraic transcription: $(3x$ in part of $x - 3) + (4x$ in part of $x - 4) = 3x \cdot (x - 4) + 4x \cdot (x - 3)$ in part of $(x - 4) \cdot (x - 3) = 7x^2 - 24x$ in part of $x^2 + 12 - 7x$.

WILBUR KNORR

TECHNIQUES OF FRACTIONS IN ANCIENT EGYPT AND GREECE

The subject-matter of [logistic] is all things which can be numbered; its branches are the methods called Greek and Egyptian for multiplication and division and the summation and separation of parts (moria).[1]

In this and several other such passages Greek writers acknowledged their debt to the ancient Egyptian arithmetical methods[2]. The reference here to "parts" (*moria*) is noteworthy, in that by using the standard term for proper measuring parts, or submultiples of the unit, the writer appears to allude to the manipulation of unit-fractions. A technique of precisely this sort dominates the computations with fractions found in the ancient Egyptian papyri—most notably, the Rhind Mathematical Papyrus—and despite the passage of over two millennia is still to be found in late Greek papyri from the Graeco-Roman and Byzantine periods. This is perhaps the most striking of several
| evidences of the continuity of the Egyptian and Greek technical traditions, all 135
the more impressive in view of the fact that the availability of the Mesopotamian sexagesimal mode, from at least the second century B.C. on, failed to displace the cumbersome and limited Egyptian mode, save in the context of astronomical computation[3]. Thus, despite the undeniable influence of Mesopotamian techniques, especially in the fields of elementary and metrical geometry and astronomy, our sources must be taken seriously when they affirm their debt to the Egyptian tradition[4].

As far as the use of unit-fraction methods is concerned, the agreement between the Egyptian and Greek approaches is apparent. Yet rare is the study of these methods which attempts to probe the relation of the two traditions, to determine, for instance, whether they shared the same underlying computational procedure, or whether the later texts reveal signs of greater technical sophistication[5]. Complicating an investigation of this type is the fact that several recent studies have proposed a variety of rationales for the unit-fraction computations of P. Rhind. In effect, their authors seek to articulate quasi-esthetic criteria which the ancient scribes followed in selecting from among the dozens of possible alternative solutions[6]. A far more plausible approach has been adopted by B. L. van der Waerden, who bases his account on the computational procedures evident in the papyrus itself[7].

Comparisons with the later Greek papyri confirm the latter approach. Not only has much the same procedure led to the values in the Greek documents, but also the presence of two or more different expressions for the same fractional value seriously undermines the hypothesis that esthetic considerations were important for the scribes' hitting upon the listed values. What I shall do in the present study is review briefly the computational methods of P. Rhind and then show how these persist with certain modifications and elaborations in the Greek papyri. My findings are recapitulated in

J. Christianidis (ed.), Classics in the History of Greek Mathematics, 337–365.

Section V, and extended extracts of numerical data from the principal ancient documents are listed in the tables in the Appendix.

I. THE SO-CALLED 2/*n* TABLE IN P. RHIND

Interpretations of the opening section (commonly called the "2/*n* table") of the Rhind Papyrus have drawn R. J. Gillings into a debate with critics [Gillings 1972, Chaps. 6-7, 10; 1978; Bruchheimer & Salomon 1977; Bruins 1975]. The issue centers on the method by which the scribe of the Papyrus expressed fractions of the form 2/*n* (for *n* odd) as sums of unit-fractions. Since there are in all cases many ways of doing
136 this, what con|siderations led to the choice of the value preserved in the Papyrus? Expressed in this way, the question leads readily to the formulation of rules of some sort affecting the choice, so that the scribe, presumably aware of many possible decompositions, was thereby able to single out the "best" one.

I believe this misconstrues the nature of the document and its production. The numbers in the Papyrus do not form a *table*, but rather are a set of division problems in which 2 is to be divided in succession by the odd numbers from 3 to 101. It results from the method of division used that the quotient is given as a sum of unit-fractional parts. Moreover, enough detail is provided in most instances for us to see precisely what the computational procedure is.

There are two methods, depending on whether *n* has proper divisors or not. For prime *n*, the scribe first sets down two-thirds or half of *n*, then takes successive halves of this part until the quotient has become less than 2. (An exceptional case is 2/59, where the lead term is 36′; also, the scribe later admits division by 10 (see below).) He thus obtains as the first term in his answer 3′ (that is, 1/3), 6′, 12′, 24′, etc., or alternatively, 4′, 8′, etc.* He then seeks those additional fractional parts which must be added to this quotient to bring it up to 2. In Table I in the Appendix I have given the final quotient for each case. The following examples for the division by 5, 7, 13, and 19 reproduce the steps actually listed in the Papyrus and so reveal the method of computation[8]:

1	5	
3″	3	3′
/ 3′	1	3″
/15′	3′	

1	7	
½	3	½
/ 4′	1	½ 4′
/28′	4′	

1	13	
½	6	½
4′	3	4′
/ 8′	1	½ 8′
/ 52′	4′	
/104′	8′	

1	19	
3″	12	3″
3′	6	3′
6′	3	6′
/ 12′	1	½ 12′
/ 76′	4′	
/114′	6′	

In the process of "complementation" of the quotient up to 2, the scribe utilizes a variety of unit-fraction identities, such as 1/2 1/2 = 1, 4′ 4′ = 1/2, 8′ 8′ = 4′, etc.; 3″ = 1/2 6′ and from it 3′ = 4′ 12′, 6′ = 8′ 24′, etc.; 1/2 = 3′ 6′ and from it 4′ = 6′ 12′, etc. Among other identities implicit in his computations are 4′ = 5′ 20′, 5′ = 6′ 30′,

* In writing 5′, 6′, etc., for 1/5, 1/6, etc., I use a notation borrowed from the Greek papyri, but identical in spirit with the Egyptian. The papyri use special symbols for one-half and two-thirds, for which I adopt the notations 1/2, 3″, respectively. The sum of such terms is indicated merely by juxtaposition; e.g., 1/2 4′ signifies 1/2 + 1/4, or 3/4. See Heath [1921 I, 41–44] for details.

6′ = 7′ 42′, 7′ = 8′ 56′, and others derived from these (e.g., 8′ = 10′ 40′). A leather roll preserved from ancient Egypt lists 26 such identities, many | agreeing with those 137
implicit here in P. Rhind, so that this view of the scribe's method of complementation is confirmed[9]. (See Table II.)

Once the fractional parts needed to complement the partial quotient to 2 have been determined, the corresponding parts of the unit may be produced to complete the quotient. For instance, in the computation of the parts of 7 the scribe needs 4′ to complement 1 1/2 4′ (which is 4′ of 7) up to 2; the corresponding part of the unit is 28′ to yield the complete quotient 4′ 28′. Similarly, in the case of 13 the partial product 1 1/2 8′ is complemented to 2 by the addition of 4′ 8′, corresponding to the parts of the unit 52′ 104′, respectively.

Beginning with the computation for 31, and commonly thereafter, the scribe varies the procedure by initiating the sequence with the tenth part, then taking the half or the third of this, followed by successive halves. In this way he obtains 20′, 40′ or 30′, 60′ as the leading term in his quotient. In the following examples I have set in square brackets those steps not appearing in the Papyrus:

1	31
[10′	3 10′]
/ 20′	1 ½ 20′
/124′	4′
/155′	5′

1	47
[10′	4 ½ 5′]
/ 30′	1 ½ 15′
/141′	3′
/470′	10′

1	71
[10′	7 10′]
/ 40′	1 ½ 4′ 40′
/568′	8′
/710′	10′

In these instances, the complementation is evident through use of the identities 20′ 5′ = 4′, 15′ 10′ = 6′ and 40′ 10′ = 8′, respectively.

The cases for 43 and 97 are unusual in that their leading terms (42′ and 56′, respectively) do not fall in the primary sequence, but signify an intermediate division by 7. The texts in P. Rhind are as follows:

1	43
/ 42′	1 42′
/ 86′	2′
/129′	3′
/301′	7′

1	97
/ 56′	1 ½ 8′ 14′ 28′
/679′	7′
/776′	8′

[1	97
8′	12 8′
56′	1 ½ 7′ 14′ 56′
	1 ½ 8′ 14′ 28′]

[1	97
4′	24 4′
28′	3 7′ 4′ 28′ 28′
	3 4′ 7′ 14′
56′	1 ½ 8′ 14′ 28]

In the case of 43 complementation is easily effected via the identity 42′ 7′ = 6′. The procedure for 97 is a bit more complicated. In the bracketed section I show how 56′ of 97 might be obtained by taking (a) the seventh of its eighth; or (b) the half of the seventh of its fourth. Both result through the standard Egyptian method of division; but (a) requires adjust|ment via the identity 7′ = 8′ 56′, while (b) produces the text 138
value directly. In either case, complementation is effected via the identity 7′ 14′ 28′ = 4′[10]. In this way the scribe's computational procedure is evident, even if his decision to divide by 7 is not. Application of the standard method to 43 produces the quotient 24′ 344′ 516′ (or, alternatively, 30′ 86′ 645′), so that it is difficult to perceive any special advantage in the variant method adopted by the scribe. Similarly, as the cases of 73, 79, 83, and 89 each work toward the leading term 60′, one would have expected the same in the case of 97. This yields the quotient 60′ 291′ 1940′ (or,

alternatively, 60′ 485′ 970′ 1164′) via the addition of 3′ 20′ (or 5′ 10′ 12′) in complementation. Neither is as compact as the scribe's value 56′ 679′ 776′, so that his ploy of dividing by 7 seems to have paid off here.

Only the procedure for the division of 2 by 101 is entirely exceptional. The scribe's result of 101′ 202′ 303′ 606′ is both obvious and disappointing. Following this pattern, one could express *any* such division as a four-term sum. If the scribe had persisted in the usual method, he would have worked out the quotient 60′ 404′ 1515′ (via the addition of 4′ 15′ to complement). One might suppose that he was reluctant to admit such large denominators. But if this is the case, he is fighting the inevitable. For as the divisors increase in size, any restriction to small numbers (say, below 1000) will eventually become untenable.

For the cases of composite *n*, a modification of the procedure is introduced: one of the proper parts (usually the greatest) is taken before the standard sequences of parts are computed. Thus, the division here is in effect reduced to one of those already listed. For instance,

1	15
[5′	3]
/10′	1 ½
/30′	½

1	21
[7′	3]
/14′	1 ½
/42′	½

1	49
[7′	7]
/ 28′	1 ½ 4′
/196′	4′

1	65
[13′	5]
/ 39′	1 3″
/195′	3′

1	95
[5′	19]
/ 60′	1 ½ 12′
/380′	4′
/570′	6′

Thus, for 65 the scribe first removes the factor of 13 to obtain 5. From here he can follow the procedure already used in the division of 2 by 5: he takes the third part of 13′ (= 39′), which results in the partial quotient 1 3″, and then complements this to 2 by the addition of 3′ (corresponding to 195′). I believe it is such a *computational* correspondence which accounts for the compatibility of the results for 5 and 65, rather than through the application of a *formulaic* procedure, like 2/65 = 13′ (2/5) = 13′ (3′ 15′).

139 | A minor deviation occurs with 95. Here the scribe removes the smaller factor of 5, completing the computation according to the procedure for 19. But we should have expected him to remove the *larger* factor 19, as he does in the other cases, and then complete the process as for 5, to obtain the quotient 19′ 57′ 285′. But either way, the procedure is evident. By contrast, Gillings finds this case especially problematic, as his proposed criteria for the selection of solutions make 60′ 228′ the expected entry [Gillings 1972, 68]. (Note, similarly, that 2/55 is given as 30′ 330′ (reduced to 2/11) instead of as 33′ 165′ (by reduction to 2/5).) From what we have seen, however, it is apparent that no simple manner of complementation could produce the term 228′ in this case.

Only two cases adopt a different procedure:

1	35
/30′	1 6′
/42′	3″ 6′

1	91
/ 70′	1 5′ 10′
/130′	3″ 30′

The scribe explicates his method in the former case: "For 35′ applied to 210 gives 6, and 2 times 6 is 12, or 7 and 5, which are 30′ and 42′ of 210" [*Rhind Mathematical*

Papyrus, Chace and Manning, eds., 1927 II, 53]. Apparently, he begins with the factors of 35, observing that the sum of 7 and 5 is 12, and that the sixth part of this yields the desired value of 2. Thus, the quotient must be the sum of the sixths of the respective parts, i.e., 30′ and 42′. His derivations of the latter entail taking 7′ and 5′ of 210 (= 6 × 35). By analogy, the procedure for 91 involves recognizing that its factors 7 and 13 sum to 20, from which the desired value of 2 would result from taking the tenth. He would thus form 910, of which 13 and 7 are 70′ and 130′, respectively. In principle, this variant procedure makes accessible a wide range of alternatives in the case of composite divisors. We shall see a number of interesting elaborations on it in the Greek papyri. Indeed, one wonders that the Egyptian scribe did not invoke it more frequently. For instance, in the case of 95, whose divisors 5 and 19 sum to 24, one may introduce the auxiliary factor 12 × 95 to obtain the quotient 60′ 228′. The fact that he does not proceed in this way here indicates, not a failure to apply esthetic criteria correctly, as Gillings would maintain [Gillings 1972, 68], but rather the availability of a number of different approaches to follow, of which any one will produce a viable result.

Thus, the divisions presented in P. Rhind reveal, not the application of esthetic rules for choosing among solutions, but the consistent use of a particular computational procedure. Among the prime divisors, only 43, 97 (where an unusual division | by 7 is intro- 140
duced), and 101 deviate from the standard pattern, and of the composite divisors, only 35 and 91. The procedure does not yield unique solutions, but rather allows a certain degree of latitude. For instance, one has at least four sequences from which to choose the leading term of the quotient (e.g., 3″, 3′, 6′, 12′, etc.; 1/2, 4′, 8′, etc.; 10′, 20′, 40′, etc.; and 10′, 30′, 60′, etc.). Indeed, alternative forms have survived; for instance, Neugebauer reports an ostracon on which the division of 2 by 7 is given as 6′ 14′ 21′ [Neugebauer 1957, 92-93]; here the scribe found that 6′ of 7 is 1 6′, and then complemented to 2 by adding 1/2 3′, corresponding to 14′ 21′. This contrasts with P. Rhind, where the scribe first works out 4′ of 7 and then completes the quotient by the addition of 28′. We have seen a similar flexibility in the treatment of composite divisors. Any of these could have been solved by the same method used for the prime divisors; but instead a variation is introduced involving the removal of factors, so that the computation is markedly simplified.

It is precisely in his repeated application of the computational procedure that the scribe reveals his distance from a modern view of the problem. Had he been searching for the most efficient or elegant way of expressing the quotients, he would surely have come to recognize the relation $2/m = 1/n + 1/nm$ (for $m = 2n - 1$), producing a two-term value in all cases. As it happens, his values agree with this relation only for $m = 5, 7, 11$, and 23, and then, of course, only by accident. The cases of 19, 29, 31, 47, 59, 71, and 79 are thus of particular interest, as this rule yields here values not only simpler than those given in the Papyrus, but also accessible (indeed, expected) under the application of the scribe's procedure; for in each case, the leading terms (e.g., 10′, 15′, 16′, 24′, etc.) fall within the sequences of parts he considers. Thus, the scribe's project is best understood as the working out of a set of problems in accordance with a general computational procedure, rather than as the search for an ideal solution for this particular set of problems.

II. UNIT-FRACTIONS IN GREEK PAPYRI

Representing the late Greek tradition of elementary arithmetic are three noted papyri: the Papyrus Akhmîm and two others, now held in the Michigan Collection.

141

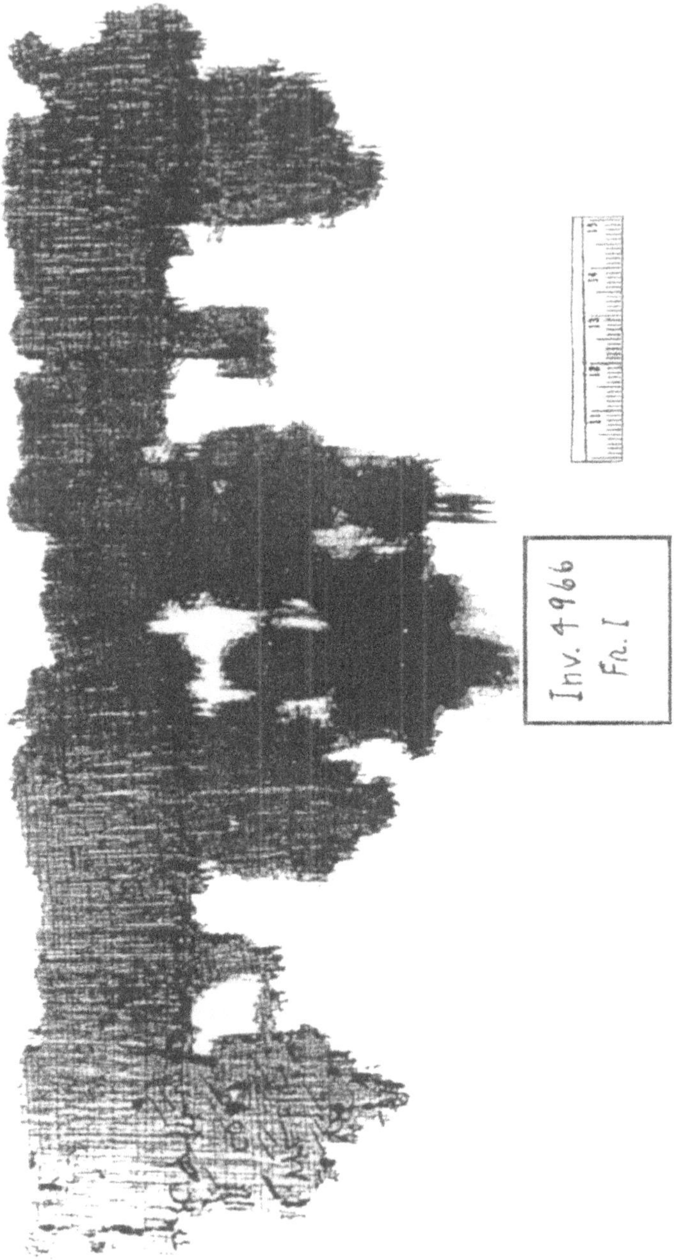

Plate I. From P. Michigan 145: fragments of a table of twenty-thirds (left) and of twenty-ninths (center). For a reconstruction, see Table III. (Printed by permission of the Library of the University of Michigan.)

Recovered from sites in Egypt, the papyri have been dated on paleographical grounds to the second and fourth centuries A.D. in the case of the Michigan papyri and to a much later period, probably the seventh or eighth century, in the case of P. Akhmîm[11]. In all three, divisions effected via unit-fractions play a promi|nent part. Indeed, 142
P. Mich. 146 consists entirely of a fragment of a table of quotients, duplicated almost entry for entry in the opening section of P. Akhmîm. The latter goes on to present fifty arithmetic problems, many solved via techniques comparable to those implicit in the table, although the values actually worked out in the problems do not always agree with those listed in the table. (See Tables III and IV in the Appendix.)

P. Mich. 145, like the other two, also contains a table listing the quotients of division in terms of unit-fractions. But it is highly fragmentary and does not allow of direct comparison with the others, since it is unique in its considering cases with denominators 23 and 29 alone. Nevertheless, it betrays a computational method comparable to that used in P. Rhind. For instance, its values for 23′ of 2, 3, 4, and 5 have leading terms, respectively, of 12′, 10′, 6′, and 6′, which, as we have seen, all fall within the primary sequences examined by the Egyptian scribe. Moreover, in both the Egyptian and Greek cases the computer tends to seek the largest such unit-fraction possible (that is, the smallest possible denominator in the sequence) to start off, although exceptional cases are to be found[12]. Of the rest of the table in the Greek papyrus, only six entries from the sequence of twenty-ninths remain: from 12 through 17. Here again, the lead terms 4′, 3′, and 1/2 all occur within the primary sequences, as well as being in all cases but one the greatest possible primary term. The exception occurs at 14, where the lead term is 4′ instead of 3′, the lead term of the preceding entry. Further, one must surely be amazed by the remarkably inefficient manner of computation implicit in his values for 13 and 14. I produce below specimen computations leading to the values he gives for these two cases, and then by 13* and 14* two procedures leading to simpler alternative values. (One should note that the scribe lists only the results, without specific indication of the computing procedure.)

13:	1	29								
	½	14	½							
	/ 3′	9	½	6′		R:	3	3′		
	/ 15′	1	3″	5′	15′	R:	1	3′	15′	
	/ 29′	1								
	/ 87′	3′								
	/435′	15′								
| 14:	1	29								143
	½	14	½							
	/ 4′	7	4′			R:	6	½	4′	
	/ 5′	5	½	5′	10′	R:	½	4′	5′	
	/ 58′	½								
	/116′	4′								
	/145′	5′								
13*:	1	29								
	/ 3′	9	½	6′		R:	3	3′		
	/ 9′	3	6′	18′		R:	9′			
	/261′	9								

14*:	1	29	
	/ 3′	9 ½ 6′	R: 4 3′
	/ 8′	3 ½ 8′	R: 3′ 4′ 8′
	/ 87′	3′	
	/116′	4′	
	/232′	8′	

The computational procedure underlying P. Rhind thus leads naturally to the values listed in P. Mich. 145, even if the scribe's reasons for carrying the calculation through in the manner he did are not entirely clear. One notes that the entries appear to have resulted through such a computational procedure, for *each* separately, rather than through a method of adding a unit-fractional part (here, 23′ or 29′) to one entry in order to obtain the next after a combination of terms. For *both* the entries for 12 and 13 contain 29′ among the parts. Moreover, the mere addition of 29′ to the entry for 13 could hardly have yielded a value for 14 in which the lead term 4′ was *less* than the term 3′ which leads 13.

The introductory section of P. Akhmîm is a table of the results of divisions expressed in the unit-fractional mode. It opens with a list of two-thirds of each number in the series 6000, 1, 2, 3, . . . , 10, 20, 30, . . . , 1000, 2000, 3000, . . . , 10,000[13]. The next block lists the thirds of the same numbers; the next their fourths; and so on through their tenths. At this point the structure of the table changes: the elevenths are listed only for 6000, 1, 2, 3, . . . , 11; the twelfths only for 6000, 1, 2, 3, . . . , 12; and so on through the twentieths. P. Mich. 146 presents the same table, albeit in a truncated version; for it begins with the seventh of 1000 and, having then run through the same sequence as P. Akhmîm, ends abruptly with the end of the eighteenths column and the
144 heading of the nineteenths. Otherwise, the two papyri agree not only | in their structure, but also in the actual values given for each entry. Robbins lists only 33 discrepancies between them, of which about half represent correct alternative values, the others being scribal errors in one or the other papyrus [Robbins 1936, 58].

Table IVa lists from P. Akhmîm the entries m/n for $m = 2$ through $n - 1$ and $n = 4$ through 20, and Table IVb indicates where it differs from P. Mich. 146. Of the missing portions from the latter, the values for the sevenths may be recovered from the extant entries for the sevenths of 1000 through 6000 (e.g., 7′ of 1000 is given as 142 1/2 3′ 42′, so that 7′ of 6 is 1/2 3′ 42′); the values so obtained are consistent with those listed for the fourteenths of the even integers. Similarly, the fifths may be recovered by examination of the entries for tenths and fifteenths, these again being consistent with each other. In general, when P. Akhmîm considers a ratio not in lowest terms, the listed value agrees with that for the ratio in reduced terms (the only exception being 14/18, whose listed value of 1/2 4′ 36′ differs from that for 7/9, viz. 3″ 9′). The same applies in the case of P. Mich. 146 (where, here, the values for 14/18 and 7/9 are identical, viz. 3″ 9′)[14]. It is significant that this is true even where the two papyri happen to differ from each other. For instance, in P. Akhmîm the entries for 4/5, 8/10, and 12/15 are all 1/2 4′ 20′, while in P. Mich. they are all 3″ 10′ 30′[15]. Similarly, P. Akhmîm lists 3′ 14′ 42′ as the value for both 3/7 and 6/14, while P. Mich. lists them both as 3′ 15′ 35′. Thus, the papyri attest independently that entries not in lowest terms were found not by a new computation, but by reference to the prior appearances.

The implicit computational procedure of the papyri agrees with what we have seen from the survey of P. Rhind and P. Mich. 145. Entries are produced by separate computations in which the first term of each is the largest possible standard part (e.g., 10′, 8′, 6′, 4′, 3′, 1/2, 3″, or 1/2 4′) and the rest are worked out through the process of complementation. Among the few exceptions one may note 3/14 (P. Mich.: 7′ 14′; P. Akhmîm: 5′ 70′) and 2/13 (7′ 91′ in both)[16]. The latter is of interest in that the Greek papyri here diverge from P. Rhind, which lists 8′ 52′ 104′, the value we should

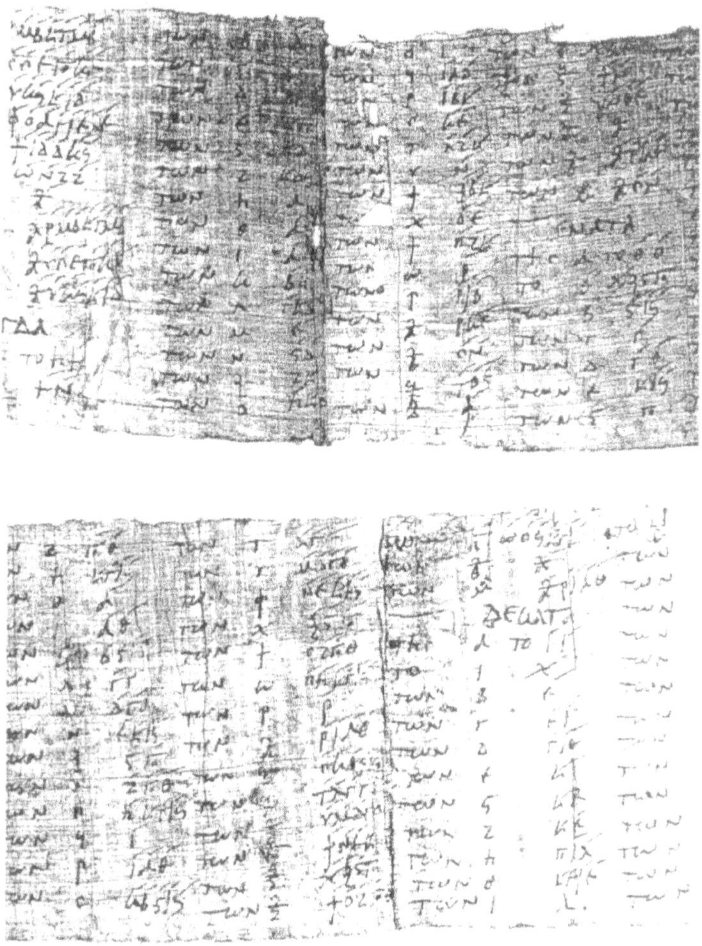

Plate II. From P. Michigan 146: portions of a table of unit-fraction expressions. The first column (only partially visible) lists sevenths; the next two columns eighths; the next three columns ninths; and the last tenths. For selected entries, see Table IVb. (Printed by permission of the Library of the University of Michigan.)

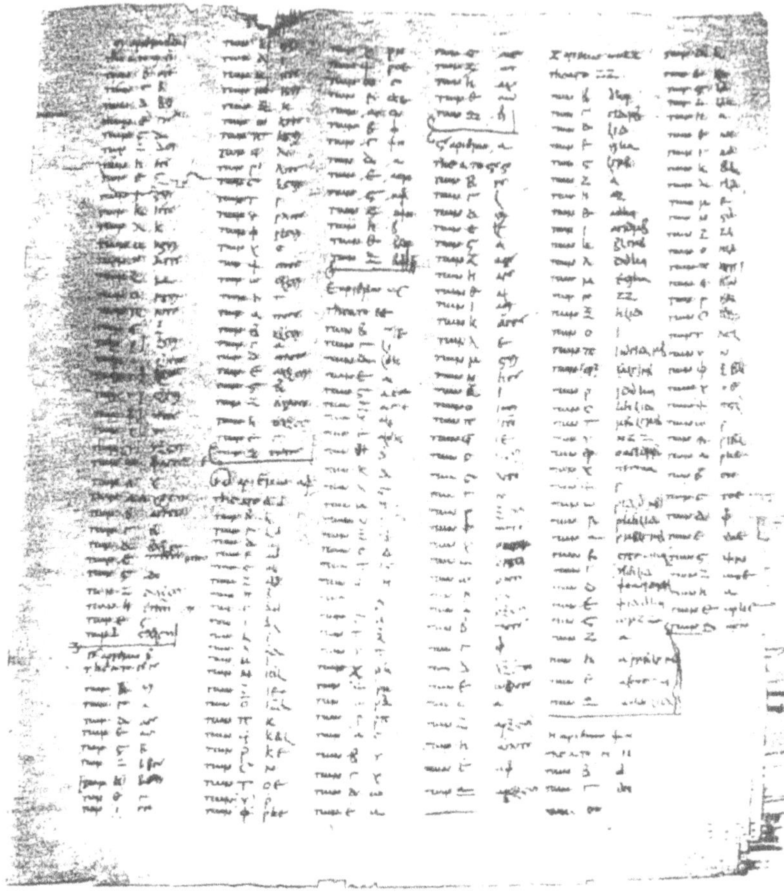

Plate III. A section of P. Akhmîm [Baillet 1892, Plate II] listing successive multiples of 3″ (col. 1), 3′ (cols. 1-2), 4′ (cols. 2-3), 5′ (cols. 3-4), 6′ (col. 4), 7′ (col. 5) and 8′ (cols. 5-6). For selected entries, see Table IVa.

have expected from the standard method[17]. Another discrepancy occurs in the case of 2/19, where the Greek papyri give 10′ 190′ in contrast with the Egyptian value of 12′ 76′ 114′; here, both reveal application of the standard method of computation, since both 10′ and 12′ are in the primary sequence of lead terms. Otherwise, in the values for the division of 2 by 5, 7, 9, 11, 15, and 17, the three papyri are in agreement. Save for mere scribal error, the remaining discrepancies between the two Greek papyri
147 result either from the failure of | P. Mich. to choose the largest lead term, instead

carrying forward the lead term of the preceding entry (i.e., 3/10, 3/14, 5/14, 4/15, 8/15, 3/18), or the tendency of P. Mich. to prefer the lead term 3″ (= 1/2 6′) where P. Akhmîm starts with 1/2 4′ (i.e., 4/5, 9/12, 10/13, 14/18)[18]. The sequence of tenths enables comparisons among the three papyri[19]: they agree for 2, 4, 5, and 6; the Greek agree on 7/10 (= 1/2 5′) and 9/10 (= 1/2 3′ 15′) in contrast with the Egyptian (3″ 30′ and 3″ 5′ 30′, respectively); but in the cases of 3 and 8, P. Rhind and P. Mich. agree (5′ 10′ and 3″ 10′ 30′, respectively) in contrast with P. Akhmîm (4′ 20′ and 1/2 4′ 20′, respectively). These comparisons point to a certain small degree of latitude within a tightly structured computational mode.

An anomaly occurs at the entry in P. Akhmîm for 3/19. One would have expected a value with a lead term of 8′ under the standard method (e.g., 8′ 38′ 152′); instead, the value 15′ 20′ 57′ 76′ 95′ is given, despite the appearance of the smaller lead term 10′ in the entry for 2. Similar anomalies emerge in the entries for the seventeenths of 3 and 4. Both Greek papyri list 12′ 17′ 51′ 68′ for 3/17, which is readily understood as influenced by the computation for 2/17 (i.e., 12′ 51′ 68′), although the value 6′ 102′ would be expected (indeed, the value 3′ 51′ for 6/17 later reflects the standard procedure). The pattern persists at 4, where the listed entry in P. Akhmîm is 12′ 15′ 17′ 68′ 85′[20]. Despite the appearance of the term 17′, this has not been formed from 3/17 by the addition of 17′, since the entry for 3/17 itself contains the term 17′[21]. It must thus result from a separate computation. But why the scribe did not follow the usual procedure, which would have produced the value 6′ 17′ 102′, for instance, is not clear.

More striking than the appearance of these anomalies at 3/17 and 4/17 is the fact that the two Greek papyri here plainly *agree*. Thus, despite their wide separation both in time and, presumably, also in place, the two papyri represent efforts within the same *textual* tradition, as well as within a uniform *computational* tradition. This was already evident to a large degree in the identical organizations of the tables of fractions in the papyri. But the agreement in detail, almost entry for entry, is now seen to follow through a textual dependence of some sort. The later papyrus, P. Akhmîm, cannot be viewed merely as a recopying of texts derived ultimately from P. Mich.; for the two nowhere agree on an incorrect entry, while each has its own share of scribal errors. Now, in several instances, the incorrect values in P. Mich. suggest computational errors: e.g., at 8/13, 12/13, 11/14, 10/15, 11/15, 4/17, 11/17, and 15/17. While listing correct values in these instances, P. Akhmîm appears to follow the computational procedure used in P. Mich., even where that might be unusual, as we saw for 4/17.
148 Thus, it seems to me probable that P. Akhmîm was produced as a | corrected version of a table quite close to that in P. Mich. The fact that the two papyri occasionally present different, but correct, values for the same fraction indicates that P. Akhmîm is working from a modified document, as compared with P. Mich. One cannot specify the source or motive of these modifications: were they part of a document antedating P. Mich. or, perhaps more likely, the result of changes by later editors, for instance, by the scribe of P. Akhmîm himself? This much is clear: the scope of such editorial intervention is extremely slight. The two Greek papyri attest to a remarkable continuity of textual and computational tradition spanning at least three centuries in late antiquity and founded on the arithmetic procedures already well established among the Egyptians almost two millennia earlier.

III. FURTHER EVIDENCE IN P. AKHMÎM

Beyond the table of divisions we have just examined, P. Akhmîm presents a series of computational problems, some dealing with elementary relations of geometric figures, others considering situations of applied arithmetic (e.g., in problems of interest), and others requesting the solution of purely computational problems. Of the fifty problems in this part of the papyrus, twenty-six fall within this third category, and of these all entail computations of unit-fractions[22]. The procedures for solving these problems typically refer to the results listed in the preceding table of fractional values, but never actually perform a decomposition according to what we have called the standard procedure. Instead, new decompositions are worked out by means of free manipulations of the factors of the divisors.

Consider, for instance, problem 23: to multiply 5′ by 4′ 28′. The scribe first rewrites the second multiplier as 2/7 (through reference to the table) and then reduces the problem to that of dividing 2 by 35. He notes that 7′ and 5′ of 35 sum to 12, of which 6′ yields the desired value 2; thus, the answer is 42′ 30′. One cannot fail to recognize here a method based on that underlying the value for the same problem 2/35 (as well as for 2/91) in P. Rhind. While in the context of the Egyptian papyrus, these values were unusual, it is clear that the method used to obtain them persisted in the ancient arithmetic tradition. Indeed, many of the other problems in P. Akhmîm depend on elaborations of this procedure.

One class of problems in P. Akhmîm poses that the calculator decompose (*chōrison*) a given fraction into a specified number of unit-fractional parts (*moria*). (The procedure of *chōrismos* is posed in problems 16, 19, 20, 50.) For instance, in No. 16 the fraction 22′ is to be expressed as the sum of three parts The scribe raises the
149 terms, seeking instead to express 5 as | parts of 110. Since 55′ of 110 is 2, the remainder to be expressed as parts is 3. Now, 10′ and 11′ of 110 sum to 21, of which 7′ is the desired value 3. Thus, the solution of the problem is found to be 55′ 70′ 77′.

A virtuoso variation of this type of problem appears in No. 20: to express the division of 75 by 323 as a sum of eight parts. The scribe takes 17′ and 19′ of the divisor, observing that their sum is 36. Of this sum, the half is 18, the third is 12, and the fourth is 9. Since these four numbers sum to the desired value of 75, he can state the solution 17′ 19′ 34′ 38′ 51′ 57′ 68′ 76′. The element of prearrangement in this problem is hardly to be missed.

In similar fashion, No. 19 requires the expression of 55′ 56′ 70′ as a sum of four parts. Using the multiple 3080 (= 55 × 56), the scribe reduces the given sum to the division of 155 by 3080, that is, of 31 divided by 616. Since 88′ of 616 is 7, he must thus express the remainder 24 as parts of 616, that is, of 3 by 77. He next takes 77′, or 1, of this, thus being left with 2. The sum of 11′ and 7′ of 77 is 18, of which 9′ produces the needed value of 2. The final answer thus becomes 63′ 77′ 88′ 99′.

Two more-elaborate applications of this procedure of manipulating factors appear in problems 39 and 40. The former seeks the result of division of 3 1/2 by 88, that is, of 7 by 176. The scribe observes that 16′ of 176 is 11, of which the triple is 33, while 11′ of 176 is 16. Since 33 and 16 sum to 49, of which 7′ produces the desired value of 7, the answer will be 77′ together with 7′ of three 16′. The latter term entails the division of 3 by 112. Now, 16′ of 112 is 7, of which the double is 14, while 7′ of 112 is 16. As 16 and 14 sum to 30, of which 10′ produces the desired value of 3, the quotient becomes 70′ together with 10′ of twice 16′, the latter equaling twice 160′, or 80′. Thus, the final answer is 70′ 77′ 80′. Similarly, in No. 40 one must divide 9 3″ by 119, that is 29 by 357. Now 51′ of 357 is 7, of which 30 times is 210. To the latter the scribe adds 7′ of 357, that is, 51, to obtain 261. Since 9′ of 261 gives the desired value of 29, the answer will be 63′ together with 9′ of 30 times 51′, the latter equaling the quotient of 10 by 459, or 10 by 153. The scribe leaves the computation in this incomplete state.

The modern editor of P. Akhmîm, J. Baillet, has summarized these procedures into two formulas for the expression of terms of form *ab/c* as sums of two terms, one or both of which will be unit-fractions ([Baillet 1892, 39, 42f]; cf. [Heath 1921 II, 543-545]). But the survey we have just given makes fully clear that the scribe is not resorting to anything like an explicit *formula* for the solution of these problems; he is applying a computational *procedure* involving manipulations of the factors of the divisors. Recognizing this, one becomes aware of the | significant fact that the Greek 150
scribe's procedure of "decomposition" is but an extension and elaboration of a procedure employed in P. Rhind centuries earlier. Thus, the continuity of the ancient tradition of elementary arithmetic runs far more deeply than Baillet perceives[23].

Several of the above problems suggest another aspect of the scribe's technique in P. Akhmîm: the utter artificiality of the unit-fractional mode. Although the problems are invariably expressed in terms of unit-fractions and the final solutions are given in this same mode, the actual execution of the arithmetic operations first introduces their conversion to terms of the form *a/b*. For instance, in No. 40 the mixed term 9 3″ is first changed to 29/3 before computation continues; in No. 23 the given divisor is changed to 2/7; in No. 19 the given term 55′ 56′ 70′ is first converted to 155/3080 via the common multiple 55×56, and then after the common factor of 5 is removed, reduced to 31/616. Consider problem 30, one of ten problems posing the subtraction of terms expressed in the unit-fraction mode (subtractions are posed in problems 6-9, 12, 14, 29-32): to subtract 4′ 44′ from 1/2 4′ the scribe first converts the terms to 3/11 and 3/4, respectively, then recasts the problem as the division of $3 \times 11 - 4 \times 3$ (that is, $33 - 12$, or 21) by 4×11 (that is 44). Without presenting further steps, he states the value of 21/44 as 3″ 11′ 33′ 44′. In another ten problems dealing with the division by terms expressed as unit-fraction sums, the procedure is the same (cf. No. 23) (multiplications or divisions are posed in problems 13, 17, 18, 21-25, 38-40): first reexpress the terms in the general fractional mode, carry out the arithmetic operations (as one now would do in modern school arithmetic), then state the solution after conversion back to the unit-fractional mode. The most elaborate instance of this approach is to be found in problem 12: to subtract from 3″ the nineteen-term sum of unit-fractions 10′ 11′ 20′ 22′ … 90′ 99′ 100′ 110′. (The term 80′ is omitted from the twenty-term sequence indicated.) Combining terms by introducing the multiple 110, the scribe recasts the subtrahend as 110′ of 60 10′ 30′. The problem is then worked out as the difference of 3″ of 110 and 60 10′ 30′ (= 13 5′) divided by 110, that is, of 66 divided by 550. Consideration of the factors of 550 then leads to the final answer: 10′ 50′.

Thus, the scribes in the arithmetic tradition of late antiquity present themselves as *virtuosi* in the art of manipulating unit-fractions; yet their very methods reveal this to be a superfluous art. How was it possible that they failed to perceive this, embellishing these techniques and so distracting from the teaching and development of the more general techniques of fractions employed within their computations? It may be that

these three papyri (and the other documents to be considered below) are not fully
151 typical of the ancient tradition in this | respect. But I see little reason to doubt that they are, and that their practice reflects the strong element of conservatism implicit in a technical tradition. The unit-fractional mode, if unwieldy, was surely adequate for solving arithmetic problems of the type given in the papyri. To neglect training in this mode would leave the scribe ill prepared to cope with the received texts used in the teaching and application of his discipline.

IV. THE HERONIAN METRICAL TRADITION

The work of Hero of Alexandria (latter part of the first century A.D.) consists mainly of introductions of a practical sort into the fields of geometry and mechanics (cf. Hero's *Mechanica* [*Opera* II] and *Metrica* and *Dioptra* [*Opera* III]). The *Metrica* is a compendium of geometric results on the measurement of plane and curvilinear surfaces (Book I) and solids (Book II), and on the manner of dividing such figures in given proportions Book III). Addressing the needs of his audience of practitioners, like surveyors and architects, Hero provides statements of geometric relationships, sometimes (but rarely) with proof, illustrated by problems worked out in full. Reflecting the utility of his manner of presenting this material, derivative versions appeared in which Hero's rules were illustrated by additional problems[24]. Throughout these works computations involving fractions arise, and one frequently finds these solved by means of the unit-fractional methods.

In the *Metrica* itself the unit-fractional mode is of minimal importance. It figures prominently in only six passages: the square root of 63 is approximated as 7 1/2 4′ 8′ 16′ (I, 9); the difference of 169 and 72 5′ is given as 96 1/2 5′ 10′ (I, 14; Cf. 15); via the Archimedean rule that "11 squares of the diameter of the circle are very nearly equal to 14 circles" (cf. *dimension of the Circle*, prop. 2) the area of the circle of diameter 10 is given as 78 1/2 14′ (I, 26), while that of diameter 17 1/2 is given as 240 1/2 8′ (I, 33); by way of determining the volume of a pyramid, the third of 1333 3′ is given as 444 3′ 9′ (II, 5). In four other passages one meets the notations 1/2 3′ (I, 8; II, 16) and 1/2 4′ (II, 6; III, 2) instead of 5/6 and 3/4, respectively. But these appear to be of only *notational*, rather than *computational*, significance. For in III, 2, Hero expresses the approximate square root of 126 1/2 4′ as 11 4′; multiplies of this by 15 to obtain 168 1/2 4′; but then writes the thirteenth of this product as 12 51/52. Later in the same section he finds the division of 100 4/5 by 10 22/65 to be 9 1/2 4′. Thus, the unit-fractions do not actually enter into the computational procedure in these instances.

152 | In order to facilitate computations with fractions, or even to avoid them, Hero often resorts to proportions. For instance, in I, 20, he replaces the ratio 14 3′:7 by its equal, 43:21; in I, 22, he replaces 12:76 1/2 by 24:153, that is, 8:51; and so on. Nevertheless, Hero possesses a general notation for fractions and makes frequent use of it in his computations. We have already seen an instance of this in the computations in III, 2 above. In all, seventeen sections of the *Metrica* adopt the general mode for fractions; only three of these fall within Book I (Chaps. 16, 17, 24), six appear in II (Chaps. 1, 2, 10, 11, 13, 18) and eight in III (Chaps. 2, 4, 7, 8, 9, 20, 21, 22). Interestingly, this pattern reverses that seen for the appearance of the unit-fraction mode, where five passages in Book I use it, but only one in II, and none in III. One suspects that unit-fractions figured more prominently in Hero's elementary sources;

but as he turned to more advanced sources, based primarily on results proved by Archimedes, the computations in the latter part of the *Metrica* required use of the more serviceable general mode of fractions.

In contrast with the *Metrica*, the derivative writings collected as the *Geometrica* and *Stereometrica* make extensive use of the unit-fractional mode. An effective way to see this is through consideration of the values they present for the approximation of square roots[25]. Of the forty-five instances extracted by Hofmann from these works, seven come from the *Metrica*, and of these only one (i.e., the approximation of $\sqrt{63}$ by 7 1/2 4′ 8′ 16′ in I, 9) employs the unit-mode, although in three others one meets the notations 1/2 3′ and 1/2 4′, which we have argued are not of computational significance. A further instance is drawn from Hero's *Dioptra* (Chap. 28): here the radicand is given as 68 1/2 14′, but the root as 8 2/7. This indicates that Hero has recast the radicand as 68 4/7 and then worked out the root as 8 + (4 4/7)/16, that is, 8 2/7. Thus, the unit-mode in this case also lacks computational significance.

All of the remaining thirty-seven cases come from the Hero-based writings. Of these, five are nearest-integer approximations, while another seven introduce only a single unit-fractional part or 2/3 or 1/2 4′, so that these cases do not distinguish between the two fractional modes. But of the twenty-five cases left, *all* are clear instances of the unit-mode. Only two of these are hybrid in form. For instance, the root of 2460 15/16 is given as 49 1/2 17′ 34′ 51′ (No. 38). Through consideration of the context, Tannery has shown how this value results as the difference between the roots of 6300 (given as 79 3′ 34′ 102′ in No. 45) and 886 − 16′ (given as 29 1/2 4′ 68′ in No. 17) [Hofmann 1934, 111]; the expected value through direct computation would be expected to be about 49 1/2 10′.

Even in these cases where the radicand or the root or both are expressed in the unit-fractional mode, we might ask whether | this mode figured within the actual process 153
of computation, or whether it was first converted to the general mode, as we saw above in the example from the *Dioptra*. In most of his discussion, Hofmann adopts the latter approach, and this may indeed represent the actual method followed. But in several instances I believe a better account is possible on the assumption that the unit-fractional mode was maintained throughout, and this appears especially useful in capturing the scribe's method of approximating fractions. For instance, for the root of 43 1/2 4′ (No. 16) we obtain 6 1/2 + (1 1/2)/13, or 6 1/2 13′ 26′, without the intermediate appearance of 3/26. Similarly, for the root of 43 1/2 4′ 9′ (No. 34) we obtain 6 1/2 + (1 1/2 9′)/13; via the standard method the division of 1 1/2 9′ by 13 leads to the quotient 9′ 78′, from which the given value of 6 1/2 9′ readily follows by neglecting the term 78′. Again, for the root of 593 (No. 29) we obtain 24 + 17/48, simplify to 24 4′ + 5/48, and then round off to 24 4′ 8′. (One notes here the possibility of the closer value 24 3′ 48′.) The root of 356 18′ (No. 35) will be 18 1/2 + (14 18′)/37; the remainder term may be simplified to 14/36 and then expressed as 4′ 9′, so as to produce the given value 18 1/2 4′ 9′. Further, the root of 444 3′ 9′ (No. 25) will be 21 + (3 3′ 9′)/42; noting that 12′ of 42 is 3 1/2 (that is, 3 3′ 6′), one is led to the approximation of the text, 21 12′.

These examples suggest that the unit-mode for fractions has played a role within the computations, as well as being the preferred notational mode for expressing results. In another case, the unit-mode can assist us in restoring a defective text.

The root of 32 (No. 37) is given as 5 1/2 14′, a remarkably poor approximation, since the correct value is only slightly short of 5 1/2 6′. Indeed, using the standard Heronian procedure for roots, we should have expected the latter value, via the computation of 6 – 4/12. Now, the Heronian procedure is recursive; given any initial approximation, a second is found by dividing the radicand by the first, whereupon a much better approximation is obtained by taking the arithmetic mean of these first two[26]. In the present case, if we take the initial value to be 5 1/2 6′ (as derived above), division produces the second value 5 1/2 7′ 238′, from which follows the new value 5 1/2 12′ 14′ 476′. Neglecting the small last term, we obtain the approximation 5 1/2 12′ 14′. The value actually given in the text, 5 1/2 14′, thus appears to have resulted through a scribal omission[27].

These observations on the interplay of the two modes of expressing fractions are confirmed through a general survey of the Hero-based metrical writings. In the *Geometrica* about fifty pages of Heiberg's edition contain computations relying on expensive use of fractions; of these about thirty pages introduce the unit-mode alone,
154 in twenty both modes appear, but in none is | the general mode used exclusively. Sometimes a result is expressed in both modes, one immediately after the other; for instance, the square root of 43 1/2 4′ is stated as "6 1/2 13′ 26′, or six units and remainder 13′ 13′ eight," whereupon the latter, a notation for 6 + 8/13, is at once reintroduced as "6 and 8 13′ 13′" for the next computation[28].

Typical of these mixed usages is a set of computations of the area of the circle: (a) First, one is to find the perimeter when the diameter is "16 3′ 15′ *schoinia*, that is, 16 *schoinia* and 5′ 5′ two"; this is converted to 82/5, multiplied by 3 7′ to obtain 257/5 plus (5/7)/5, which he then states in the form 51 3′ 7′ 15′. (b) The scribe then presents in their turn five variant computations of the area. Although the answer in each case is stated as 211 4′ 25′ 28′, the operands are given in the general notation (e.g., as 16 2/5 and 51 19/35 for the diameter and the perimeter, respectively). In the first of these one can view the answer in its stated form as actually derived via the procedure given; but in the other four it is difficult to do so. Thus, the unit-mode is even further removed from the effective computational procedure [Hero IV, 346-351][29].

These examples thus show that, in contrast with Hero's own writing, the *Metrica*, those derived from it employ the unit-mode as the standard way of expressing fractions. Yet even in the latter, the general mode for fractions is familiar, being in fact the mode through which the computations are actually performed. This is precisely the ambivalence we saw in the case of the problems in P. Akhmîm: there, as here, problems were typically phrased in the unit-mode, then converted to the general mode for computational purposes, after which the solution was converted back into unit-mode. If in rare instances one perceives that the unit-mode might be a convenient way to execute a computation: (for example, in multiplying 5 1/4 by itself to produce 27 1/2 16′), for all but such simple cases, computation demands use of the general mode. Thus, the unit-mode serves almost exclusively as a notation for the recording of numbers, rather than for operating with them.

V. THE CONTINUITY OF THE ANCIENT COMPUTATIONAL TRADITION

This survey of the arithmetic of unit-fractions has revealed a striking uniformity in the techniques used by the Egyptian scribes of well before the middle of the second millennium B.C. and the Greek scribes throughout the first half-millennium A.D. Filling the gap is a set of demotic (Egyptian) papyri from the Hellenistic and early

Greco-Roman periods (third century B.C. onward). In the edition by R. A. Parker, this comprehends seventy-two problems preserved in five documents estimated to | range from the third century B.C. to the second century A.D. Like the other 155
papyri we have considered, these are devoted to detailed arithmetical manipulations in contexts like solving commercial problems or finding areas and volumes according to the rules of metrical geometry. Considerable interest centers on operations involving fractions, and in these cases we encounter the same mixture of methods, dominated by the use of unit-fractions, that appear in P. Rhind and P. Akhmîm.

Implicit in these demotic problems is access to tables of unit-fraction expressions like those we have seen. For instance, when the scribe of No. 41 multiplies 1 5/6 42′ times itself, he must know that the given terms equal 1 6/7, for the result is stated as 3 22/49. Again, in Nos. 57-61 the given terms are 3′ 15′ and 3″ 21′; but the scribe knows to convert them to 2/5 (or 14/35) and 5/7 (or 25/35) before the computations begin. Thus, in No. 57 their product is worked out as 10/35 and then written as 4′ 28′; in No. 58 their quotient is known to be 14/25, written as 1/2 25′ 50′; in No. 60 the difference is first found to be 11/35, then converted to 4′ 28′ 35′; in No. 61 their sum is 39/35, then written as 1 10′ 70′. Clearly, the implied procedure of expressing the result of a division in terms of unit-fractions is the same as that used in P. Rhind and in the Greek papyri. But one also sees how this mode might sometimes perform merely a notational, rather than a computational, role. This break, however, is by no means as evident as in P. Akhmîm or the Heronian metrical writings. The demotic scribes typically retain the unit-mode throughout. Indeed, even in the examples just cited, no general notation for fractions is used; for instance, the scribe still conceives of 10/35 as a division: "make 10 a part of 35." Only rarely does a general notation enter (as in Nos. 2, 3, 10, 13), and there in the case of mixed fractions as well as proper. In No. 13, for example, the implied division of 60 by 131 is written as 3′ 15′ (7 1/2 10′)/131. One is reminded of problems in P. Rhind and P. Akhmîm where the combination of fractions is assisted by first raising the terms, but the new denominator happens not to be a common multiple of the given ones[30].

By virtue of their adherence to the same procedure of division, the demotic scribes present many of the same values found in the other papyri. A surprising instance is the expression of 2/35 as 30′ 42′, just as in P. Rhind and P. Akhmîm. Indeed, its appearance in No. 56 is worked out precisely as in P. Akhmîm No. 23. For further comparisons, one may note in No. 46 the value 1 − 6′ written via a special notation for 5/6 (analogous to 3″ 6′ elsewhere)[31]; in No. 47 the value 1 − 7′ is 5/6 42′ (cf. P. Akhmîm: 3″ 6′ 42′); but in No. 48 the value 1 − 8′ is 3″ 12′ 8′ (P. Akhmîm: 1/2 4′ 8′) and in No. 49 the value 1 − 9′ is 5/6 30′ 45′ (P. Akhmîm: 3″ 6′ 18′); in No. 50 the value 1 − 10′ is 5/6 15′, in agreement with P. Akhmîm (3″ 6′ 15′). The | unusual order of the 156
terms in No. 48 suggests that its 3″ 12′ resulted from partition of 1/2 4′ as in P. Akhmîm. As for No. 49, its very unusual value for 8/9 is consistent with a value for 8/90 (i.e., 15′ 45′) listed in a table of 90ths in No. 66. We may thus have additional indication of use of tables in these computations. In Nos. 66 and 67 portions of such tables are preserved, for 90ths and 150ths, respectively (see Table V). In format these hardly differ from those in the other papyri, but there are idiosyncracies in the implied computational procedure. The value for 2/150 is not 75′, for instance, but 90′ 450′; that for 3/150 is not 50′, but 60′ 300′; and so on. The scribe may first have taken 30′

of 150, so reducing these cases to 2/5 and 3/5, respectively; if so, his reasons for doing this must lie with the special purposes intended for the table[32].

An interesting link between the demotic papyri and the later Greek computational tradition is evident in the method they both use for estimating square roots. For in all dozen instances in the papyri the value is derived via the "Heronian" or Babylonian rule, in which the square root of $a^2 + b$ is taken as $a + b/2a$[33]. In No. 62, for example, the root of 10 is worked out step by step according to this rule, resulting in the value 3 6′. In No. 35 the roots of 345 and 105 are given as 18 1/2 12′ and 10 4′, respectively, where reliance on the same procedure is clear, even though the steps are not set out. Often a rounding off is involved. For instance, in Nos. 7 and 15 the root of 1500 is given as 38 3″ 20′. The rule would yield 38 56/76 or 38 14/19; separation into unit-fractions might lead to the form 38 3″ 19′ 57′ (as in the P. Akhmîm), from which the given estimate follows. In No. 32 the root of 133 3′ is given as 11 1/2 20′, where the rule yields 11 1/2 22′ 66′. In No. 18 the root of 1000 appears as 31 1/2 10′ 30′. The rule yields 31 39/62 or 31 1/2 8/62; I suspect that the scribe then rounded off 8/62 to 8/60 to obtain 10′ 30′. An interesting case is No. 37 where the root of 450 is given as 21 5′ 60′. The rule yields 21 9/42 or 21 3/14, so that consulting a table (like that in P. Akhmîm) would lead to the value 21 5′ 70′. As the scribe persists in the stated value with 60′, instead of the expected value with 70′, one would gather either that he has miscomputed the reduction of the fraction or drawn it from a table with an error in that entry[34].

Through these examples one can see that the demotic papyri serve as a bridge between the ancient Egyptian and Mesopotamian scribal methods on the one hand and the later Greek papyri and metrical writings on the other. Unfortunately, the Hellenistic provenance of the demotic papyri does not exactly secure this view. It is in theory possible, for instance, that the "Heronian" methods they use derived from interaction with an older Greek computational tradition, leading to an advancement of the
157 | native Egyptian technique beginning from the third century B.C. But the latter view is quite unconvincing in the light of other features of the papyri. In particular, circle measurement invariably uses 3 as the value for π, just as in the older Mesopotamian tradition [Parker 1972, No. 32 f, 36-38][35]. By contrast, the Greek metrical tradition always uses the Archimedean value 3 1/7; thus one would surely suppose that any contact with Greek methods would lead to its adoption. Similarly, a set of demotic problems measures segments of circles via the rule $(h/2)(b + h)$, for height h and base b [Parker 1972, No. 36-38]. The rule is known to Hero, who ascribes it to unnamed "ancients" and points out that in the special case of the semicircle (where $b = 2h$) it implies the value 3 for π (*Metrica* I, 30); he adds that this "rather careless" method was made more accurate by others who added on a fourteenth part of the square on half the base (*Metrica* I, 31). Surely it would be odd for the demotic scribes to latch onto the crude rule at just the time when their Greek counterparts were engaged in its correction. On the other hand, it would be quite natural for the Greek computers, upon learning of such a rule from demotic sources, to recognize its shortcomings in the light of Archimedes' findings on the circle-measurement.

Under this view, the demotic papyri present to us a late phase of the native Egyptian computational tradition. This is in keeping with certain conjectures by Parker [1972, 5 f., 8-10], notably that the invention of the general mode for fractions as an outgrowth of techniques (like the raising of terms) already seen in the Rhind

Papyrus, although of course even in the demotic papyri its application is minimal. This native tradition could borrow from Mesopotamia certain methods, like the rule for square roots and the value of 3 for π[36]. When this importation occurred cannot directly be ascertained, owing to the relatively recent dating of our sources. But a persuasive circumstantial case can be made for the late sixth and fifth centuries B.C., when Egypt was under the administration of the Persians. In proposing this view, Parker calls attention to a demotic astrological papyrus which indicates the introduction of Mesopotamian modes of calendary and divination into Egypt during this same period[37]. With respect to the Egyptians' adoption of new mathematical techniques, their acutely selective attitude is not really surprising. The scribes would certainly wish to avail themselves of techniques not held in their own tradition, for instance, the method for estimating roots. But already possessing an arithmetical notation and elaborate procedures for effecting the arithmetical operations, including the methods of unit-fractions, they would understandably resist adoption of the alternative Mesopotamian methods, despite the enormous gain in flexibility entailed by the sexagesimal place-notation[38].

| These considerations lead me to view the Greek computational tradition as con- 158
sisting of two largely independent branches. Implicit in the formal treatment of mathematics by Euclid is a basis of metrical and arithmetical techniques oriented toward the practical needs of fields like commerce and surveying. Some part—and doubtless no small part—of these techniques could be imported through contact with the older traditions, as indeed the Greeks themselves insist[39]. Seeing that the Greeks look toward Egypt, rather than Mesopotamia, as the source of mathematical technique, I infer that the Egyptian contacts were the more significant during this formative phase and that knowledge of Mesopotamian methods, such as the "geometric algebra" implicit in Euclid's Book II, might come through the mediation of the Egyptian scribes[40]. This is the pattern of transmission we have already argued for some of the techniques used in the demotic papyri, and the Greeks could hardly be blamed if they failed to perceive the Mesopotamian origins of such features of the contemporaneous Egyptian tradition. This also helps us to understand how the Greeks, in the infancy of their own technical tradition, could have failed to adopt the superior Mesopotamian computational techniques, if indeed they had access to the Mesopotamian tradition directly[41].

This practical tradition in Greek served as the technical basis for the formal corpus of geometry as compiled by Euclid and extended by Archimedes. As the practical complement of the formal geometry, it profited in its turn from the incorporation of findings due to the advances in geometric theory. Especially prominent in this movement is Hero of Alexandria, a writer fully sensitive to the aims of theory, but concerned with addressing an audience of trained mathematical practitioners, like surveyors and military engineers. Thus, in a work like his *Metrica* he begins with the basic rules familiar in the elementary metrical tradition, but continues well beyond these in the presentation of more advanced materials, sometimes with derivations and proofs. He so brings in Archimedean results on the measurement of plane figures like circular and parabolic segments, and of solid figures like segments of spheres and cylindrical sections; he presents the familiar rule for square roots in a more general form, as well as a rule for estimating cubic roots; elsewhere, he gives a geometric solution with proof of the problem of duplicating the cube[42]. These examples reveal that the level of expertise implicit in Hero's writings is far higher than that indicated in the mathematical papyri.

The tradition represented in papyri like P. Mich. and P. Akhmîm, however, appears to be independent of the one underlying Hero's writings. To be sure, these papyri display greater dexterity in technical manipulations than one finds in older documents like P. Rhind and the demotic papyri. But we have seen that the germ of these refinements
159 was already present in the | earlier tradition. The Greek papyri show little, if any, sign of access to the more sophisticated sources exploited by Hero. To the contrary, they adhere to an elementary technical level with phenomenal tenacity. For instance, the virtual identity of the fraction tables in P. Mich. 146 and P. Akhmîm reveals a continuity spanning several centuries in the transmission, not only of certain *techniques* of unit-fractions, but even of the *text* of a specific table of computed values. In contrast to Hero's audience, the users of the papyri are quite limited in technical competence, operating within a narrow range of practical commercial contexts hardly different from that of the earlier Egyptian scribes. Indeed, the Greek papyri are best viewed as an extension of the Egyptian tradition. I would suppose that as Greek gradually supplanted demotic Egyptian on the popular level in the course of the Hellenistic and Greco-Roman periods, training in the technical methods for dealing with everyday situations as provided in the demotic papyri would have to be made available in Greek as well. In this way, the Greek mathematical papyri would originate effectively as a translation literature.

Intermediate between Hero's writings and the papyri is the metrical corpus called the *Geometrica* and the *Stereometrica* by its modern editor, Heiberg. Its dependence on Hero's *Metrica* is manifest throughout. But this is a hybrid literature, seeking to present Hero's procedures in a form appropriate for less expert users, like those of the papyri. Thus, rules are presented not with proofs or derivations, but rather in the guise of case after case of fully worked out numerical examples. Sometimes the same problem, framed around the same data, is worked out according to several alternative sequences of the arithmetical operations; this would surely indicate that the editor considered the exposition of basic arithmetic among the principal objectives of his teaching. As we have seen, a subtle index of the lower technical level of the derivative metrical collections is their retention of the unit-mode for fractions as a computational method; by contrast, in Hero's writings this has all but yielded entirely to the general mode, entering only as a notation for the expression of the answers.

How can one explain the longevity of the methods of unit-fractions? A hint may lie in the practical writers' preoccupation with units of weight and measure. In contrast with Hero, who omits mention of the specific units measuring the magnitudes given in his problems, the scribes invariably specify so many *schoinia*, feet, spans or daktyls of length, so many acres of planar measure, and so on. One comes upon extensive lists of units and their diverse subdivisions in accordance with the wide variety of systems associated with different countries and chronological periods. Units of weight
160 and currency are also of major importance, and in a representative | passage we learn that "the Roman *dēnarion* has 1252 parts, … the *assarion* is divided into the 1/2, 3′, 4′, 6′, 8′, 9′, 10′, 12′, 16′, 18′, 24′, 36′, 40′, 50′, 72′, and these parts have their own names among the Roman computers"[43]. Thus afflicted, the ancient scribe had better have mastered the art of converting the fractional parts of units into subunits and sub-subunits, and so on. Exercises in the manipulation of unit-fractions might well be an effective preparation for this kind of practical situation.

But there is another factor which should not be overlooked. The methods of the Egyptian scribes were part of a tradition already more than two thousand years old when first encountered by the Greeks in the sixth or fifth century B.C. Despite their limited range and unwieldy implementation, these ancient techniques were quite feasible, and their great antiquity would render their modification or displacement extremely difficult. In this respect, then, the conservatism of the ancient computational tradition would conform to a general pattern of transmission, whether of texts or techniques, in the traditions of many other literary and technical fields.

NOTES

1 Scholium to Plato's *Charmides* 165e; cf. [Thomas 1957 I, 16–19].

2 For Greek expressions of this debt to Egyptian mathematics, see [Thomas 1957, 8; Heath 1921 II, 440].

3 The 20th-century study of the mathematical and astronomical cuneiform texts has revealed the profound debt owed by Greek science to its *Babylonian* precursors. This is evident in the use of sexagesimals for astronomical computation, while Babylonian data and even parameters are crucial for Hellenistic astronomy. In mathematics O. Neugebauer has pointed to Babylonian sources for the Greek metrical tradition, as seen in such instances as the methods for approximating square roots (the so-called "Heronian rule"; see below, note 25), and the techniques of examining quadratic geometric relations, the so-called "geometric algebra" (cf. [Neugebauer 1957, Chap. VI, esp. pp. 144–151]).

4 On the basis of the technical evidence from Egypt, Mesopotamia, and Greece, in the light of literary sources, I have proposed a transmission of Mesopotamian mathematical techniques to the Greeks through Egypt after the Persian occupation of the late sixth and early fifth centuries B.C. The details of this argument are given in an unpublished paper, "The Greeks Learn Geometry," a version of which was presented to the seminar on the history of mathematics of the Courant Institute (New York University) in October 1975.

| 5 Baillet [1892] makes a few passing comparisons between the Egyptian and Greek texts (see note 161
23 below). Hultsch [1901; see note 17 below] and Tannery [1884] attempt to interpret the Greek texts in the light of the Egyptian, but both are severely handicapped by clumsy and improbable views on the methods followed by the Egyptian scribes.

6 See [Gillings 1972, Chaps. 6–7, 10; 1978; Bruchheimer & Salomon 1977; Bruins 1975]. Among older accounts seeking to articulate esthetic criteria underlying the ancient arithmetic are Hultsch' studies of the Egyptian [1901, drawing upon his massive study of P. Rhind in 1895] and Baillet's of the Greek [1892, 22; see Section II below]. One may note that the literature on the "2/*n* table" in P. Rhind is immense, including beyond those cited above dissertations by Neugebauer (1926) and by Vogel (1929) and lengthy studies by Tannery [1884] and by Hultsch' (1895). For references see the bibliographies in [Gillings 1972] and [Archibald 1927; in *The Rhind Mathematical Papyrus*, Chace and Manning, eds., Vol. I].

7 [van der Waerden 1980]; cf. also [van der Waerden 1954, 23–26] for an extremely clear and useful short resume of the computational methods and [van der Waerden 1938] for a detailed account.

8 Compare the editions of P. Rhind by Peet (1923) and Chace and Manning (1927). In each computation the terms in the left-hand column are the parts of the unit, those in the right-hand column the corresponding parts of the divisor. The slash (/) before a row indicates that that entry is to be included in the sum yielding the final quotient; the sum of the corresponding terms on the right will be the number being divided (here, 2).

9 For discussions of this document, see [Gillings 1972, Chap. 9; van der Waerden 1954, 21 ff and Plate 3]. For those of its entries which are not obvious, it seems to me that a method of raising terms, followed by separation and reduction, may have been employed (cf. the procedure of "*chōrismos*" used in P. Akhmîm, discussed in Section III below). For instance, $8' = 5/40 = 40' + 4/40 = 40'\ 10'$. Again, $8' = 25/200 = 200' + 24/200 = 200' + 3/25 = 200' + 25' + 2/25 = 200'\ 25'\ 15'\ 75'$ (the last step using the value for 2/25 given in P. Rhind). The entry for $13'$ is quite incorrect. Gillings [1972, 99] suggests that it be emended to $26'\ 39'\ 78'$, on the pattern of the entries just preceding. But, of course, it bears little resemblance to the value actually given, and

seems unlikely to have resulted either through textual or computational errors. I believe that the scribe may have used a method like the above, but instead of changing 13′ to 3/39, he made it 3/49. (The implied error in the computation of 3 × 13 would amount to a difference of only one stroke in the tens′ grouping.) Then, 3/49 = 49′ + 2/49 = 49′ 28′ 196′ (this last using the value
162 | listed in P. Rhind), the entry given on the leather roll. This view of the method underlying the table in the leather roll points to a continuity of computational technique linking the Egyptian and Greek traditions; see the discussions in Sections II and III below.

10 The identity 4′ = 7′ 14′ 28′ is given in the leather roll (see note 9). As it happens, the other two, 6′ = 7′ 42′ and 7′ = 8′ 56′, are not listed there. But they follow the same pattern of formation illustrated by the two entries 3′ = 4′ 12′ and 4′ = 5′ 20′ which do appear there.

11 P. Mich. 145: F. E. Robbins in *Papyri ... Michigan*, III, pp. 34 ff.; P. Mich. 146: *ibid.*, pp. 52 ff. (cf. also [Robbins 1922]); P. Akhmîm: Baillet [1892, 3–4].

12 Half of the entries for cases of prime divisor in P. Rhind conform to this pattern. The scribe chooses smaller initial terms in 17 and 19 (12′ instead of the possible 10′), in 37 (24′ instead of 20′), in 59 (36′ instead of 30′). In the latter parts of the series he seems to prefer a leading term from the tenths sequence (cf. the cases of 47, 61, 67, 71, 83) to alternatives. Other instances are: 29 (lead term of 24′ instead of 16′); 31 (20′ instead of 16′); 43 (42′ instead of 24′); 97 (56′ instead of 48′). Of course, the pattern does not apply to his values when the divisor is composite.

13 After the first term, these are, in accordance with the Greek alphabetic-numeral notation: α, β, γ, ..., ι, κ, λ, ..., ρ, σ, τ, ..., $/\alpha$, $/\beta$, $/\gamma$, ..., $\overset{\alpha}{\mu}$. (On this notational system, see [Heath 1921 I, 31–40].) Thus, to find two-thirds of 427, say, one merely adds the entries for 400, 20, and 7. Baillet [1892, 20] explains that the initial entry listing the part of 6000 may have been motivated by the fact that the basic unit of currency, the talent, consisted of 6000 drachmas.

14 P. Mich. 146 lists 3/18 as 9′ 18′; in P. Akhmîm this is 6′, as expected. P. Mich. gives different values for 9/12 and 12/16 (namely, 3″ 12′ and 1/2 4′, respectively); its values for 3/4 and 15/20 are lost. In P. Akhmîm all four have the same value of 1/2 4′.

15 P. Akhmîm gives 16/20 as 1/2 4′ 20′, consistent with its prior values; the value for this term has been lost from P. Mich.

16 Also, in the table of sixteenths only the parts 1/2, 4′, 8′, 16′ appear; thus, 3/16 is given as 8′ 16′ rather than as 6′ 48′. Other exceptions arise at 3/17, 4/17 and 3/19, to be discussed below.

17 Interestingly, an Egyptian papyrus edited by E. Revillout in 1895 and dated by him to the mid-second millennium B.C. adopts the value 7′ 91′, thus at variance with P. Rhind, but in agreement with the later Greek papyri. This document contains the fragment of a table of fractions,
163 presenting some three dozen entries from the sequence of sevenths, eighths, ..., fifteenths. | It thus establishes that tables of the sort preserved in the Greek papyri Akhmîm and Michigan 145 and 146 had firm precedents in the older Egyptian tradition. Hultsch [1901] has examined the entries in Revillout's papyrus in comparison with those in P. Rhind and P. Akhmîm; but his own view of the methods underlying P. Rhind is entirely unpersuasive and so prevents him from perceiving the essential unity of computational technique linking all of these documents.

18 For 9/14 P. Akhmîm has the expected 1/2 7′; P. Mich. has 1/2 8′ 56′.

19 For the table of tenths in P. Rhind, see [Chace and Manning 1927 II, 60].

20 Here P. Mich. gives the incorrect value 12′ 15′ 17′ 34′ 51′; but the scribe clearly is adhering to the same unusual computational pattern as that in P. Akhmîm.

21 Implicit in the passage from 3 to 4 and from 7 to 8 in the table of seventeenths is the identity 51′ 17′ = 15′ 85′. But it is difficult to see why the scribe would have made the use of such an identity part of his procedure here.

22 Problems 6–9, 12–25, 29–32, 38–40, and 50; Nos. 28 and 49 also manipulate unit-fractions, but in the context of problems of distribution, rather than of pure computations.

23 In his discussion of the comparisons between P. Rhind and P. Akhmîm, Baillet [1892, 59–62] perceives agreement in pedagogical motives, in the classes of problems posed (e.g., interest, distributions, etc.), and sometimes, if not always, in the specific numerical values given (e.g., in the values for 2/*n*, save for 13 and 19). But he views the Greek procedure of *chōrismos* as an entirely new factor [Baillet 1892, 62] and otherwise fails to sense the unity of computational *procedure* linking the two texts.

24 The compilations of metrical problems are collected as the *Geometrica* and *Stereometrica* [*Opera* IV, V].

25 In most cases Hero uses $a \pm b/2a$ to approximate the square root of $a^2 \pm b$. A review of these materials is given by Hofmann [1934] with extensive references to the analyses by Hunrath, Tannery, and others. I believe Hofmann's views are generally quite sound; but in the details of several difficult cases, better alternatives are possible (see note 27, for instance), as I show in a separate paper currently in progress.

26 In fact, the rule stated in note 25 above is derived from this more general form. For if a is taken as the initial approximation to the root of $a^2 \pm b$, the quotient after division by a will yield $a \pm b/a$ as the second approximation, and the arithmetic mean will be $a \pm b/2a$. Hero is aware that his procedure is recursive (cf. *Metrica* I, 8).

27 Hofmann's attempt here to save the text is strained [1934, 111]; Bruins' attempt to emend (*Codex* III, p. 93) is vitiated by a substitution error in the computation of the lower bound (read "12 − 2" for "12 + 2").

| 28 *Geometrica* IV, 290 ff.; cf. also pp. 297, 323. This style of expressing the denominator doubly seems idiosyncratic of some manuscripts of the *Geometrica;* for it is not used in the *Metrica*, for instance. For a discussion, see [Heath 1921 I, 43]. 164

29 For other examples where alternative executions of a problem merely restate the answer in its earlier form, rather than working out a new form, see *Geometrica* IV, 290–297, 344–347.

30 On Egyptian precedents for the method of common denominators, see [van der Waerden 1954, 27; Parker 1972, 8–10].

31 Parker [1972, 8] notes that the scribe uses a ligature based on 3″ 6′ to denote 5/6.

32 Parker [1972, 73] offers no firm reason for the scribe's factoring out 30′ in this manner. One may observe, however, that the opaquely motivated choices for the 150ths are entirely consistent with the choices for the entries in the table of 90ths, where they are quite natural. I suggest that the scribe might hold in view a base of 360 for its association with the number of days in a year. Then 90 corresponds to the days in a 3-month period, 150 to those in a 5-month period, and the entries in the tables give the fraction of the base period which each of the days in a decanal (10-day) period amounts to. Such tables might be useful for the computation of interest on short-term loans.

33 Parker accounts for several cases via the alternative subtractive rule: that $a - b/2a$ approximates the root of $a^2 - b$. But in only two instances does the latter form yield the scribe's value exactly where the additive form does not. Thus, in No. 33 the root of 13 3′ is 3 3″ (via 16 − 2 3″) and in No. 38 the root of 75 is 8 3″ (via 81 − 6). But in these cases the additive rule would result in 13 3″ 18′ and 8 11/16, respectively, and these might easily have been rounded off to produce the scribe's values. In other cases, the scribe prefers the additive form, even when the alternative would give a closer answer. Thus, in No. 18 the root of 1000 is given as 31 1/2 10′ 30′ via 961 + 39, rather than as 31 1/2 8′ via 1024 − 24; and in No. 35 the root of 345 is given as 18 1/2 12′ via 324 + 21, rather than as 18 1/2 19′ 38′ via 361 − 16. Since the scribal methods of unit-fractions do not employ subtractive techniques—indeed, they are extremely rare even in Hero—I am inclined to doubt that subtractive forms were introduced in the special context of square roots.

34 That the papyri are not immune from miscomputation and scribal error should be clear from the preceding sections. See also Table IVb; Parker [1972, *passim*]; and the critical apparatus to the editions of Hero's metrical writings.

| 35 See [Neugebauer 1957, 46 f.] on the Mesopotamian value; he notes that one set of tablets indicates an alternative value of 3 1/8. 165

36 It is odd that the procedure for finding the area of the circle used in Rhind Papyrus 48 yields the closer value 256/81, or 3 13/81, that is, a bit less than 3 1/6. Either this was not an established procedure in the Egyptian tradition at large, or else the scribes of the demotic papyri accepted the value 3 for its obviously greater convenience in computations.

37 Parker [1972, 6], who cites his edition of a Vienna demotic papyrus on eclipse omens (Brown Egyptological Studies, 2, 1959). I advocate much the same view of the interaction of these traditions in the paper cited in note 4 above.

38 On the Mesopotamian computational methods, see [Neugebauer 1957, Chap. II].

39 Herodotus assigns to the Egyptians the origination of geometry for the purposes of land-measurement (*Histories* II, 109), and this view is maintained by several later Greek witnesses (for citations, see [Heath 1921 I, 121 f.]). Neugebauer wishes to discount all such testimonia of Egyptian contributions to the early Greek technical tradition [1957, 72, 80, 151]. While, admittedly, one might exaggerate the importance of Egyptian technique, as the later writers often do,

the respect which earlier writers like Plato and Aristotle accord to the Egyptian tradition of learning, including mathematics, is clear and apparently sincere (see, for instance, *Philebus* 274 c; *Laws* 819 b; *Metaphysics* I, 1). The stories of visits to Egypt by several of the Presocratics, by Plato, by Eudoxus and others are hardly unreasonable. Although the technical range represented in the demotic papyri is quite limited, it embraces most of what we can infer of the technical base of the fifth-century Greek tradition.

40 For brief accounts of these techniques and their Mesopotamian parallels, see [Neugebauer 1957, 41 f., 149 f.; van der Waerden 1954, 63–73, 118–126].

41 I here refer of course only to the computational methods in the metrical and papyrus traditions. When the Greeks gained direct access to the Mesopotamian astronomical methods, beginning in the third or second century B.C., they assimilated with these the sexagesimal computational methods. These methods thus became a fixture of the Greek astronomical tradition and were transmitted with it into later antiquity, the Middle Ages and the Renaissance. But among the Greeks, their use remained restricted to the astronomical field.

42 *Metrica* I, 32–35, 38; II, 11–15; on roots, see I, 8 and III, 20; on the cube-duplication, see *Mechanica* I, 11 and *Belopoeica* (Wescher, 114–119). For a survey account, see [Heath 1921 II, 316–344].

166 | 43 Hero [1912 IV, 410 f]. Robbins perceives a commercial motivation, when he writes of the fraction tables in the Michigan Papyri, “It is very probable that documents of this class were intended for reference in the course of ordinary business” [1936, 54].

APPENDIX: TABLES

TABLE I. Values of $2/n$ in the Rhind Papyrus

Prime divisors					Composite divisors			
3	3″				9	(3)	6′	18′
5	3′	15′			15	(3)	10′	30′
7	4′	28′			21	(3)	14′	42′
11	6′	66′			25	(5)	15′	75′
13	8′	52′	104′		27	(3)	18′	54′
17	12′	51′	68′		33	(3)	22′	66′
19	12′	76′	114′		35	(*)	30′	42′
23	12′	276′			39	(3)	26′	78′
29	24′	58′	174′	232′	45	(3)	30′	90′
31	20′	124′	155′		49	(7)	28′	196′
37	24′	111′	296′		51	(3)	34′	102′
41	24′	246′	328′		55	(11)	30′	330′
43	42′	86′	129′	301′	57	(3)	38′	114′
47	30′	141′	470′		63	(3)	42′	126′
53	30′	318′	795′		65	(5)	39′	195′
59	36′	236′	531′		69	(3)	46′	138′
61	40′	244′	488′	610′	75	(3)	50′	150′
67	40′	335′	536′		77	(7)	44′	308′
71	40′	568′	710′		81	(3)	54′	162′
73	60′	219′	292′	365′	85	(5)	51′	255′
79	60′	237′	316′	790′	87	(3)	58′	174′

(continued)

TABLE I. *(continued)*

Prime divisors					Composite divisors				
83	60′	332′	415′	498′	91	(*)	70′	130′	
89	60′	356′	534′	890′	93	(3)	62′	186′	
97	56′	679′	776′		95	(19)	60′	380′	570′
101	101′	202′	303′	606′	99	(3)	66′	198′	

Results of the division of 2 by each of the odd numbers from 3 through 101 as presented in the Rhind Papyrus (see Chace and Manning 1927 I, 21–22, 50–60). Under the column of composite divisors, the number indicated in parentheses is the divisor to which the computation is reduced, after removal of one of the factors; e.g., the computation of 2 ÷ 15 is reduced to that of 2 ÷ 3 after removing the factor 5.

TABLE II. Fraction Values in the Mathematical Leather Roll

8′	=	10′	40′			7′	=	14′	21′	42′
4′		5′	20′			9′		18′	27′	54′
3′		4′	12′			11′		22′	33′	66′
5′		10′	10′			13′		28′	49′	196′
3′		6′	6′			15′		30′	45′	90′
½		6′	6′	6′		16′		24′	48′	
3″		3′	3′			12′		18′	36′	
8′		25′	15′	75′	200′	14′		21′	42′	
16′		50′	30′	150′	400′	30′		45′	90′	
15′		25′	50′	150′		20′		30′	60′	
6′		9′	18′			10′		15′	30′	
4′		7′	14′	28′		32′		48′	96′	
8′		12′	24′			64′		96′	192′	

Decompositions of unit-fractions, as listed in the Egyptian Mathematical Leather Roll. For sources, see [Gillings 1972, Chap. 9] and the remarks in the present paper, note 9. 167

TABLE III. Table of Fractions in P. Michigan 145

[I, i]				[I, ii]					
of 1	23′			of 12	4′	8′	29′	232′	
[of 2]	12′	276′		of 13	3′	15′	[29′]	87′	435′
[of 3]	10′	46′	115′	of 14	4′	5′	58′	116′	145′
[of 4	6′]	138′		of 15	½	58′			
[of 5	6′	23′]	138′	of 16	[½]	29′	58′		
...		...		[of 17	½	12′	348′		
				...			...		

Fragment of a table of fractions from P. Mich. 145. The first column lists consecutive quotients of division by 23; the second column, of division by 29. Bracketed entries have been supplied by Robbins [1936, 43]. For discussion, see the present paper, Section II.

167–168 TABLE IVa. Fractions Listed in P. Akhmîm

Fourths	
2 =	½
3	½ 4′
Fifths	
2 =	3′ 15′
3	½ 10′
4	½ 4′ 20′
Sixths	
2 =	3′
3	½
4	3′
5	½ 3′
Sevenths	
2 =	4′ 28′
3	3′ 14′ 42′
4	½ 14′
5	3″ 21′
6	½ 3′ 42′
Eighths	
2 =	4′
3	4′ 8′
4	½
5	½ 8′
6	½ 4′
7	½ 4′ 8′
Ninths	
2 =	6′ 18′
3	3′
4	3′ 9′
5	½ 18′
6	3″
7	3″ 9′
8	½ 3′ 18′
Tenths	
2 =	5′
3	4′ 20′
4	3′ 15′
5	½
6	½ 10′
7	½ 5′
8	½ 4′ 20′
9	½ 3′ 15′
Elevenths	
2 =	6′ 66′
3	4′ 44′
4	3′ 33′
5	3′ 11′ 33′
6	½ 22′
7	½ 11′ 22′
8	3″ 22′ 66′
9	½ 4′ 22′ 44′
10	½ 3′ 22′ 33′
Twelfths	
2 =	6′
3	4′
4	3′
5	3′ 12′
6	½
7	½ 12′
8	3″
9	½ 4′
10	½ 3′
11	½ 3′ 12′
Thirteenths	
2 =	7′ 91′
3	6′ 26′ 39′
4	4′ 26′ 52′
5	3′ 26′ 78′
6	3′ 13′ 26′ 78′
7	½ 26′
8	½ 13′ 26′
9	3″ 39′
10	½ 4′ 52′
11	½ 3′ 78′
12	½ 3′ 13′ 78′
Fourteenths	
2 =	7′
3	5′ 70′
4	4′ 28′
5	3′ 42′
6	3′ 14′ 42′
7	½
8	½ 14′
9	½ 8′ 56′
10	3″ 21′
11	½ 4′ 28′
12	½ 3′ 42′
13	½ 3′ 14′ 42′
Fifteenths	
2 =	10′ 30′
3	5′
4	4′ 60′
5	3′
6	3′ 15′
7	3′ 10′ 30′
8	½ 30′
9	½ 10′
10	3″
11	3″ 15′
12	½ 4′ 20′
13	½ 3′ 30′
14	½ 3′ 10′
Sixteenths	
2 =	8′
3	8′ 16′
4	4′
5	4′ 16′
6	4′ 8′
7	4′ 8′ 16′
8	½
9	½ 16′
10	½ 8′
11	½ 8′ 16′
12	½ 4′
13	½ 4′ 16′
14	½ 4′ 8′
15	½ 4′ 8′ 16′
Seventeenths	
2 =	12′ 51′ 68′
3	12′ 17′ 51′ 68′
4	12′ 15′ 17′ 68′ 85′
5	4′ 34′ 68′
6	3′ 51′
7	3′ 17′ 51′
8	3′ 15′ 17′ 85′
9	½ 34′
10	½ 17′ 34′
11	½ 12′ 34′ 51′ 68′
12	½ 12′ 17′ 34′ 51′ 68′
13	½ 4′ 68′
14	½ 4′ 17′ 68′
15	½ 3′ 34′ 51′
16	½ 3′ 17′ 34′ 51′
Eighteenths	
2 =	9′
3	6′
4	6′ 18′

(continued)

TABLE IVa. *(continued)*

5		4′	36′			
6		3′				
7		3′	18′			
8		3′	9′			
9		½				
10		½	18′			
11		½	12′	36′		
12		3″				
13		3″	18′			
14		½	4′	36′		
15		½	3′			
16		½	3′	18′		
17		½	3′	9′		
		Nineteenths				
2	=	10′	190′			
3		15′	20′	57′	76′	95′
4		5′	95′			
5		4′	76′			
6		4′	19′	76′		
7		3′	38′	114′		
8		3′	30′	38′	57′	95′
9		3′	12′	38′	57′	76′
10		½	38′			
11		½	19′	38′		
12		½	12′	38′	76′	114′
13		3″	57′			
14		3″	19′	57′		
15		½	4′	38′	76′	
16		½	4′	19′	38′	76′
17		½	3′	30′	57′	95′
18		½	3′	12′	57′	76′
		Twentieths				
2	=	10′				
3		10′	20′			
4		5′				
5		4′				
6		4′	20′			
7		3′	60′			
8		3′	15′			
9		3′	10′	60′		
10		½				
11		½	20′			
12		½	10′			
13		½	10′	20′		
14		½	5′			
15		½	4′			
16		½	4′	20′		
17		½	3′	60′		
18		½	3′	15′		
19		½	3′	10′	60′	

TABLE IVb. Alternative Values in P. Michigan 146 169

(i) Correct alternative

4/5	=	3″	10′	30′	3/14	=	7′	14′		12/15	=	3″	10′	30′
3/7		3′	15′	35′	5/14		4′	14′	28′	3/18		9′	18′	
3/10		5′	10′		6/14		3′	15′	35′	5/18		6′	9′	
8/10		3″	10′	30′	9/14		½	7′		11/18		½	9′	
9/12		3″	12′		4/15		5′	15′		14/18		3″	9′	
10/13		3″	13′	39′	8/15		3′	5′						

(ii) Incorrect alternatives

8/13	≠	½	13′	26′	78′		11/15	≠	½	3′	15′			
12/13		½	3′	13′	26′	78′	4/17			12′	15′	17′	34′	51′
11/14		3″	14′	28′			11/17			½	12′	17′	34′	51′
10/15		½	3′				15/17			½	4′	17′	34′	68′

TABLE IVa: Values for fractions listed in P. Akhmîm; I have excerpted from the complete list given by Baillet [1892, 24–31]. See Section II of the present paper for discussion.

TABLE IVb: Values for fractions as listed in P. Mich. 146, where these disagree with the entry in P. Akhmîm. The alternatives given in (i) are correct, but those in (ii) represent errors on the part of the scribe of P. Mich. For the complete table in P. Mich. 146 and a full listing of alternatives, see [Robbins 1936, 52-58]. (I have omitted those entries which involve scribal error in P. Akhmîm.) Note that P. Mich. gives only the last part of the sevenths table, the complete tables for eights through eighteenths, and then breaks off; but the entries for the earlier parts up to and including the sevenths can be reconstructed from entries in the extant portion.

TABLE V. Two Lists of Fractions in a Demotic Papyrus

	90ths			150ths		
[1	90′		1	150′		
2	45′		2	90′	450′	
3	30′		3	60′	300′	
4	30′	90′	4	45′	2[25′]	
5	30′	45′	5	30′		
6	15′]		6	30′	[150′]	
7	15′	90′	7	30′	90′	[450′]
8	15′	45′	8	20′	300′	
9	10′		9	30′	[45′	225′]
10	10′	90′	10	1[5′]		

Table V: Values for fractions as listed in P. British Museum 10794; see [Parker 1972, Nos. 66 – 67, pp. 72 f., Plate 24]. Dating and provenance of the papyrus are uncertain [Parker 1972, 2]. Bracketed figures follow Parker's restorations.

170 BIBLIOGRAPHY

I. Primary Sources

Codex Constantinopolitanus, E. M. Bruins, ed., 3 vols, Leyden, 1964 (I: facsimile; II: Greek text; III: translation (English) and commentary).

Hero of Alexandria, *Opera*, 5 vols, Leipzig (III, 1903: *Metrica* and *Dioptra*, H. Schöne, ed.; IV/V, 1912/1914: *Geometrica/Stereometrica*, J. L. Heiberg, ed.).

Papyri in the University of Michigan Collection, J. G. Winter, ed., Vol. III, Ann Arbor, 1936 (P. Michigan 145, F. E. Robbins, ed., pp. 34–52; P. Michigan 146, F. E. Robbins, ed., pp. 52–59; on the latter, cf. also the same editor in *Isis*, 1922, 5, pp. 20–25).

Le Papyrus mathématique d'Akhmîm, J. Baillet, ed., in *Mémoires ... de la Mission Archéologique française au Caire*, Vol. 9, Fasc. 1, Paris, 1892.

The Rhind Mathematical Papyrus (British Museum Nos. 10057, 10058), T. E. Peet, ed., London, 1923; A. B. Chace and H. P. Manning, eds., 2 vols., Oberlin, Ohio, 1927 (the latter with a "Bibliography of Egyptian Mathematics," by R. C. Archibald).

II. Discussions

Baillet, J. 1892. See *Le Papyrus Akhmîm*.

Bruchheimer, M. & Salomon, Y. 1977: Some comments on R. J. Gillings' analysis of the 2/*n* table in the Rhind Papyrus. *Historia Mathematica*, 4, 445–452.

Bruins, E. M. 1975. The part in ancient Egyptian mathematics. *Centaurus* 19, 241–251.

Gillings, R. J. 1972. *Mathematics in the time of the Pharaohs*. Cambridge, Mass.

—— 1978. Response to "Some comments. ...," *Historia Mathematica* 5, 221–227.

Heath, T. L. 1921. *A history of Greek mathematics*. 2 vols. Oxford.
Hofmann, J. E. 1934. Über die Annäherung von Quadratwurzeln bei Archimedes und Heron. *Jahresbericht der Deutschen Mathematiker-Vereinigung* 43, 187–210. (Reprinted in O. Becker, ed., *Zur Geschichte der griechischen Mathematik*. Darmstadt, 1965.)
Hultsch, F. 1901. Neue Beiträge zur ägyptischen Teilungsrechnung. *Bibliotheca Mathematica* 23, 177–184.
Neugebauer, O. 1957. *The exact sciences in antiquity*. Providence, R.I.
| Parker, R. A. 1972. *Demotic mathematical papyri*. Providence, R.I. 171
Robbins, F. E. 1922; 1936: see *Papyri in the ... Michigan Collection*.
Tannery, P. 1884. Questions héroniennes. *Bulletin des sciences mathématiques* 8_2 (= 19), 329–344, 359–376.
Thomas, I. 1957. *Selections illustrating the history of Greek mathematics*. 2 vols. London/Cambridge, Mass. (Reprint of edition of 1939–1941.)
van der Waerden, B. L. 1938. Die Entstehung der aegyptischen Bruchrechnung. *Quellen und Studien B* 4, 359–382.
—— 1954. *Science awakening*, Groningen (Reprinted, New York, 1963.)
—— 1980. The (2:*n*) table in the Rhind Papyrus. *Centaurus* 24, 259–274.

DAVID H. FOWLER

LOGISTIC AND FRACTIONS IN EARLY GREEK MATHEMATICS: A NEW INTERPRETATION

SINCE THE popularisation of techniques of decimal fractions at the end of the sixteenth century, western mathematics has drawn inspiration from the fluent manipulations of more and more general, more and more abstract, kinds of numbers.[1] I shall refer to mathematics that uses, like this, some system of numbers sufficiently general to include fractional quantities *and their arithmetic* as 'arithmetised mathematics'. For example, Mesopotamian mathematics and astronomy is arithmetised; but we have no unambiguous evidence of the influence of this Mesopotamian mathematics on Greek mathematics before the second century BC; and thereafter, in Greek texts, Mesopotamian sexagesimal numbers are found only in astronomical contexts. I am exploring a novel interpretation of early Greek mathematics in which Mesopotamian influence on Greek mathematics is minimal, even non-existent, and will not consider this sexagesimal evidence further here.

I propose that early Greek mathematics and astronomy, up to and including Archimedes, was *not* arithmetised. The argument proceeds on the following three fronts, of which I shall only deal with the second here:[2]

| (i) The development of proposals for different kinds of ratio theory ('theoretical *logistikē*') associated with mathematics, astronomy, and music theory, and an exploration of the relations between them, the whole being set against our evidence concerning early Greek mathematics and carried out using the techniques found there. (The principal texts are the curriculum in Plato's *Republic* VII; Euclid's *Elements*, Books II, IV, X, and XIII; the *parapēgma* tradition in astronomy; and the Euclidean *Sectio Canonis*.) This re-interpretation persuades me that there is no need to import Mesopotamian or our ideas of general kinds of numbers and common fractions into the interpretation of early Greek mathematics; indeed, any such interpretation is fraught with difficulties.[3] 134

(ii) An examination of our early evidence, on papyrus, of various kinds of texts involving numerical, and especially fractional, manipulations. At the outset, it

I would like to thank Wilbur Knorr for his corrections to and comments on an earlier version of this article. This acknowledgment is all the more heartfelt since his own conclusions about this topic are very different from those presented here; see KNORR 1982.

1 For example, the same people—in particular Viète, Stevin, and Harriot—are involved in popularising or using decimal fractions and in developing the then new and now ubiquitous symbolic, 'algebraic', notations of mathematics. I believe that this is much more than a coincidence; see FOWLER 1985 and KLEIN 1934–6.

2 See FOWLER 1987; Parts 1, 2, and 3 of this book correspond to the three parts of the argument.

3 For a brief illustrations of several different kinds of ratio and proportion theories and their relevance or not to early Greek mathematics, see FOWLER 1989 and 1991.

J. Christianidis (ed.), Classics in the History of Greek Mathematics, 367–380.

must be emphasised that most of this material is fragmentary and comes from commercial calculations ('practical *logistikē*') so may be irrelevant to mathematical practice,[4] but nowhere do I find any convincing evidence for the proposal that 'the Greeks[5]' used anything like our notations for common fractions and our ways of performing fractional arithmetic.[6]

(iii) A re-examination and re-interpretation of the sources and transmission of the received interpretation of early Greek mathematics, because we find a striking absence from our earliest sources of stories and interpretations that later become ubiquitous and crucial.[7]

Almost all of our *written* evidence about Greek culture has passed via Egypt, and almost all of it has been later rewritten, from the ninth century AD onwards, in a modernised Byzantine script. Numerical material in these Byzantine manuscripts is liable to have been modernised and uniformised in what might then have been considered to be unimportant ways—this applies, in particular, to the treatment of numbers and fractions. (One needs only to look at modern editions and translations, even by the most scrupulous of scholars, to see similar processes at work today.[8]) Our only earlier direct
135 | written evidence is in the form of extensive but fragmentary papyrus texts, almost all of them from the Græco-Roman administration of Egypt from the end of the fourth century BC onwards, and so first appearing only towards the end of the creative period of Greek mathematics. We should distinguish between these two different types of evidence.

We should also distinguish between theoretical mathematics and commercial material. Very little of the high, theoretical mathematical tradition is found in papyrus texts; almost none of the commercial, practical tradition appears in Byzantine manuscripts. But the middle ground between the two, represented by the Heronian corpus, is found in both kinds of texts; see, for example, the papyrus *pM.P.E.R., N.S. i 1.*

In 'theoretical *arithmētikē*', formal Greek mathematics investigates and uses the *arithmoi*. These should be conceived in some very concrete sense, for example as

(the solo), the duet, the trio, the quartet, . . .,

or in other grammatical forms, such as the adverbial 'repetition' numbers

once, twice, three-times, four-times, . . .

---

4 However I think it worth stating that everyday commercial practice can have a decisive effect on the development of mathematics. This was certainly the case for the emerging Italian mathematics of the fourteenth and fifteenth centuries, and I believe that the introduction of decimal fractions [FOWLER 1987] had an important influence on the way mathematicians now conceive of their very abstract 'real' numbers. Stevin dedicated *De Thiende*, his pamphlet of 1585 on decimal fractions, to "Astronomers, Land-meters, Measurers of Tapestry, Gaugers, Stereometers in general, Money-Masters, and to all Merchants"!

5 'Greek' means 'written in Greek', a description which takes in much of the eastern Mediterranean and beyond for more than two thousand years. It is just about as precise as use of 'Latin' for all of western Mediterranean and northern European culture would be.

6 This opinion is controversial. For another recent investigation of this same issue which arrives at very different conclusions, see KNORR 1982.

7 For example, the role of ruler-and-compass constructions; and the circumstances and impact of the discovery of incommensurability and difficulties arising out of this discovery.

8 For illustrations, see the two Documents *infra*.

So, for example, a Greek multiplication table is best described as something like:

> the duet once is the duet, the solo twice is the duet;
> the duet twice is the quartet,
> the duet three-times is the sextet, the trio twice is the sextet;
> ......

and one style of the surviving multiplication tables from the practical tradition is to commute the entries like this—not, I believe, because commutativity was problematic, as is sometimes asserted, but because what was being learned was an automatic response to the different sound patterns.

Fractional quantities are *always* expressed in practical *logistikē* using the *merē* or *moria*, 'parts', also known as 'unit fractions', 'Egyptian fractions', '*quantièmes*', etc.; I shall use the Greek transliteration '*merē*' or the translation 'parts'. After a special sign for the half, these were originally written by appending a long 'fraction indicator' to the corresponding letter numeral, thus:

the half (∠), the third (*Γ́*), the quarter (Δ́), the fifth (*É*), ...

and are now best abbreviated, I suggest, as:

$$\acute{2}, \acute{3}, \acute{4}, \acute{5}, \ldots$$

As with Egyptian fractions, we also find *ta duo merē*, 'the two parts' [Β́], our $\frac{2}{3}$; here I suggest the abbreviation $\acute{\acute{3}}$. Traces of these parts are also found, in a limited way, in the theoretical tradition.[9] Neither the Greek words nor the Greek notations contain those features which lead easily to our conception of common fractions.[10]

| Even though the idea of ratio in Greek mathematics might sometimes seem to be 136
making reference to these *merē*, mathematical ratios never seem to be conceived or manipulated arithmetically. In theoretical *logistikē*, ratios are never added, and the 'compounding', or multiplication of ratios plays a very curious and minor role in Greek mathematics before the works of Apollonius.[11] By contrast the commercial tradition is full of arithmetical calculations. The main operation which gives rise to fractional quantities is division, as in, for example:

τῶν	*ιβ*	*τὸ*	*ιζ́*	/	*∠ιβ́ιζ́λδ́νά χή*
of the	12	the	17th	is	$\acute{2}\,\acute{12}\,\acute{17}\,\acute{34}\,\acute{51}\,\acute{68}$

for what we would now[12] write as:

$$\frac{12}{17} = \frac{1}{2} + \frac{1}{12} + \frac{1}{17} + \frac{1}{34} + \frac{1}{51} + \frac{1}{68}\,.$$

---

9 See, for example, EUCLID *Elements*, Book VII, Df. 20 (Heath) = Df. 21 (Heiberg); Prop. 4–13, 37–39, etc.

10 For a clear illustration of Greek practice, see Document 1 *infra*.

11 But early music theory is built around compounding ratios. It may be difficult now for us to conceive that compounding ratios—which we assimilate into the multiplication of common fractions—may pose problems of great difficulty to mathematicians of insight. For an attempt to explain and illustrate this, and a discussion of compounding in Euclid's *Elements*, based on MUELLER 1981, see FOWLER 1987, p. 139–153.

12 The unspaced, unfriendly, majuscule script of early Greek papyri, frequently riddled with all manner of errors, contrasts strongly with the convenient, legible, compact style of Byzantine

These expressions would have been memorised, evaluated, or looked up in division tables, of which many examples have now been published.[13] The only manipulations of this basic operation that are then found are very restricted: "of *m* the nth" is seen to be the same as "of *km* the kn^{th}", and "of *m* the nth" and "of *p* the nth" can be added or subtracted to give "of $(m \pm p)$ the nth", where all of the divisions, to repeat, are expressed and still seem to be conceived as sums of *merē*. Nowhere, to my knowledge, do we get an example where two general expressions "of *m* the nth" and "of *p* the qth" are directly combined without going through some sequence of these basic manipulations.

A simple example will illustrate the point I am making. A Greek addition of the 6th of 5 and the 9th of 4 would almost always be embedded in the further information giving answers to the questions: "What *is* the 6th of 5?"—answer 2́ 3́—and "What *is* the
137 9th of 4?"—answer 3́ 9́.[14] Then 2́ 3́ and 3́ 9́ make 3̋ | 2́ 9́ which clearly is greater than 1 so, expressing 2́ as 3́ 6́, we get the answer 1 6́ 9́. A more sophisticated Greek procedure would employ the manipulations that the 6th of 5 is also equal to the 18th of 15; the 9th of 4 is also equal to the 18th of 8; so their sum will be the 18th of 23, which is 1 6́ 9́. But almost always the division are expressed and seem to be conceived as sums of *merē*.

This standard and ubiquitous phrase *tōn m to ń*, "of the *m* the nth", used to describe division is, in our papyrus texts, very occasionally abbreviated as $^{\acute{n}}_{m}$, and this is frequently described as a notation for common fractions.[15] To be quite precise, there are five known Greek or Egyptian papyrus texts *only* in which this abbreviated notation is found: *pLondon ii 265*, p. 257; *pM.P.E.R., N.S. i 1*; and three demotic texts published in *DMP*, Problems 2, 3, 10, 13, 51 and 72.[16] But in each instance, the calculations are being carried out throughout in *merē*, and in every case, the explanation can be more plausibly offered that the scribe could not remember or work out, or did not have available a table for, the particular expression *tōn m to ń*, which he then leaves unresolved and abbreviated. Then later, in the Byzantine manuscript tradition, what are

---

manuscripts and the care that is taken today over well-chosen notations, clear lay-out, and careful proof-reading. I have pointed to this contrast here by deliberately doing nothing to help the reader's interpretation of this Greek fraction beyond trying to get the letters and accents right. One can take this observation further: early written texts appear to act as *aides mémoires* to the normal substrate of communication, which was oral. For our scholarly activity today, the balance is reversed.

13 A catalogue of papyrus division tables in FOWLER 1987, p. 271–9, lists 45 division tables, 3 in hieratic, 3 in demotic, and 39 in Greek, with several more in course of publication.

14 E. M. Bruins has repeatedly and forcibly argued that expressions for the nth of *m* which contain the part *ń* should not be permitted in theoretical *logistikē* so the 9th of 4 should not be expressed as 3́ 9́, but rather as something like 4́ 6́ 3́6́. But our division tables and other calculations show little or no sign of following this rule; for example, this expression for the 9th of 4 and the earlier expression for the 17th of 12 are both taken from such division tables.

15 See, for example, the very influential HEATH 1921, vol. 1, p. 44–45:

> In this system the numerator of any fraction is written in the line, with the denominator *above* it, without accents or other marks . . .; the method is therefore simply the reverse of ours, but equally convenient.

In fact, in every known instance of this abbreviation in early Greek papyrus texts, the 'denominator' does have a fraction indicator which the editors did not transcribe. For a full discussion, see FOWLER 1987, p. 248–68.

16 See also the discussion *DMP*, p. 8–10.

again interpreted as notations from common fractions may also originally have been abbreviations for this same sterotyped phrase *tōn m to ń*, perhaps now with a relaxation of the hitherto rigorously imposed convention that these divisions should always be resolved as a sum of *merē*. With this, the earlier practice of expressing a quotient as a sum of *merē* can shade off into the later and more sophisticated tradition in which the question (*sc.* the interrogative *quotiens*, "How many times?") can be treated and manipulated exactly as if it were its own answer—our common fractions.[17]

Two final remarks will elaborate on what I meant at the beginning of this article when I said that the underlying intuitions of Greek mathematics were not arithmetised. Our answer "$\frac{4}{9}$" to the question "What is the 9th of 4?" conveys, by itself, very little further information; what adds flesh to it are the manipulations, especially the arithmetical operations, that we can perform on these common fractions.[18] On the one
hand, these arithmetical operations | seem to play almost no role in early Greek math- 138
ematics, and are found in this common fractional form neither in the theoretical nor the practical tradition; on the other hand, an important operation in the reconstruction of my proposed Greek ratio theory (or theoretical *logistikē*) is the very *un*arithmetical operation of combining $m{:}n$ and $p{:}q$ to give the mean $(m + p){:}(n + q)$. Second, I give in FOWLER 1987 some very detailed descriptions of ways of thinking about ratio that do not depend on ideas of fractional numbers. The two most important examples described there are 'anthyphairetic ratios' (based on the so-called Euclidean algorithm) and 'astronomical ratios' (based on the addition procedure underlying *Elements* Book V, Df. 5, the Eudoxan definition of proportionality); and each of these definitions provides both a clear description of the concept of ratio from which propositions can then be proved deductively, and also a practical base which enables us to perform manipulations like finding good approximations to different kinds of geometrical ratios (for example ratios of sides of squares, or cubes, or the perimeter and diameter of a circle). However the procedure for performing the arithmetical operations is not obvious for either of these examples. Hence the way towards a route in which geometry is arithmetised by assigning to each line a 'length' (its ratio to an assigned line), then to each plane region an 'area' (its ratio to the square on the assigned line) in such a way that the area of a rectangle is also equal to the product of the length of its sides; and then going further to translate the geometry into arithmetic; then ultimately to abstract this arithmetic into algebra—this route does not present itself for

[17] An exception should be made in this, as in many other, respects for the case of Diophantus, who seems to be able to handle division in a way very close to our concept of common fractions. But, judging from the manuscript of the Arabic translation of Diophantus, which seems to be closer to Diophantus' own text than our Byzantine Greek manuscripts, Diophantus does not seem to use any 'fractional notation' for division.

[18] I can no longer remember my own struggle to understand common fractions, so my children's early efforts were all the more surprising and vivid. There was no difficulty in explaining or understanding fractional notation ("$\frac{4}{9}$?"—"That's four ninths. Take a pie and cut it into nine pieces ..."), but I do not think I ever clearly explained to them or they ever successfully understood from anybody just what $\frac{4}{9} \times \frac{5}{6}$ was, let alone $\frac{4}{9} \div \frac{5}{6}$. Of course, they wanted, at that stage, to learn how to perform the manipulations correctly, and quickly stopped bothering about what they might mean!

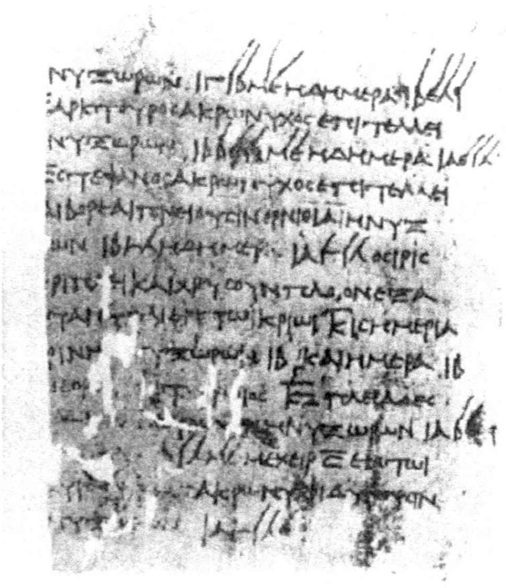

Fr. (*b*), 2nd hand. Col. iv. PLATE VIII.

ἡ νὺξ ὡρῶν ιγι´β´μ´ε´, ἡ δ´ ἡμέρα ιβ´ε´λ´q̇.
ι]ς Ἀρκτοῦρος ἀκρόνυχος ἐπιτέλλει,
ἡ] νὺξ ὡρῶν ιββ´ι̣´ε̣´μ´ε´, ἡ δ´ ἡμέρα ιαθ´ι´λ´.
κ]ς Στέφανος ἀκρόνυχος ἐπιτέλλει
κ]αὶ βερίαι πνείουσιν ὀρνιθίαι. ἡ νὺξ
ὡρ]ῶν ιβ∠λ´, ἡ δ´ ἡμέρα ιαγ´ι´λ´. Ὄσιρις
π]εριπλεῖ καὶ χρυσοῦν πλοῖον ἐξά-
γε]ται. Τῦβι (ε) ἐν τῶι Κριῶι. κ ἰσημερία
ἐα]ρινή, ἡ νὺξ ὡρῶν ιβ καὶ ἡμέρα ιβ,
[κ]α̣ὶ ἑορ[τ]ὴ Φ̣ιτω̣ράιος. κζ Πλειάδες
ἀκρόνυχ[οι, δύνου[σ]ιν, ἡ νὺξ ὡρῶν ιαβ´ς´q̇,
[ἡ] δ´ ἡμέρα [ι]βι´λ´μ´ε´. Μεχεὶρ ς ἐν τῶι
[Τ]αύρωι. Ὑάδες ἀκρόνυχοι δύνουσιν,
[ἡ] νὺξ ὡρῶν ια∠ι´λ´ε´,

55. l. ιε for ιβ. 57. ιε corr. from ε. 65. ς corr. 68. l. λ´q for λ´ε.

55–205. '(Choiak 1st:) . . . The night is $13\frac{4}{45}$ hours, the day $10\frac{41}{45}$. 16th, Arcturus rises in the evening. The night is $12\frac{34}{45}$ hours, the day $11\frac{11}{45}$. 26th, Corona rises in the evening, and the north winds blow which bring the birds. The night is $12\frac{8}{15}$ hours and the day $11\frac{7}{15}$. Osiris circumnavigates, and the golden boat is brought out.

'Tubi 5th, the sun enters Aries. 20th, spring equinox. The night is 12 hours and the day 12 hours. Feast of Phitorois. 27th, Pleiades set in the evening. The night is $11\frac{38}{45}$ hours, the day $12\frac{7}{45}$.

Figure VIII.1. Parapēgma (c. -300)—Papyrus Hibeh, i 27; a facsimile; *b* transcription; *c* translation

consideration since the very first step of performing arithmetic on these ratios seems, even to many mathematicians today, to be intractable.[19]

ANNEXE

Document 1. pHibeh i 27*: An early example of Greek fractional notation*

This papyrus is perhaps our earliest physically surviving Greek scientific or semi-scientific text. It is a *parapēgma*, a characteristic kind of early Greek calendar, with an introduction that is loosely associated with Eudoxus' astronomy, and it can be dated from internal evidence to c. 300 BC. It has easily accessible and full descriptions in its *editio princeps* [GRENFELL & HUNT 1906], and in [FOWLER & TURNER 1983] and [FOWLER 1987], so I shall restrict my description here to a brief summary of some of its features. Most of these can be seen in the first line of this extract [Figure VIII. 1 *a–c*], which reads in translation:

> "The night is 13 1́2́ 4́5́ hours, the day 10 3̋ 5́ 3́0 9́0."

| We see: 140

- The distinctive fraction indicators, and the absence of any other diacritical signs anywhere in the text apart from the superior horizontal bars, not transcribed by the editors, which are used to distinguish the dates from the other numerals and other kinds of text (clear examples on lines 62, 64, and 66).
- The letter *beta* used to denote 2, as in *Ί Β́* for 1́2́ in the first fraction, and exactly the same *Β́* used for 3̋ (i.e. $\frac{2}{3}$, *ta duo merē*), as in the second fraction.
- The undoubted error in the first fraction, which should read *ΙΓ ΊΈ Μ́Έ* for 13 1́5́ 4́5́. (This can be deduced from the simple arithmetical structure of the calendar and from the fact that the day and night together should add up to 24 hours.) Errors are very common in numerical texts.

Document 2. Archimedes' Measurement of a Circle: *The Evolution of a Text*

These excerpts will illustrate how a text has been corrupted, then refined, corrected, and modernised during its transmission down to us. During such a process of transmission, the notations and conventions for handling numbers and, especially, fractions, and the ways of setting out formal mathematical arguments, are very susceptible to irreversible evolution.

We possess no original versions of any Greek mathematical text, and most texts survive only in the form of Byzantine minuscule copies made from the mid-ninth century AD onwards. Our principal source of Archimedes' works is such a Byzantine copy, made in the ninth or tenth century but lost in the sixteenth century, and its contents have been reconstructed from four surviving copies of it made between 1450 and 1564. One folio of one of these copies (Bibliothèque Nationale gr. 2360, f° 23 v°:

---

[19] I say "seems" because, despite the common opinion today, there is an easy algorithm, due to R. W. Gosper, for adding and multiplying continued fraction expansions (today's equivalent of anthyphairetic ratios). It is described in FOWLER 1987, p. 354–360.

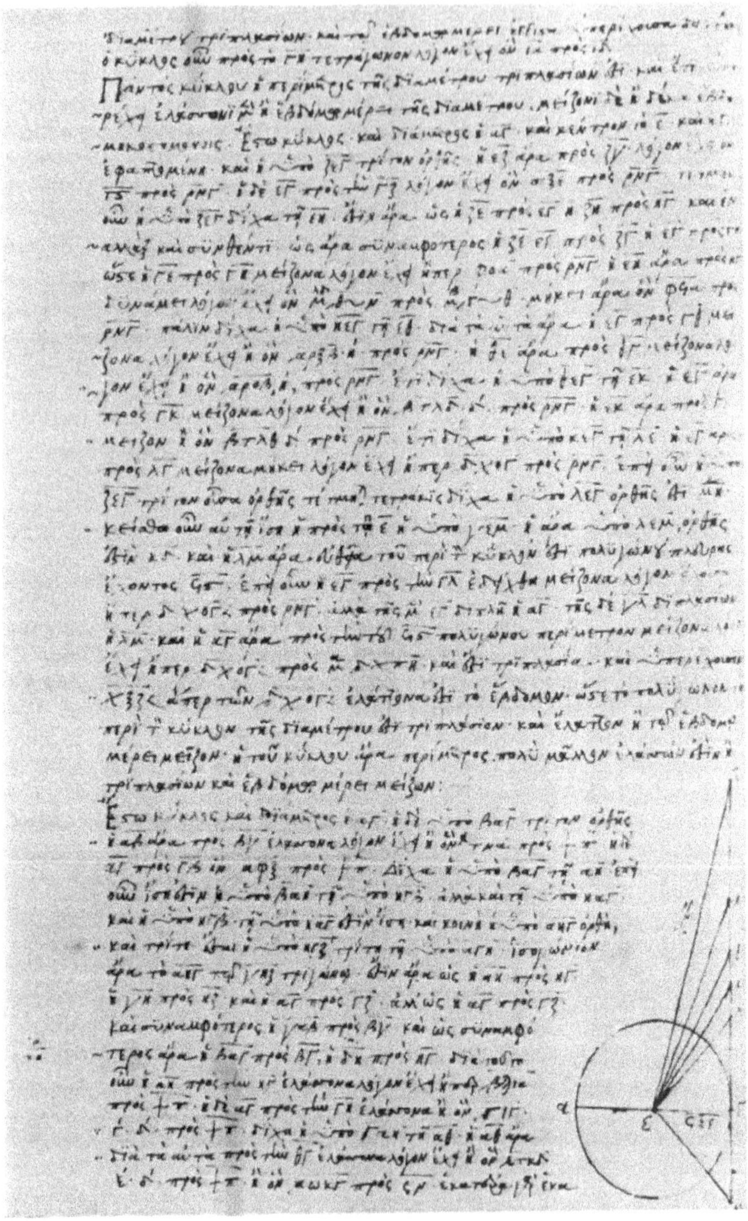

Figure VIII.2.a. Archimedes, La Mesure du cercle, 3: 10–20
Paris, Bib. nat. gr. 2360: f° 23v°

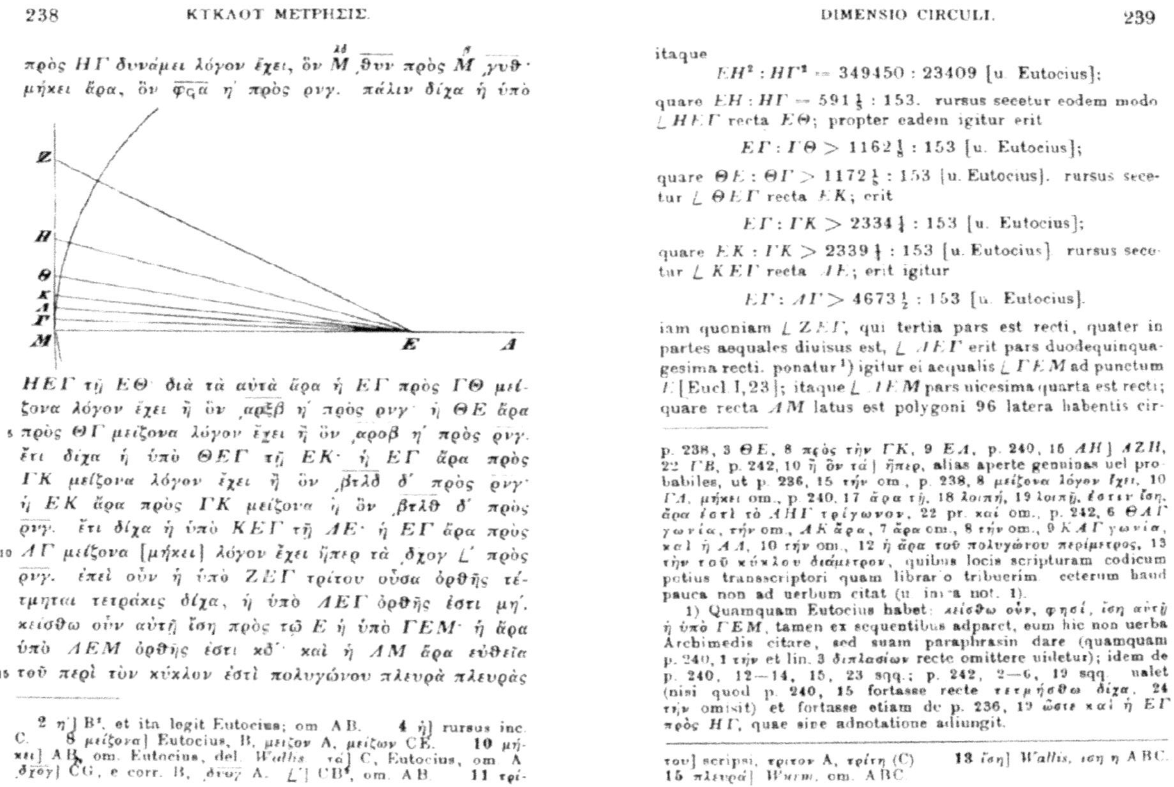

238 ΚΥΚΛΟΥ ΜΕΤΡΗΣΙΣ.

πρὸς ΗΓ δυνάμει λόγον ἔχει, ὃν $\overset{\iota\delta}{M}$,θυν πρὸς $\overset{\beta}{M}$,γυθ· μήκει ἄρα, ὃν ,φϙα η′ πρὸς ϱνγ. πάλιν δίχα ἡ ὑπὸ

ΗΕΓ τῇ ΕΘ· διὰ τὰ αὐτὰ ἄρα ἡ ΕΓ πρὸς ΓΘ μείζονα λόγον ἔχει ἢ ὃν ,αϱξβ η′ πρὸς ϱνγ· ἡ ΘΕ ἄρα πρὸς ΘΓ μείζονα λόγον ἔχει ἢ ὃν ,αϱοβ η′ πρὸς ϱνγ. ἔτι δίχα ἡ ὑπὸ ΘΕΓ τῇ ΕΚ· ἡ ΕΓ ἄρα πρὸς ΓΚ μείζονα λόγον ἔχει ἢ ὃν ,βτλδ δ′ πρὸς ϱνγ· ἡ ΕΚ ἄρα πρὸς ΓΚ μείζονα ἢ ὃν ,βτλθ δ′ πρὸς ϱνγ. ἔτι δίχα ἡ ὑπὸ ΚΕΓ τῇ ΛΕ· ἡ ΕΓ ἄρα πρὸς ΛΓ μείζονα [μήκει] λόγον ἔχει ἤπερ τὰ ,δχογ L′ πρὸς ϱνγ. ἐπεὶ οὖν ἡ ὑπὸ ΖΕΓ τρίτου οὖσα ὀρθῆς τέτμηται τετράκις δίχα, ἡ ὑπὸ ΛΕΓ ὀρθῆς ἐστι μη′. κείσθω οὖν αὐτῇ ἴση πρὸς τῷ Ε ἡ ὑπὸ ΓΕΜ· ἡ ἄρα ὑπὸ ΛΕΜ ὀρθῆς ἐστι κδ′· καὶ ἡ ΛΜ ἄρα εὐθεῖα τοῦ περὶ τὸν κύκλον ἐστὶ πολυγώνου πλευρὰ πλευρὰς

2 η′] B², et ita legit Eutocius; om. AB. 4 ἡ] rursus inc. C. 8 μείζονα] Eutocius, B, μειζον A, μείζων CE. 10 μήκει] AB, om. Eutocius, del. Wallis. τά] C, Eutocius, om. A. ,δχογ] CG, e corr. B, ,δυογ A. L′] CB², om. AB. 11 τρί-

DIMENSIO CIRCULI. 239

itaque

$EH^2 : H\Gamma^2 = 349450 : 23409$ [u. Eutocius];

quare $EH : H\Gamma = 591\frac{1}{8} : 153$. rursus secetur eodem modo $\angle H E \Gamma$ recta $E\Theta$; propter eadem igitur erit

$E\Gamma : \Gamma\Theta > 1162\frac{1}{8} : 153$ [u. Eutocius];

quare $\Theta E : \Theta\Gamma > 1172\frac{1}{8} : 153$ [u. Eutocius]. rursus secetur $\angle \Theta E \Gamma$ recta EK; erit

$E\Gamma : \Gamma K > 2334\frac{1}{4} : 153$ [u. Eutocius];

quare $EK : \Gamma K > 2339\frac{1}{4} : 153$ [u. Eutocius]. rursus secetur $\angle K E \Gamma$ recta ΛE; erit igitur

$E\Gamma : \Lambda\Gamma > 4673\frac{1}{2} : 153$ [u. Eutocius].

iam quoniam $\angle Z E \Gamma$, qui tertia pars est recti, quater in partes aequales diuisus est, $\angle \Lambda E \Gamma$ erit pars duodequinquagesima recti. ponatur[1]) igitur ei aequalis $\angle \Gamma E M$ ad punctum E [Eucl. I, 23]; itaque $\angle \Lambda E M$ pars uicesima quarta est recti; quare recta ΛM latus est polygoni 96 latera habentis cir-

p. 238, 3 ΘΕ, 8 πρὸς τὴν ΓΚ, 9 ΕΛ, p. 240, 16 ΛΗ] ΛΖΗ, 22 ΓΒ, p. 242, 10 ἢ ὃν τά] ἤπερ, alias aperte genuinas uel probabiles, ut p. 236, 15 τήν om., p. 238, 8 μείζονα λόγον ἔχει, 10 ΓΛ, μήκει om., p. 240, 17 ἄρα τῇ, 18 λοιπῇ, 19 λοιπῇ, ἔστιν ἴση, ἄρα ἐστὶ τὸ ΛΗΓ τρίγωνον, 22 pr. καί om., p. 242, 6 ΘΑΓ γωνία, τήν om., ΑΚ ἄρα, 7 ἄρα om., 8 τήν om., 9 ΚΑΓ γωνία, καὶ ἡ ΑΛ, 10 τήν om., 12 ἡ ἄρα τοῦ πολυγώνου περίμετρος, 13 τὴν τοῦ κύκλου διάμετρον, quibus locis scripturam codicum potius transscriptori quam librario tribuerim. ceterum haud pauca non ad uerbum citat (u. infra not. 1).

1) Quamquam Eutocius habet: κείσθω οὖν, φησί, ἴση αὐτῇ ἡ ὑπὸ ΓΕΜ, tamen ex sequentibus adparet, eum hic non uerba Archimedis citare, sed suam paraphrasin dare (quamquam p. 240, 1 τήν et lin. 3 διπλασίων recte omittere uidetur); idem de p. 240, 12—14, 15, 23 sqq.; p. 242, 2—6, 19 sqq. ualet (nisi quod p. 240, 15 fortasse recte τετμήσθω δίχα, 24 τήν omisit) et fortasse etiam de p. 236, 19 ὥστε καὶ ἡ ΕΓ πρὸς ΗΓ, quae sine adnotatione adiungit.

του] scripsi, τριτον A, τρίτη (C) 13 ἴση] Wallis, ιση η ABC. 15 πλευρά] Wurm, om. ABC.

Figure VIII.2.b. Archimedes, Measurement of a circle, 3: 10–20. Heiberg edition

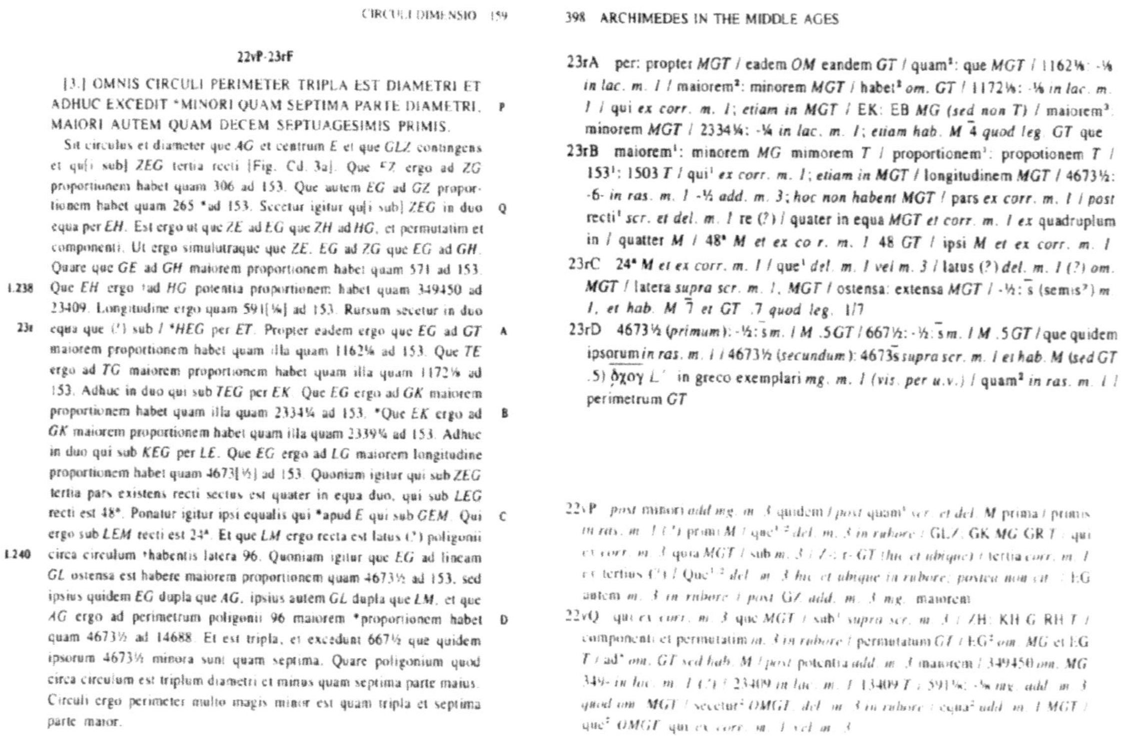

CIRCULI DIMENSIO 159

22vP-23rF

[3.] OMNIS CIRCULI PERIMETER TRIPLA EST DIAMETRI ET
ADHUC EXCEDIT *MINORI QUAM SEPTIMA PARTE DIAMETRI, P
MAIORI AUTEM QUAM DECEM SEPTUAGESIMIS PRIMIS.

Sit circulus et diameter que *AG* et centrum *E* et que *GLZ* contingens
et qu[i sub] *ZEG* tertia recti [Fig. Cd. 3a]. Que *EZ* ergo ad *ZG*
proportionem habet quam 306 ad 153. Que autem *EG* ad *GZ* propor-
tionem habet quam 265 *ad 153. Secetur igitur qu[i sub] *ZEG* in duo Q
equa per *EH*. Est ergo ut que *ZE* ad *EG* que *ZH* ad *HG*, et permutatim et
componenti. Ut ergo simulutraque que *ZE*, *EG* ad *ZG* que *EG* ad *GH*.
Quare que *GE* ad *GH* maiorem proportionem habet quam 571 ad 153.
1.238 Que *EH* ergo †ad *HG* potentia proportionem habet quam 349450 ad
23409. Longitudine ergo quam 591[⅛] ad 153. Rursum secetur in duo
23r equa que (?) sub / *HEG per *ET*. Propter eadem ergo que *EG* ad *GT* A
maiorem proportionem habet quam illa quam 1162⅛ ad 153. Que *TE*
ergo ad *TG* maiorem proportionem habet quam illa quam 1172⅛ ad
153. Adhuc in duo qui sub *TEG* per *EK*. Que *EG* ergo ad *GK* maiorem
proportionem habet quam illa quam 2334¼ ad 153. *Que *EK* ergo ad B
GK maiorem proportionem habet quam illa quam 2339¼ ad 153. Adhuc
in duo qui sub *KEG* per *LE*. Que *EG* ergo ad *LG* maiorem longitudine
proportionem habet quam 4673[½] ad 153. Quoniam igitur qui sub *ZEG*
tertia pars existens recti sectus est quater in equa duo, qui sub *LEG*
recti est 48ª. Ponatur igitur ipsi equalis qui *apud *E* qui sub *GEM*. Qui C
ergo sub *LEM* recti est 24ª. Et que *LM* ergo recta est latus (?) poligonii
1.240 circa circulum †habentis latera 96. Quoniam igitur que *EG* ad lineam
GL ostensa est habere maiorem proportionem quam 4673½ ad 153, sed
ipsius quidem *EG* dupla que *AG*, ipsius autem *GL* dupla que *LM*, et que
AG ergo ad perimetrum poligonii 96 maiorem *proportionem habet D
quam 4673½ ad 14688. Et est tripla, et excedunt 667½ que quidem
ipsorum 4673½ minora sunt quam septima. Quare poligonium quod
circa circulum est triplum diametri et minus quam septima parte maius.
Circuli ergo perimeter multo magis minor est quam tripla et septima
parte maior.

398 ARCHIMEDES IN THE MIDDLE AGES

23rA per: propter *MGT* / eadem *OM* eandem *GT* / quam¹: que *MGT* / 1162⅛: -⅛ *in lac. m. 1* / maiorem²: minorem *MGT* / habet² *om. GT* / 1172⅛: -⅛ *in lac. m. 1* / qui *ex corr. m. 1*; *etiam in MGT* / EK: EB *MG (sed non T)* / maiorem³: minorem *MGT* / 2334¼: -¼ *in lac. m. 1*; *etiam hab. M* 4̄ *quod leg. GT* que

23rB maiorem¹: minorem *MG* mimorem *T* / proportionem¹: propotionem *T* / 153¹: 1503 *T* / qui¹ *ex corr. m. 1*; *etiam in MGT* / longitudinem *MGT* / 4673½: -6- *in ras. m. 1* -½ *add. m. 3*; *hoc non habent MGT* / pars *ex corr. m. 1* / *post* recti¹ *scr. et del. m. 1* re (?) / quater in equa *MGT et corr. m. 1 ex* quadruplum in / quatter *M* / 48ª *M et ex corr. m. 1* 48 *GT* / ipsi *M et ex corr. m. 1*

23rC 24ª *M et ex corr. m. 1* / que¹ *del. m. 1 vel m. 3* / latus (?) *del. m. 1 (?) om. MGT* / latera *supra scr. m. 1, MGT* / ostensa: extensa *MGT* / -½: s̄ (semis?) *m. 1, et hab. M* 7̄ *et GT* .7 *quod leg.* 1/7

23rD 4673½ (*primum*): -½: s̄ *m. 1 M* .5 *GT* / 667½: -½: s̄ *m. 1 M* .5 *GT* / que quidem ipsorum *in ras. m. 1* / 4673½ (*secundum*): 4673s̄ *supra scr. m. 1 et hab. M (sed GT* .5) δχογ L′ in greco exemplari *mg. m. 1 (vis. per u.v.)* / quam² *in ras. m. 1* / perimetrum *GT*

22vP *post* minori *add mg. m. 3* quidem / *post* quam¹ *scr. et del. M* prima / primis *in ras. m. 1 (?)* primi *M* / que¹·² *del. m. 3 in rubore* / GLZ: GK *MG* GR *T* / qui *ex corr. m. 1* quia *MGT* / sub *m. 3* / Z-: t- *GT (hic et ubique)* / tertia *corr. m. 1 ex* tertius (?) / Que¹·² *del. m. 3 hic et ubique in rubore; postea non cit.* / EG autem *m. 3 in rubore* / *post* GZ *add. m. 3 mg.* maiorem

22vQ qui *ex corr. m. 3* que *MGT* / sub¹ *supra scr. m. 3* / ZH: KH *G* RH *T* / componenti et permutatim *m. 3 in rubore* / permutatum *GT* / EG² *om. MG et* EG *T* / ad² *om. GT sed hab. M* / *post* potentia *add. m. 3* maiorem / 349450 *om. MG* 349- *in lac. m. 1 (?)* / 23409 *in lac. m. 1* 13409 *T* / 591⅛: -⅛ *mg. add. m. 3 quod om. MGT* / secetur² *OMGT, del. m. 3 in rubore* / equa² *add. m. 1 MGT* / que² *OMGT* qui *ex corr. m. 1 vel m. 3*

Figure VIII. 2c. Archimedes, Measurement of a circle. Translation by William of Moerbeke, l. 3–28

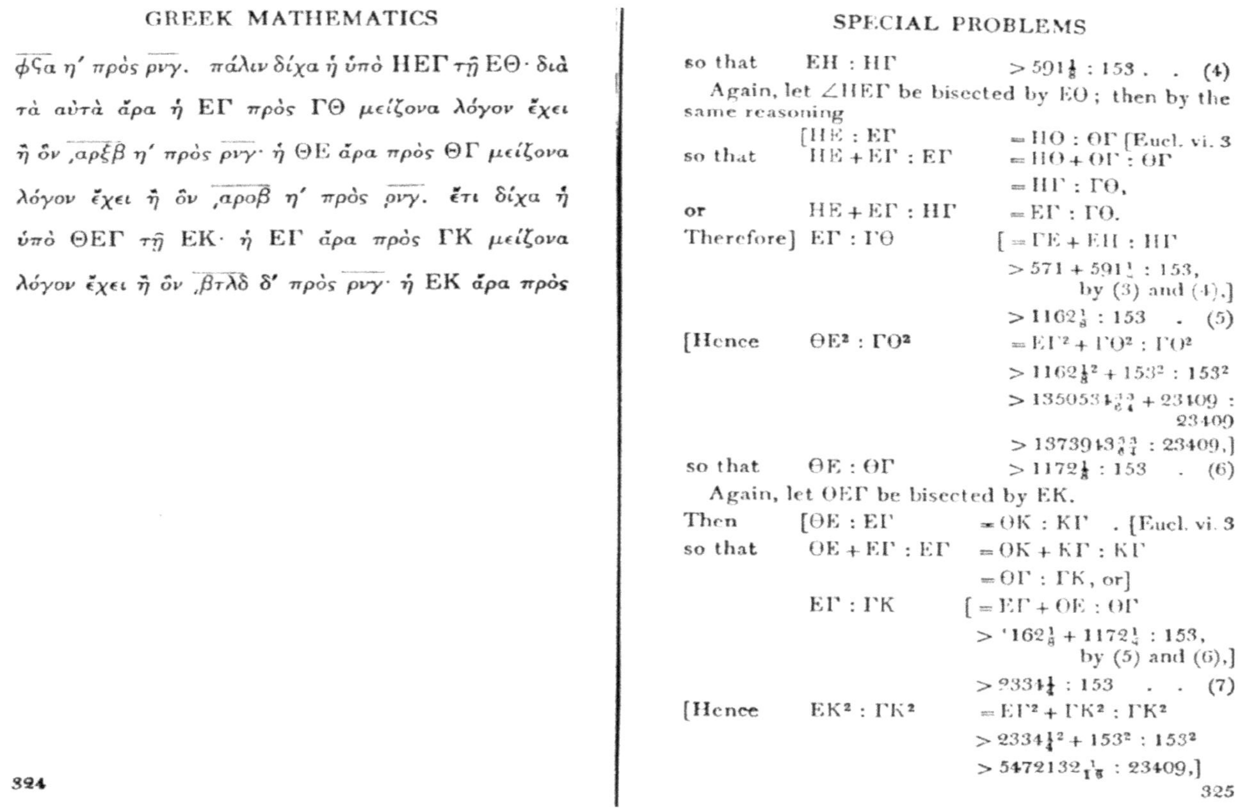

GREEK MATHEMATICS

$\overline{\phi\varsigma\alpha}$ η′ πρὸς $\overline{\rho\nu\gamma}$. πάλιν δίχα ἡ ὑπὸ ΗΕΓ τῇ ΕΘ· διὰ τὰ αὐτὰ ἄρα ἡ ΕΓ πρὸς ΓΘ μείζονα λόγον ἔχει ἢ ὃν $\overline{,\alpha\rho\xi\beta}$ η′ πρὸς $\overline{\rho\nu\gamma}$· ἡ ΘΕ ἄρα πρὸς ΘΓ μείζονα λόγον ἔχει ἢ ὃν $\overline{,\alpha\rho o\beta}$ η′ πρὸς $\overline{\rho\nu\gamma}$. ἔτι δίχα ἡ ὑπὸ ΘΕΓ τῇ ΕΚ· ἡ ΕΓ ἄρα πρὸς ΓΚ μείζονα λόγον ἔχει ἢ ὃν $\overline{,\beta\tau\lambda\delta}$ δ′ πρὸς $\overline{\rho\nu\gamma}$· ἡ ΕΚ ἄρα πρὸς

324

SPECIAL PROBLEMS

so that ΕΗ : ΗΓ $> 591\frac{1}{8} : 153$ . . (4)

Again, let ∠ΗΕΓ be bisected by ΕΘ; then by the same reasoning

[ΗΕ : ΕΓ = ΗΘ : ΘΓ [Eucl. vi. 3

so that ΗΕ + ΕΓ : ΕΓ = ΗΘ + ΘΓ : ΘΓ

= ΗΓ : ΓΘ,

or ΗΕ + ΕΓ : ΗΓ = ΕΓ : ΓΘ.

Therefore] ΕΓ : ΓΘ [= ΓΕ + ΕΗ : ΗΓ

$> 571 + 591\frac{1}{8} : 153$, by (3) and (4).]

$> 1162\frac{1}{8} : 153$. (5)

[Hence ΘΕ² : ΓΘ² = ΕΓ² + ΓΘ² : ΓΘ²

$> 1162\frac{1}{8}^2 + 153^2 : 153^2$

$> 1350534\frac{33}{64} + 23409 : 23409$

$> 1373943\frac{33}{64} : 23409$,]

so that ΘΕ : ΘΓ $> 1172\frac{1}{8} : 153$. (6)

Again, let ΘΕΓ be bisected by ΕΚ.

Then [ΘΕ : ΕΓ = ΘΚ : ΚΓ . [Eucl. vi. 3

so that ΘΕ + ΕΓ : ΕΓ = ΘΚ + ΚΓ : ΚΓ

= ΘΓ : ΓΚ, or]

ΕΓ : ΓΚ [= ΕΓ + ΘΕ : ΘΓ

$> 1162\frac{1}{8} + 1172\frac{1}{8} : 153$, by (5) and (6),]

$> 2334\frac{1}{4} : 153$ . . (7)

[Hence ΕΚ² : ΓΚ² = ΕΓ² + ΓΚ² : ΓΚ²

$> 2334\frac{1}{4}^2 + 153^2 : 153^2$

$> 5472132\frac{1}{16} : 23409$,]

325

Figure VIII.2.d. Archimedes, Measurement of a circle, 3: 10–20. Ivor Thomas, *Greek Mathematical Works*

Figure VIII.2a) is shown together with a page of Heiberg's edition of Archimedes (*Opera Omnia*. . .: Figure VIII.2b), in the apparatus of which the reconstructed manuscript is referred to as A. In addition, there is the now inaccessible Constantinople palimpsest which contains the *Method*; Heiberg incorporated some readings from this (manuscript C), and more could probably now be read using ultraviolet light. A careful literal Latin translation of the text was made by William of Moerbeke in 1269 (Heiberg's manuscript B, Clagett's manuscript *O*). This has now been re-edited in [CLAGETT 1964–84, vol. 2], and part of one page of his edition, with its apparatus, is also reproduced [Figure VIII.2c]. William appeared to have great trouble with the numbers in this text so sometimes he left a gap when he wrote the main text (referred to in Clagett's apparatus as *m.1*) and wrote the Greek in the margin; he would then write the Latin in translation in the text (as *m.2*) and erase the marginal Greek—but it can still be read under ultraviolet light (Clagett's *vis. per u.v.*; there are no examples in this excerpt). Clagett's apparatus also gives variant reading of later copies of William's manuscript (manuscripts *G, M*, and *T*); see his text for details of these.

All of these manuscripts are manifestly corrupt (for example, the propositions are in
141 the wrong order and numbered incorrectly); the text must differ | from that written by Archimedes (for example it shows no trace of Archimedes' Doric dialect); and several of the 'correct' numbers are only found in sixteenth century annotations to William's translation (Clagett's *m.3*). Moreover, Archimedes' translation is usually presented along with Eutocius' commentary of the sixth century AD. See, for example, how Heiberg repeatedly refers to Eutocius in his translation, and how, in the third extract, from the Loeb Classical Library's *Selections Illustrating* … *Greek Mathematics* [THOMAS 1939], six lines of Archimedes' text have been expanded to twenty-six lines of translation by inserting material, set off by square brackets, based on Eutocius [Figure VIII.2d]. Also observe how, conversely, the phrase *meizona logon ekhei ē hon* has three times, in this short extract, been abbreviated to an inequality sign >; and how the first fraction on this page, φϙα ή [591 8′] on line 2, has been silently corrected in the Greek. This fractional part ή does not appear in any of the manuscripts used by Heiberg (see the apparatus of the Budé edition edited by Mugler), nor is it in the Latin translation by William (where it has been added by the sixteenth century annotator). It is not in the Latin translation by Plato of Tivoli, so it may not have been in the Arabic version he worked from either, though it is in the Latin translation by Gerard of Cremona which was based on a more elaborate Arabic text than our surviving Greek text [CLAGETT 1964–84, vol. I, chapter II].[20] Also Eutocius' commentary includes it. I observe, without committing myself to any further comment here, that 591 7′ is a better approximation of the right kind than 591 8′.

Such corrections and modernisations are indispensible for making texts accessible and comprehensible, but they may distort our perception of the intentions, methods, and techniques of their original authors.[21]

[20] I would like to thank Wilbur Knorr for pointing this out.

[21] Further discussion of Archimedes' calculations can be found in FOWLER 1987, p. 54–57; 240–246. There is also an annotated bibliography by Knorr on Archimedes in the recently reprinted DIJKSTERHUIS 1929, and the entire corpus of manuscripts of Archimedes' *Measurement of a Circle*—in Greek, Arabic, Hebrew and Latin—is analyzed in detail in KNORR 1989.

REFERENCES 146

Primary Sources

ARCHIMEDES

Measurement of a Circle [Edition: J. L. HEIBERG, *Opera Omnia cum Commentariis Eutocii*, revised ed., (3 vols.), Teubner, Leipzig, 1910–15, vol. I, p. 232–242 (Reprint: Teubner, Stuttgart, 1972)—see also CLAGETT 1964–84 and THOMAS 1939.]

EUCLID

Elements [Edition: J. L. HEIBERG; revision: E. S. Stamatis, *Euclides Elementa* (5 vols. in 6 parts), Teubner, Leipzig, 1969–1977]

DMP

R. A. PARKER *Demotic Mathematical Papyri* (Brown Egyptological Studies 7), Lund Humphries (for Brown University Press), London, 1972

pHibeh i 27

[Edition: B. P. GRENFELL & A. S. HUNT, *The Hibeh Papyrus I,* Egypt Exploration Society, London, 1906, p. 138–157]

pLondon ii 265

[Edition: F. G. KENYON, H. I. BELL, *et al., Greek Papyri in the British Museum* (7 vols. to 1974), British Museum, London, 1893–, p. 257–264]

pM.P.E.R., N.S. i 1

[Edition: H. GERSTINGER & K. VOGEL, *Mitteilungen aus der Nationalbibliothek in Wien: Papyrus Erzherzog Rainer* (Neue Series) 1, Österreischischen Staatsdruckerei, Vienna, 1932, p. 11–76]

Secondary Sources

CLAGETT (Marshall)

1964–84 *Archimedes in the Middle Ages* [esp. vols. I: *The Arabo-Latin Tradition* and II: *Translations from the Greek by William of Moerbeke*], American Philosophical Society, Philadelphia

DIJKSTERHUIS (E. J.)

1929 *Archimedes*, Munksgaard, Amsterdam [Reprint with an appendix: "Archimedes after Dijksterhuis: A Guide to Recent Studies" by W. R. Knorr, Princeton University Press, 1987]

FOWLER (David H.) 147

1985 "400 Years of Decimal Fractions", *Mathematics Teaching* 110, p. 20–21; 111, p. 30–31

1987 *The Mathematics of Plato's Academy; A New Interpretation*, Clarendon Press, Oxford

1989 "*Logos* (Ratio) and *Analogon* (Proportion) in Plato, Aristotle, and Euclid" in J. PETITOT (ed.), *Logos et Théorie des Catastrophes* Patiño, Geneva, p. 443–72

1991 "Ratio and Proportion in Early Greek Mathematics" in A.C. BOWEN (ed.), *Science and Philosophy in Classical Greece*, Institute for Research in Classical Philosophy and Science, Pittsburgh

FOWLER (David H.) & TURNER (Eric G.)

1983 "Hibeh Papyrus i 27: An Early Example of Greek Arithmetical Notation", *Historia Mathematica* 10, p. 344–359

HEATH (Thomas L.)

1921 *A History of Greek Mathematics* (2 vols.), Clarendon Press, Oxford [Reprint: Dover Publications, New York, 1981]

KLEIN (Jacob)

1934–6 "Die griechische Logistik und die Enstehung der Algebra", *Quellen und Studien zur Geschichte der Mathematik* (Abteilung B) 3/1, p. 18–105; 3/2, p. 122–235 [Quoted from the English translation: *Greek Mathematical Thought and the Origin of Algebra*, (translator E. Brann) with an Appendix containing VIETA, *Introduction to the Analytical Art* (translator J. Winfree Smith), M.I.T. Press, Cambridge (USA), 1968]

KNORR (Wilbur R.)
1982 "Techniques of Fractions in Ancient Greece and Egypt", *Historia Mathematica* 9, p. 133–171
1989 *Textual Studies in Ancient and Medieval Geometry*, Birkhäuser, Basel/Boston/Berlin

MUELLER (Ian)
1981 *Philosophy of Mathematics and Deductive Structure in Euclid's* Elements, M.I.T. Press, Cambridge (USA)

THOMAS (Ivor)
1939 *Selections Illustrating the History of Greek Mathematics*, Vol. 1, Loeb Classical Library, London/Cambridge (USA)

PART 6

METHODOLOGICAL ISSUES IN THE HISTORIOGRAPHY OF GREEK MATHEMATICS

Texts selected and introduced by Sabetai Unguru

SABETAI UNGURU

INTRODUCTION

The chapter on Methodological Issues surveys summarily the methodological debate initiated with the publication in 1975 of Sabetai Unguru's article "On the Need to Rewrite the History of Greek Mathematics" in the *Archive for History of Exact Sciences*. It includes, in addition to that article, the responses, in the same journal, of B.L. van der Waerden and André Weil, as well as Unguru's rejoinder to those, and other, responses (for example, Hans Freudenthal's), published in *Isis*.

The central issue raised by Unguru's article is that of the need to deal historically with past mathematics in general, and with Greek mathematics in particular, by resisting the common temptation of emptying, as it were, the "pure" mathematical content of ancient texts from their ancient receptacles, a disgorging which, allegedly, leaves the content unaffected by merely purifying it of its formal dejecta. That this is an historically and philosophically faulty procedure, distortive of the true character of ancient mathematics, was shown, at great length, in "On the Need" article. As it was to be expected, prominent representatives of the traditional school of historians of mathematics reacted violently to the call to rewrite the history of ancient Greek mathematics on sane historical foundations, a methodology, commonly accepted in other fields of the history of ideas. Their responses brought into relief once more all that is wrong with the traditional way of writing the history of ancient mathematics. Unguru's rejoinder in *Isis*, the editor of the *Archive*, Clifford Truesdell, having refused strenuously to publish anything else from Unguru's pen, pointed out the weaknesses of the historiographic position of the traditionalists, and their failure to engage specifically and frontally the arguments put forward in the "On the Need" article.

In a nutshell, the issue between the traditionalists, who barge into foreign mathematical worlds through the mathematical door, and the new historians of mathematics, who insinuate themselves into ancient mathematical cultures through what one can call the historical door, is the position of the former that form and content are independent variables in the mathematical domain that can be separated arbitrarily without thereby damaging the identity and wholeness of ancient texts, while the latter question this arbitrary separation, pointing out the errors and distortions to which it necessarily leads and exposing the blatant anachronism that is its inseparable companion.

It turns out that the mathematical and historical approaches are mutually antagonistic and that no compromise is possible between the mathematical and historical methodological principles. Adopting one or the other has fateful consequences for one's research, effectively determining the nature of the results reached and the tenor of the inferences used in reading them.

Tel Aviv University

J. Christianidis (ed.), Classics in the History of Greek Mathematics, 383–384.

REFERENCES

Unguru, S.: "On the Need to Rewrite the History of Greek Mathematics," *Archive for History of Exact Sciences*, 15 (1975), pp. 67-114.

Unguru, S.: "History of Ancient Mathematics: Some Reflections on the State of the Art," *Isis*, 70 (1979), pp. 555-565.

Van der Waerden, B.L.: "Defence of a 'Shocking' Point of View," *Archive for History of Exact Sciences*, 15 (1976), pp. 199-210.

Weil, A.: "Who Betrayed Euclid? (Extract from a Letter to the Editor)", *Archive for History of Exact Sciences*, 19 (1978), pp. 91-93.

SABETAI UNGURU

ON THE NEED TO REWRITE THE HISTORY OF GREEK MATHEMATICS

History is the most fundamental science, for there is no human knowledge which cannot lose its scientific character when men forget the conditions under which it originated, the questions which it answered, and the function it was created to serve. A great part of the mysticism and superstition of educated men consists of knowledge which has broken loose from its historical moorings.

BENJAMIN FARRINGTON[1]

It would not occur to the modern mathematician, who uses algebraic symbols, that one type of geometrical progression [*i.e.*, 1, 2, 4, 8] could be more perfect or better deserving of the name than another. For this reason algebraic symbols should not be employed in interpreting such a passage as ours [*i.e.*, Plato, *Timaeus*, 32A, B].

FRANCIS M. CORNFORD[2]

Any historian of mathematics conscious of the perils and pitfalls of Whig history quickly discovers that the translation of past mathematics into modern symbolism and terminology represents the greatest danger of all. The symbols and terms of modern mathematics are the bearers of its concepts and methods. Their application to historical material always involves the risk of imposing on that material, a content it does not in fact possess.

MICHAEL S. MAHONEY[3]

The previous string of quotations is (most certainly) **not** illustrative of the ways in which the history of mathematics has traditionally been written. The authors of the quotations themselves have not always practiced what they occa|sionally preached.[4] 68
Indeed, the discipline is exceedingly rich in works written (as it were) as a living illustration of P. W. BRIDGMAN'S exhortation:

> ... the past has meaning only in terms of the present. The impartial recovery of the past, uncontaminated by the influence of the present, is held up as a professional ideal, and a criterion of technical competence is the degree to which this ideal is reached. This ideal is, I believe, impossible of attainment, and cannot even be formulated without involvement with meaningless verbalisms.[5]

---

1 *Greek Science Its Meaning For Us* (Harmondsworth: Penguin Books, 1953), 311.

2 *Plato's Cosmology* (New York: The Liberal Arts Press, 1957), 49.

3 *The Mathematical Career of Pierre de Fermat (1601–1665)* (Princeton, N.J.: Princeton University Press, 1973), XII–XIII.

4 Ironically, the very works of FARRINGTON and MAHONEY mentioned above are cases in point for the very popular syndrome referred to by HEINE in the following phrase: 'Sie predigen öffentlich Wasser, Und drinken heimlich Wein'; the difference being, however, that, in this instance, both the 'preaching' and the 'drinking' take place openly, in the public domain. For an analysis of FARRINGTON'S work see LUDWIG EDELSTEIN, 'Recent Trends in the Interpretation of Ancient Science', *Roots of Scientific Thought A Cultural Perspective*, P.P. WIENER & A. NOLAND (eds.) (New York: Basic Books, 1957), 90–121; as to MAHONEY'S book, I will be dealing with it in a future essay review in *FRANCIA-Forschungen zur westeuropäischen Geschichte*, the journal of the *Institut Historique Allemand* in Paris.

5 'Impertinent Reflections on the History of Science', *Philosophy of Science*, **17** (1950), 63–73, at 64; it is also there that BRIDGMAN says (among other things): 'It seems to me that there is a very real danger in a too assiduous devotion to the historical point of view ...' (*ibid.*, 72). Without

J. Christianidis (ed.), Classics in the History of Greek Mathematics, 385–431.

The situation is particularly scandalous in the history of ancient and medieval mathematics. It is in truth deplorable and sad when a student of ancient or medieval culture and ideas must familiarize himself first with the notions and operations of *modern* mathematics in order to grasp the meaning and intent of modern commentators dealing with ancient and medieval mathematical texts. With very few and notable exceptions, Whig history *is* **history** in the domain of the history of mathematics; indeed, it is still, largely speaking, the standard, acceptable, respectable, 'normal' kind of history, continuing to appear in professional journals and scholarly monographs. It is *the* way to write the history of mathematics. And since this is the case, one is faced with the awkward predicament of having to learn the language, techniques, and ways of expression of the modern *mathematician* (typically the manufacturer of 'historical' studies) if one is interested in the *historical* exegesis of *pre-modern* mathematics; for it is a fact that the representative audience of the mathematician fathering 'historical' studies consists of historians (or people who identify themselves as historians) rather than mathematicians. The latter look condescendingly upon their (usually older) colleagues in their new and somewhat strange hypostasis which seems to indicate to the working mathematician an implicit, but public, confession of professional (*i.e.*, mathematical) impotence.

As to the goal of these so-called 'historical' studies, it can easily be stated in one
69 sentence: to show how past mathematicians hid their modern ideas and proce|dures under the ungainly, *gauche*, and embarrassing cloak of antiquated and out-of-fashion ways of expression; in other words, the purpose of the historian of mathematics is to unravel and disentangle past mathematical texts and transcribe them into the modern language of mathematics, making them thus easily available to all those interested.

If the preceding description seems unconvincing and written by a reckless partisan of hyperbole, the balance of this paper should correct this mistaken impression. Indeed it is the purpose of this paper to show what is historically wrong with the traditional way the history of ancient Greek mathematics has been written and to call to the new generation of historians of Greek mathematics to rewrite that history on a new and historically sane basis.

I

One of the central concepts for the understanding of ancient Greek mathematics has customarily been, at least since the time of PAUL TANNERY and HYERONIMUS GEORG ZEUTHEN, the concept of 'geometric algebra'.[6] What it amounts to is the view

---

denying the pregnant philosophical problems stemming from the reconstruction of the past, and accepting the obvious conclusion that 'the impartial recovery of the past', *etc.* is indeed an impossible ideal, it does not follow that abandoning irrevocably this unattainable ideal is tantamount to an abandonment of the historical method. Indeed, to repeat a truism, the fact the historian knows that it is in principle impossible to relive the past and that his reconstruction is inherently deficient and inadequate represents for him the utmost challenge *to try* and look at the past through sympathetic and understanding eyes and to achieve a reconstruction which does no patent violence to that which is to be reconstructed. That there is something to be reconstructed and understood is taken for granted by any mentally healthy historian worth his salt.

6 *Cf.*, for instance PAUL TANNERY, *Mémoires Scientifiques*, **1**, *Sciences Exactes Dans L'Antiquité*, J.-L. HEIBERG & H.G. ZEUTHEN (eds.) (Toulouse: Edouard Privat and Paris: Gauthier-Villars,

that Greek mathematics, especially after the discovery of the 'irrational' by the Pythagorean school, is *algebra* dressed up, primarily for the sake of rigor, in geometrical garb. The reasoning of Greek mathematics, the line of attack of its various problems, the solutions provided to those problems, *etc.* all are essentially *algebraic*, though, to be sure, for reasons that have never been fully elaborated, attired in geometrical accouterments. We are, then, not only authorized to look for the algebraic 'subtext' (so to speak) of any geometrical proof, but it is indeed wise (historically!) always to transcribe the geometrical content of any proposition in the symbolic language of modern algebra, especially when the former is particularly cumbersome and awkward, while the recourse to the latter always makes the logical structure of the proof clear and convincing, without thereby losing | anything not only in generality 70
but also in any possible *sui generis* features of the ancient way of doing things.[7]

1912), 254–280. Characteristically, the title of this study is "De la solution géométrique des problèmes du second degré avant Euclide"! See also, *Mémoires Scientifiques*, **3** (1915), 158–187 and 244–250. For Zeuthen's views see his *Die Lehre von den Kegelschnitten im Altertum* (Hildesheim: Georg Olms, 1966, being a photographic reproduction of the Copenhagen, 1886, edition), 1–38 and *Geschichte der Mathematik im Altertum und Mittelalter* (Copenhagen: Andr. Fred. Höst & Sön, 1896), 32–64.

P. Tannery & H.G. Zeuthen were not the originators of the concept of 'geometric algebra'. Pierre de la Ramée seems to deserve the doubtful credit for this invention. It was he, apparently, who 'discerned' that the algebraic art must underlie some parts of Euclid's *Elements* (Books II and VI) and, perhaps, also Greek analysis. (*Cf.* Michael S. Mahoney, 'Die Anfänge der algebraischen Denkweise im 17. Jahrhundert', *Rete*, **1** (1971), 15–31, especially p. 25.)

It is quite interesting (and, as will become clear later, strongly supportive of one argument of this paper) that practically all the founders of modern mathematics (Viète, Descartes, and Fermat) followed Ramus in his belief that algebra lies at the root of Greek analysis! Remarkably enough, William Oughtred in the seventeenth century, in his most famous mathematical textbook, *Clavis mathematicae*, is also of the opinion that algebra can serve as a means of understanding difficult problems in Euclid, Archimedes, Apollonios, and Diophantos (*cf. ibid.*, n. 49, 28). Our nineteenth and twentieth century historians of mathematics can indeed be proud of their lengthy and aristocratic mathematical lineage; in truth they have made Oughtred's 'insight' the keystone of their methodological and interpretive approach!

7 Thus spake Tannery: 'Je veux parler de tout le livre *X* d'Euclide et de la théorie des irrationnelles qui s'y trouve renfermée... Ce n'est, rien moins que le détail complet de la solution géométrique de l'équation bicarrée et le commencement de celle de l'équation tricarée, avec l'invention d'une nomenclature destinée à suppléer au défaut de notations' (*op. cit.*, **1** 263). In a historical appendix written for his brother's (Jules) *Notions de Mathématiques* (Paris: Ch. Delagrave, 1903), Paul emphatically states: 'Quoique leurs [*i.e.*, the ancients] procédés d'exposition aient toujours présenté, par rapport aux nôtres, des différences essentielles, leur méthode zététique était au fond beaucoup plus voisine de la nôtre qu'on n'est porté à le croire au premier abord. C'est que, tandis que leur symbolisme algébrique [*sic* !] se développait péniblement, ils en avaient, dès le quatrième siècle avant notre ère, constitué un pour la géométrie, ... Ce langage présentait en même temps tous les avantages de l'emploi des lettres dans l'analyse de Viète [!] au moins pour les puissances 2 et 3. *Ils avaient dès lors pu constituer, probablement dès le temps des premiers pythagoriciens, une véritable algèbre géométrique pour les premiers degrés, avec la conscience très nette qu'elle correspondait exactement à des opérations numeriques'* (*op. cit.*, **3**, 167, my italics). Then he goes on to say: 'Quoiqu'ils [*i.e.*, the Greeks] ne se soient pas élevés... au concept général des coordonées, leur façon de considérer les coniques est tout à fait analogue à celle de notre géométrie analytique [!]... *L'équation qu'ils établissent* [!] revient à la forme générale moderne: $y^2 = p x \pm p/a\, x^2$ ...

Les procédés de transformation des coordonnées chez les anciens sont imparfaits, par suite du défaut de conception générale du problème... Mais ces procédés n'en existent pas moins' (*ibid.*, 168, my italics). Such examples could be multiplied *ad nauseam*.

71 | In other words, there is nothing unique and (ontologically) idiosyncratic concerning the way in which ancient Greek mathematicians went about their proofs, which might be lost in the process of translation from the geometrical to the algebraic

---

H. G. Zeuthen has stated his views repeatedly and in various places. The most cogent and complete statement, however, appears in the two works quoted in the previous note. Thus, in *Die Lehre von den Kegelschnitten im Altertum*, Zeuthen entitles his first chapter 'Voraussetzungen und Hülfsmittel; Proportionen und geometrische Algebra' (*op. cit.*, 1). It is there that, in a paragraph remarkable for its *non sequiturs*, Zeuthen says that though the Greeks did not possess the concept of a system of coordinates, they nevertheless used 'rechtwinklige und schiefwinklige Koordinaten', and though Algebra was unknown to them, the historian must establish what they used in its stead (*op. cit.*, 2)! He continues by saying that the Greek theory of proportions '... enthielt Sätze, welche es ermöglichen die wichtigsten algebraischen Operationen ... auszuführen' (*ibid.*, 4) and 'Auf diese Weise hat man einen Apparat, mit Hülfe dessen man die Zusammensetzung algebraischen Grössen ausdrücken kann' (*ibid.*). Furthermore, after the discovery of incommensurability by Pythagoras or one of his disciples, '... wurde der unmittelbaren Anwendung von Zahlen und daran geknüpften Proportionen in der Geometrie, welche Anspruch auf Stringenz sollte erheben dürfen, ein Halt geboten ... Indessen konnte es nicht fehlen, dass man praktisch Zahlen und Proportionen auch auf die Geometrie anwandte, wenn auch mit dem Bewusstsein, dass man, um die gewonnenen Resultate anerkannt zu sehen, dieselben hinterher [!] auf einem anderen Wege beweisen müsse' (*ibid.*, 5).

But, the modern manipulation of proportions is a direct outgrowth of the existence of a symbolic mathematical language in which the symbols themselves are manipulated and operated on. On the other hand, 'Das Altertum hatte allerdings keine Zeichensprache, aber ein Hülfsmittel *zur Veranschaulichung dieser sowie anderer Operationen besass man in der geometrischen Darstellung und Behandlung allgemeiner Grössen und der mit ihnen vorzunehmenden Operationen*' (*ibid.*, 6). And now comes the pregnant statement:

> In dieser Weise entwickelte sich eine geometrische Algebra, wie man sie nennen kann, da dieselbe als Algebra teils allgemeine Grössen, irrationale sowohl wie rationale, behandelt, teils andere Mittel als die gewöhnliche Sprache benutzt um ihr Verfahren anschaulich zu machen und dem Gedächtnisse einzuprägen. Diese geometrische Algebra hatte zu Euklids Zeiten eine solche Entwicklung erreicht, dass sie dieselben Aufgaben bewältigen konnte *wie unsere Algebra solange diese nicht über die Behandlung von Ausdrücken zweiten Grades hinausgeht*, ein Gebiet, welches sie auch, ... in ihrer Anwendung auf die Lehre von den Kegelschnitten ausgefüllt hat. Eine solche Anwendung entspricht der Anwendung unserer Algebra in der analytischen Geometrie (*ibid.*, 7).

Having dealt with the ancient theory of proportions, Zeuthen passes on to 'geometric algebra' proper and establishes that the Greeks had the means to represent the equation $\alpha x + \beta y + \gamma z + \cdots = d$ as follows:

> ... auf einer Geraden neben einander Stücke abgetragen würden, die in den Verhältnissen α, β, γ ... zu x, y, z ... stehen. Der Abstand zwischen dem Anfangspunkt und dem Punkt, den man durch successives Abtragen der Stücke erreicht, wird dann d sein. Auf ähnliche Weise kann man verfahren, wenn andere Vorzeichen in der Gleichung [!] vorkommen. Ebenso wie wir bei der jetzt gebräuchlichen Darstellung im Gedächtnis behalten müssen, was jeder einzelne von unseren Buchstaben bedeutet, ebenso mussten die Alten behalten, was das für Stücke waren, die man abgetragen hatte; *dann aber hatten die Alten ebenso wie wir eine Darstellung der Gleichung* ...
>
> Mit Hülfe einer solchen Darstellung werden Gleichungen ersten Grades auf Wegen gelöst, welche viel mit unserer algebraischen Behandlung gemeinsam haben (*ibid.*, 10, my italics).

This should suffice here. More about Tannery's and Zeuthen's views on 'geometric algebra' will come to the fore in the balance of this study.

language; the main reason for this being that the ancient mathematical reasonings and structures *are* indeed substantially algebraic. As B.L. van der Waerden put it:

> Theaetetus and Apollonius were at bottom algebraists, they thought algebraically even though they put their reasoning in a geometric dress.
>
> Greek algebra was a geometric algebra, a theory of line segments and of areas, not of numbers. And this was unavoidable as long as the requirements of strict logic were maintained. For "numbers" were integral or, at most, fractional, but at any rate rational numbers [?], while the ratio of two incommensurable line segments cannot be represented by rational numbers. It does honor to Greek mathematics that it adhered inexorably to such logical consistency.[8]

Nevertheless, adopting such a procedure necessarily implied imposing very stringent limitations upon the kind of problems one could solve and, therefore, upon the results one could achieve. Van der Waerden, following in the footsteps of his illustrious predecessors but adding pinch, sharpness, and pungency to their sometimes (by comparison) mild, moderate, and gentle statements, goes on to ascribe to the ancient Greek mathematicians (and he is **not** referring to Diophantos here) the solution of '*equations*' in their geometrical propositions:

> Equations of the first and second degree can be expressed clearly in the language of geometric algebra, and, if necessary, also those of the third degree. But to get beyond this point, one has to have recourse to the bothersome [?] tool of proportions.
>
> | Hippocrates, for instance, reduced the cubic equation [!] $x^3 = V$ to the proportion 72
>
> $$a:x = x:y = y:b,$$
>
> and Archimedes wrote the cubic [?]
>
> $$x^2(a - x) = bc^2$$
>
> in the form
>
> $$(a - x):b = c^2:x^2$$
>
> In this manner one can get [*Who* can get?] to equations of the fourth degree; ... But one can not get any farther; besides, one has to be a mathematician of genius, thoroughly versed in transforming proportions with the aid of geometric figures, to obtain results by this extremely cumbersome method [?]. Any one can use our algebraic notation, but only a gifted mathematician can deal with the Greek theory of proportions and with geometric algebra.[9]

---

[8] B. L. van der Waerden, *Science Awakening* (New York: John Wiley & Sons, Inc., 1963), 265–66. In this instance too, as in so many others (*cf.* footnote 15, *passim*), van der Waerden mirrors Otto Neugebauer's views on the fundamentally algebraic character of Apollonius' *Conics*. Thus, Neugebauer thinks that '... auch in der scheinbar rein geometrischen Theorie der Kegelschnitte vieles steckt, das uns Aufschlüsse geben kann, über die sozusagen latente algebraische Komponente in der klassischen griechischen Mathematik' ('Apollonius-Studien', 216; full reference in footnote 15). And, speaking of the structure of the *Conics*, Neugebauer says: 'Die Behandlung des Evolutenproblems ohne jede Benutzung von Infinitesimal methoden *aus rein algebraischen Betrachtungen* ist überhaupt ein besonderes Glanzstück des ganzen Werkes. *Ebenso ist das ganze Arsenal von Identitäten und zugehörigen Ungleichungen aus Buch VII ... rein algebraischer Natur'* (*ibid.*, 218, n. 4, my emphasis).

[9] *Op. cit.*, 266. It is clear that *not* 'any one can use our algebraic notation'. For somebody to use it, he must *have* such a notation at his disposal in the first place and he must know to use it, *i.e.*, he

73 | What are the grounds for such a view and what are its underlying assumptions? Let me state from the outset that I cannot find any *historically* gratifying basis for this generally accepted view, which, I think, owes its origin, in part, to the fact that those who have been writing the history of mathematics in general, and that of ancient mathematics (including Greek) in particular, have typically been mathematicians, abreast of the modern developments of their discipline, who have been largely unable to relinquish and discard their laboriously acquired mathematical competence when dealing with periods in history during which such competence is historically irrelevant and (I dare say) outright anachronistic. Such an approach, furthermore, stems from the unstated assumption that mathematics is a *scientia universalis*, an algebra of thought containing universal ways of inference, everlasting structures, and timeless, ideal patterns of investigation which can be identified throughout the history of civilized man and which are *completely independent of the form in which they happen to appear at a particular juncture in time*. In other words, such an interpretation takes it for granted that *form* and *content* do not constitute an integral whole in mathematics, that, as a matter of fact, **content** is independent of **form**, and that one can, therefore, transcribe

---

must be aware of and conversant with the algebraic way of thinking! The position exemplified by the above quotation is also (though, perhaps, not always presented with the same bluntness) that of PAUL TANNERY and ZEUTHEN. Thus, TANNERY begins his study on the geometrical solution of second degree problems before EUCLID with the following statement: 'Si nous nous proposons de parler de la solution géométrique des problèmes du second degré avant Euclide, il est clair cependant que ce n'est que dans l'oeuvre de ce dernier que nous pouvons trouver l'exposition de cette solution' (*Mém. Scient.*, **1**, 254) and ZEUTHEN, who, according to his own confession, adopts the point of view of TANNERY (*Die Lehre*, note 1, 5), says: 'Um ... zu erfahren, wie weit die Bekanntschaft der Alten mit gemischten quadratischen Gleichungen und deren Lösung oder Reduktion auf rein quadratische Gleichungen sich erstreckte, wird es zweckmässig sein zu prüfen, welche Gestalt die quadratische Gleichung in der Sprache der geometrischen Algebra annehmen musste ...' (*ibid.*, 15). ZEUTHEN also categorically proclaims: 'Wir sehen also, dass die Alten alle Formen der Gleichung zweiten Grades behandelt haben ...' (*Gesch. der Math. im Alt. und Mittel.*, 50). This is also he position of NEUGEBAUER in his 'Apollonius Studien' when playing havoc among APOLLONIUS' geometrical propositions, by transcribing the latter's rhetorical descriptions into the language of algebraic, manipulative symbolism. There is very little of APOLLONIUS in NEUGEBAUER'S transcriptions as even a glance at HEIBERG'S edition (or VER EECKE'S translation) will show. 'Bei den vorangehend geschilderten Überlegungen', says NEUGEBAUER, 'bin ich nirgends anders von den Apolloniusschen Text abgegangen als durch die äussere Form' (*op. cit.*, 250). As if this is not precisely the supreme sin a historian of mathematics may perform! (More on the relation between form and content in mathematics, below.) Furthermore, this statement is not even true, since NEUGEBAUER has not respected (among other things) APOLLONIUS' division into propositions. NEUGEBAUER goes on: 'Es wäre selbstverständlich auch bei der griechischen Ausdrucksweise der Beweise ohne weiteres möglich gewesen [?] bei analogen Beweisen gleiche Bezeichnungen einzuführen. [This is retrospective, hindsightful history! It is obvious for us that identical notation in analogous proofs is preferable to arbitrary notation, only because for us matters of notation are more than mere name-calling, baptizing! We *operate on* our notations; for the Greeks this was utterly inconceivable.] Dass die antike Mathematik so gänzlich unempfindlich gegen diese uns so sehr lästige Unsystematik gewesen ist, zeigt, dass man sehr vorsichtig damit sein muss, wenn man behauptet, die Unübersichtlichkeit der Beweise habe ihre Weiterentwicklung schliesslich verhindert. Offenbar [?] überblickte man das Buchstabengewirr einer Konstruktion mit derselben Selbstverständlichkeit vie wir heute komplizierte Formeln' (*ibid.*, 250, n. 28). *This is incredible!* It presupposes that formulae exist somehow independently of their actual, *i.e., written* presence, that they are 'hidden' within the Greek notational chaos, to be merely disentangled by the penetrating eye of the modern historian of mathematics!

with impunity ancient mathematical texts by means of modern symbolic algebraic notation in order to gain an 'insight' into their otherwise 'cumbersome' content.

Furthermore, exactly because this content (like the inert gases) is essentially unaffected by its formal surroundings, the ability of the modern mathematician to uncover it and give it a 'palatable' (*i.e.*, modern) form constitutes not only the best *modern* reading of ancient 'burdensome' and 'oppressive' mathematical texts but also the only *correct* reading and, at the same time, *the* proof that this is what the ancient mathematician had in mind when he put down (in an awkward fashion, to be sure) for posterity his mathematical thoughts. Thus, if *we* see in a number of EUCLIDEAN propositions in the *Elements* quadratic equations, then this is what EUCLID had in mind when he enunciated and proved those propositions geometrically; if *we* can identify equations of the fourth degree in APOLLONIUS,[10] this is what APOLLONIUS had in mind, though this identification of the algebraic kernel of APOLLONIAN thought is not always easy and requires, obviously, modern mathematical training:

> Reading a proof in Apollonius requires extended and concentrated study. Instead of a concise algebraic formula, one finds a long sentence, in which each line segment is indicated by two letters which have to be located in the figure. *To understand the line of thought, one is compelled to transcribe these sentences in modern concise formulas.*[11]

Besides, and as a natural corollary of such a view, when the mathematician succeeds in showing that two apparently unrelated mathematical texts, belonging to two alien cultures and to two different time periods have the same algebraic[12] content, in spite of their totally different formal outlook and context – like, say, a Babylonian tablet involving lists and manipulations of numbers and the Greek | geometrical 74
propositions dealing with application of areas –, it becomes legitimate to inquire into possible influences, questions of priority, ways of transmission, *etc.* This is precisely the way followed by OTTO NEUGEBAUER in his *Vorgriechische Mathematik*,[13] which has recently been reprinted, and (in greater detail and cogency) by VAN DER WAERDEN in *Science Awakening*,[14] where it is argued, exclusively on the basis of this type of reasoning, that Greek 'geometric algebra' is nothing but 'Babylonian algebra' in geometrical attire![15]

---

10 VAN DER WAERDEN, *ibid.*

11 *Ibid.*, italics provided.

12 This qualifier is actually superfluous since this is the only possible content according to the view expounded here.

13 *Vorlesungen über Geschichte der antiken mathematischen Wissenschaften*, Band I: *Vorgriechische Mathematik* (Berlin-Heidelberg-New York: Springer-Verlag, 1969); this is an unrevised reprint of a book first published in 1934.

14 *Cf. op. cit.*, 82–147.

15 *Ibid.*, 118–124. VAN DER WAERDEN has essentially adopted *in toto* NEUGEBAUER'S approach and findings in the latter's three "Studien zur Geschichte der antiken Algebra'; the first study (I) appeared in *Quellen und Studien zur Geschichte der Mathematik, Astronomie und Physik*, **2** (1932), Abteilung B: *Studien*, 1–27; the second (II), carrying the additional title 'Apollonius-Studien', came out in the same volume, same section (*Studien*) of the same journal, pp. 215–254; finally, the third (III), entitled 'Zur geometrischen Algebra', saw the light of the day in volume **3** (1936) of the same journal, same section, pp. 245–259. NEUGEBAUER summarized his well-known views on Greek 'geometric algebra' in his *The Exact Sciences in Antiquity* (Princeton, N.J.: Princeton University Press, 1952); I have used the second edition of this work (New York: Dover

75 | Is this an acceptable position? As a historian, I must answer this question with an emphatic '***no***'! *This position*, I happen to believe, *is historically unacceptable*. Among other obvious drawbacks, it fails to answer the most stringent and manifest question, *viz.*, why did Greek mathematics stick throughout its development to the 'cumbersome', 'awkward', 'highly difficult' method of 'geometric algebra' with its application of areas, transformation of proportions by means of geometrical figures, *etc.*?

Publications, Inc., 1969). In *The Exact Sciences*, NEUGEBAUER confesses that there is no documentary evidence for what he calls 'Oriental influence' on theoretical Greek mathematics (p. 147). His 'working hypothesis', however, is: 'the theory of irrational quantities and the related theory of integration [?] are of purely Greek origin, but the contents of the "geometrical algebra" utilize results known in Mesopotamia' (*ibid.*). The only evidence for this mathematically beautiful 'working hypothesis' that NEUGEBAUER is able to produce is the fact that both the Babylonian numerical-arithmetical material and some Greek geometrical propositions lend themselves rather easily to an algebraic rendering which, when performed, shows them to be identical. There is no question, indeed, about their identity for NEUGEBAUER (and any modern mathematician) who has at his disposal the algebraic language and the rules of translations *into* it. The real question is: Did the ancients (Babylonians and Greeks) know the algebraic language and the rules of translating *it* into either number manipulations or geometric propositions? Pointing out that the problem of application of areas, which he calls 'the central problem of the geometrical algebra' (p. 149) is 'rather difficult to motivate' (*ibid.*) in any other way than by translating it into the language of equations (the same procedure as that followed in the transcription and solution of so-called Babylonian 'problems of second degree'), NEUGEBAUER is really telling us something about his own motivations, idiosyncrasy, and background rather than anything significant about the ancients. Why is the problem of the application of areas a 'strange geometrical problem' (*ibid.*, 150)? What is *strange* about it? Was it strange for the Greeks? Why does it need any motivation? Why is the Babylonian method of solution by reduction to the 'normal form' not in need of any motivation? NEUGEBAUER acknowledges that attempts to explain the problem of application of areas independently of algebraic translations have been made, but he claims that the algebraic explanation is 'by far the most simple and direct explanation' (*ibid.*). Fully aware that simplicity does not amount to historical proof, NEUGEBAUER rests his case on the plausibility of his algebraic interpretation and on the historical likelihood of contacts between the Babylonian and the Greek civilizations in Hellenistic times (*ibid.*, 150–151). NEUGEBAUER has dealt at great length with the historical problem of the alleged relations between Babylonian and Greek mathematics in the 'Schlussbemerkungen' to his third study on ancient algebra (*q.v.*). It is there that after stating that in the realm of elementary geometry as well as in the realms of the theory of proportions and the theory of equations (!), Babylonian mathematics contains the entire substantive material on which Greek mathematics continued to erect its structures, NEUGEBAUER points out that, in spite of the total lack of explicit citations of sources, he is convinced of the indubitable influence of Babylonian on Greek mathematics. His conviction is based on the following three factors: 1) The specific evidence of the relation between the two (by which he means their identity when submitted to the same algebraic treatment); 2) The historical fact of a widely spread Hellenistic culture reaching the 'Orient'; and, finally, 3) The numerous Greek citations referring to Greeks having studied in the 'Orient' (*loc. cit.*, 258). According to NEUGEBAUER, the period during which contacts between Greek and Babylonian mathematics took place should be taken as the period from PLATO to HIPPARCHUS. A notable result of these contacts is the Greek geometric algebra, which was later applied to conic sections, achieving there its most remarkable results (*ibid.*) A few questions naturally arise. If the Greeks were so smart to take over 'Babylonian algebra' and geometrize it, why did they adopt the Babylonian dainties rather selectively? Specifically, why did they not adopt a positional number system from the Babylonians rather than clinging to a dreadful one? Why did they fail to see the great 'advantages' of the Babylonian approach to astronomy, sticking exclusively to geometrical models rather than to arithmetical sequences? Why did they not deal with the 'irrational' like the Babylonians? (There would not have been then any 'crisis of the irrational'!) The above is by no means an exhaustive list of troublesome questions stemming from NEUGEBAUER'S hypothesis.

This question gains even more in acuity when one keeps in mind that the perpetrators of the view embodied in the concept of 'geometric algebra' presume without any qualms (and rest assured) that there has been an underlying algebraic edifice to Greek geometry throughout its development. Why, then, did this algebraic framework remain all the time in the background, hidden, camouflaged, concealed?

Answering by pointing out that the Greek system of numeration employed the letters of the Greek alphabet as number symbols and thus made those letters unavailable to the mathematician to serve him as algebraic symbols, leaving at his disposal 'only' the geometrical representation, is missing the point entirely by begging the question.[16] If a necessary ingredient of the algebraic way of thinking is the existence of an operational symbolism, and if the Greeks were thinking algebraically, then, they possessed such an operational symbolism. The graphical shape of the symbols is immaterial; if the letters of the alphabet could not be used (and this is far from clear), then some *other* symbols *had to be* used for an algebraic mode of reasoning to become reality. The geometrical diagrams which we encounter in Greek mathematical texts most certainly are ***not*** algebraic symbols in the proper sense of the word; besides, they are not brought into play *operationally*.

So the question remains unanswered: If thinking algebraically simplifies things, as everybody would agree, and if the great Greek mathematical geniuses were algebraists at heart, then why did they put their relatively simple algebraic reasonings in the clumsy and unwieldy molds of geometrical form? Furthermore, if they thought algebraically, and if the most fundamental difference between the | algebraic and the geo- 76
metric mode of reasoning lies, as I think it does, in the distinction between *symbolic* and *extensive* (*i.e., spatial*) magnitude, then why did they systematically fail to use any algebraic symbolism whatever in their writings? How can one reasonably explain such a failure? Is the unwarranted assumption of such mathematical schizophrenia accountable in any convincing historico-rational manner?

II

In the previous paragraph, I touched on the characteristic features of geometry and algebra. Let us pursue this matter a little further. What are the most fundamental traits of geometrical thinking? Geometry is thinking about space and its properties; furthermore, it is thinking embodied in, fused with graphic, diagrammatic representation. If the diagrams of geometry are its 'symbols', then these 'geometrical symbols'

---

[16] This is precisely Abel Rey's point of view in *La Science dans l'Antiquité*, **3** (*La Maturité de la Pensée Scientifique en Grèce*), (Paris: Albin Michel, 1939), note 1, 391. In that note Rey seems to imply that there was in existence an 'algebra numerosa' and that the ancient Greeks had, therefore, a real choice between this algebra and the geometrical symbolism of 'geometrical algebra' and that, furthermore, they preferred (wisely) the latter! But if this is the case, where are the traces of this 'algebra numerosa' during the first six centuries B.C.? There are no such traces, and this is for a very simple reason, mentioned by A. Rey on a previous page of his book: 'Le mathématicien grec est un geomètre. Il n'arrivera à l'algèbre numérique, et bien imparfaitement encore, qu'à l'extrême fin de la période gréco-romaine, au IVe siècle après J.C.' (*ibid.*, 349). Moreover, this point of view of A. Rey clearly contradicts what its author said elsewhere in this work (and in other works) about the inherent fundamental differences between geometry and algebra, the latter requiring a new way of thinking, *etc.* (More about this, below.)

display a feature which is totally absent from a true (algebraic) symbol: they are inherently extended because space, which they represent, is extended; they appeal to the eye of the geometer and to his spatial intuition; they are indeed, in a very real sense, the hypostatization of the geometer's spatial intuition. 'True' geometry (not analytical geometry) is inconceivable without diagrams and geometrical constructions. These diagrams are the characters in which the geometrical language is written: no diagrams, no geometrical way of thinking. Though it is true that these diagrams are only poor and imperfect copies of the real geometrical objects and relations, it is only through them that the geometer can pursue those lengthy and involved chains of reasoning which constitute the beauty and the glory of geometry.

Though diagrams constitute an integral and inseparable part of geometrical thinking, they are not its only ingredient. They must usually be accompanied by a rhetorical component, the proof, the most important function of which is to introduce the time parameter necessary in obtaining the finished, polished, wholesome diagrams through all the required intermediary, manipulative steps leading to the desired conclusion. In other words, if there is an operational, manipulative aspect in geometrical thinking, and I think there is, it takes place not at the level of the 'geometric symbol', the diagram (at least not in the written tradition, which is our only concern here), but at the rhetorical, descriptive, hortative level of the actual proof. Simply put, if one wishes to ascribe status of symbol to geometric diagrams (and it is far from clear that this is an entirely legitimate ascription), he will necessarily realize that the 'symbolism' thus constituted is, most certainly, not *operational* symbolism, when compared, say, with modern algebraic symbolism which is truly operational.[17]

77 | This brings us to the characteristics of the algebraic mode of thinking as they constituted themselves in the course of the historical development of algebra. According to a recent study,[18] the main features of the algebraic way of thinking are: 1. Operational symbolism; 2. The preoccupation with mathematical relations rather than with mathematical objects, which relations determine the structures constituting the subject-matter of modern algebra. The algebraic mode of thinking is based, then, on relational rather than on predicate logic; 3. Freedom from any ontological questions

---

[17] *Cf.* what A. REY has to say concerning the characteristics of the geometrical method: 'La construction géométrique ... nécessite ... une intuition plus singulière que les formules de l'algèbre ... D'abord elle a besoin d'une intuition concrète. L'esprit comme dira Descartes, y est asservi aux lignes, aux angles, et aux figures, aux agencements complexes de leurs traces et, comme les Grecs le professaient, à la règle et au compas. Il y a là effort pénible prétendra encore Descartes, nous ajouterons *limitatif*, d'imagination: limitatif parce que l'image est quelque chose de limité et de singulier en face de l'acte conceptuel, de la relation saisie toute nue.

Ensuite, ... [1]a construction géométrique est une synthèse où chaque pas prépare le suivant, où les inventions se lient et se commandent. Mais dans l'invention elle-même, chaque construction nécessite encore un tour de main, un biais, une intuition, une finesse particulière ...

Le symbolisme géométrique reste toujours en deça du symbolisme algébrique. Il faut, pour atteindre les articulations de pensée dans l'algèbre s'affranchir de la nécessité de construire, de même que la «construction» permettait de s'affranchir de la necéssité de compter et de calculer, et du spécifisme qui es [*sic*] affectait' (ABEL REY, *Les Mathématiques en Grèce au Milieu du* Ve *Siècle* (Paris: Hermann & Cie, 1935), 55–56).

[18] MICHAEL S. MAHONEY, 'Die Anfänge der algebraischen Denkweise im 17. Jahrhundert', *Rete*, **1** (1971), 15–31.

and commitments and, connected with this, abstractness rather than intuitiveness.[19] It seems, therefore, that the algebraic way of reasoning is *different* from the geometric one. It is completely abstract, free from dependency on perceptional, spatial considerations, it is manipulative, the entities it manipulates are themselves completely abstract, mere signs, it is analytical, functional, it possesses a universality of application missing in geometrical reasoning, and it is, at least to a certain extent, mechanical in the rules of manipulation of its symbols.[20]

III

Let us return now to the concept of 'geometric algebra'. It would seem, from what was said above alone, that it is a monstrous, hybrid creature, a contradiction in terms, a logical impossibility. Indeed it is. And, as we shall see, it is also an historical impossibility. The argument, to be sure, is very simple and straightforward. To | have an *Y*-like *X* 78
presupposes the prior or concomitant existence of some *X*, with respect to which alone departures from *X*-ness make sense and could be assessed. If there is not now and there never has been in the past an *X, Y*-like *X*'s are impossible creatures both logically and actually. In the same fashion, to speak of 'geometric algebra' in Greek antiquity makes good sense only if contemporaneously or formerly there existed an algebra from which the Greeks departed in certain ways. The fact is that (in spite of many historically unsubstantiated claims to the contrary on behalf of an alleged Egyptian, Babylonian, or even PYTHAGOREAN algebra) there has never been an algebra in the pre-Christian era.[21] Consequently, there could not have been any 'geometric algebra' either.

---

19 *Op. cit.*, 16–17. There is a slight variation of this characterization, appearing in an enlightening essay review of the reprint of NEUGEBAUER'S *Vorgriechische Mathematik*: MICHAEL S. MAHONEY, 'Babylonian Algebra: Form Vs. Content', *Studies in History and Philosophy of Science*, **1** (1971), 369–380; see particularly p. 372.

20 ABEL REY, in his highly illuminating analysis of the features of algebraic thinking (which makes some of his other hackneyed and erroneous conclusions stand out like the proverbial sore thumb), is in substantial agreement with the shorter characterization of MAHONEY. To illustrate: 'La condition *sine qua non* d'une algèbre sera ... un système de symboles et de règles mécaniques pour agencer ces symboles. C'est une *spécieuse* universelle, et c'est par ce mot que Viète l'a distinguée du calcul numérique. Son signe éminent c'est l'évasion hors de tout concret dans le pur abstrait. Il faut donc y faire abstraction des nombres et du calcul numérique, opérer sur des termes qui en soient des substituts universels, à l'aide d'un symbolisme opératoire. Les inconnues ont par là-même la même nature, et jouent le même rôle dans l'opératoire que les quantités connues' (*Les Math. en Grèce*, 38). '... les signes opératoires ... se substituent en algèbre aux articulations du raisonnement' (*ibid.*). [En algèbre] On n'opère plus sur des nombres, sur des quantités, des valeurs des termes. On opère sur des *relations*. Les termes ici sont déjà des relations, car ils sont imbriqués les uns avec les autres, et pour employer le mot dans un sens très général mais qui prélude à son sens technique moderne, ils sont *fonction* les uns des autres' (*ibid.*, 40). 'Le besoin du symbole et sa création montrent que la pensée ne peut plus, pour l'objectif qu'elle vise et qu'elle trouve, utiliser une représentation concrète et particulière. Le saut, le voilà ...' (*ibid.*, 45). 'L'algèbre seule peut permettre de transcender l'espace de la perception' (*ibid.*, 56).

21 See M. MAHONEY, 'Babylonian Algebra: Form Vs. Content' and 'Die Anfänge der algebraischen Denkweise im 17. Jahrhundert'; ABEL REY, *Les Math. en Grèce*, 30, 32, 34, 36–37, 41, 44, *passim*; also LÉON RODET, *Sur les Notations Numériques et Algébriques antérieurement au XVI*[e] *Siècle* (Paris: Ernest Leroux, 1881), *passim*, especially 69–70; JACOB KLEIN, *Greek Mathematical Thought and the Origin of Algebra* (Cambridge, Mass.: M.I.T. Press, 1968), *passim*; Á. SZABÓ, *Anfänge der griechischen Mathematik* (Müchen-Wien: R. Oldenburg, 1969) *passim*, but especially

If in spite of the preceding, however, various authors speak of Greek 'geometric algebra', this is due exclusively to the fact that these authors happen to live in a period after the invention of algebra and its application to geometry (analytical geometry) and assume, therefore, unwarrantedly and ahistorically that the symmetric case, *i.e.*, the application of *geometry* to *algebra*, has also taken place. This conclusion, however, is historically inadmissible. There is (broadly speaking) in the historical development of mathematics an *arithmetical* stage (Egyptian and Babylonian mathematics) in which the reasoning is largely that of elementary arithmetic or based on empirically paradigmatic rules derived from successful trials taken as a prototype,[22] a *geometrical* stage, exemplified by and culminating in classical Greek mathematics, characterized by rigorous deductive reasoning presented in the form of the postulatory-deductive method, and an *algebraic* stage, the first traces of which could be found in DIOPHANTOS' *Arithmetic* and in AL-KHWARIZMI'S *Hisab al-jabr w'al muqàbalah*, but which did not reach the beginning of its full potentiality of development before the sixteenth century in Western Europe;[23] it is characterized, as we saw, by its supreme
79 degree of abstract|ness, by its operational symbolism of universal applicability, and by its preoccupation with relations and structures.

28, 34, 35–36, and primarily the 'Anhang' appearing on 455–488; PAUL-HENRI MICHEL, *De Pythagore à Euclide: Contributions à l'histoire des mathématiques Préeuclidiens* (Paris: Les Belles Lettres, 1950), 639–646; G. A. MILLER, 'Weak Points in Greek Mathematics', *Scientia*, **39** (1926), 317–322.

Some of the works cited here establish the point solidly and unambiguously (SZABÓ and to a lesser extent KLEIN); some of them are, at best, ambiguous (MAHONEY and REY) succeeding in determining at one and the same time the ontological incommensurability of the geometric and the algebraic way of thinking and, yet, accepting (openly or implicitly), without realizing the contradiction involved, the historical legitimacy of the concept 'geometric algebra'; finally, some, though presenting a less clearcut point of view (RODET), or an unacceptable, ahistorical point of view (MICHEL and, especially, MILLER), enable the astute eye of the historically minded reader to reach easily a conclusion opposite to that presented by the author.

22 *Cf.* A. REY, *Les Math. en Grèce*, 34, 41.

23 *Ibid.*, 43, 45, 91–92. Against this view, for NEUGEBAUER, it seems, mathematics has always historically been algebra in various disguises and shapes. Thus, the first stage in the development of mathematics (algebra) was represented by the Babylonian sexagesimal place-value system and the operations with numbers made possible by the existence of such a system ('Zur geometrischen Algebra', 247). The second stage was represented by 'Babylonian algebra' proper, in which problems are reduced to quadratics of the normal form; the third stage is illustrated by the translation of algebraic techniques into the language of geometry – Greek mathematics or geometrical algebra (*ibid.*), and, finally, the fourth stage in the development of mathematics (algebra) is '... [die] Periode der neuerlichen Rückübersetzung der geometrischen Algebra in eine "algebraische" Algebra' (*ibid.*, 249). Besides, for NEUGEBAUER, geometry has always had secondary, derivative character: 'Die grossen Fortschritte der Geometrie sind in allen Phasen [!] immer unlösbar mit der Entwicklung anderer Disziplinen verknüpft (analytische Geometrie und elementare Algebra, Differentialgeometrie und Analysis, Topologie und Riemannsche Flächen + abstrakte Algebra), so dass das Geometrische an sich immer erst nachträglich [!] wieder aus dieser Verknüpfung gelöst werden musste [?]. Für die Frühgeschichte der Mathematik ist eine „reine" („synthetische") Geometrie viel zu schwierig [Why should this be so? Is there any substantiation for this unqualified claim, except algebraic hindsight?]. Das primäre Hilfsmittel ist hier die Verknüpfung mit dem Bereich der (rationalen) Zahlen und ein wesentlicher Fortschritt der Geometrie ist immer erst möglich, wenn die ungeometrischen Hilfsmittel weit genug entwickelt sind' (*op. cit.*, 246). It is true that NEUGEBAUER knows very well the importance of symbolism for the development of mathematics (*cf. op. cit.*, 246–247); (after all this should not

It was only after the instauration of the algebraic stage that 'algebraic geometry' (*i.e.*, analytical geometry) could and did occur. The symmetric counterpart of this 'algebraic geometry', *i.e.*, 'geometric algebra' is not a historical entity, but only the fruit of the mathematico-historical lucubrations of mathematicians born during the algebraic stage in the development of mathematics. It is a figment of their mathematical imaginations, rather than anything real. It is an invention of the modern mathematician reading ancient texts through modern glasses, *i.e.*, an immediate and net outcome of the modern mathematician's ability to read geometry algebraically, to transcribe geometrical propositions into the language of algebraic equations, and to assume 'therefore' that this is what geometry is all about always, everywhere.

Additionally, if it is possible to supplement the above argument by showing that the assumption (for the sake of the argument) of a 'geometric algebra' leads to absurdities in the *specific* analysis of ancient mathematical texts, this should finally dispose of this monstrous concept and lead urgently to its demise. This is exactly what we plan to do by using the great classic of Greek mathematics, EUCLID'S *Elements*, as our source of illustrations; let us hope, therefore, that the hours of 'geometric algebra', this arbitrary and aberrant concept, are indeed numbered!

IV

It is at least in principle possible that a partisan of the view embodied in the concept of 'geometric algebra' might counter the argument expounded above in the following fashion: 'I will grant you (though reluctantly!) that since there are no *explicit* instances of algebraic texts in Egyptian and Babylonian mathematical | sources, there was no pre- 80
Hellenic algebra. The situation is, however, totally different with Greek mathematics. It is clear to the shrewd and trained eye that Greek geometry is nothing but geometrically clad algebra. So the Greeks must be taken as the inventors of algebra.[24] However, for reasons that are immaterial to our issue, they decided not to use the standard type of algebraic symbolism, but to dress their algebraic formulas in geometrical outfits. So the very existence of Greek geometry is the best proof for the existence of an ancient algebra, Greek algebra. As to the alleged irreconcilableness and incommensurability of the two ways of thinking (geometrical and algebraic), I do not buy this, modern mathematics doesn't buy it, and, obviously, the ancient Greeks didn't buy it either!'

What does one answer to such an interlocutor? I happen to believe that his 'argument' is really no argument and that it has been taken care of already in what was said

---

come as a surprise from the part of somebody who, for all practical purposes, identifies mathematics with algebra!) But he draws from this awareness what seems to me to be unwarranted conclusions. Even if he is right about the precedence of computational techniques in pre-Greek civilizations over geometrical considerations, it does *not* follow that algebra preceded geometry in the Hellenic civilization. Logistic is most certainly not algebra, and quoting ARCHYTAS (p. 245) to the effect that logistic takes precedence over the arts, including geometry (DIELS-KRANZ, *Die Fragmente der Vorsokratiker*, 5th ed., 47B4) does not prove that algebra preceded geometry in Greek mathematics!

24 Indeed the nineteenth century originators of the concept of 'geometric algebra', ZEUTHEN and TANNERY, wrote before NEUGEBAUER (the 'discoverer' of Babylonian algebra) and VAN DER WAERDEN (the articulate spokesman for the view that 'Babylonian algebra' became Greek 'geometric algebra') and yet, did not hesitate to speak freely of Greek 'geometrical algebra' when they encountered in the *Elements* propositions which seemed to them out of place, unwieldy, incongruous!

above. However, in a more substantive fashion and in order to adapt the general analysis presented above to the *ad hoc*-ness of the interlocutor's alleged counter-argument, the following reply is in place: *Language* is the immediate reality of *Thought*. The differences between the two ways of thinking are real differences, which could not be dismissed off hand, rooted as they are in the features of perceptible space on the one hand (geometry) and the universal denotativeness of the supremely abstract manipulable symbol (algebra) on the other hand. Different ways of thinking imply different ways of expression. It is, therefore, impossible for a system of mathematical thought (like Greek mathematics) to display such a discrepancy between its alleged underlying *algebraic* character and its *purely geometric* mode of expression. Furthermore, why did the ancient Greeks hide their ways of reasoning? What was there to hide?

That this last question is pertinent indeed – and not at all gratuitous as it may, *prima facie*, seem – is shown by the following quotation from VAN DER WAERDEN in which the learned author discusses Book X of EUCLID'S *Elements:*

> Up to X 28 it goes fairly well, but when the existence proofs start with X 29 ... one does not see very well what purpose all of this is to serve. *The author succeeded admirably in hiding his line of thought* by starting with his constructions, even before having introduced the concept of binomial which does throw some light on the purpose of these constructions, and by placing at a still later point the division into 6 types of binomials.[25]

81 | Lest the readers believe that TANNERY, ZEUTHEN, NEUGEBAUER, and VAN DER WAERDEN are the only 'villains', I would like to state that practically 'anybody who counts' in the writing of the history of Greek mathematics in modern times has adopted the stand that 'geometric algebra' provides one with an illuminating insight into the inner workings of Greek mathematics.[26] This certainly is true of contemporary writers

---

25 *Sci. Awak.*, 172, my italics. Once more, VAN DER WAERDEN follows in the footsteps of O. NEUGEBAUER. It was NEUGEBAUER who, in his 'Apollonius Studien (Studien zur Geschichte der antiken Algebra II.)', already stated that, though there are recognizable structures and algorithms all over the *Conics*, which the trained eye of the mathematician can disentangle, these structures, algorithms, and methods of proof have been subsequently completely hidden, camouflaged ('... nachträglich völlig erdeckt' ...; see *loc. cit.* (footnote 15), 253). Furthermore, speaking of his own analytical transcriptions and manipulations of APOLLONIUS' geometrical rhetoric, NEUGEBAUER says: '*Diese höchst einfache Schlussweise gibt den Schlüssel zu sämtlichen hier zusammengestellten Konstruktionen.* Bei Apollonius ist nur alles mit grosser sorgfalt auf den Kopf gestellt und verschleiert' (*ibid.*, 251).

26 NEUGEBAUER'S unmitigated enthusiasm for geometric algebra (for which he erroneously takes ZEUTHEN as the originator) is typical: 'Zeuthen verdankt man die für das Verständnis der ganzen griechischen Mathematik grundlegende Einsicht, dass es sich insbes. in den Büchern II und VI von Euklids Elementen um eine geometrische Ausdrucksweise eigentlich algebraischer Probleme handelt. Insbesondere hat er an vielen Stellen darauf hingewiesen, dass in den „Flächenanlegungs"-Aufgaben von Buch VI und zugehöriger Sätze der Data die vollständige Diskussion der Gleichungen zweiten Grades steckt. Er hat dann weiter gezeigt, wie diese „geometrische Algebra" die Basis für die „analytische Geometrie" der Kegelschnitte des Apollonius bildet, deren Bezeichnungen „Ellipse", „Hyperbel", „Parabel" noch heute auf die Fundamentalfälle der „Flächenanlegung" zurückwiesen' ('Zur geometrischen Algebra', *Q.U.S.*, **3** B (1936), 249). There are hardly any unambiguous, clear-cut exceptions to the rule. Even those who, for one reason or another, began doubting the inherited interpretation (and this 'doubting' got under way only in recent times) did not, as a rule, abandon the concept of 'geometric algebra'. A case in point is represented by MICHAEL MAHONEY. (*Cf.*, for instance, his otherwise enlightening article, 'Another Look at Greek Geometrical Analysis', *Archive for History of Exact*

of textbooks like CARL B. BOYER and HOWARD EVES;[27] it is true of J.F. SCOTT, DIRK STRUIK, FLORIAN CAJORI, DAVID EUGENE SMITH, EDNA E. KRAMER, and so forth.[28] And, of course, it is true of the greatest writer | on the history of Greek mathematics in the 82
English language in modern times, Sir THOMAS LITTLE HEATH.

HEATH interests us here since he is the author of (among other things) the English editions of the writings of the great classics of Greek mathematics, EUCLID, ARCHIMEDES, and APOLLONIUS.[29] His translations (when he is satisfied to limit himself to the role of translator![30]) are considered reliable and insightful. Since we shall largely

---

Sciences, **5** (1968), 318–48, where he says: 'For example, Proposition VI, 28 [of the *Elements*], which is part of the "geometric algebra" of the Greeks ...' (328) or 'The earliest techniques of analysis evolved from the researches of the ... Pythagoreans, and are brought together in the major contribution of this mathematico-philosophical school to Greek mathematics: geometrical algebra. Geometrical algebra was one of the basic tools of the mathematical analyst. In the *Data*... Euclid gave prominent place to the doctrine of the application of areas, which is the essence of Greek geometrical algebra' (*ibid*., 330–31); even in his cogent and powerful criticism of NEUGEBAUER'S ahistorical procedures ('Babylonian Algebra: Form *vs.* Content'), MAHONEY somehow considers the ahistorical concept of 'geometrical algebra' as legitimate, since he says that 'Greek geometrical algebra' could construct '... a quadratic system of equations in two unknowns from the values of those unknowns ...' (376), but could not construct '... a single quadratic equation from its two roots ...' (*ibid*).). Another, earlier instance of the same syndrome is illustrated by ABEL REY'S writings quoted above. (Incidentally, 'Chapitre IX' of 'Livre III' of *La Maturité de la Pensée Scientifique en Grèce* reproduces verbatim, *in toto*, 'Chapitre IV' ('Arithmétique et Système Métrique Algèbre, Géométrie et Algèbre Géométrique') of *Les Mathématiques en Grèce au Milieu du Ve Siècle*, without any hint whatever to the reader!) To my knowledge, it is only ÁRPÁD SZABÓ, who, in the introduction and (primarily) in an appendix appearing in *op. cit.*, 455–88 (about which more will be said below), unequivocally and forcefully calls attention to what is wrong with the concept of 'geometric algebra' and asks for its abandonment. I had arrived at my ideas concerning the historical unsoundness of the notion of 'geometrical algebra' independently, while, as a graduate student, I immersed myself in reading Greek mathematical texts and the modern commentaries on them. I reached my final conviction about the necessity to discard and repudiate 'geometric algebra' as an explanatory device in the study of the history of Greek mathematics and about the need, growing out of this rejection, to rewrite that history on a sound basis, while teaching a graduate seminar on EUCLID'S *Elements* at the University of Oklahoma in the fall of 1972. I gave a talk on this topic at the Hebrew University of Jerusalem in the late fall of the same year, which got (so far as I can judge) a mixed reception: historians and the (very few) historically-minded mathematicians present seemed to like its conclusions, while the mathematicians (to put it mildly) remained unconvinced.

27 See CARL B. BOYER, *A History of Mathematics* (New York-London-Sydney: John Wiley & Sons, Inc., 1968), 85–87, 114–15, 121–131, *passim* and HOWARD EVES, *An Introduction to the History of Mathematics* (New York: Holt, Rinehart and Winston, 1964, rev. ed.), 64–69, *passim*.

28 See J.F. SCOTT, *A History of Mathematics* (London: Taylor & Francis Ltd., 1960), 23, *passim*; DIRK J. STRUIK, *A Concise History of Mathematics* (New York: Dover Publications, Inc., 1948, 2nd rev. ed.), 58–60, *passim*; FLORIAN CAJORI, *A History of Mathematics* (New York: The MacMillan Co., 1919), 32–33, 39; DAVID EUGENE SMITH, *History of Mathematics*, 2 vols. (New York: Dover Publications, Inc., 1958), **1**, 106 and **2**, 290; EDNA E. KRAMER, *The Nature and Growth of Modern Mathematics*, 2 vols. (Greenwich Connecticut: Fawcett Publ. Inc., 1974), **1**, 108, 137–40, 146.

29 *The Works of Archimedes* (Cambridge: At the University Press, 1897), *Apollonius of Perga Treatise On Conic Sections* (Cambridge: W Heffer & Sons Ltd., 1961), *The Thirteen Books of Euclid's Elements*, 3 vols. (Cambridge: At the University Press, 1908); HEATH'S edition of EUCLID will be referred to in the future as EUCLID, *Elements*.

30 *Cf.*, in this context, SZABÓ'S remark: 'Es wurde also eben betont, dass man auf die Übersetzungen der Quellen – vom Gesichtspunkt der Mathematikgeschichte aus – sich häufig nicht verlassen kann, auch dann nicht, wenn die fraglichen Übersetzungen manchmal philologisch so gut wie *tadellos* sind' (*op. cit.*, 16).

confine our discussion in what follows to EUCLID'S *Elements*, let us see what HEATH'S views on 'geometric algebra' are, as they pertain to the *Elements*. HEATH thinks that after the discovery of the irrational, '... it was possible to advance from a geometrical arithmetic to a geometrical *algebra*,[31] which indeed by EUCLID'S time (and probably long before) had reached such a stage of development that it could solve the same problems as our algebra so far as they do not involve the manipulation of expressions of a degree higher than the second.'[32] HEATH goes on to say that '... Book II gives the geometrical proofs of a number of algebraical formulae [!]'[33] and then, without apparently grasping the inconsistency involved, continues:

> It is important however to bear in mind that the whole procedure of Book II is *geometrical;* rectangles and squares are shown in the figures, and the equality of certain combinations to other combinations is proved by those figures. We gather that this was the classical or standard method or proving such propositions, and that the algebraical method of proving them, with no figure except a line with points marked thereon,[34] was a later introduction.[35]

Finally, HEATH finishes his introductory remarks to Book II of the *Elements* with a description of what he calls '... the geometrical equivalent of the algebraical operations'[36] allegedly undertaken by Greek geometers in their *geometrical* treatises, noting, among other things, that 'The division of a product of two quantities by a third is represented in the geometrical algebra by the finding of a rectangle with one side of a given length and equal to a given rectangle or square. This is the problem of *application of areas* ...'[37]

83 | Let us try, then, to sum up the views of those who see in Greek geometry (at least in some crucial parts of it) a 'Geometric Algebra' by referring (copiously) to VAN DER WAERDEN, not because I particularly pick on him as my 'bouc émissaire', but for the simple reason that his assertions are among the most shocking in their bluntness and outspokenness, and because his book is one of the most recent pronouncements on the issue, and (in addition) is easily available to the interested (but needy student) in paperback.[38]

'When one opens Book II of the Elements', says VAN DER WAERDEN, 'one finds a sequence of propositions, *which are nothing but geometric formulations of algebraic rules. ... We have here, so to speak, the start of an algebra textbook, dressed up in geometrical form.*'[39]

---

[31] This is also ZEUTHEN'S view. *Cf.*, for instance, *Gesch. der Math. im Alt. und Mittel.*, 42.

[32] EUCLID, *Elements*, **1**, 372.

[33] *Ibid. Cf.* also ZEUTHEN, *Die Lehre von den Kegel.*, 12.

[34] Why is such a procedure '*algebraical*'?

[35] *Elements*, **1**, 373.

[36] *Ibid.*, 374.

[37] *Ibid. Cf.* also ZEUTHEN, *Die Lehre*, 14 and TANNERY, *Mém. Scient.* **1**, 256–57.

[38] *Science Awakening* (which was originally published in Dutch as *Ontwakende Wetenschap* (Groningen, 1950) appeared first in English translation at Groningen in 1954; since then the scholarly world was supplied with a paperback edition (New York: John Wiley & Sons, 1963), used in this study, in which the beautiful illustrations of the hard cover edition are marred by imperfect typographical reproductive processes, and, very recently (what a dream for a publishing house!), with a new hardcover 'Third Edition' in English (Groningen: Noordhoff, n.d.).

[39] *Sci. Awaken.*, 118, my italics. *Cf.* also, ZEUTHEN, *Die Lehre*, 12–13. Interestingly, G.H.F. NESSELMANN in his *Die Algebra der Griechen* (Berlin, 1842) – a photographic reprint (Frankfurt:

And,

> Quite properly, Zeuthen speaks in this connection of a "geometric algebra." Throughout Greek mathematics, one finds numerous applications of this "algebra." *The line of thought is always algebraic, the formulation geometric*. The greater part of the theory of polygons and polyhedra is based on this method; the entire theory of conic sections depends on it. Theaetetus in the 4th century, Archimedes and Apollonius in the 3rd are perfect virtuosos [*sic*!] on this instrument.[40]

| For VAN DER WAERDEN, Greek '... geometric algebra is the continuation of 84
Babylonian algebra.'[41] However, the Greeks, unlike their Mesopotamian forerunners, translated *everything* into geometric terminology.

Minerva, 1969) is available – considers Book II as *arithmetical* (*not* algebraic) in character: "Jedenfalls ... müsen wir ... das zweite Buch ... zu den arithmetischen zählen, da von seinen vierzehn Sätzen die ersten zehn gleichfalls nur geometrisch ausgesprochene und bewiesene, aber ihrem Wesen nach [?] lauter arithmetische Wahrheiten enthalten' (*op. cit.*, 154). NESSELMANN (about whose book L. RODET remarked that a better title would be *'le calcul chez les Grecs'* (*op. cit.*, 57)) then goes on to transcribe the first ten propositions in algebraic symbolism! Incidentally, we do possess an *arithmetical* translation of these ten propositions dating from the 14th century by a Byzantine monk, BARLAAM, entitled *ἀριθμητικὴ ἀπόδειξις τῶν γραμμικῶς ἐν τῷ δευτέρῳ τῶν στοιχείων ἀποδειχθέντων* and another *arithmetical* translation by CONRAD DASYPODIUS published with the original Book II of EUCLID in 1564. The proofs in these arithmetical translations are patterned after those appearing in the so-called 'arithmetical Books' of the *Elements* (VII–IX). For an example of BARLAAM'S translation and proofs, see NESSELMANN, *op. cit.*, 155, where the proposition dealt with is II. 4.

40 *Op. Cit.*, 119, my italics. Again, NEUGEBAUER espoused similar views long before VAN DER WAERDEN. Thus, describing the contents of the *Conics*, NEUGEBAUER said: 'Im ersten Buch werden die *Grundgleichungen* [!] der Kurven und ihrer Tangenten entwickelt ... ' ('Apollonius Studien', 218, italics added). Referring in a more detailed fashion to the contents of Book I, NEUGEBAUER again spoke of, *'Gewinnung der Grundgleichung* a) *zunächst in unmittelbar geometrischen Form*, b) Umformung in eine solche Gestalt, wie sie für die Anwendung bequemer ist ... Schliesslich wird gezeigt: ist ein Kegelschnitt durch seine *Gleichung* gegeben, so gibt es auch ... einen Kegel ... auf dem er liegt. *Zusammen mit der ursprünglichen Gewinnung der Gleichung aus dem räumlichen Schnitt ist damit die volle Äquivalenz von räumlicher und analytischer Darstellung bewiesen*' (*ibid.*, 219, my emphasis).

41 *Ibid.* Needless to say, the originator of this view is OTTO NEUGEBAUER. Thus, in his 'Zur geometrischen Algebra', speaking of the title he chose for this study, NEUGEBAUER confesses that, although it may be too narrow for his purposes, it was selected, ' ... um anzudeuten welchen Punkt ich für den eigentlichen Schlussstein für das Verständnis des Verhältnisses der griechischen Mathematik zur babylonischen halte' (*op. cit.*, 246). Having accepted ZEUTHEN'S views on the nature of 'geometric algebra' *in toto*, NEUGEBAUER goes on to ask what was the historical origin of the problem of application of areas, a question left unanswered by ZEUTHEN. NEUGEBAUER'S answer runs as follows: 'Die Antwort ... liegt einerseits in der aus der Entdeckung der irrationalen Grössen folgenden Forderung der Griechen der Mathematik ihre Allgemeingültigkeit zu sichern durch Übergang vom Bereich der rationalen Zahlen zum Bereich der allgemeinen Grössenverhältnisse, andererseits in der daraus resultierenden Notwendigkeit [Is this *logical* necessity or *historical* necessity? Clearly, the former! And so, NEUGEBAUER has sinned once more against history, by substituting logical for historical criteria in his analysis.], *auch die Ergebnisse der vorgriechischen "algebraischen" Algebra in eine "geometrischie" Algebra zu übersetzen'* (*ibid.*, 250). Is there any historical proof for the above italicized statement? As NEUGEBAUER would ask: 'Ist diese mächtige Behauptung textlich belegt?' **No!** What, then, is NEUGEBAUER'S basis for making such a statement? He tells us in what immediately follows the above passage: 'Hat man das Problem einmal in dieser Weise formuliert, so ist alles Weitere vollständig trivial und liefert *den glatten Anschluss der babylonischen Algebra an die Formulierungen bei Euklid*

85 *'But since | it is indeed a translation which occurs here and the line of thought is algebraic, there is no danger of misrepresentation, if we reconvert the derivations into algebraic language and use modern notations.'*[42]

VAN DER WAERDEN goes on to say that we can use, without conscience qualms, modern algebraic symbolism, '... provided we take good care, *not to use algebraic transformations, which can not immediately be reformulated in the Greek terminology.*'[43] After doing exactly that, VAN DER WAERDEN obtains (no surprise, since he is the prestigious author of *Moderne Algebra*) for some of the propositions in Book II what he calls '... normalized forms of systems of equations ...,'[44] identical to those equations

---

(ibid.) In other words, NEUGEBAUER begins with what one would normally expect the historian to conclude (namely, that the Greeks knew the Babylonian stuff and 'translated' it into geometrical language), and from this historically totally unfounded assumption, by transcribing both the Greek geometrical propositions and the Babylonian numerical manipulations into algebraic symbolism, 'manages to show' that they are both the same and 'therefore' the Greeks copied the Babylonians. The vicious circle of his reasoning is obvious! Having thus shown the complete mathematical equivalence between the Babylonian 'normal form' and the simplest case of application of areas, NEUGEBAUER then exclaims in pleasant amazement: 'Das ist aber genau die einfachste Formulierung der Flächenanlegungsaufgabe des „elliptischen" Falles, wie sie bei Euklid VI, 28 steht ... Euklid VI, 29 steht dann die Übersetzung der Normalform (2), d.h. der „hyperbolische" Fall' (*ibid.*). What does this prove? To my mind, nothing else than the fact that if one performs the historically impermissible translation of the Babylonian and Greek mathematical stuff into algebraic symbolism, one can see that they are the same. *It certainly does not prove that the Greeks knew the Babylonian stuff!* But all this is not enough, since NEUGEBAUER goes on: 'Damit ist gezeigt, dass die ganze Flächenanlegung nichts anderes ist, als die mathematisch evidente geometrische Formulierung der babylonischen Normalformen quadratischer Aufgaben' (*ibid.*, 251). Aber ist was ist mathematisch evident auch historisch evident? Das scheint mir nicht! Neugebauer continues in the same vein by showing '... dass auch die griechische *Lösungsmethode* nicht anderes ist als die wörtliche Übersetzung der babylonischen Formel ... $\left.\begin{matrix}x\\y\end{matrix}\right\} = \frac{b}{2} \pm \sqrt{(\frac{b}{2})^2 - c^2}$ ..., (*ibid.*). Having done this, he remarks: 'Die einzig neue Überlegung ist hier die Bemerkung über die Grösse der Gnomon figur, also etwas, was so nahe liegt, dass es gewiss keiner besonderen Motivierung bedarf, *wenn man den Ausgangspunkt so wählt, wie es hier geschehen ist, nämlich in der Aufgabe, die algebraische Formel* (2) $\left[\text{i.e., } \left.\begin{matrix}x\\y\end{matrix}\right\} = \frac{b}{2} \pm \sqrt{(\frac{b}{2})^2 - c^2}\right]$ *ins Geometrische zu übersetzen.* Hat man aber (2) nicht zur Verfügung, d.h. müsste man auf jede algebraische Formulierung verzichten, so ist gar nicht einzusehen, wie man auf eine derartige Konstruktion verfallen kann' (*ibid.* my emphasis). In other words, it is impossible to understand Greek geometry without seeing it as derivative, secondary, illustrative of a *truly* algebraic background and motivation! Indeed NEUGEBAUER procceds to show by the same historically indefensible method (*i.e.*, by transcription of both Babylonian number computations and Greek geometrical propositions into algebraic manipulative symbolism) that: ... *die ganze Flächenanlegungsaufgabe wird sowohl hinsichtlich Fragestellung wie hinsichtlich Lösungsmethode unmittelbar verständlich wenn man sie nur als die sinngemässe Übersetzung der babylonischen Methode in die Sprache der geometrischen Algebra auffasst' (ibid.*, 252).

[42] *Ibid.*, my italics.
[43] *Ibid.*
[44] *Ibid.*, 124.

he has previously obtained by a similar procedure from Babylonian cuneiform tablets, and concludes that:

> *Apparently the Pythagoreans formulated and proved geometrically the Babylonian rules for the solution of these systems.*[45]

Further:

> Thus we conclude, that *all the Babylonian normalized equations have, without exception, left their trace*[46] *in the arithmetic and the geometry of the Pythagoreans.* It is out of the question to attribute this to mere chance.[47] What could only be surmised before, has now become certainty [!], namely that the Babylonian tradition supplied the material which the Greeks, the Pythagoreans in particular, used in constructing their mathematics.[48]

I think this should be more than enough!

As I already intimated above, I believe such a view is offensive, naive, and historically untenable. It is certainly indefensible on the basis of the historical | record, *i.e.*, 86
on the basis of a study of the documents of Greek mathematics, undertaken *not* from the point of view of the achievements, results, and methods of modern mathematics, which, it should be unequivocally understood, are completely irrelevant in attempting to understand Greek mathematics for its own sake, but from the standpoint adopted by the ancient Greek mathematicians themselves, inasmuch as this standpoint could be grasped by a modern mind. To read ancient mathematical texts with modern mathematics in mind is the safest method for misunderstanding the character of ancient mathematics, in which philosophical presuppositions and metaphysical commitments played a much more fundamental and decisive role than they play in modern mathematics. To assume that one can apply automatically and indiscriminately to any mathematical content the modern manipulative techniques of algebraic symbols is the surest way to fail to understand the inherent differences built into the mathematics of different eras.

---

45 *Ibid.*

46 The reader may like being reminded that, strictly speaking, *there are no Babylonian equations*, normalized or 'abnormalized'. There are only tablets containing numbers and operations executed on these (specific) numbers, which the modern mathematician can translate, if he so wishes, into equations. On this whole issue, I urge the reader to peruse MAHONEY'S 'Babylonian Algebra: Form *Vs.* Content'. It is there that MAHONEY says: 'All that Babylonian texts contain is series of arithmetical operations that lead to (usually) correct results. The rest is interpretation by the historian. The Babylonians state the problem and compute the solution; the derivation of that solution is the work of the historian, and one may question whether the derivation tells us more about the historian's mathematics than about Babylonian mathematics' (*op. cit.*, 375). Indeed if NEUGEBAUER was able at all to speak of a 'Babylonian Algebra' and to tie it in with Greek geometry, this was due to his rather radical methodological innovation. As he tells us in his 'Studien zur Geschichte der antiken Algebra I' (full reference in footnote 15): 'Dabei verstehe ich unter „antiker Algebra" einen wesentlich weiteren Problemkreis als dies üblicherweise der Fall ist. Einerseits fasse ich das Wort „Algebra" sachlich möglichst weit, d.h. *ich ziehe auch stark geometrisch betonte Probleme mit in Betracht, wenn sie mir nur auf dem Wege zu einen letztlich „algebraisch" zu nennenden formalen Operieren mit Grössen zu liegen scheinen*. Andererseits gehe ich zeitlich weit über das übliche Mass hinaus, ...' (*op. cit.*, 1, italics added).

47 Indeed it is far from chancy! It is due to the conscious *premeditated* talents of the modern mathematician, turned historian, who has managed to translate ('traduttore traditore') both Babylonian *numerical* manipulations and Greek *geometrical* propositions into algebraic language.

48 *Ibid.*

'Mathematics is a reflection of culture ...'[49] It is, clearly, not immune to the intellectual and cultural environment in which it grows. Nothing is. This is the most fundamental reason why we have an *Egyptian*, a *Babylonian*, and a *Greek* mathematics (to limit ourselves to Antiquity only), and not just *Ancient* mathematics. It is indeed a truism that in some very substantive and irreducible aspects, Egyptian mathematics is *not* Babylonian mathematics, and Babylonian mathematics is *not* Greek mathematics. They become practically indistinguishable only if one commits the deadliest of sins a historian may be tempted to commit, namely that of inflicting upon them the ultimate historiographical insult of considering them mere adumbrations of modern mathematics and, therefore, proceed to translate them into modern algebraic symbolism.[50]

Whig history, a dead horse nowadays – one would like to believe – in most branches of history, is alive and thriving in the history of mathematics, where its dangers are no less real than in the more traditional types of intellectual history. It seems perfectly obvious to me that the ultimate implication of the historiographical view which allows one to read ancient mathematical texts through modern glasses must be that in mathematics, unlike any other domain of intellectual endeavour, the 'real stuff', the 'hard-core' mathematical content, the very essence of the discipline, its true fabric is immune to historical development and change, representing, in good Platonic fashion, a given, permanent, universal, stable structure, which man somehow grasped from the very beginning of his preoccupation with mathematical topics and which can
87 easily be identified, | recognized, and labeled by the perceptive and skilled sage, *i.e.*, by the individual trained in (modern) mathematics, who knows with certitude what *Mathematics* is all about.

In a very real sense such a view must lead one to look at the mathematics of bygone eras as *preparatory* for modern mathematics, in the sense that the essential structure is already there, the only real difference being that this structure is expressed in 'cumbersome', 'awkward', 'unnecessarily difficult' form or language. In other words, the development of mathematics becomes the (almost) exclusive development of mathematical form, the groping for the 'right' kind of language to express the universal 'truths' which were there and were apprehended all the time. This approach, I submit, is not just naive and offensive historiographically, but it undermines the very fiber of the history of mathematics as a historical discipline. In short, it is, I believe, unacceptable to us as historians and should, therefore, be relinquished.

---

49 M. MAHONEY, 'Babyl. Alg.", 370.

50 Recently, more attention is being paid to mathematics as a reflection of culture. *Cf.* in this context DAVID BLOOR, 'Wittgenstein and Mannheim on the Sociology of Mathematics', *Studies in History and Philosophy of Science*, **4** (1973), 173–91. It is there that one finds the following interesting remark: 'As evidence for the idea that mathematical notions are cultural products, consider the historical case of the concept zero. Our present concept is not the one that all cultures have used. The Babylonians, for example, used a place-value notation but had a different, though related, concept. Their nearest equivalent to zero operated in the way that ours does when we use it to distinguish, say, 204 from 24. They had nothing correspond – [*sic*!] to our use when we distinguish, say 240 from 24. As Neugerbauer [*sic!*] puts it, 'context alone decides the absolute value in Babylonian mathematics' ... *If the Babylonians used a zero which left some aspects of a calculation context dependent, then, thus far, their concept of zero differs from ours'* (*ibid.*, 186, italics provided).

I am not about to enter into an exhaustive analysis of the sociological roots of such
a scandalous situation. Let me only suggest again, however, that the fact that the his-
tory of mathematics has been typically written by mathematicians might have some-
thing to do with it; and in many instances it was not just 'broadminded'
mathematicians who engaged in such ventures;[51] on the contrary, these were mathe-
maticians who have either reached retirement age and ceased | to be productive in their 88
own specialties or became otherwise professionally sterile. However, both of these
categories had something in common: in order to serve humanity and expend
untapped remnants of scholarly energy, they decided to employ their creativity in a
field, *history of mathematics*, 'half' of which – the *history* – was too alien and exotic
to them while the other 'half' – the *mathematics* – was, alas, too familiar to them; the
underlying assumption being that history does not really require any training, its nar-
rative, reportorial methods and techniques being common-sensical and self-evident;
and since they were highly proficient in mathematics they had *all* which was required
to become successful historians of mathematics! ...

If the above suggestion is correct, then, the reader may judge for himself how wise a decision it is for a professional to start writing the history of his discipline, when his only calling lies in professional senility which bars him from encroaching on more friendly, familiar, and hospitable territory!

---

[51] As an illustration of the kind of mathematician I have in mind, I shall refer the reader to G.A. MILLER (see reference in note 21, above). His article on 'Weak Points in Greek Mathematics' is a genuine *chef-d'oeuvre* and should be read in its entirety. Space limitations, however, permit me to quote here only sparingly. Thus, after deploring the '... undue emphasis on the geometric view ...' (317) of the ancient Greeks, MILLER declares emphatically that *'the lack of emphasis on the formal algebraic side of mathematics doubtless constituted the greatest inherent weakness of Greek mathematics'* (*op. cit.*, 317–18, my italics). He illustrates this 'drawback' with 'the Greek attitude towards the solution of the quadratic equations. Not only did they solve certain quadratic equations geometrically, but ... it appears clear that they had three general formulas [! ?] for the algebraic solutions of such equations ... they failed to see the general significance of the formulas and hence ... did not succeed in obtaining a general solution of the quadratic equation in the modern sense. It seems therefore unfortunate that many writers claim that they solved the quadratic equation' (*ibid.*, 318). 'By overlooking the fact that the algebraic equation frequently gives us much more than what we explicitely put into it, *the Greeks made a blunder and failed to put into their work one of the most fruitful ideas of later mathematics'* (*ibid.*, my italics).

'The awe inspired by the immortal *Elements* ... is partly offset by the short-sightedness exhibited by the Greeks when they failed to extend the number concept so as to include the negative and the ordinary complex numbers [!]. In fact, the earlier Greek writers did not include the irrational numbers in their concept of numbers [Imagine, such nasty behaviour!]' (*ibid.*, 318–19).

'It was well that the Greeks developed the theory of conic sections *without awaiting the discovery of the usefulness of this theory in the study of our solar system* ... (*ibid.*, 320, my italics).

Finally: 'The painstaking care which the modern scientist employs in making accurate measurements was foreign to the Greek mind. They devoted their attention to the shorter and easier routes leading to scientific truths' (*ibid.*, 320–21).

I have burdened the reader with this string of quotations for two main reasons: 1. Most of the views expressed by MILLER relate to issues discussed in this study and 2. These views, though representing a much lower degree of sophistication than those embodied in the term 'geometric algebra', stem ultimately from the same condemnable approach to the history of mathematics. If they, rightly, seem offensive and simple-minded, let the reader keep in mind that they, at least, condemn Greek mathematics for not being algebraic rather than (which I think is potentially much more dangerous) *making* it algebraic and then discussing it as such.

V

In this section I shall select a few examples from EUCLID'S *Elements* and analyse them in detail, in order to show, I think peremptorily, the inherent deficiencies of the time-honoured and venerable viewpoint that Greek geometry (at least some very important parts of it) is algebra in disguise.

My examples will be taken from Books II and VI (the books, *par excellence*, containing the so-called 'geometric algebra of the Greeks'), which will enable me to say something about the characteristic features of Greek geometry, from one of the so-called 'arithmetical books' (Book IX) and from Book X, dealing with incommensurable lines and their classification. Let me state that these are just a few examples out of a luxuriant plethora of similar illustrations which 'beautify' the thirteen books of the *Elements*. I have selected them because they represent striking illustrations of my point, namely:

Greek geometry is not algebra (geometric or otherwise) but simply geometry. Clearly, since there is (and this is obvious for us) a complete isomorphism between geometry and algebra – what else if not this is the message of analytical geometry? –, one can practically always use algebraic techniques for transferring the geometrical form and structure to their algebraic, analytical counterparts. There is no quarrel about this. However this is ***not*** the crucial historiographical point! The crucial historiographical point is that in this transfer-process one does irreparable violence and inflicts unrectifiable damage to the unique, peculiar, *sui generis* traits of Greek geometry which are not, let me state this emphatically, reducible to something 'simpler', less 'clumsy', *etc*. There is nothing 'clumsy', 'awkward', 'cumbersome', and so forth about Greek mathematics when it is not taken out of its own context. It certainly was not cumbrous, unwieldy, and oppressive for EUCLID, ARCHIMEDES, and APOLLONIUS, and this is what is historically important and, from our point of view, constitutes the most decisive clincher.

This being my point of view, I shall display, in what I consider to be an irrefragable fashion, the absurd consequences of the traditional interpretation when this interpreta-
89 tion is submitted to the most important test, *i.e., the test of Greek | mathematics itself.* Let me state from the outset that, to my mind, the traditional interpretation does not withstand such a test, – indeed it collapses noisily under its own unwarranted assumptions.

I have included Book X among the sources of my examples because the book is considered by many as the crowning achievement of the *Elements*, the most 'powerful' of all the thirteen books, and because historians of mathematics have traditionally analyzed its contents in purely algebraic terms.[52] Thus, VAN DER WAERDEN thinks that, 'In X 33–35, the solution of ... equations is indicated, for various cases, by the use of geometric algebra.'[53] And, in perusing HEATH'S edition of the *Elements* one is immediately struck by the whole tenor of HEATH'S commentaries on Book X, in which, from the very beginning, he speaks freely (and abundantly) of 'quadratic equations',[54] 'roots of equations of the second degree as are incommensurable with the given

---

[52] *Cf.* TANNERY, *Mém. Scient.*, **1**, 264–67; ZEUTHEN, *Gesch. d. Math. im. Alt. u. Mittel.*, **56**, 158–161; ZEUTHEN, *Die Lehre*, 24–26; HEATH'S edition of EUCLID, *Elements*, **3**, *passim*; VAN DER WAERDEN, *Sc. Awaken.*, 168–172; BOYER, *A Hist. of Math.*, 128–29; *etc., etc.*

[53] *Op. cit.*, 170.

[54] *Elements*, **3**, 5.

magnitudes', 'a classification of ... irrational magnitudes ... arrived at by successive solution of equations of the second degree',[55] and, finally (to cut the quotations short), of the fact that '... the Greeks would write the equation leading to negative roots in another form so as to make them positive, *i.e. they would change the sign of x in the equation.*'[56] HEATH also says, to mention one last shocking example, that the *binomial* and the *apotome* (which he writes as '$\rho \pm \sqrt{k\rho}$')'... are the positive roots of the biquadratic (reducible to a quadratic) $x^4 - 2(1+k)\,\rho^2 \cdot x^2 + (1-k)^2\,\rho^4 = 0$.'[57]

***

Let us now proceed to the promised examples.

Proposition II.5 states: 'If a straight line be cut into equal and unequal segments, the rectangle contained by the unequal segments of the whole together with the square on the straight line between the points of section is equal to the square on the half.'[58]

EUCLID'S proof advances through the following stages: Let *AB* be given and bisected at *C*; let it also be divided into two unequal segments at *D*. Construct the square on *CB* and draw *BE*. Let $DG \parallel CE$. Through *H*, the point of intersection of *DG* and *BE*, let *KM* be drawn $\parallel AB$, and through *A* let *AK* be drawn $\parallel BM$. By I.43, the complement $CH =$ the complement *HF*. Consequently, $CM = DF$. But $CM = AL$ (because $AC = CB$ by hypothesis); therefore $AL = DF$. By adding *CH* to each of the preceding rectangles, it follows that $AH =$ gnomon *NOP*. But *AH* | is the rectangle *AD, DB* 90
(because $AK = DH = DB$), therefore the gnomon $NOP =$ rectangle *AD, DB*. Let *LG* (the square on *CD*) be added to each member of the previous equality. ∴ Square on *CB* = rectangle *AD, DB* + square on *CD*, q.e.d.[59]

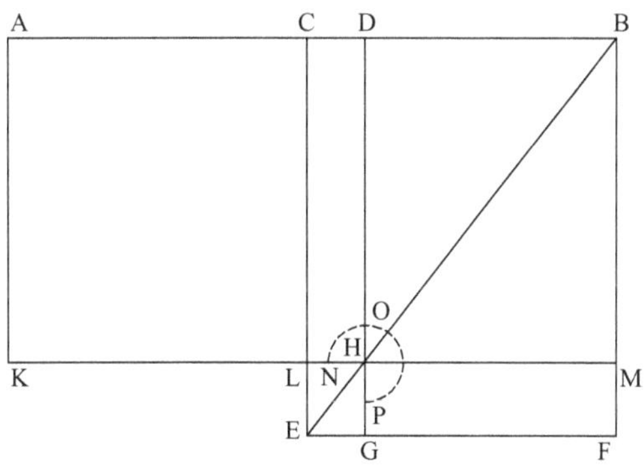

[55] *Ibid.*, 4–5; these expressions are taken over approvingly by HEATH from ZEUTHEN'S *Gesch. d. Math. im. Alt. u. Mittel.*, 56.

[56] *Ibid.*, 5, my italics.

[57] *Ibid.*, 7. By the way, Sir THOMAS avows somewhat belatedly, in a confessional slip, while discussing the character of the first five propositions of Book XIII of the *Elements*, that, '... the method of [proof of] the propositions is that of Book II., being strictly geometrical and not algebraical ...' (*ibid.*, 441).

[58] EUCLID, *Elements*, **1**, 382.

[59] *Ibid.*, 382–83.

These are Euclid's enunciation and proof. There is no trace of equations here as there is no trace of equations anywhere in Greek classical mathematics, *i.e.*, in Greek geometry. The proof is purely geometrical, constructive, intuitive (or visual), in the sense of its appeal to the eye, and it consists of a logical concatenation of statements about geometrical objects (in this case, rectangles, squares, and gnomons). There are no symbols and, consequently, there are no operations performed on symbols; the proof appeals to spatial perception rather than being abstract and it is essentially rooted in what has become known as Aristotelian predicate logic. All these are the very characteristics of Greek geometry.[60]

And yet what do we find in Heath's commentary on II.5? 'Perhaps the most important fact about II.5,6 is however their bearing on the *Geometrical Solution of a quadratic equation*.'[61] How does Heath discern such a bearing? This is very simple:

> Suppose, in the figure of II.5, that $AB = a$, $DB = x$; then
>
> $$ax - x^2 = \text{the rectangle } AH$$
> $$= \text{the gnomon } NOP.$$
>
> Thus, if the area of the gnomon is given ($= b^2$, say), and if a is given ($= AB$), the problem of solving the equation
>
> $$ax - x^2 = b^2$$
>
> is, in the language of geometry, *To a given straight line (a) to apply a rectangle which shall be equal to a given square (b^2) and shall fall short by a square figure*, i.e. to construct the rectangle AH or the gnomon NOP.[62]

91 | What does this prove? That the Greeks solved quadratic equations? *Not at all*! The only thing it proves is that Heath (and Zeuthen, and Tannery, and van der Waerden, and P.H. Michel, *etc., etc.*) can transcribe Euclid's geometry into algebraic symbolism and obtain (in this case) a quadratic equation. Does this tell us anything about the Greeks in general, and about II.5 in particular? *Nothing. There is not the slightest shred of genuine historical evidence that* Euclid (*or the other great Hellenistic mathematicians, let alone the* Pythagoreans) *ever used equations in their geometrical works. The sources do not contain equations.* This, however, does not prevent historians of mathematics from applying foreign (algebraic) techniques to Greek geometry and obtaining thus algebraic counterparts to Greek geometrical propositions, which they, then, illegitimately consider as being the genuine Greek stuff.

Paul-Henri Michel is a case in point. It seems to me highly interesting and significant that he begins his discussion of 'geometrical algebra' with the following statement: 'Pour faire comprendre comment la géométrie *pouvait* "jouer le rôle d'algèbre", nous prendrons un cas très simple.'[63] He then takes the equation $bx = c$, shows how it is

60 *Cf.* 'Babyl. Algebra', 372; see also 'Die Anfänge der algebr. Denkweise', 17–18, *passim*.

61 *Elements*, 383.

62 *Ibid.; cf.* also Zeuthen, *Die Lehre*, 19 and Tannery, *Mém. Scient.*, **1**, 257–59. Also, Zeuthen, *Geschichte d. Math. im. Alt. u. Mittel.*, 47–48 and 52.

63 *De Pythagore à Euclide*, 639, my italics.

mathematically equivalent to simple Greek geometrical techniques of application of areas, and soon claims, serenely and coolly, that the Greeks *solved the equation* by means of the application of areas! Then he says: 'Telle fut longtemps l'algèbre des Grecs.'[64] From here to the next claim there is just one step: 'Les solutions géométriques d'équations du deuxième degré abondent chez Euclide.'[65] As an example, MICHEL uses EUCLID II.5, the same proposition we discussed above. Let us see what it becomes in his skillful hands:

> Si une droite [b] est coupeé en deux parties égales [$b/2$] et en deux parties inégales [x et y], le rectangle [xy ou c] formé par les deux segments inégaux de la droite entière, plus le carré du segment placé entre les sections [$(b/2 - y)^2$] est égal au carré de la moitié de la droite entière [$(b/2)^2$].[66]

The preceding is a beautiful illustration, I think, of the despicable methods of historians of mathematics, which enable them so easily to 'discover' equations in ancient Greek mathematics. Their procedures are clearly unveiled by MICHEL'S square brackets used to transcribe EUCLID'S geometrical proposition into algebraic symbolism, symbolism which does *not* appear in the EUCLIDEAN text at all. Indeed, MICHEL goes
on to say, 'Pour démontrer ce théorème,[67] Euclide fait usage du gnomon des 92
Pythagoriciens'[68], and then he summarizes EUCLID'S geometrical | proof, but not without adding (again in square brackets) the corresponding algebraic expressions (missing in EUCLID), as if EUCLID'S procedure and the algebraic manipulations are exactly one and the same thing! Furthermore, he continues his anachronistic analysis of II.5 by saying the following about EUCLID'S diagram:

> On peut d'ailleurs constater immédiatement sur la figure... l'égalite du gnomon... et du rectangle xy, ou c; et en conséquence *deduire*:
>
> $$x = b/2 + \sqrt{(b/2)^2 - c} = \frac{b + \sqrt{b^2 - 4c}}{2}$$
>
> et
>
> $$y = b/2 - \sqrt{(b/2)^2 - c} = \frac{b - \sqrt{b^2 - 4c}}{2}.$$

---

[64] *Op. cit.*, 640. It is clear that MICHEL has a weird (if unoriginal) view of both algebra and the historical method. Thus he says: 'Nous ne considérons pas l'algèbre comme nécessairement liée à un certain système de symboles, mais, à la suite de M. Thureau-Dangin, comme "une application de la méthode analytique [des Grecs] à la résolution des problèmes numériques" ... Pour qu'il y ait algèbre (non pas algèbre "lettrée" mais algèbre "parlée") il faut mais il suffit qu'une quantité inconnue soit posée d'emblée comme connue. Dès que le mathématicien adopte cette méthode, *son discours est susceptible d'être traduit en équations, ce que nous faisons couramment pour la commodité du lecteur' (ibid.*, 641–42, my italics).

[65] *Ibid.*, 643.

[66] *Ibid.*, 643–44.

[67] This is what II.5 is: a theorem, a *geometrical* theorem and not a quadratic equation, or a problem leading to a quadratic equation!

[68] *Op. cit.*, 644.

> Nous sommes ainsi ramenés aux formules par lesquelles se traduisent les opérations et les résultats de l'algèbre numérique babylonienne.[69]

And so, having translated in good traditional fashion, on the one hand, *both* the Babylonian specific numbers and the Babylonian cookbook-recipe type of solution procedure into algebraic symbols and algebraic operations and, on the other hand, the Euclidean purely geometric procedure into the same symbolic and operational language, Michel may now (like others before him) marvel at their 'identity' and, consequently, establish the necessary historical connections and influences between the two mathematical cultures! History? Perhaps, but certainly not sane, acceptable history. As if to crown his entire discussion, P.H. Michel continues by making this profound historical statement:

> Une même équation est donc susceptible d'être résolu par deux méthodes bien différentes, sur la valeur desquelles nous n'avons pas à nous prononcer.[70]

Beautiful! History? Perhaps, but certainly not sound, acceptable history. It is rather 'logical history', *i.e.*, in more cases than not, *non-history. It is history as it should be* rather than an honest attempt to establish it as it was; it is, in other words, a logical rather than a historical reconstruction.

Noting that, historically, the geometrical treatment (coming after the 'arithmetical atomism' of the Pythagoreans) represented an advancement in Greek mathematics, P.H. Michel asserts: 'La géométrie (*tenant lieu d'algèbre*) permettait en effet la généralisation des calculs arithmétiques et l'inclusion des quantités irrationnelles dans ces calculs généralisés.'[71] Why did geometry take the place of algebra? Where is the algebra which it allegedly replaced? No trace of it is found in the known sources of Greek mathematics and this is for a very good reason: There was no algebra preceding geometry. The *arithmetic* of the Pythagoreans, which (even according to the standard treatment) was replaced by the geometrical approach, was
93 **not** algebra and cannot therefore be identified | historically with it.[72] If words have (within given historical periods) more or less settled meanings, then, most certainly, Pythagorean arithmetic (with its treatment of discrete entities), which was replaced by the geometrical approach (with its treatment of continuous magnitudes), was not algebra.

Historians of mathematics cannot have it both ways! It is logically impossible to claim at one and the same time that, on the one hand, one of the main reasons for the general decline of mathematics in the post-Hellenistic era was due to the Greek emphasis on Geometry,[73] and that, on the other hand, Greek geometry was nothing but algebra in disguise; had the latter been the case, it would have been very easy to abandon the

---

69 *Op. cit.*, 645, my italics.

70 *Ibid.*

71 *Ibid.*, 646, my italics.

72 *Cf.* the following: '... chaque type de problème arithmétique nécessite une invention de l'esprit particulière à ce problème, adaptée à sa solution, et que ne peut pas servir à d'autres types dérectement [*sic*]; car, indirectement, toute opération contribue bien à former l'esprit arithmétique et à faciliter les inventions nouvelles' (*Les Math. en Grèce*, 55–56).

73 See Michel, *op. cit.*, 646; A. Rey, *La Science dans l'Antiquité*, **3**, 388–91; G.A. Miller, *op. cit.*, *passim*.

disguise, to drop the mask, and pursue undisguised algebra, while, as is known, in reality one must wait until the sixteenth century for this to start happening.[74]

It seems to me that it is a considerably more appealing (and certainly historically more defensible) thesis that Greek mathematics, as found in the *Elements*, is an outgrowth of PYTHAGOREAN mathematics, the arithmetical discreteness of the latter (with all its accompanying inherent weaknesses) having been replaced in the former by the continuity of geometrical magnitude; thus, in EUCLID numbers are not collections of points anymore, but segments of straight lines, *etc*. This replacement enabled Greek geometry to deal 'honourably' (and vigorously) with the alleged 'scandal' generated by the discovery of the irrational.[75]

It also seems true that the 'figurative', numerical approach of the PYTHAGOREANS contained somehow in germ another possibility of generalization (and, potentially, of removal of contradictions) than that actually taken by classical Greek mathematics (*i.e.*, the purely geometric approach), and this is the possibility of distinguishing visually relations between numbers of the same kind, by means of the gnomonic differences in their punctiform representation, which relations could, perhaps, be seen retrospectively as a step in the direction of algebra proper. (But algebra did *not* develop in the sixteenth century out of this consideration!)

The only contention one can make in this context with a reasonable degree of accuracy (and it does not amount to an earth-shaking position) is that for the Greek mathematician living before the discovery of the irrational and working within the tradition of arithmetical geometry, the very way of representing numbers geometrically by points and punctiform figures contained intrinsic possibilities of grasping visually numerical relations; in other words, the PYTHAGOREAN way of representing numbers gave the PYTHAGOREAN mathematician an intuitive, visual means of generalization which, undoubtedly, contributed to the progress of mathematics.

Before leaving proposition II.5, I would like to call the attention of the reader | to 94
VAN DER WAERDEN'S discussion of II.5 and II.6 and to SZABÓ'S criticism thereof. For VAN DER WAERDEN both II.5 and II.6 are nothing but the geometrical expression of one and the same algebraic formula: $a^2 - b^2 = (a - b)(a + b)$.[76] 'But', as VAN DER WAERDEN says, 'it can not have been the sole purpose of the two propositions to give [the above] formula... a geometric dress and to prove it in that way; for, why should

---

[74] See A. REY, *Les Math. en Grèce*, 32, 45 (note 1), *passim*; MAHONEY, 'Die Anfänge der algebr. Denkweise', 18, 23, *passim*.

[75] That there was such a 'scandal' in the *mathematical* world is, at best, doubtful. *Cf.*, in this respect, Á. SZABÓ, *op. cit.*, 115.

[76] *Sci. Awaken.*, 120. ZEUTHEN, closer to EUCLID, transcribes II.5 as $(a - b)b + (\frac{1}{2}a - b)^2 = (\frac{1}{2}a)^2$ or as $(a - b)b + \left(b - \frac{1}{2a}\right)^2 = (\frac{1}{2}a)^2$ and II.6 as $(a + b)b + (\frac{1}{2}a)^2 = (\frac{1}{2}a + b)^2$ or $b(b - a) + (\frac{1}{2}a)^2 = (b - \frac{1}{2}a)^2$ (*Die Lehre*, 12). NESSELMANN, for whom, as we saw, these are we saw, these are *arithmetical* propositions, chooses another, equivalent algebraic form for II.5, $ab + \left(\frac{a - b}{2}\right)^2 = \left(\frac{a + b}{2}\right)^2$, and one of the two variants of ZEUTHEN as II.6 (*Die Algebra der Griechen*, 154). This variety and richness in transcription is, in itself, a clear-cut indication that the venerable authors are performing geometricide!

two propositions be given for *one* formula?[77] Indeed, why? Van der Waerden's answer is that the two propositions are really not propositions but '… *solutions of problems*; II 5 calls for the construction of two segments *x* and *y* of which the sum and product are given, while in II 6 the difference and the product are given.'[78]

Árpád Szabó, in one of the most effective criticisms ever leveled against the term of 'geometric algebra', thoroughly takes apart van der Waerden's interpretation, based as it is on unbridled manipulations of algebraic symbols, leading to equations, *etc*. According to Szabó, II.5 is a purely geometrical proposition, more exactly a lemma, necessary in the proof of the very important purely geometrical proposition II.14. This follows not only from modern editions of Euclid, in which during the proof of II.14 one is referred back to II.5, but also from the essentially identical wording of large parts of both propositions in the original, ancient Greek text. Indeed, the very 'clumsiness' and sluggishness of the language in which II.5 is enunciated seems to indicate that it was meant to represent a 'pre-fabricated' constitutive part, to be used ready-made in the proof of II.14.

Furthermore, this is not the only instance of such a procedure in the *Elements*. Other purely geometrical propositions, which were previously taken to illustrate Greek 'geometrical algebra', display a similar relation, in that one of a couple of propositions represents a modular unit necessary in the proof of the other member of the couple. A case in point is represented by II.6 and II.11, where II.6 constitutes such an integral 'pre-fabricated' part of II.11, both being, again, purely geometrical theorems. The reason that II.6 looks as a special case of II.5, says Szabó, is simply that II.11 (in whose proof II.6 is used as a module) is indeed a special case of II.14 (in
95 whose proof II.5 is used as a module).[79] Another case | in point (on the authority of Proclus – in his commentary on Plato's *Republic*) is provided by II.10 which represented a module for a proposition not included in the *Elements*, but which was reconstructed on the very basis of Proclus' remarks.[80]

One does not necessarily have to accept Szabó's preceding interpretation for the 'similarities' between II.5 and II.6, in order to agree with the main thrust of his argument against 'geometric algebra'. To begin with, Szabó shows that even if such a creature as 'Babylonian Algebra' ever existed (and this is rather doubtful), '… *auch dann hat man bisher noch mit gar keiner konkreten Angabe* wahrscheinlich machen können, dass die Griechen in voreuklidischer Zeit eine solche Algebra wirklich gekannt hätten, geschweige denn, dass sie dieselbe übernommen und geometrisiert hätten.

77 *Ibid.*

78 *Ibid.*, 121.

79 I wonder if this really solves the 'problem'! Let me state emphatically that there is a problem *only* if one transcribes II.5 and II.6 into modern symbolism. It is only due to this totally unacceptable procedure that van der Waerden could make his initial claim that II.5 and II.6 are nothing but the same algebraic formula! If one stays within the Euclidean realm (and this is the only admissible procedure), *i.e.*, if one does not transcend the boundaries of geometry, then, most clearly, II.5 and II.6 are **not** the same proposition. Specifically, in the language of application of areas, II.5 asks to apply a rectangle to a given line such that it will be equal to a given square and *fall short* by a square figure, while II.6 asks for the application of a rectangle to a given line such that it will be equal to a given square and *exceed* by a square figure! (*Cf.* Euclid, *Elements*, **1**, 385–86). These can be shown to be the same only by somebody who has the benefit of formulaic expression.

80 Á. Szabó, *Anfänge der griechischen Mathematik*, 458–59.

(Die Griechen haben nicht einmal die positionelle Bezeichnungsart der Zahlen von den Babyloniern übernommen!)'[81]

Furthermore, those so-called 'geometrically clad algebraic propositions' in Euclid are 'algebraic' only in the sense that we can rather easily make them algebraic. 'Aber es kann gar keine Rede davon sein dass diese Theoreme ursprünglich "*algebraische* Sätze" oder Lösungen für "*algebraische* Aufgaben" gewesen wären. Nein, diese sind alle sowohl die Sätze wie auch die Aufgaben – *rein geometrischen Ursprungs*. Auch II.5 ist ein rein geometrischer Satz. Wohl kann man diesen Satz in der modernen Interpretation mit einer "algebraischen Aufgabenlösung" *vergleichen*. Aber man nehme sich in acht, damit ein solcher Vergleich den ursprünglichen und echt geometrischen Sinn des Satzes nicht verdunkle!'[82]

Szabó points out that the problem of incommensurability itself was originally a *geometrical* problem[83], and he chooses to talk of 'Pythagorean geometry of surfaces'[84], rather than the hackneyed and wrong term of 'geometric algebra'. Summing up his criticism, Szabó says:

> Es wäre irreführend, diesen Satz [i.e., II.5] als "Lösung einer algebraischen Gleichung" aufzufassen. Die algebraische Auslegung – auch wenn sie dem Satz Euklids äquivalent ist – verdunkelt den wahren geometrischen Sinn dieses Satzes, und historisch erweckt sie den falschen Schein, als hätten die Griechen in voreuklidischer Zeit in der Tat mit "algebraischen Gleichungen" operiert.[85]

| And finally: 96

> . . . auch die übrigen Sätze der sog. "geometrischen Algebra der Pythagoreer" sich als rein geometrische Sätze erklären lassen. Dagegen hat man Spuren von echt algebraischen Gedankengängen aus der voreuklidischen und Euklidischen Überlieferung bisher *nicht* nachweisen können.[86]

***

Our next example, from a Book considered a mainstay of 'geometric algebra' is VI.28. Euclid's enunciation is:

> To a given straight line to apply a parallelogram equal to a given rectilineal figure and deficient by a parallelogramic figure similar to a given one: thus the given rectilineal figure must not be greater than the parallelogram described on the half of the straight line and similar to the defect.[87]

---

81 *Ibid.*, 457.

82 *Ibid.*, 458. Szabó aims a scathing criticism at the Tannery-Zeuthen thesis: 'H.G. Zeuthen, dessen "Verdienste" um die Entdeckung der sog. "geometrischen Algebra der Griechen" durch O. Neugebauer . . . so übertrieben hervorgehoben wurden, hat in Wirklichkeit in seinen beiden *Werken* (Die Lehre von den Kegelschnitten im Altertum, . . . und Geschichte der Mathematik im Altertum und Mittelalter, . . .) – was die "geometrische Algebra" betrifft – nur den irreführenden *Vergleich* von P. Tannery weitergebaut. (Man hätte sich nämlich erst einmal fragen müssen, inwiefern überhaupt *erlaubt* ist, im Zusammenhang mit Euklids *geometrischen Konstruktionen* über Lösungen von *algebraischen Gleichungen* zu reden!)' (*ibid.*, note 6, 457). For additional elements of this criticism, *cf. op. cit.*, 35–36, 474, 488.

83 *Ibid.*, 36.

84 *Ibid.*, 465, *passim*.

85 *Ibid.*, 487.

86 *Ibid.*

87 Euclid, *Elements*, **2**, 260.

EUCLID'S proof proceeds as follows: Let AB be the given line, C the given rectilineal figure (not greater than the parallelogram described on half of AB and similar to the defect), and D the parallelogram to which the defect is to be similar.

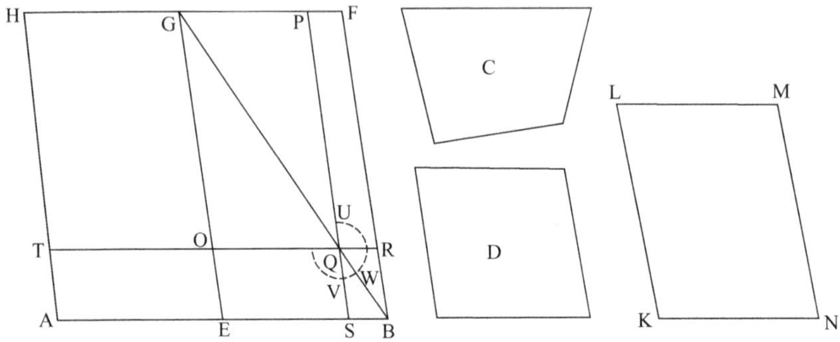

It is required to apply to AB a parallelogram equal to C and deficient by a parallelogramic figure similar to D. Bisect AB at E. On EB construct the parallelogram $EBFG$ similar and similarly situated to D. (This is done by VI.18.) Complete the parallelogram AG.

Now, if ▱ $AG = C$, then the requirement is obviously fulfilled. If however, ▱ $AG \neq C$, then the only remaining possibility (due to the διορισμός included in the enunciation) is that ▱ $AG > C$. If this is the case, then ▱ $GB > C$ since ▱ AG = ▱ GB, by construction. Let, now, ▱$KLMN$ be constructed such that it is at one and the same time equal to the excess of ▱ GB over C and similar (and similarly situated) to D. (This can be done by VI.25.)

97 | Since ▱ $GB \sim$ ▱ D, $\therefore$ ▱ $KM \sim$ ▱ D (by VI.21). Assume that in the two similar parallelograms, GB and KM, the corresponding sides are respectively GE and KL and GF and LM.

Since ▱ $GB = C$ + ▱ KM (by construction), $\therefore$ ▱ $GB >$ ▱ KM. EUCLID now concludes (and this is a tacit assumption) that

$$GE > KL \quad \text{and} \quad GF > LM$$

Let, then, $GO = KL$ and $GP = LM$ (by construction), and let the ▱ $GOPQ$ be constructed. Obviously, ▱ $GQ \cong$ ▱ KM. $\therefore$ ▱ $GQ \sim$ ▱ GB (by VI.21), and, by VI.26, $\therefore$ ▱ GQ is about the same diameter with ▱ GB. Let the common diameter GQB be described. Since ▱ $BG = C$ + ▱ KM and ▱ GQ = ▱ KM, $\therefore$ gnomon $UWV = C$. Furthermore, since ▱ PR = ▱ OS (by I.43), $\therefore$ ▱ PB = ▱ OB. But ▱ OB = ▱ TE (by I.36), since $AE = EB$. $\therefore$▱ TE = ▱ PB, $\therefore$ ▱ TS = gnomon UWV. But gnomon $UWV = C$, $\therefore$ ▱ $TS = C$, q.e.d.[88]

Again, no algebraic symbols, no equations; a typical, purely geometrical, proposition belonging to the PYTHAGOREAN geometry of surfaces. Only somebody already steeped in the modern algebraic wisdom can 'discern' the 'algebraic line of thought'

88 *Ibid.*, 260–62.

behind the traditional geometrical reasoning, transcribe the proposition into modern symbolism, and, then (if he does not discern the historical blunder involved and the *non sequitur* on which his conclusion is based) claim that this proposition is nothing but geometrically clad algebra. This is exactly what VAN DER WAERDEN[89] and T.L. HEATH are doing.

For HEATH this proposition '... is the geometrical equivalent of the solution of the quadratic equation $ax - \frac{b}{c}x^2 = S$, subject to the condition necessary to admit of a real solution, namely that $S \ngtr c/b \cdot a^2/4$.'[90] Who can see that? If you face a smart student of Greek synthetic geometry, whose mind was never exposed to the algebraic way of thinking, with HEATH'S statement quoted above, there is not the slightest doubt whatever that he would fail to understand it. Indeed, I think that EUCLID *himself would have failed to understand* HEATH'S *statement*; not because EUCLID was less smart than HEATH, but because, living when he did, he did not have at his disposal what HEATH had in the nineteenth and twentieth centuries (primarily algebra and analytical geometry) and because (and this is another way of saying the same thing) his pattern of mathematical thought was different than HEATH'S. HEATH approached mathematics *algebraically*, EUCLID (like all the Greeks) approached it *geometrically*, and never did the twain seriously meet before the sixteenth century.[91]

***

89 *Op. cit.*, 121–22; *cf.* also ZEUTHEN, *Die Lehre*, 19–20, 29–31 and *Gesch. d. Math. im Alt. u. Mittel.* 47–48.

90 *Elements*, **2**, 263.

91 HEATH goes on, in his commentary, 'To exhibit the exact correspondence between Euclid's geometrical and the ordinary algebraical method of solving the equation ...' (*ibid.*, 263). He transforms the equation in various ways and, finally reaches the following expression for the root:

$$x = \frac{c}{b} \cdot \frac{a}{2} \pm \sqrt{\frac{c}{b}\left(\frac{c}{b} \cdot \frac{a^2}{4} - S\right)}.$$

Then he manages to show how he can find for every step in EUCLID'S proof a corresponding algebraic expression, until he reaches an expression for *QS* (see figure above) identical to the expression he found previously for *x* (with *minus* before the radical). No quarrel with HEATH'S procedure as long as he does not ascribe it to EUCLID. However, that is exactly what HEATH does when he says: ' ... Euclid *really* finds *GO* from the equation $GO^2 \cdot \frac{b}{c} = \frac{c}{b} \cdot \frac{a^2}{4} - S$ (*ibid.*, 264, my italics). This is inexcusable! There are other unacceptable historical blunders in HEATH'S commentary. For instance, realizing that EUCLID'S solution 'corresponds' to only one root of HEATH'S equation, the latter makes the following remark: 'He [*i.e.*, EUCLID] cannot have failed to see [?] how to *add GO* to *GE* would give another solution' (*ibid.*); then HEATH shows how '... the other solution can be arrived at ... [!]' (*ibid.*). The root of the last blunder can be found in ZEUTHEN, *Die Lehre*, 19–21.

98 | We are now going to discuss two propositions belonging to Book IX. In proposition IX.8, EUCLID says:

> If as many numbers as we please beginning from an unit be in continued proportion, the third from the unit will be square, as will also those which successively leave out one; the fourth will be cube, as will also all those which leave out two; and the seventh will be at once cube and square, as will also those which leave out five.[92]

The EUCLIDEAN proof proceeds as follows (and my paraphrase is *faithful* to EUCLID'S way of reasoning):

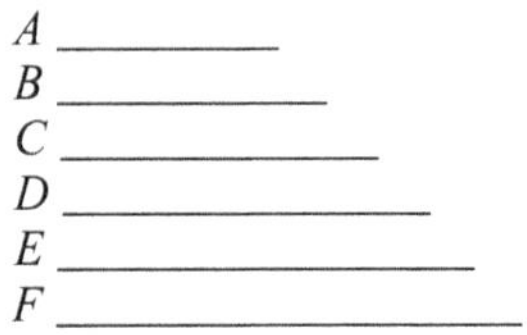

Let there be given as many numbers as one pleases, starting from the unit, like *A, B, C, D, E*, and *F*. By the definition of 'continued proportion', it follows that as the unit is to *A* so is *A* to *B*. But the unit measures *A*; so, as many times as the unit measures *A, A* measures *B* (by Definition VII.20). Now, since the unit measures *A* according to the units in *A, A* measures *B* also according to the units in *A*. Therefore *A* by multiplying itself makes *B*, and so *B* is a square. Furthermore, since *B, C, D* are in continued proportion and *B* is square it follows (by VIII.22) that *D* is also square. For the same reason, *F* is square and so are all those which leave out one.

Now since as the unit is to *A*, so is *B* to *C*, the unit measures *A* the same number of times that *B* measures. *C*. But, again, the unit measures *A* according to the units in the latter; thus *B* measures *C* according to the units in *A, i.e., A* by multiplying *B* makes *C*.

Now, since *A* by multiplying itself makes *B* and by multiplying *B* makes *C*, it follows that *C* is cube. Furthermore, since *C, D, E*, and *F* are in continued proportion
99 and *C* is cube, it follows (by VIII.23) that *F* is also cube. But *F* was | also proved to be square so that the seventh from the unit is both cube and square. In a similar fashion one can prove that all other numbers which leave out five are both cube and square, q.e.d.[93]

So much for EUCLID. This proposition can serve, I think, as a beautifully striking example of the inherent limitations and the built-in chronic inadequacies of the beloved method (practiced by mathematicians posing as historians) of automatically and ahistorically transcribing EUCLID'S language into the modern symbolism of algebra. A proposition for the proof of which EUCLID has to toil energetically (perhaps, it would be no great exaggeration to say, with might and main) and in the course of whose proof he had to rely on many previous propositions and definitions (*e.g.*, VIII.22, VIII.23, def. VII.20), becomes a trivial commonplace, which is an immediate outgrowth, a trite after-effect of our symbolic notation:

$$1, a, a^2, a^3, a^4, a^5, a^6, a^7, a^8, a^9, \ldots$$

---

92 *Elements*, 2, 390.
93 *Ibid.*, 390–91.

As a matter of fact, if we use modern algebraic symbolism, this *ceases altogether to be a proposition* and its truthfulness is an immediate and trivial application of the definition of a geometric progression in the particular case when the first member equals one and the ratio, q, is a positive integer (for EUCLID)!

Second Example

In proposition IX.9, EUCLID states that:

> If as many numbers as we please beginning from a unit be in continued proportion, and the number after the unit be square, all the rest will also be square. And, if the number after the unit be cube, all the rest will also be cube.[94]

The Euclidean Proof

Let there be given as many numbers as we please starting from an unit and in continued proportion, namely *A, B, C, D, E*, and *F*, where *A*, the number after the unit, is square. By the previously proved proposition (IX.8), *B* and all those which leave out one are square. *A, B, C* being in continued proportion and *A* being square, it follows (by VIII.22) that *C* is also square. In the same manner, one can prove that all the rest are also square.

| Now let *A* be cube. By IX.8, *C* is cube and so are all those which leave out two. 100
Since as the unit is to *A* so is *A* to *B*, the unit measures *A* as many times as *A* measures *B*. But the unit measures *A* according to the units in it; so *A* measures *B* according to the units in *A*. Consequently, *A* by multiplying itself makes *B*. But *A* is cube, and a cube by multiplying itself makes a cube (by IX.3). So *B* is cube. Now, *A, B, C, D*, are four numbers in contiunued proportion, the first of which is cube; consequently, by VIII.23, *D* is also cube. And, for exactly the same reason, *E* is also cube, and so are all the rest, q.e.d.[95] So much for EUCLID.

What happens to this proposition if we sin once more and employ the modern notation? Clearly, its 'propositional' character — if I may be allowed to use this term in such a context — vanishes and *the proposition* becomes once more a trivial consequence of the general definition of a geometric progression.

$$1, a^2, a^4, a^6, a^8, \ldots, a^{2n}, \ldots$$

$$1, a^3, a^6, a^9, a^{12}, \ldots, a^{3n}, \ldots$$

---

94 *Ibid.*, 392.
95 *Ibid.*, 392–93.

There is nothing to be proved, there is no proposition any more. Again, in the *definition* of a geometric progression, some particular values have been substituted for the first term and the ratio, and the whole thing is nothing but this particularized form of the definition!

***

Let me now switch Books and pick up some EUCLIDEAN propositions from the longest and one of the most remarkable books of the *Elements, viz.*, Book X, most of the results of which are, apparently, due to THEAETETUS. EUCLID,[96] however, as usual, systematized, made precise definitions and distinctions, and clarified.

Proposition X.9

> The squares on straight lines commensurable in length have to one another the ratio which a square number has to a square number; and squares which have to one another the ratio which a square number has to a square number will also have their sides commensurable in length. But the squares on straight lines incommensurable in length have not to one another the ratio which a square number has to a square number; and squares which have not to one another the ratio which a square number has to square number will not have their sides commensurable in length either.[97]

Now EUCLID'S proof proceeds along the following lines: First, if *A*, *B* are commensurable in length, then *A* has to *B* the ratio which a number has to a number (by X.5). Let this ratio be equal to the ratio of *C* to *D*, i.e., *A* is to *B* as *C* is to *D*.

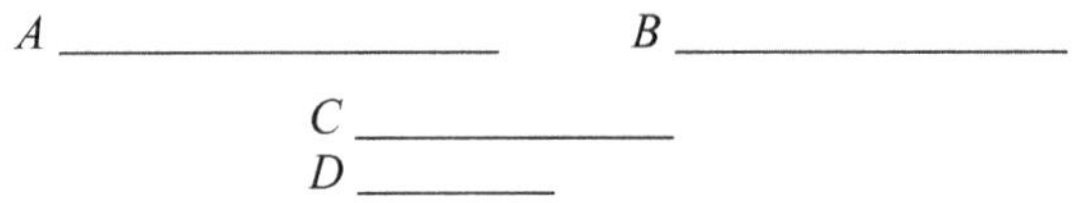

101

But the square on *A* has to the square on *B* a ratio which is the duplicate of the ratio which *A* has to *B*, since similar figures are in the duplicate ratio of their corresponding sides (by VI.20 Porism); and the ratio of the square on *C* to the square on *D* is the duplicate of the ratio of *C* to *D*, for between two square numbers there is one mean proportional number, and, by VIII.11, the square number has to the square number a ratio duplicate of that which the side has to the side. Consequently, as the square on *A* is to the square on *B*, so is the square on *C* to the square on *D*.

Next, let us assume that as the square on *A* is to the square on *B* so is the square on *C* to the square on *D*. We must prove that *A* is commensurable in length with *B*.

From the hypothesis it follows, says EUCLID, that as *A* is to *B* so is *C* to *D*.[98] Consequently, *A* has to *B* the ratio which a number (*C*) has to a number (*D*), *i.e.*, *A* is commensurable in length with *B* (by X.6).

---

96 This follows from a scholium to X.9 and from PAPPUS' commentary on Book X, preserved in Arabic; *cf.*, however, concerning the veracity of these sources, Á. SZABÓ, *op. cit.*, 100–111.

97 *Elements*, **3**, 28.

98 Incidentally, EUCLID takes this for granted, *i.e.*, without further ado, he assumes that ratios the duplicates of which are equal are themselves also equal; the converse of this assumption was employed in the preceding stage of the proof.

Next, assume that A is incommensurable in length with B. (The proof proceeds by *reductio ad absurdum*.) If the square on A has to the square on B the ratio which a square number has to a square number, then, by the immediately preceding, it would follow that A is *commensurable* in length with B, which it is not; therefore the square on A cannot have to the square on B the ratio which a square number has to a square number.

Again, assume now that the square on A has *not* to the square on B the ratio which a square number has to a square number. If A were commensurable in length with B, then, by the preceding, the square on A would have to the square on B the ratio which a square number has to a square number, which is not the case; consequently, A is not commensurable in length with B, q.e.d.[99]

What does this theorem become when one throws away the geometrical négligé which barely covers its algebraic nudity ..., in order to uncover the hidden charms of the latter?

If A, B be straight lines and C, D be numbers, then, if $A/B = C/D$, $A^2/B^2 = C^2/D^2$ and conversely. That is it! Is *this* what EUCLID says? Is this what EUCLID hid from our view in his geometrical dishabille? Were one to believe the partisans of 'geometric algebra', the answer to these questions should be an unequivocal yes. And yet the very serious deficiencies of such an interpretation are, I submit, self-evident in this case. To be sure, even HEATH himself, in his commentary on this proposition, says:

> This inference, which looks so easy when ... symbolically expressed, was no means so easy for Euclid owing to the fact that $\underline{a}$, $\underline{b}$ are straight lines, | and $\underline{m}$, $\underline{n}$ numbers.[100] 102
> He has to pass from $\underline{a}:\underline{b}$ to $a^2:b^2$ by means of VI.20, Por. through the duplicate ratio; the square on $\underline{a}$ is to the square on $\underline{b}$ in the duplicate ratio of the corresponding sides $\underline{a}$, $\underline{b}$. On the other hand, $\underline{m}$, $\underline{n}$ being *numbers*, it is VIII.11 which has to be used to show that $m^2:n^2$ is the ratio duplicate of $\underline{m}:\underline{n}$.[101]

What HEATH says is, in effect, an unwitting confession of the ahistoricity lying at the very root of the concept of 'geometric algebra' (which, incidentally, does not prevent Sir THOMAS from using indiscriminately modern symbolism in the very same commentary, a few paragraphs below the above quotation)![102]

I shall not belabour this point anymore. Let us now go over to some other examples culled from Book X. The lemma before proposition X.22 states that:

> If there be two straight lines, then, as the first is to the second, so is the square on the first to the rectangle contained by the two straight lines.[103]

EUCLID proves this in the following manner:
FE and EG being two straight lines, as FE is to EG so is the square on FE to the rectangle FE, EG. Let us describe on FE the square FD, and let GD be completed. By VI.1, it follows that as FE is to EG so is FD to GD. But FD is the square on FE and GD is the rectangle GE, DE, *i.e.*, the rectangle GE, FE. Hence, as FE is to EG so is

---

[99] *Ibid.*, 28–30. There is a porism (and a lemma) after this proposition; they do not interest us here.

[100] This is how HEATH transcribed EUCLID'S enunciation: 'If a, b be straight lines, and $a : b = m : n$, where m, n are numbers, then $a^2 : b^2 = m^2 : n^2$ and conversely' (*ibid.*, 30).

[101] *Ibid.*, 31.

[102] *Ibid.*

[103] *Ibid.*, 50.

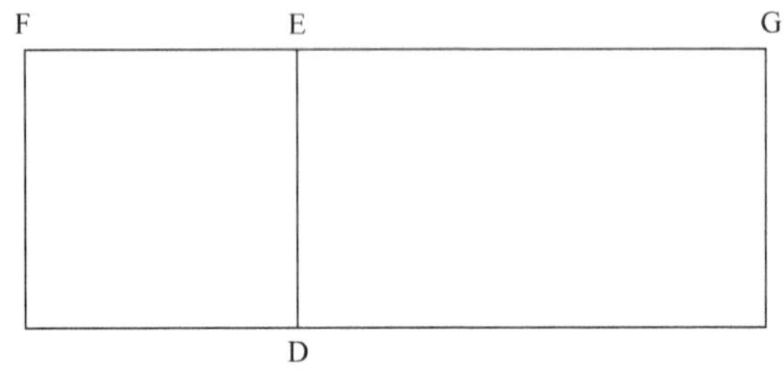

the square on *FE* to the rectangle *FE*, *GE*. Similarly, EUCLID goes on, as the rectangle *GE*, *FE* is to the square on *FE*, that is, as *GD* is to *FD*, so is *GE* to *EF*, q.e.d.[104]

Compare the above proof with the algebraic content of the lemma, which says that $a/b = a^2/ab$. In its algebraic form, the triviality of the entire enterprise becomes striking. The lemma becomes nothing but an inane, vapid, banal illustration of the simplification of fractions!

A similar instance is provided by the lemma after proposition X.53. In it EUCLID shows that if *AB*, *BC* are two squares, placed as they are in the accompanying diagram, and if the
103 parallelogram *AC* be completed, then *AC* is a square, | *DG* is a mean proportional between *AB* and *BC*, and, finally, *DC* is a mean proportional between *AC* and *CB*.[105]

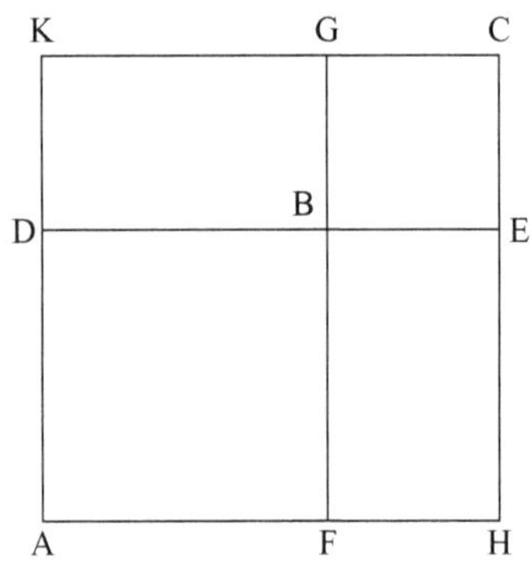

EUCLID'S proof contains the following steps:

$$DB = BF$$

$$BE = BG$$

$$\therefore DE = FG.$$

But $DE = AH = KC$, and $FG = AK = HC$, by I.34.

$\therefore AH = KC = AK = HC$, hence the ▱ *AC* is equilateral. It is also rectangular; therefore it is a square.

104 *Ibid.*, 50–51.
105 *Ibid.*, 115.

Now since *FB* is to *BG* as *DB* is to *BE* and

$$\left.\begin{array}{l}\text{as } FB \text{ is to } BG\text{, so is } AB \text{ to } DG\text{, and}\\ \text{as } DB \text{ is to } BE\text{, so is } DG \text{ to } BC\text{,}\end{array}\right\} \quad \text{(by VI.1),}$$

then it follows that as *AB* is to *DG*, so is *DG* to *BC* (by V.11). Consequently, *DG* is a mean proportional between *AB* and *BC*.

Next, since as *AD* is to *DK*, so is *KG* to *GC* (for they are respectively equal) and, *componendo*, as *AK* is to *KD*, so is *KC* to *GC* (by V.18), while,

$$\text{(by VI.1)} \quad \left\{\begin{array}{l}\text{as } AK \text{ is to } KD\text{, so is } AC \text{ to } DC\text{, and}\\ \text{as } KC \text{ is to } GC\text{, so is } DC \text{ to } BC\text{, it follows}\end{array}\right.$$

that as *AC* is to *DC*, so is *DC* to *BC* (by V.11), *i.e.*, that *DC* is a mean proportional between *AC* and *BC*, q.e.d.[106]

In algebraic notation, this lemma asserts that

$$x^2/xy = xy/y^2 \quad \text{and} \quad (x+y)^2/(x+y)y = (x+y)y/y^2.$$

Not only is this a beautiful example of the inadequacies inherent in transcribing geometrical propositions into algebraic notation, which transform once more a | rather involved – 104
though straightforward – proposition into a trivial matter of simplifying fractions, but, what is even more significant, the first half of the proposition becomes *obviously* (in algebraic form) a repetition of something already proved by EUCLID during his proof of X.25![107] Uneasy about this fact, HEATH suggests that the lemma may not be genuine! ...[108] There are some real dangers lurking behind the back of 'geometric algebra' ...

Passing now to more complicated propositions, let me mention – without reproducing the proof, which is long and rather difficult – proposition X.92. EUCLID'S enunciation reads:

> If an area be contained by a rational straight line and a second apotome, the "side" of the area is a first apotome of a medial straight line.[109]

---

[106] *Ibid.*, 115–16.

[107] *Cf. ibid.*, 56–57, especially the beginning of 57.

[108] *Ibid.*, 116.

[109] *Ibid.*, 194. To enable the reader to grasp the meaning of the proposition and, at the same time, to give him an inkling of its complexity, I shall define the crucial concepts appearing in it: An *apotome* is an irrational straight line obtained by subtracting from a rational straight line another rational straight line, the two rational straight lines being commensurable in square only.

The *Annex* is the straight line which, when added to a compound irrational straight line obtained by subtraction (like an *apotome*) makes up the greater term, *i.e.* the *annex* is the negative term in an apotome.

A *second apotome* is an apotome having the following characteristics: Given a rational straight line and an apotome, if the square on the whole be greater than the square on the annex by the square on a straight line commensurable in length with the whole, and the annex be commensurable in length with the rational straight line set out, the apotome is called a *second apotome.*

A *medial* straight line is a mean proportional between two rational straight lines commensurable in square only.

A *first apotome of a medial* straight line is an irrational line obtained by subtracting from a medial straight line another medial straight line commensurable with the former in square only and which contains with it a rational rectangle.

EUCLID'S proof is, as usual, *completely* geometrical in character (relying on *two* diagrams and on *many* previous propositions), and in its course EUCLID uses the method of application of areas and the EUDOXEAN theory of proportions developed in Book V.[110] Clearly, there is not, and there could not be, any talk of 'equations', 'square roots', *etc.*, and no algebraic symbolism whatever is used. Yet HEATH, in his lengthy commentary,[111] starts by saying:

> This proposition amounts to finding and classifying
>
> $$\sqrt{\rho\left(\frac{k\rho}{\sqrt{1-\lambda^2}} - k\rho\right)}.$$
>
> The method is that of the last proposition. Euclid solves, first, the equations
>
> $$u + v = \frac{k\rho}{\sqrt{1-\lambda^2}}$$
> $$uv = \tfrac{1}{4} k^2 \rho^2. \qquad (1)$$
>
> Then, using the values of u, v so found, he puts

105

> $$x^2 = \rho u$$
> $$y^2 = \rho v \qquad (2)$$
>
> and $(x - y)$ is the square root required.[112]

A greater discrepancy than that between what EUCLID is doing and HEATH'S 'translation' of it is indeed hard to come by, although, from the point of view of modern mathematics, what HEATH is doing is correct, and the two ways of approaching the proposition are mathematically equivalent. Historically, however, there is an unbridgeable chasm between EUCLID'S *way* and HEATH'S *way!* Sir THOMAS' procedure is, I think, a vivid exemplification of what the Italians must have meant when they came up with the phrase 'traduttore traditore'!

For whatever it is worth, let me note that in a graduate seminar on EUCLID'S *Elements*, the students and I have found that one of the greatest difficulties in the study and understanding of Book X – typically considered among the most difficult, if not **the** most difficult in the *Elements* by historians of mathematics – consists in HEATH'S modern interpretation of it and in the dangerous exercises in intellectual equilibrium required by the continuous adjustment to the two incommensurable ways of thinking when switching from EUCLID to HEATH and *vice versa*. To read Book X **not** through modern eyes, it would appear, removes the brunt from such an undertaking. A good edition for such a purpose is the copy of the *Elements* in Great Books of the Western World, which contains HEATH'S translation without his commentaries! ...

I shall not burden the patience of the reader with many more examples of the type I have been lastly discussing. Let me only note that beautiful instances of the historical incompatibility between EUCLID'S *geometry* and HEATH'S *algebra* are offered by propositions X.100, X.101, X.102, and X.103.[113]

---

110 *Ibid.*, 194–97.
111 *Ibid.*, 197–98.
112 *Ibid.*

What I would like to do now is to call the reader's attention to a string of three consecutive propositions in Book X,[114] namely, 112, 113, and 114, which, I believe, show beyond any reasonable doubt that what EUCLID is doing is **not** algebra, but geometry. These are all long and (relatively) involved propositions, which I do not intend to prove here (for obvious reasons). I must, however, reproduce the enunciations so that the reader can get the flavour of EUCLID's ideas and, then, see my point more easily.

Proposition X.112

> The square on a rational straight line applied to the binomial straight line[115] produces as breadth an apotome the terms of which are commensurable with the terms of the binomial and moreover in the same ratio; and further | the apotome so arising will have the same order as the binomial straight line.[116]

106

Proposition X.113

> The square on a rational straight line, if applied to an apotome, produces as breadth the binomial straight line the terms of which are commensurable with the terms of the apotome and in the same ratio; and further the binomial so arising has the same order as the apotome.[117]

Proposition X.114

> If an area be contained by an apotome and the binomial straight line the terms of which are commensurable with the terms of the apotome and in the same ratio, the "side" of the area is rational.[118]

It is easy for us to discern the complete symmetry of these three enunciations. There is no symmetry, however (and this is of utmost significance), in the three proofs. About X.112 HEATH says that '... it is the equivalent of rationalising the denominators of the fractions $\frac{c^2}{\sqrt{A}+\sqrt{B}}$, $\frac{c^2}{a+\sqrt{B}}$, by multiplying numerator and denominator by $\sqrt{A}-\sqrt{B}$ and $a-\sqrt{B}$ respectively [!].'[119] HEATH goes on to say that 'Euclid proves that $\frac{\sigma^2}{\rho+\sqrt{k\rho}} = \lambda\rho - \sqrt{k}\cdot\lambda\rho\,(k < 1)$, and his method enables us to see that $\lambda = \sigma^2/(\rho^2 - k\rho^2)$.'[120] In the continuation of his commentary, HEATH

---

[113] *Ibid.*, 221–231.
[114] *Ibid.*, 243–53.
[115] A *binomial* straight line is an irrational line obtained from the addition of two rational straight lines commensurable in square only.
[116] *Ibid.*, 243.
[117] *Ibid.*, 248.
[118] *Ibid.*, 252.
[119] *Ibid.*, 246. Isn't this a 'faithful' translation? ...

considers it a certainty that '... the Greeks must have had some analytical method which suggested the steps of such proofs',[121] the entire burden of the statement seeming to be that 'analytical' is used here as synonymous to 'algebraic'.[122]

| Now I happen to think that this is not at all certain. It is certain only (and I am not referring specifically to HEATH, whose serious scholarly contributions to the history of mathematics are firmly established) for somebody who has grown arrogant of

---

120 *Ibid.*

121 *Ibid.*

122 On 'genuine' Greek analysis see R. ROBINSON, 'Analysis in Greek Geometry', *Mind*, N.S., **45** (1936), 464–73 and M. MAHONEY, 'Another Look at Greek Geometrical Analysis' (full reference in note 26, above). MAHONEY'S article is, on the whole (as are his other studies), perceptive, penetrating and insightful. Concerning his attitude toward 'geometrical algebra', however, these qualities are lacking. Thus, he says: 'As they are given in the *Data*, however, the theorems pertaining to geometrical algebra are *cumbersome* [?], involving as they do the intricate construction of plane figures. Working mathematicians used a simpler form of geometrical algebra, an algebra of line lengths ... Although it is an example of theoretical, rather than problematical, analysis, the analysis of Euclid XIII, 1 ... illustrates the use of the simplified algebra of line lengths ...' (*op. cit.*, 331, my italics).

Where is there any concrete, specific proof for the use of 'geometric algebra' in pre-EUCLIDEAN, or even EUCLIDEAN, times? *There is none!* The reference to the scholiast's interpolation to XIII.1 misses the point, I think, since the exact date of the interpolation is unknown;
107 furthermore, one of the few | positive statements one can make about this interpolation (and others) is that it is *spurious*, or, as HEATH put it, '... altogether alien from the plan and manner of the *Elements'* (*Elements*, **3**, 442). It was interpolated perhaps as late as 500 years after the writing of the *Elements* (HEATH says that all the interpolations to XIII.1–5 '... took place before Theon's time ...' (*ibid.*)–*i.e.*, fourth century A.D.), and the method of proof it displays is totally foreign to classical Greek mathematics. There is not one shred of reliable historical evidence to support the speculations of BREITSCHNEIDER, HEIBERG, *etc.* that this method represents '... a relic of analytical investigations by Theaetetus or Eudoxus ...' *(ibid.)*; indeed the whole history of Greek mathematics seems to exclude such an inference. But even if one believes HEIBERG'S later dating (in 'Paralipomena zu Euklid', *Hermes*, **38** (1903), 46–74, 161–201, 321–356), namely that the author of these interpolations is HERON OF ALEXANDRIA, this would still make these additions some 400 years younger than EUCLID and would place them comfortably (to the exclusion of PAPPUS and DIOPHANTUS) *after* the decline of classical Greek mathematics.

MAHONEY also speaks of the '... increased use of an informal, but subtle and penetrating, *algebra of line lengths*' ... (*op. cit.*, 337, my italics) in the works of the post-EUCLIDEAN mathematicians of the 3rd century, APOLLONIUS and ARCHIMEDES. Then he goes on saying, 'ARCHIMEDES provides an example of these analyses' (*ibid.*). His reference to proposition II.1 of *On the Sphere and the Cylinder*, however, does not warrant any allusion to ARCHIMEDES' proof as an 'algebra of line lengths.' (*Cf.* J.L. HEIBERG, ed., *Archimedes Opera Omnia*, 2nd ed. (Leipzig: Teubner, 1910) **1**, 170–74.) The proof is still essentially geometrical in the best tradition of EUCLID'S *Elements!* MAHONEY, then, proceeds to give a detailed example of '... Greek geometrical analysis in action, *one which proceeds by an algebra-like manipulation of line lengths ...*' (*ibid.*, my italics). His chosen example is proposition II.4 of ARCHIMEDES' *On the Sphere and the Cylinder*. I must say, however, that I am unconvinced. Again, what ARCHIMEDES is doing in II.4 of *On the Sphere and the Cylinder* is very much like what EUCLID is doing in the *Elements* (though there are obviously differences, some of which do point toward a freer manipulation of lines; interestingly, however, in both examples given by MAHONEY the line lengths *are closely associated* with two or three-dimensional figures!); I cannot see how somebody whose mind was not 'corrupted' by algebraic reasoning and manipulations can describe ARCHIMEDES' proof as 'algebra-like manipulation of line lengths', though, to be sure, this name is less offensive than 'geometrical algebra'.

the past and cannot consequently think anymore, when complicated geometrical ques- 108
tions are involved, but in analytical terms. In other words, it is certain for somebody who knows how to get out of geometrical difficulties by translating them into analytical terms. What I am saying, I guess, is that if **we** do not see any other way, it does **not** mean that the *Greeks*, who obviously did not have our algebra, did not see any either! So the Greeks did **not** use '... geometry as the equivalent of our algebra' – this is infatuation of the twentieth century with its own great achievements and it certainly is anachronistic – *they used geometry*.' It is **we** who are using algebra, with remarkable dexterity, I must confess, as the equivalent of *their geometry!*

Furthermore, in a more substantive fashion (and in a less polemical vein), if Euclid's *lines* were *general algebraic symbols* (which they are not), which could be manipulated like such symbols, then the essence of X.112 could be expressed as follows: If $R^2 = B \cdot A$, where R is a rational line and B is a binomial, then A is a corresponding apotome. Under such circumstances, X.113 would follow immediately and trivially from X.112, as a consequence of the unicity of algebraic operations and the commutativity of multiplication, since X.113 states only that

> If $R^2 = A \cdot B$, where R is rational and A an apotome, then B is a corresponding binomial.

In such a setting, all of Euclid's efforts to prove X.113[123] would have been in vain, and therefore incomprehensible. Indeed, under such circumstances, no proof at all of

---

One more remark. Speaking of Euclid's *Porisms*, Mahoney says that it represents '... the best example of the sort of treatise included in the Treasury of Analysis. It also illustrates well Tannery's remark that the Greeks lacked not so much the methods as the language to express them' (*ibid.*, 343–44). There is another laudatory reference to Tannery's saying at the end of the article. According to Mahoney, the *Porisms* indicates '... why the lack of a suitable mode of exposition – such as symbolic algebra – prevented the Greeks from pursuing geometrical analysis further and from being able to express clearly what they had accomplished [!]. In the realm of geometrical analysis in particular, Tannery's remark holds true; the Greeks did not so much lack methods of mathematics as means to express them' (*ibid.*, 348). Finally the motto itself of Mahoney's article is, once more, Tannery's original saying: 'Ce qui manque aux mathématiciens grecques [*sic*] ce sont moins les méthodes ... que des formules propres à l'exposition des méthodes' (*ibid.*, 318). Mahoney is not alone in praising this famous 'fliegende Wort' of Tannery; so do Zeuthen, Heath, *etc.* And yet, is not this famous saying an unwitting confession that 'geometric algebra' is a pernicious and historically stillborn concept to use? Furthermore, is it not absurd to talk of *the methods* when *the means to express them*, i.e., *to use them*, are not available? How could one use a method which is *de facto* inexpressible, *i.e.*, unthinkable? Within the given limits of coherence of a mathematical culture, the *methods* available to that culture are exactly those by means of which the culture reached and *expressed* its mathematical achievements. The methods are contained in the tangible products of that mathematical culture. In the absence of treatises on the methodology of mathematics, the methods are those embodied in and displayed by the actual mathematical works available to the historian. The question is really very simple: To what extent does one possess the method if he lacks the means to put it to use? And the answer seems to me obvious. 'Wovon man nicht sprechen kann darüber muss man schweigen' has not only hortatory and prescriptive consequences; it is also, *mutatis mutandis*, a correct description of the *historical* state of affairs in intellectual history: 'Wovon man nicht sprechen kann darüber schweigt man'. If a culture (any culture!) cannot speak it does **not** speak. It remains silent. It certainly does not hide its impotence. Part of being ignorant of something is being ignorant of your ignorance. If you know that you are ignorant, your ignorance *stricto sensu* has disappeared. And the Greeks, clearly, did not know that they did not know algebra. So they did not hide their ignorance behind a geometrical screen. There is nothing lurking in hiding behind Greek geometry!

X.113 would have been necessary and X.113 would have become, *at best*, a Porism[124] and not an independent proposition. But this is certainly not the case in the *Elements*, and this is, I believe, a beautiful substantiation and corroboration of my view: *Greek geometry is geometry!* It is not algebra (without qualification), *i.e.*, it is not even 'geometrical algebra' if the term is understood as it has been traditionally understood since TANNERY and ZEUTHEN.

Moreover, proposition X.114 is another case in point. In the same notation I used above, it merely states that

> If $A \cdot B = R^2$, where A is an apotome and B is a corresponding binomial, then R is rational,

which, once more, *algebraically* is nothing but X.112, or for that matter, X.113 read in the opposite direction so to speak. Had algebra been in the background of EUCLID'S mind, he would **not** have spent great intellectual energies to prove *thrice exactly the same thing*. The conclusion is clear: For EUCLID, who did not think algebraically, the triplet of propositions I discussed did not represent one and the same proposition; and, indeed, *geometrically*, they are different. It would be enough to go laboriously through the proofs to convince oneself of the truth of this, my last assertion.

VI

| The last topic I want to deal with in this paper is the question of its originality. How original is it? In its form, the thrust of its argument, most of the examples cited to sub-
109 stantiate different points, its 'radicalism', and in its main conceptual emphasis the paper is original. Still I know very well that 'there is nothing new under the sun'. This is why I was not completely surprised when, in the course of my research, I encountered isolated and sporadic ideas which I had, naively, considered my exclusive intellectual property. Thus, in the 1930's, two lengthy articles appeared under the title 'Die griechische Logistik und die Entstehung der Algebra'.[125] Their author was JACOB KLEIN, who, recently, had the articles published in book form in an English translation by EVA BRANN.[126] This is a book which I personally consider to be one of the most substantial contributions to the literature on the history of mathematics; by the same token, it seems to be one of the least influential. In this book, KLEIN deals primarily

---

123 *Elements, ibid.*, 248–50.

124 In its sense of 'corollary'.

125 *Quellen und Studien zur Geschichte der Mathematik, Astronomie und Physik* (Abteilung B: Studien), **3** (fasc. 1, 1934), 18–105 and **3** (fasc. 2, 1936), 122–235. Ironically, KLEIN'S insightful articles were followed in each case by NEUGEBAUER'S unbridled transcriptions of ancient mathematical texts into algebraic language; thus, in fasc. 1, NEUGEBAUER published 'Serientexte in der babylonischen Mathematik' (*ibid.*, 106–114), while in fasc. 2 KLEIN'S article was succeeded by 'Zur geometrischen Algebra'! To make things even more piquant, the article immediately following the last part of KLEIN'S study was 'Eudoxos Studien III. Spuren eines Stetigkeitaxioms in der Art des Dedekind'schen zur Zeit des Eudoxos' (*op. cit.*, 236–244), by OSKAR BECKER!

126 *Greek Mathematical Thought and the Origin of Algebra* (Cambridge, Mass.: 1968, M.I.T. Press). The book includes an 'Appendix', containing VIETA'S *In artem analyticem* [*sic*] *isagoge* (Tours, 1591), translated into English by the Reverend J. WINFREE SMITH. (Parenthetically, let me urge those readers who have a choice and wish to read KLEIN'S highly interesting study to refer back to the original German articles: somehow the pomposity, stuffiness, and turgidity of the author's

with the differences between the Greek and the modern concept of number, and his conclusions completely match mine. I shall return to KLEIN'S ideas below.

Two other authors whose works contain interpretations similar to mine are ABEL REY and MICHAEL S. MAHONEY. Strangely, however, each of them adheres to a view which accepts the legitimacy of the term 'geometrical algebra'. Nevertheless this does not make those of their ideas which are strongly supportive of my interpretation less interesting. Thus, ABEL REY questioned the propriety of ZEUTHEN'S interpretation of PYTHAGOREAN and EUCLIDEAN mathematics.[127] Both he and MAHONEY pointed out forecefully and, I think, convincingly (and in this they also agree with KLEIN), that algebra is a product of modern times, starting its full flowering with the sixteenth and seventeenth centuries.[128] Both, as we saw, emphasized the differences between the geometrical and the algebraic way of thinking,[129] and both attacked the view which identified pre-Hellenic mathematics as algebraical.[130]

| The only scholar (so far as I know) who, in a book remarkable for its solid, penetrating, and far-sighted analysis, rejected peremptorily the historical value of the concept of 'geometric algebra' is the Hungarian philologist ÁRPÁD SZABÓ; the book, to 110
which we already referred before, is *Anfänge der griechischen Mathematik.* Though his remarks about 'geometric algebra' are a mere aside to the main endeavour of his analysis,[131] they fit *one* of the main messages of his book, *viz.*, that essentially, fundamentally, Greek mathematics became *very* early in its historical development Greek *geometry*, and that it grew and matured in very close nearness to PYTHAGOREAN musical theory. The main issues of Greek mathematics were *geometrical* issues, not the least of which is the issue of incommensurability.[132] These issues were *Greek* issues, not borrowed and disguised ones. Indeed,

> Jene Vermutungen die die "geometrische Algebra der Pythagoreer" als Übernahme bzw. griechische Weiterentwicklung von ursprünglich babylonischen Gedankengängen auffassen wollten, waren voreilig. Der Zusammenhang dieser Art Kentnisse mit der "babylonischen Wissenschaft" ist in Wirklichkeit *nirgends erwiesen.* Im Gegenteil! Man hat eher den Eindruck, dass die hier behandelte 'Flächengeometrie der Pythagoreer' eine *rein griechische Errungenschaft* war.[133]

---

style are better accomodated by the Teutonic cadences than by the more friendly sounds of the perfidious Albion ...)

[127] *Les Math. en Grèce*, 30.

[128] REY, *op. cit.*, 32, 44, 45, 48, 91; MAHONEY, 'Die Anfänge der algebr. Denkweise', *passim.*

[129] This makes their acceptance of the legitimacy of the term 'geometric algebra' the more difficult to understand. Instances of this acceptance in REY can be found in *op. cit.*, 33, 46, 49–51, 52 ('Le théorème de Pythagore lui-même est une résolution intuitive de l'equation $x^2 + y^2 = z^2$), 56–57; also, *La Science dans l'Antiquité*, 352; for examples of MAHONEY'S acceptance see note 26 above.

[130] MAHONEY, 'Babyl. Algebra', *passim*; Rey, *op. cit.*, 34, 36–37, 41, 91–92. The last reference in REY is a scathing attack against NEUGEBAUER'S interpretation of Babylonian mathematics. It is there that REY says: 'Du reste la preuve convaincante c'est que si, bien avant l'ère chrétienne et surtout avant Diophante, on avait eu l'idée algébrique des équations et, peu on prou, la pensée algébrique, toute la face de la mathématique en eût été changée' (*ibid.*, 91).

[131] Most of them appear in an appendix to the book.

[132] *Op. cit.*, 28, 36, *passim.*

[133] *Op. cit.*, 488.

Algebra is not geometry and, therefore, algebraic transcriptions of non-algebraic mathematical texts are historically inadmissible.[134] Besides, there are no traces in the Greek mathematical tradition (of both the pre-EUCLIDEAN and the EUCLIDEAN period) of any genuine algebraic ways of thinking.[135] This is why I think ABEL REY was right to state:

> Elle [la mathématique greque] sera géométrique. Lorsqu'elle ne le sera plus – lorsqu'elle tendra à devenir calculante, sous l'influence orientale sans doute, donc par l'affaiblissement même de ses forces créatrices, voire chez le plus puissant de ses derniers représentants, Diophante, à la fin de la civilization antique – elle sera tout près aussi de ne plus être.[136]

It is now time to return to JACOB KLEIN'S book. *Number* for the Greeks meant *positive integer. Numbers* are represented by EUCLID as *line segments*. After the discovery of the irrational, it became obvious that it is not the case that 'all *line segments* could be associated with numbers'; however, it does not follow from here that the converse
111 statement is also false, and, indeed, EUCLID associates *numbers* | with *line segments* throughout his so-called 'arithmetical books', *i.e.*, Books VII, VIII, and IX of the *Elements*.

The Greek concept of **arithmos** (number), *i.e.*, a 'number of things' (what KLEIN calls *Anzahl*), was replaced in the sixteenth century by a *new* concept of number as an abstract symbol. Instrumental in this change was FRANCOIS VIÈTE (VIETA), 1540–1603, who transformed the concept of *arithmos* into the modern concept. This transformation marks the beginnings of *modern* mathematics. Greek **arithmos** and modern **number** do *not* mean the same thing. As KLEIN has it, the two concepts differ in 'Begrifflichkeit', *i.e., conceptualization* and *intentionality*. (By the latter KLEIN understands '... the mode in which our thought, and also our words, signify or intend their objects.'[137]) For the Greeks, *arithmos* always meant a *number of things* (Anzahl), although 'things' did not have to be mentioned explicitly; for modern mathematics after VIÈTE, on the other hand, *Number is a concept*; it is *the concept of quantity*! As *numbers* come to be regarded as *abstract* and *symbolic* entities, a 'new' mathematics (and by the same token, a 'new' science on the long run) came into being, the mathematics in which the *symbolic form* of a statement is inseparable from its content; indeed, if I may put it this way, **the form is the content!** *Mutatis mutandis*, this separation is also, to a very great extent, impossible in modern (physical) science, where *mathematical form* and *physical content* are irreducibly intertwined and hopelessly enmeshed.

With VIÉTE and his successors (STEVIN, DESCARTES, WALLIS, *etc.*), then, a radical conceptual change has occurred. It is, therefore, historically unwarranted to apply mechanically to Greek mathematics the manipulations and jugglings of modern mathematical symbolism. Historians of mathematics, however, have been doing exactly this. Being themselves immersed in modern ways of thought, they have been reading Greek mathematical texts through modern glasses and, to nobody's surprise, were

---

134 *Ibid.*, note 21, 487. It seems to me, therefore, a concession to the prevailing mode of writing the history of mathematics, which SZABÓ so eloquently denounced, when he himself starts, somewhat indiscriminately, using algebraic notation in his geometrical discussions (*cf. op. cit.*, 483).

135 *Op. cit.*, 472–73.

136 *La science dans l'antiquité*, 390.

137 *Op. cit.*, 118.

rather successful in identifying in these texts an *inexistent* Greek algebra.[138] They could achieve such a fantastic result only by | betraying Greek mathematics, only by 112
applying to it foreign categories of post-Renaissance mathematical thinking.

One should not apply modern symbolism to Greek mathematics with impunity, as if modern symbolism were nothing but a temporally universal (*i.e.*, historically indifferent) means for organizing and simplifying *any* given conceptual content. The fact that it is *modern* symbolism that one applies is, in itself, the best evidence for the ahistoricity of such a procedure. Klein has shown, and I think successfully, that '... symbolic formalism is at the *core* of the modern concept of number, and that to translate Greek mathematics into its terms obscures completely both the meaning of the Greek concept and the genuine Greek achievement in the theory of number.'[139]

In his commentary on proposition VIII. 4, Heath talks '... of the cumbrousness of the Greek method of dealing with *non-determinate numbers*. The proof in fact is not easy to follow', he goes on, 'without the help of modern symbolical notation. If this be used, the reasoning can be made clear enough.'[140] The question, however, is: Did the Greeks in general, and Euclid in particular, ever use 'non-determinate' numbers?

In his *Die Algebra der Griechen*, G.H.F. Nesselmann produced a since famous trichotomous classification of the historical development of algebra.[141] The *three* stages distinguished by Nesselmann are: *Rhetorical, Syncopated*, and *Symbolic* Algebra. In Nesselmann's classification, Diophantus' *Arithmetica* fell in the second category (syncopated).[142] Nesselmann's analysis, however, is very approximative and, at best,

[138] Modern algebra (the only true algebra) is a creation of the sixteenth and seventeenth centuries. Its great protagonists are Viète, Descartes, and Fermat. It marks the passage from an old way of thinking in mathematics (the geometrical way, the *mos geometricus*) to a new way (the symbolic way, the *mos per symbola*). Its historical development is rightly connected with the reintroduction into the West of the great works of classical Greek mathematics which, however, contained the old way of thinking, to be discarded by modern mathematics. With Viète algebra becomes the very language of mathematics; in Diophantus' *Arithmetica*, on the other hand, we possess merely a refined auxiliary tool for the solution of arithmetical problems (*cf.* M. Mahoney, 'Die Anfänge der algebr. Denkweise', *passim*). In the seventeenth century, algebra was called *ars analytica*, a pregnant name indeed. It shows the difference between the Greek approach and that of the seventeenth century. For the Greeks, mathematics was not an *art*, a manipulative technique *(techne)* but a science *(episteme, scientia)*. Furthermore, for the Greeks *analysis* was merely a means of discovery, a heuristic tool. Mathematics, *episteme*, was limited to *synthesis*. In the seventeenth century, on the other hand, one is faced with algebraical analysis without any synthesis. This new approach meant (among other things) a certain loosening of the Greek strictures of rigor and a new mathematical style. Mahoney identifies the necessarily *external* factors which led to this development as Petrus Ramus' pedagogical endeavours and the search for a universal symbolism *(characteristica universalis)* starting with Ramon Lull in the thirteenth century. These two factors were united in Ramus, who contributed to a separation of the universal symbolism from its ties with magic *via* the *ars memoriae*. According to Mahoney, Ramus seems to have been the first to demand more respectability and status for the algebraic *art*, practiced, as a rule, outside the walls of the university establishment (*cf. ibid.*, 25).

[139] From the dust jacket of Klein's book.

[140] *Elements*, **2**, 353, my italics.

[141] *Op. cit.*, 301–303.

[142] Klein, I think rightly, sees Diophantus' *Arithmetica* as an exercise in *theoretical logistic* (*cf. op. cit.*, 127–149, *passim*).

faulty.[143] Léon Rodet evolved another, dichotomous classification: 1. The 'algebra of abbreviations and *given* numbers' and 2. Symbolic algebra (*i.e.*, the only true algebra, algebra proper).[144]

113 | According to Rodet even Diophantus' algebra belongs to the first type.[145] Modern algebra '... n'a pris naissance que lorsqu'on eut l'idée

> de représenter les données du problème sous forme générale par un symbole, de symboliser également les opérations chacune par un signe spécial, et d'arriver ainsi non plus à resoudre avec plus on moins de facilité un problème particulier, mais à trouver des formules donnant la solution de tous les problèmes d'une même espèce, et, parce qu'elle servait à caractériser chaque espece de problème, servant à exprimer les propriétés générales de certaines catégories des nombres, de certaines familles de figures, où a formuler les lois de certaines classes de phénomènes naturels.[146]

Do we find, then, any *algebra* in Euclid? I doubt it! Euclid's *numbers* are *given* line-segments, no abstract symbols, and Euclid's presentation is *not* symbolic. *It always deals with determinate numbers* of units of measurement which are **not** seen as representing specific illustrations, instances of a *concept of general magnitude*.[147] From here on, allow me to quote Jacob klein:

> In *illustrating* each determinate number of units of measurement by measures of distance it [*i.e.*, the Euclidean presentation] does *not* do two things which constitute the heart of the symbolic procedure: It does *not* identify the object represented with the means of its representation, and it does *not* replace the real determinateness of an object with a *possibility* of making it determinate, such as would be expressed by a sign which, instead of *illustrating* a determinate object, would *signify* possible determinacy ... when in the arithmetical books an arithmetical, or more exactly a logistical proposition is

---

143 Léon Rodet in *op. cit.* (see note 21 above for full reference) demolishes Nesselmann's taxonomy. It is there that Rodet says: '... il faut reconnaître que cette distinction des trois étapes successives du langage algébrique a quelque chose de séduisant. Il n'y a qu'un malheur: c'est qu'elle est bâtie uniquement sur un échafaudage d'inexactitudes ...' (*ibid.*, 56). Rodet points out that even admitting the truthfulness of Nesselmann's classification, *it is wrong to call it historical!* The three stages do not correspond to three *historically successive* stages even on Nesselmann's own account, since the lowest rank of this classification is occupied by the Arabs and by Italian mathematicians writing between the Crusades and the sixteenth century, while Diophantus (3rd century A.D.) corresponds to the middle stage and the Hindus, reported masters of the Arabs, are occupying the highest rank, *i.e.*, the same spot as modern symbolic algebra! Rodet destroys especially this characterization of Hindu mathematics and reveals its absolute historical falsehood due to Nesselmann's ignorance of '... les notations algébriques des Indiens' (*ibid.*, 57). Speaking of Hindu 'algebraic notation', Léon Rodet says: 'Il lui manque, pour être mise en parallèle avec la nôtre, deux choses essentielles: des signes spéciaux pour les deux opérations directes de l'addition et de la multiplication, et le moyen de représenter autrement que par des nombres particuliers les *paramètres* qui entrent, simultanément aux *variables* proprement dites, dans nos expressions algébriques. Enfin, comme chez Diophante, les symboles qu'elle emploie ne sont que les initiales des noms des quantités qu'elle veut représenter. L'algèbre Indienne mérite tout autant que celle des Grecs et des Européens entre le XIIe et le XVIIe siécles, le nom d' *Algébre* syncopée ...' (*ibid.*, 60).

144 *Op. cit.*, 69–70.

145 *Cf.*, in this connection, Michel's statement: 'D'une façon générale, le vocabulaire de Diophante reste imprégné de géométrie, comme en témoignent ces énoncés de problémes' (*op. cit.*, 641); also, Nesselmann: 'So finden wir wirklich selbst bei Diophant Beispiele von gänzlicher Vernachlässigung des Gebrauches der Abbreviaturen ..., die also ganz der rhetorischen Stufe angehören' (*op. cit.*, note 15, 304).

146 *Ibid.*

147 Klein, *op. cit.*, 123.

> demonstrated *generally* with the aid of lines, this does not in the least mean that there exists either a general number or the concept of a "general," i.e., indeterminate, number corresponding to this general proof ... the *general* "linear approach" ... intends only *determinate* numbers ... Since ... in Euclid ... the single illustrative lines are additionally identified by a letter, the possibility arises of representing the numbers intended by those letters. This does not, however, in the least amount to the introduction of symbolic designations. Letters for indicating magnitudes and numbers seem to have been used already by Archytas ... [As TANNERY put it, however,] *the letter does not symbolize the value of a number, and does not lend itself to being operated on.* Aristotle, too made use of such mathematical letters, e.g. in the *Physics* and in *On the Heavens;* and he even introduced them into his "logical" and ethical investigations. But such a letter is never a "symbol" in the sense that that which is signified by the symbol is in itself a "general" object.[148]

It simply cannot be said any better!

| Now, *Symbolic Algebra* (*i.e.*, algebra proper) was not born, as RODET has 114
shown,[149] 'before someone had the idea of representing what is given in a problem in a general form by means of a symbol, and of similarly symbolizing each of the operations by a special sign.'

Such an idea, so far as I am aware, certainly does **not** appear in the *Elements*, in which EUCLID, according to PROCLUS, collected '... many of the theorems of Eudoxus, perfecting many others by Theaetetus, and bringing to irrefragable demonstration the things which had only been somewhat loosely proved by his predecessors.'[150]

I acknowledge the financial support of the College of Arts and Sciences of the University of Oklahoma, by way of a Summer Fellowship, in the writing of this paper. I am also very thankful to Professor WILLY HARTNER for his helpful and encouraging editorial comments. A drastically curtailed version of the argumentive essence of this paper was presented at the XIVth International Congress of the History of Science, Section II (History of Mathematics), Tokyo, August 1974. (*Cf.* Proceedings of the Congress.)

148 *Op. cit.*, 123–24.

149 See text to note 145 above; *cf.* also, KLEIN, *op. cit.*, 146–47.

150 EUCLID, *Elements*, **1**, 37. I have striven in this paper to demolish the validity of the concept of 'geometric algebra' as a useful historiographic term. In this, if POPPER is right, I must have achieved the highest level of understanding of the true underpinnings of that concept ... According to Sir KARL, there are three levels of understanding: 1. *The lowest* represented by the pleasant feeling of having grasped the argument 2. *The medium level*, represented by the ability to repeat the argument 3. *The highest level*, represented by the ability to refute the argument (*Cf.* IMRE LAKATOS, 'Proofs and Refutations (II)', *British Journal for the Philosophy of Science*, **14** (1963–64), 120–139, at 131.)

B. L. VAN DER WAERDEN

DEFENCE OF A "SHOCKING" POINT OF VIEW

1. INTRODUCTION

In his paper "On the Need to Rewrite the History of Greek Mathematics" SABETAI UNGURU severely criticizes the views of TANNERY, ZEUTHEN, NEUGEBAUER and myself on the "Geometrical Algebra" of the Greeks. UNGURU summarizes our position in one sentence: "Greek 'geometric algebra' is nothing but 'Babylonian algebra' in geometrical attire", and he starts to prove that this position is historically unacceptable. UNGURU states his objections very clearly. The object of the present paper is to defend our position against this emphatic attack.

2. ALGEBRAIC THINKING

After having summarized our views, UNGURU starts his discussion by summing up the characteristic features of geometric and algebraic thinking. According to UNGURU (quoting MAHONEY), the main features of algebraic thinking are:

1. Operational symbolism,
2. The preoccupation with mathematical relations rather than with mathematical objects, which relations determine the structures constituting the subject-matter of modern algebra . . .,
3. Freedom from any ontological questions and commitments and, connected with this, abstractness rather than intuitiveness[1].

If this definition of "algebraic thinking" is accepted, then indeed UNGURU is right in concluding that "there has never been an algebra in the pre-Christian era", and that Babylonian algebra never existed, and that all assertions of TANNERY, ZEUTHEN, NEUGEBAUER and myself concerning "Geometric algebra" are complete nonsense.

Of course, this was not our definition of algebraic thinking. When I speak of Babylonian or Greek or Arab algebra, I mean algebra in the sense of AL-KHWĀRIZMĪ, or in the sense of CARDANO's *Ars magna*, or in the sense of our school algebra. Algebra, then, is:

> *the art of handling algebraic expressions like* $(a + b)^2$ *and of solving equations like* $x^2 + ax = b$.

[1] This definition was taken from M.S. MAHONEY: Die Anfänge der algebraischen Denkweise im 17. Jahrhundert, *Rete* 1 (1971), pp. 15–31. However, I don't think MAHONEY meant to give a general definition of "Algebraic Thinking". He was concerned with modern tendencies in algebra, which first manifested themselves in the 17th century. He certainly did not include the algebraic thinking of AL-KHWĀRIZMĪ and CARDANO, to which the characteristic features 1.2.3 do not apply.

J. Christianidis (ed.), Classics in the History of Greek Mathematics,, 433–445.

200 | If this definition is applied to any Babylonian or Arab text it is unimportant what symbolism the text uses. Our relation

$$(a + b)^2 = a^2 + b^2 + 2ab$$

can be stated in words thus:

"The square of a sum is the sum of the squares of the terms and twice their product."

The statement in words says exactly the same thing as the formula. Instead of "product" one may also say "area" (of a rectangle), as the Babylonians did, or just "rectangle", as the Greeks did.

Let us now look into history, and see where we find Algebra in the sense of our definition.

3. ARAB ALGEBRA AND CARDANO'S *ARS MAGNA*

The word "al-jabr" in the title of AL-KHWĀRIZMĪ's Algebra is a part of the full expression *al-jabr wa'l-muqābala*, which ROSEN translates as *Completion and Reduction*. The Arab words denote two simple operations necessary for solving equations. AL-KHWĀRIZMĪ's treatise deals mainly with the art of solving equations.

Similarly, CARDANO's *Ars magna* (Great Art) is mainly concerned with the art of solving linear, quadratic, cubic and biquadratic equations.

At school we learn how to handle expressions like $(a + b)^2$ and how to solve linear and quadratic equations. This subject is called "Algebra". So the definition of Algebra just given is in full accordance with standard usage from 800 A.D. up to the present day.

In what follows, the word Algebra will be used only in this sense.

4. BABYLONIAN ALGEBRA

UNGURU denies the existence of Babylonian algebra. Instead he speaks, quoting ABEL REY, of

> an *arithmetical* stage (Egyptian and Babylonian mathematics), in which the reasoning is largely that of elementary arithmetic or based on empirically paradigmatic rules derived from successful trials taken as a prototype.

I have no idea on what kind of texts this statement is based. For me, this is history-writing in its worst form: quoting opinions of other authors and treating them as if they were established facts, without quoting texts.

Let us stick to facts and quote a cuneiform text BM 13901 dealing with the solution of quadratic equations. Problem 2 of this text reads:[2]

> I have subtracted the (side) of the square from the area, and 14,30 is it.

The statement of the problem is completely clear: It is not necessary to translate it into modern symbolism. If we do translate it, we obtain the equation

$$x^2 - x = 870.$$

---

[2] O. NEUGEBAUER: *Mathematische Keilschrifttexte* III, p. 6.

The solution given in the text reads:

> | Take 1, the coefficient (of the unknown side). Divide 1 into two equal parts: 0;30 201
> times 0;30 is 0;15. Add this to 14,30, and (the result) 14,30; 15 has 29;30 as a square root. Add the 0;30 which you have multiplied by itself to 29;30, and 30 is the (side of the) square.

This is the same method of solution we learn at school. According to our definition, this is algebra.

5. AL-KHWĀRIZMĪ

AL-KHWĀRIZMĪ teaches the same method as the Babylonians. Let me give an example from page (5) of KHWĀRIZMĪ's treatise (p. 8 of ROSEN's translation).

> *Roots and Squares are equal to numbers*; for instance "one square, and ten roots of the same amount to thirty-nine dirhems"; that is to say, what must be the square which, when increased by ten of its own roots, amounts to thirty-nine? The solution is this: you halve the number of the roots, which in the present instance yields five. This you multiply by itself; the product is twenty-five. Add this to thirty-nine; the sum is sixty-four. Now take the root of this, which is eight, and subtract from it half the number of the roots, which is five; the remainder is three. This is the root of the square which you sought for; the square itself is nine.

There is a striking similarity between the Babylonian way of treating quadratic equations and KHWĀRIZMĪ's, a similarity not only in content but also in form.

Unlike the Babylonians, KHWĀRIZMĪ also gives proofs. For solving quadratic equations of the form

$$x^2 + ax = b$$

he needs the rule of computation

$$(x + \tfrac{1}{2}\,a)^2 = x^2 + ax + (\tfrac{1}{2}\,a)^2$$

and he proves it by means of drawings. First he gives a complicated proof by means of a square surrounded by four rectangles and four small squares, but later on he presents a simplified diagram (see Fig. 1), which looks just like the diagram accompanying EUCLID's proposition II 4.

Fig. 1. Diagram from the Algebra of AL-KHWĀRIZMĪ, p. 16 in ROSEN's translation.

6. GEOMETRY AND ALGEBRA

The diagram in KHWĀRIZMĪ's Algebra is typical of a general tendency we can observe in Greek arithmetics as well as in Babylonian and Arab algebra, namely the tendency to illustrate algebraic notions and methods by means of diagrams.

Let us first consider Babylonian algebra. In problems with two unknowns, these
202 are often called *length* and *width*, and their product *area*. The product of a | number
by itself is always called its *square*. In problems with three unknowns, these are sometimes called *length, width* and *height*, and their product *volume*.

In Greek arithmetics, the product of a number by itself is always called its *square*. This term is found in all texts from PLATON to DIOPHANTOS. Numbers of the form mn with $m \neq n$ were called *oblong numbers*, and two products mn and pq were called *similar*, if m is to n as p is to q. This means that products mn were visualized as rectangles. Just so, numbers of the form $n^2(n \pm 1)$ were interpreted as volumes of rectangular parallelepipeda and called "Arithmoi paramekepipedoi"[3].

In passing we may note that there are Babylonian tables of these numbers $n^2(n \pm 1)$, and that these tables were used to solve cubic equations of the form

$$x^2(x \pm a) = b.$$

Thus we see that the tendency to translate algebraic or arithmetical notions into geometric terminology was common to the Babylonians, Greeks and Arabs.

7. SIDE- AND DIAGONAL-NUMBERS

A good example of how the Greeks used to translate arithmetical operations and theorems into the language of geometry and conversely is offered by the Pythagorean theory of Side-Numbers and Diagonal-Numbers[4]. They were pairs of numbers s_n and d_n satisfying the relation

$$d_n^2 = s_n^2 \pm 1. \tag{1}$$

Starting with $s_1 = d_1 = 1$, these numbers were defined recursively by the relations

$$s_{n+1} = s_n + d_n, \quad d_{n+1} = 2s_n + d_n, \tag{2}$$

or, if one prefers a definition in words:

"If a side-number is added to the corresponding diagonal-number, the next side-number is obtained. If the side-number is added twice to the diagonal-number, the next diagonal-number is obtained".

This sentence in words is completely equivalent to the formulae (2), so there is no danger in using the formulae, contrary to MAHONEY's opinion quoted by UNGURU at the beginning of his paper.

PROKLOS informs us that the Pythagoreans proved (1) by means of Proposition II 10 of EUCLID. This means: they transformed the purely arithmetical statements (1) and

---

3 O. BECKER: ΑΡΙΘΜΟΙ ΠΑΡΑΜΗΚΕΠΙΠΕΔΟΙ, *Quellen und Studien zur Geschichte der Math.*, B 4, p. 181 (1938).

4 See PROKLOS: *Commentary on Platon's Republic II*, Chapters 23 and 27, or VAN DER WAERDEN: *Science Awakening I*, p. 126.

(2) into the language of geometry, representing the numbers s_n and d_n by line segments and their squares by geometrical squares. To these segments and squares they applied II 10, thus obtaining the desired relation (1). This example shows that the Pythagoreans were able to translate arithmetical statements like (1) and (2) into the language of geometry, and conversely.

8. GEOMETRICAL ALGEBRA 203

Now we are sufficiently prepared to discuss Greek "Geometrical Algebra". Algebra, as we had defined it, is an art which can be applied to numbers as well as to line segments and areas, and in fact the Babylonians already applied it to numbers as well as to line segments. Now if algebra is restricted to line segments and their products (*i.e.* rectangles and squares), one obtains a restricted algebra which may be formulated in purely geometric terms, and which may well be called "geometric algebra". Thus, if the formula

$$(a + b)^2 = a^2 + b^2 + 2ab \tag{3}$$

is restated in words and restricted to line segments a and b, one obtains just the theorem II 4 of EUCLID's elements.

Thus, "geometric algebra" is by no means a *contradictio in terminis*, as UNGURU claims, but it is a reality. It is algebra restricted to line segments and areas, and hence a part of algebra, but it is also a part of geometry, namely a set of geometrical theorems and solutions of problems, in which only line segments, rectangles and orthogonal parallelepipeda are considered. Examples of geometrical algebra are the propositions II 1–6 and II 9–10 of EUCLID.

9. TWO ROADS TO GEOMETRICAL ALGEBRA

As we have seen, geometrical algebra is a part of algebra as well as a part of geometry. It follows that one can arrive at geometrical algebra by two different roads: One can either start with geometrical problems concerning rectangles and squares, and solve these problems by means of theorems, or one can start with algebraic problems such as the solution of quadratic equations and reformulate them in geometrical language, writing "rectangle" instead of "product".

The Greeks, and in particular the Pythagoreans, were perfectly able to follow either road. They called the product of a number by itself "square", and they solved a purely arithmetical problem concerning Side-Numbers and Diagonal-Numbers by means of the geometrical theorem II 10. The question is now: What road did the Greeks actually follow? Did they start with algebraical (or arithmetical) or with geometrical problems?

We (ZEUTHEN and his followers) feel that the Greeks started with algebraic problems and translated them into geometric language. UNGURU thinks that we argued like this: We found that the theorems of EUCLID II can be translated into modern algebraic formalism, and that they are easier to understand if thus translated, and this we took as "*the* proof that this is what the ancient mathematician had in mind". Of course, this is nonsense. We are not so weak in logical thinking! The fact that a theorem can be translated into another notation does not prove a thing about what the author of the theorem had in mind.

No, our line of thought was quite different. We studied the wording of the theorems and tried to reconstruct the original ideas of the author. We found it *evident* that these theorems did not arise out of geometrical problems. We were not able to find any interesting geometrical problem that would give rise to theorems like II 1–4. On the other
204 hand, we found that the explanation of these | theorems as arising from algebra worked well. Therefore we adopted the latter explanation.

Now it turns out, to my great surprise, that what we, working mathematicians, found evident, is not evident to UNGURU. Therefore I shall state more clearly the reasons why I feel that theorems like EUCLID II 1–4 did not arise from geometrical considerations.

10. THE ORIGIN OF EUCLID II, PROPOSITIONS 1–4

Proposition 1 reads in the translation of HEATH:

> If there be two straight lines, and one of them be cut into any number of segments, the rectangle contained by the two straight lines is equal to the rectangles contained by the uncut straight line and each of the segments.

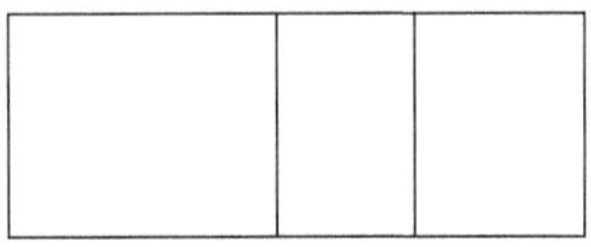

Fig. 2. Diagram to EUCLID's Prop. II. 1.

Geometrically, this theorem just means that every rectangle can be cut into rectangles by lines parallel to one of the sides. This is evident: everyone sees it by just looking at the diagram. Within the framework of geometry there is no need for such a theorem: EUCLID never makes use of it in his first four books.

However, if one starts with the algebraic operations of addition and multiplication of numbers and asks: how does one multiply a sum by a quantity a? the answer is: Multiply the terms of the sum by a and add the results. In elementary arithmetics, this rule is needed all the time. If this rule of computation is translated into the language of geometry, Proposition II 1 results. In other words: Proposition II 1 furnishes a geometrical proof of an algebraic rule of computation.

II 2 and II 3 are just special cases of II 1. Once more, from the point of view of geometry there is no reason to formulate these trivialities as theorems.

II 4 says:

> If a straight line be cut at random, the square on the whole is equal to the squares on the segments and twice the rectangle contained by the segments.

Geometrically, this means: If we take a point Z on the diagonal of a square and draw lines through Z parallel to the sides of the square, the square will be divided into two squares and two rectangles. This is trivial.

205 | As we have seen, the same diagram of a square divided into two squares and two rectangles (without the diagonal, which is not necessary) also appears in AL-KHWĀRIZMĪ's treatise. Here it occurs in its natural place: The author needs it to justify

II 4 reads:

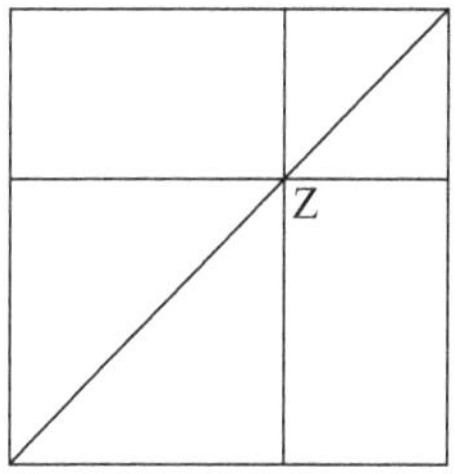

Fig. 3. Diagram to Euclid, Prop. II 4.

his method of solving quadratic equations. In this case we can see why AL-KHWĀRIZMĪ; inserted the diagram. If we assume that the author of Book II also started with an algebraic tradition, to which a rule for squaring a sum belonged, we can understand why he formulated the theorems 1–4 just as he did, but if he came from geometry, we cannot understand his line of thought.

11. A TREATISE OF THĀBIT IBN QURRA

THĀBIT IBN QURRA, a contemporary of AL-KHWĀRIZMĪ, was an excellent geometer and astronomer, fully conversant with the work of EUCLID. In a little known treatise[5], THĀBIT pointed out that the solution of the three types of quadratic equations according to "the Algebra people" is equivalent to the "Application of areas with excess or defect" as presented by EUCLID.

The example of THĀBIT shows that UNGURU is completely wrong in thinking that mathematicians like ZEUTHEN came to their opinions about Greek geometric algebra only because they translated EUCLID's propositions into modern algebraic symbolism. It is true that ZEUTHEN was able to use modern symbolism, but THĀBIT was not, and yet he arrived at the same conclusion as ZEUTHEN, namely that AL-KHWĀRIZMĪ's solution of quadratic equations is equivalent to EUCLID's procedure.

UNGURU, like many non-mathematicians, grossly overestimates the importance of symbolism in mathematics. These people see our papers full of formulae, and they think that these formulae are an essential part of mathematical thinking. We, working mathematicians, know that in many cases the formulae are not at all essential, only convenient. The treatise of THĀBIT offers a good illustration of this thesis.

12. THE APPLICATION OF AREAS

In his Commentary to EUCLID I, Prop. 44, PROKLOS informs us:

> These things, says Eudemos, are ancient and are discoveries of the Muse of the Pythagoreans, I mean the application (παραβολή) of areas, their exceeding

---

[5] See P. LUCKEY: Tābit b. Qurra über den geometrischen Richtigkeitsnachweis der Auflösung der quadratischen Gleichungen. Sitzungsberichte Sächsische Gesellschaft der Wiss. Leipzig 1941, pp. 93–114.

> (ὑπερβολή) and their falling short (ἔλλειψις). It was from the Pythagoreans that later geometers took the names, which they again transferred to the so-called conic lines … whereas those godlike men of old (the Pythagoreans) saw the things signified by these names in the construction, in a plane, of areas upon a finite straight line. For, when you have a straight line set out and lay the given line exactly alongside the whole of the straight line, then they say that you *apply* the said area; when however you make the length of the area greater than the straight line itself, it is said to *exceed*, and when you make it less, in which case, after the area has been drawn. There is some part of the straight line extending beyond it, it is said to *fall short*. (Translation by HEATH, The Thirteen Books of Euclid's Elements I, p. 343).

The same terms *application, exceeding* and *falling short* are used in EUCLID VI 28–29:

206 | 28. To a given straight line to apply a parallelogram equal to a given rectilinear figure and deficient by a parallelogram similar to a given one.
29. To a given straight line to apply a parallelogram equal to a given rectilinear figure and exceeding by a parallelogram similar to a given one.

E. NEUENSCHWANDER[6] has shown that the term "parallelogram" was introduced into geometry at the time of EUDOXOS, long after the time of the Pythagoreans. In book 2, which is due to the Pythagoreans, only squares and rectangles occur, and the notions "Proportion" and "Similarity" do not occur. So, when the Pythagoreans invented their application of areas with defect or excess, the defect or excess was probably required to be just a square, not a parallelogram similar to a given one. Now if EUCLID's diagrams to VI 28 and VI 29 are simplified by assuming a square excess or defect, the resulting diagrams are essentially the same as EUCLID's diagrams to II 5 and II 6. Also, the single steps in the proofs of VI 28 and VI 29 are just generalizations of the single steps in the proofs of II 5 and II 6. Thus, one sees:

II 5 and II 6 were just the theorems the Pythagoreans needed for the solution of their problems of application of areas with defect or excess.

The application of a given area C to a line segment AB with a square defect or excess may be illustrated by Figures 4 and 5 below. In both cases, the rectangle AQ is required to be equal to a given area C, and the defect or excess BQ is required to be a square.

In modern notation, the two problems may either be written as pairs of equations with two unknowns

$$\text{(I)} \quad \begin{cases} x + y = a \\ xy = C \end{cases} \qquad \text{(II)} \quad \begin{cases} x - y = a \\ xy = C \end{cases}$$

(compare EUCLID, Data 85 and 86), or as single equations with one unknown x or y

$$x(a - x) = C \quad \text{or} \quad x^2 + C = ax, \tag{4}$$

$$x(a + x) = C \quad \text{or} \quad x^2 + ax = C, \tag{5}$$

---

6 E. NEUENSCHWANDER: Die ersten vier Bücher der Elemente Euklids. *Archive for History of Exact Sciences* 9 (1973), pp. 325–380.

(6) $$y(y-a) = C \quad \text{or} \quad y^2 = ay + C.$$

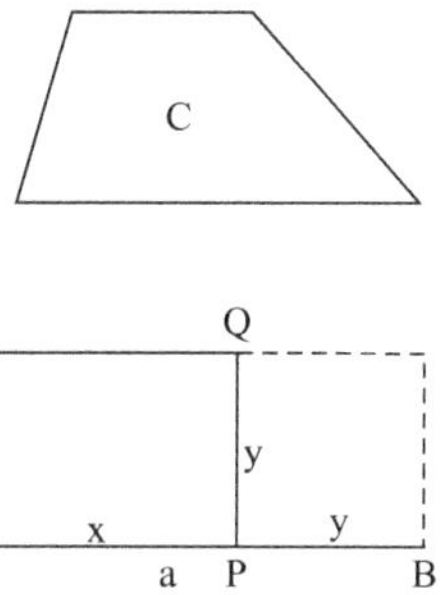

Fig. 4. Application of an area C to a line segment AB with square defect.

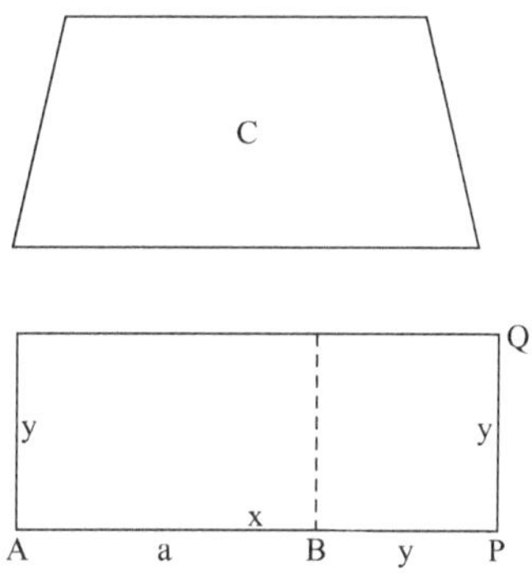

Fig. 5. Application of an area C to a line segment AB with square excess

| The equations (4), (5), (6) are just the three types of mixed quadratic equations 207
treated by KHWĀRIZMĪ, as THĀBIT rightly noted.

13. THE APPLICATION OF II 6 TO THE PROBLEM II 11

In the preceding section we have seen that one of the purposes of the pair of propositions II 5–6 was to justify the solution of the problems of application of areas with defect or excess. A confirmation of this view is obtained by considering the application of the theorem II 6 to the solution of the problem II 11.

SZABÓ and UNGURU have noted that II 6 was used in the solution of the problem II 11. For once, I fully agree with them. Let us now examine how it was used.

The problem II 11 reads:

> To cut a given straight line so that the rectangle contained by the whole and one of its segments is equal to the square of the remaining segment.

If the given straight line is called a and the "remaining segment" x, the problem can be formulated as an equation

$$a(a - x) = x^2 \tag{7}$$

or, which is equivalent,

$$x^2 + ax = a^2. \tag{8}$$

This is an equation of type (5), and indeed it is solved by applying an area a^2 to the line segment a with a square excess. In Fig. 6 I have reproduced EUCLID's diagram. The perpendicular lines AB and AC are made equal to the given line segment a. The additional segment $x = AF$ is constructed in such a way that the rectangle FK is equal to the square AD. The text first describes the construction of the segment AF and next proves, by means of II 6, that the rectangle FK is indeed equal to the square AD. Thus, equation (8) is solved.

Now the text proceeds to show that the segment x just constructed also satisfies the original condition (7). In modern notation one would say: Subtract ax from both sides of (8), and you obtain (7). The text proceeds just so:

> ... *FK* is equal to *AD*. Let *AK* be subtracted from each; therefore *FH* which remains is equal to *HD*.

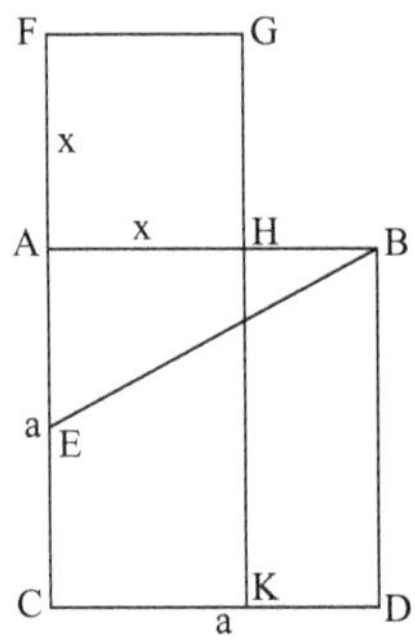

Fig. 6. Diagram to EUCLID II 11. From HEATH, The 13 Books of Euclid's Elements, p. 402

208 | The operation here applied is just what KHWĀRIZMĪ; calls "reduction" (*muqābala*): one and the same term is subtracted from both sides of an equation.

Quite apart from terminology, our analysis shows that II 6 was in fact used by EUCLID in order to solve a problem of application of an area with a square excess. This is what I wanted to prove.

14. EUCLID'S METHOD OF SOLUTION

EUCLID's method of solving a pair of equations like

$$\text{(I)} \quad \begin{cases} x + y = a \\ xy = C \end{cases}$$

is exactly the same as the Babylonian method. EUCLID halves the line segment a and erects a square on $\frac{1}{2}a$. The given area C is taken away from this square, and the remainder is converted into a square of equal area. The side of this square is the geometrical equivalent of the Babylonian square root. Added to $\frac{1}{2}a$, it yields the line segment x, and subtracted, it yields y.

I claim that this is a cumbersome method, and that from the geometrical point of view other methods would be easier and more natural.

For instance: The Greeks knew how to convert a given polygon C into a square of equal area h^2. They also knew: If a line segment $h = CD$ is drawn in a semicircle, perpendicular to the diameter AB, and if the segments AC and CB are called x and y, then xy is equal to h^2 (Prop. II 14). Now if $AB = a$ and h are given, one can draw a line parallel to AB in a distance h, and take one of the two points of intersection, say D. Dropping a perpendicular from D, one obtains the required line segments x and y.

There are other, still simpler geometrical solutions. "Simpler" means: Requiring fewer circles and straight lines in the construction. I shall give an example of a simple solution of Problem II:

$$\text{(II)} \quad \begin{cases} x - y = a \\ xy = C. \end{cases}$$

| Let bc be the given area C, and let b be larger than c. In a sufficiently large circle 209
one can construct a chord AB equal to $b-c$. On the production of AB a segment BC can be made equal to c; then AC is equal to b. Construct another chord DE equal to a. Draw

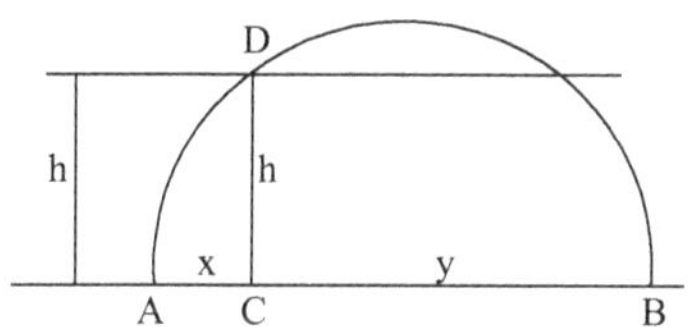

Fig. 7. Solution of (I) by means of a semicircle.

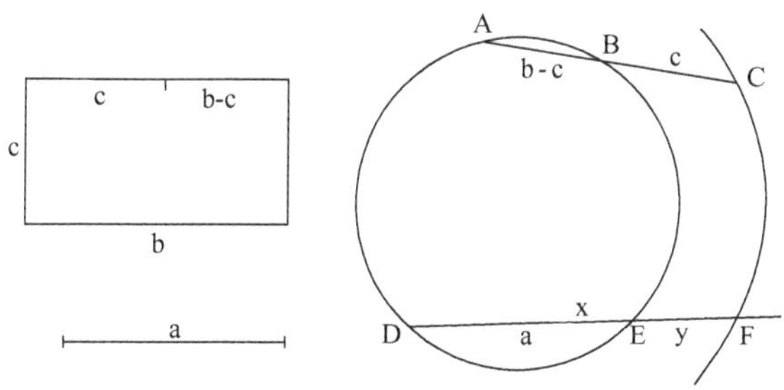

Fig. 8. Solution of Problem (II) by chords in a circle.

a concentric circle through C, which intersects the prolongation of DE in F. Making use of EUCLID III 35, one sees easily that $DF = x$ and $EF = y$ satisfy the conditions

$$x - y = a,$$
$$xy = bc.$$

If b happens to be equal to c, one can draw a tangent instead of the chord AB and apply III 36 instead of III 35. The same method of construction can also be applied to solve Problem I.

Note that III 35–36 were known to the Pythagoreans, as NEUENSCHWANDER[6] has shown. It follows that the construction just given was well within the reach of the Pythagoreans.

Why did the Pythagoreans (or why did EUCLID) leave aside these simple solutions, which were well within their reach? My explanation is: Their starting point was not geometry but algebra, and they translated the algebraic solution, step by step, into the language of geometry.

If this hypothesis is accepted, the question arises: How did the Pythagoreans learn algebra? Did they invent it anew or did they learn it from the Babylonians? To answer this question, it will be best not to restrict ourselves to algebra.

15. THE RELATIONS BETWEEN BABYLONIAN AND GREEK MATHEMATICS

As I have shown in my book, *Science Awakening I*, there are several points of contact between Babylonian and Greek mathematics. I shall now, once more, enumerate these points, referring to my book for further details.

1. The Babylonians as well as the Pythagoreans knew the "Theorem of PYTHAGORAS".
2. They had methods of constructing "Pythagorean Triples", *i.e.* integer solution of the equation

$$x^2 + y^2 = z^2.$$

3. Both were interested in solving systems of linear equations in several unknowns.
4. Both were interested in numbers of the form $n^2(n + 1)$ or $n^2(n-1)$, which the Greeks called *Arithmoi paramekepipedoi*.
5. Both knew how to solve quadratic equations. The Greek method of solution is the same as the Babylonian one, but in geometric attire.
6. The Babylonians had four standard types of linear and quadratic equations with two unknowns:

$$\text{(I)} \quad \begin{cases} x + y = a \\ xy = C, \end{cases} \qquad \text{(II)} \quad \begin{cases} x - y = a \\ xy = C, \end{cases}$$

$$\text{(III)} \quad \begin{cases} x + y = a \\ x^2 + y^2 = S, \end{cases} \qquad \text{(IV)} \quad \begin{cases} x - y = a \\ x^2 + y^2 = S. \end{cases}$$

| The Greeks formulated four theorems II 5–6 and II 9–10, by means of which these types can be solved. The solutions thus obtained are the same as the Babylonian solutions, but in geometric language. They differ from all simpler geometrical solutions. 210

7. S. GANDZ[7] has indicated a remarkable similarity between methods of DIOPHANTOS and Babylonian methods. When DIOPHANTOS wants to find two numbers x and y whose sum a is given, he often puts

$$x = \tfrac{1}{2}\, a + s \quad \text{and} \quad y = \tfrac{1}{2}\, a - s,$$

and when the difference $x - y = d$ is given, he sometimes puts

$$x = s + \tfrac{1}{2}\, d \quad \text{and} \quad y = s - \tfrac{1}{2}\, d,$$

s being a new unknown. The Babylonians applied the same "plus-and-minus method" in a number of cases, including the standard cases (I)–(IV).

8. The Babylonians solved equations of type $x^3 = V$ by means of tables of cube roots. The Greeks solved the same equations by geometric constructions.

I feel the evidence is overwhelming, even if one leaves aside the Pythagorean traditions about the instruction in the science of numbers and the other mathematical sciences which PYTHAGORAS is said to have received in Babylon (IAMBLICHOS, *Vita Pyth.* 19, and PORPHYRIOS, *Vita Pyth.* 11).

---

7 S. GANDZ, *Osiris* 3 (1938), p. 405.

ANDRÉ WEIL

WHO BETRAYED EUCLID?
(Extract from a Letter to the Editor)

Some time ago your *Archive* printed a paper on Greek mathematics which, in tone and style as well as in content, fell significantly below the usual standards of that journal. As it has already quite adequately (if perhaps too gently) been refuted there by V.D. WAERDEN and by FREUDENTHAL, there is no need for referring to it by name. My only purpose in this letter is to point out that we have here almost a textbook illustration of the very thesis which the author (let us call him Z) sought to discredit, *viz*., that it is well to know mathematics before concerning oneself with its history; just as it is well to know Greek before dealing with Greek mathematics.

Z discusses a number of examples from EUCLID; I shall examine only the simplest one, which raises no side-issue; it is taken from EUCLID IX.8. As this consists of parallel statements about squares and about cubes, I may, for brevity, consider only the former. The paper quotes that proposition as follows (in HEATH's literal translation):

> "If as many numbers as we please beginning from an unit be in continued proportion, the third from the unit will be a square, as will also those which successively leave out one."

As the proof would show if necessary, the latter clause means the fifth, the seventh, *etc*. "Numbers" (VII, def. 2) means integers other than the unit or "monad". "In continued proportion" (ἑξῆς ἀνάλογον; VII, def. 21) means that the ratio of each integer to the next remains the same throughout the sequence. A number is called "a square" (VII, def. 19) if it is equal to some number multiplied with itself.

HEATH also gives an alternative translation of the same statement, equally faithful but in shorthand:

> "If 1, a, a_2, a_3, . . be a geometrical progression [*i.e.*, as explained later on, if $1{:}a = a{:}a_2 = a_2{:}a_3 = . .$], then a_2, a_4, a_6, . . are squares".

In EUCLID's proof, the "numbers" in the proposition are denoted by Greek capitals, A, B, Γ, Δ, E, Z; HEATH gives a literal translation of the proof, followed again by a transcript in shorthand, and ends up with the remark:

> | "The whole result is of course obvious if the geometrical progression is written, with 92
> our notation, as 1, a, a^2, a^3, . . ,a^n".

This gives occasion to Z, after pouring totally unwarranted obloquy upon HEATH, to make this pronouncement:

> "If we use modern algebraic symbolism, this ceases altogether to be a proposition and its truthfulness is an immediate and trivial application of the definition of a geometric progression".

When HEATH (imprudently, perhaps) wrote "obvious", he was not writing for laymen. He meant that the result is obvious for one who, having at least learned school-algebra,

J. Christianidis (ed.), Classics in the History of Greek Mathematics, 447–449.

will recognize in it the special case $q = 2$ of the rule $a^{pq} = (a^p)^q$. He knew that any mathematician would make the distinction (a subtle one to the layman) between the obvious and the trivial. Mathematicians are trained to know the difference between a definition, a notation and a theorem.

Perhaps the *modern* mathematician finds it easier, in this case, to perceive the truth of the matter, because nowadays the exponential notation x^α is used in many situations where x, α are not numbers. For instance, the exponent α may be taken from a non-commutative group; some care is then needed in the choice of definitions and notations if the rule $x^{\alpha\beta} = (x^\alpha)^\beta$ is still to hold true. However that may be, one who thinks that the rules governing the use of the exponential notation are trivial must be lacking, not only in mathematical understanding, but also in historical sense. Let him read EUCLID's book IX, then ARCHIMEDES' *Sandreckoner*, then pages 132 to 166 of J. TROPFKE's excellent *Geschichte der Elementar-Mathematik*, volume II. There he will learn that the development of the exponential *notation* and the realization of its *properties* went hand in hand for almost twenty centuries before they reached perfection. If now our notation allows schoolchildren to use the properties of exponentiation without ever being conscious of them, this does them no harm; they may then imagine that this makes those properties "trivial consequences of the definition", but we know better.

To berate HEATH and others for betraying EUCLID when all they do is to use a certain amount of notation to clarify the contents of his writings does not merely indicate a lack of mathematical sense; it argues a deficiency in logic. As everyone knows, words, too, are symbols. The content of a theorem does not change greatly, whether it is expressed in words or in formulas; the choice, as we all know, is mostly a matter of taste and of style. "Euclid's *numbers*", we read in Z's article, "are *given* line-segments, no abstract symbols" (his italics). What are A, B, Γ, Δ, E, Z in the proof of IX.8, if not symbols?

As to "numbers" being "line-segments", every reader of EUCLID knows how punctilious he is in distinguishing between line-segments (εὐθεῖαι), magnitudes (μεγέθη) and numbers (ἀριθμοί). Where, in IX.8 or indeed in the whole text of books VII, VIII and IX, is there a mention of line-segments? The layman may be misled by the diagrams in the margins; but a mere glance, for instance at the proof of IX.8, will show that the diagram contributes nothing to our understanding of the text, which carries no reference to it. If the unit had been thought of as a unit of length, it would appear in
93 the diagram, but it does not. It is open | to question whether such diagrams belong to the "tradition", *i.e.* whether they go back to EUCLID; even if we assume that they do, it is clear to the mathematician's eye that they are no more than a partial visualization of a piece of abstract reasoning. HASSE and his school used diagrams to illustrate the mutual relationships between algebraic number-fields; that did not make their subject into geometry. In EUCLID's books VII, VIII and IX, there is no trace of geometry, nor even of so-called "geometrical algebra". According to our modern classifications, those books are mostly algebra pure and simple (the algebra of the ring of integers); the balance, which is far deeper and more interesting, is pure number-theory. Of course it is more practical to carry out algebraic operations as we do, with the help of our algebraic symbolism, than in words as EUCLID did; just as it is more practical to perform arithmetical operations in the decimal (or, as computers do, in the dyadic) system, rather that as ARCHIMEDES did; this does not affect the substance of the matter. Who, one may ask, has been betraying EUCLID?

One point more deserves touching upon. EUCLID is the first extant mathematical text where the concept of proof is identified with a *gapless* chain of reasoning; this, and for good reasons, is still our view of the matter. Often it compels one to include, so to say for the record, much laborious routine; those who take shortcuts do so at their peril. The trained mathematician has learnt to discern, and indeed to skip, such passages, while the would-be historian concludes (in Z's words) that the writer has had "to toil energetically", little imagining that the poor wretch was just cursing the dullness of his self-inflicted task. It is not always easy, in a given historical context, to distinguish between mere routine and creative reasoning; there can be no worthwhile history of mathematics unless this is done.

To conclude: when a discipline, intermediary in some sense between two already existing ones (say A and B) becomes newly established, this often makes room for the proliferation of parasites, equally ignorant of both A and B, who seek to thrive by intimating to practitioners of A that they do not understand B, and vice versa. We see this happening now, alas, in the history of mathematics. Let us try to stop the disease before it proves fatal.

SABETAI UNGURU

HISTORY OF ANCIENT MATHEMATICS: SOME REFLECTIONS ON THE STATE OF THE ART

Denn eben wo Begriffe fehlen
Da stellt ein Wort zur rechten Zeit sich ein.
—JOHANN WOLFGANG VON GOETHE[1]

THE HISTORY OF MATHEMATICS typically has been written as if to illustrate the adage "anachronism is no vice." Most contemporary historians of mathematics, being mathematicians by training, assume tacitly or explicitly that mathematical entities reside in the world of Platonic ideas where they wait patiently to be discovered by the genius of the working mathematician. Mathematical concepts, constructive as well as computational, are seen as eternal, unchanging, unaffected by the idiosyncratic features of the culture in which they appear, each one clearly identifiable in its various historical occurrences, since these occurrences represent different clothings of the same Platonic hypostasis.

Various forms of the same mathematical concept or operation are not considered merely mathematically equivalent but also historically equivalent. Indeed mathematical equivalence is taken to represent historical equivalence. Since the mathematical Forms are eternal and since in their works mathematicians of all ages share in the expression of the same Forms, the specific mathematical idiom used by a mathematician has no bearing on the content of his thought. Mathematical language is at best a secondary appurtenance of the mathematical culture of any epoch. The mathematical kernel is untouched by the peculiar language used, since all mathematical languages lead back to the same ideal Forms. This makes the various casts in which the same mathematical truth has been expressed throughout the centuries completely equivalent. As one of my colleagues put it: "Under such an ontology, the object of the history of mathematics becomes the task of identifying the ideal forms present in the work of each historical author and apportioning out proper credit to that mathematician who first gave expression to one of these eternal forms, i.e., who first brought it out of the eternal Platonic realm into the world of human consciousness." This is precisely the task performed traditionally by the historian of mathematics.

But if scholars continue to neglect the peculiar specificities of a given mathematical | culture, whether as a result of explicitly stated or implicitly taken-for-granted 556

This paper is dedicated to my parents, Zeida and Ghiza Unguru. In its present form it owes a lot to the criticisms of Willy Hartner and Matthias Schramm, whom I thank for their assistance. I am also grateful to the anonymous referees and to many friends and colleagues for their critical suggestions. I am exclusively responsible for the views expressed here.

1 *Faust* I (Mephistopheles speaks).

J. Christianidis (ed.), Classics in the History of Greek Mathematics, 451–461.

assumptions, then by definition their work is ahistorical and should be recognized as such by the community of historians. History, as Aristotle knew, focuses on the idiosyncratic rather than the nomothetic.[2] It is impossible for modern man to think like an ancient Greek. Historical understanding, however, involves the attempt at faithful reconstruction of the past. In intellectual history this necessarily means the avoidance of conceptual pitfalls and interpretive anachronisms. Though it is impossible to think like Euclid, it is rather facile to think obtrusively unlike him. We cannot know what went through Euclid's mind when he wrote the *Elements*. But we can determine what Euclid could not have thought when he compiled his great work. He, most likely, did not employ concepts or operations for which there is no genuine evidence either in his time or in the works of his predecessors. This much is safe to conclude. Furthermore, he clearly could not have foreseen what mathematicians and historians of mathematics were going to do in the long run to his *Elements*; he could not have used mathematical devices and procedures which were invented many hundreds of years after his death. This much is obvious too. Given that we cannot think like Euclid, we should, nevertheless, strive to avoid thinking unlike him when elucidating and commenting on his writings. This is (and must remain) the historian's goal. One way of thinking *unlike* Euclid is to use the algebraic approach in interpreting his works.

It continues to be habitual among some historians of mathematics to say that what really lies behind Euclid's geometrically couched statements are algebraic reasonings, appearing in geometrical garb because of the lack of an appropriate algebraic symbolism. This strikes me as both inaccurate and unilluminating. To see this, let us ask the following question: how illuminating would it be to propose that Euclid really thought in Sanskrit but because of his ignorance of the Sanskrit alphabet, had to use the Greek one and consequently expressed himself in Greek? Greek mathematics must be understood in its own right. This can be done by refusing to apply to its analysis foreign, anachronistic criteria. The only acceptable meta-language for a historically sympathetic investigation and comprehension of Greek mathematics seems to be ordinary language, not algebra.

However, many scholars and in particular B. L. van der Waerden and Hans Freudenthal do not endorse these ideas.[3] Instead, they argue that mathematicians can easily discern in Greek mathematics its underlying algebraic basis and interpret it accordingly. Who would deny this manifest truth? The real question is: how accurate and factual is the mathematicians' interpretation? Is the treatment to which mathematicians have submitted Greek mathematics historically adequate? In fact, van der Waerden and Freudenthal argue the cogency and the completeness of the mathematicians' interpretation and treatment of Greek mathematics in algebraic form. But there is a serious problem here. The fact that modern mathematicians can interpret Greek mathematics algebraically is one thing. The conclusion that therefore the train of thought of the Greek mathematicians was algebraic is an entirely different matter. The step from the former statement to the latter is both a logical and a historical *non*

2 Aristotle, *Poetica* 1451b 1–19.

3 B. L. van der Waerden, "Defence of a 'Shocking' Point of View," *Archive for History of Exact Sciences*, 1975, 15:199–210 and Hans Freudenthal, "What Is Algebra and What Has It Been in History," *Arch. Hist. Exact Sci.*, 1977, 16:189–200, both written in response to my article "On the Need To Rewrite the History of Greek Mathematics," *Arch. Hist. Exact Sci.*, 1975, 15:67–114.

sequitur. "Logical," for obvious reasons; "historical," because it is possible to show | 557
by an analysis of Greek mathematical texts that the assumption of an underlying algebraic foundation for Greek mathematics leads to insoluble dilemmas and dreadful quandaries.[4]

In brief, van der Waerden defines algebra as "*the art of handling algebraic expressions like* $(a+b)^2$ *and of solving equations like* $x^2+ax=b$."[5] But no algebra exists in Babylonian and pre-Diophantian Greek mathematical sources. "Babylonian and Greek algebra" came into being only after the specific, numerical Babylonian examples and the Greek geometrical propositions had been transcribed into algebraic language; only as a result of the mathematician's elucidation of the texts was "algebra" brought into existence.[6] But the text itself did not present this elucidation; the imaginative creation of the interpreter did so. And there is a mathematical imagination and a historical imagination, and they typically run on different tracks. Finally, continuing allegiance to the algebraic interpretation of Babylonian mathematics is rendered still more untenable by the following rather damaging confession: "In den eigentlich mathematischen Texten ... die meisten Beispiele sicherlich von ihrem Resultat aus hergerichtet sind."[7] If this is true (and it seems to be, for most answers are "nice," even numbers), then what is the basis for the claim that the Babylonians solved equations? Do we as a rule solve equations by starting with the answer?

Those who perceive an algebraic substructure bolstering Greek mathematics claim that the Greeks started with algebraic problems but, then, translated them into a geometric format. They have reached this conclusion, according to van der Waerden, by studying "the wording of the theorems" and by trying "to reconstruct the original ideas of the author. We found it *evident* that these theorems did not arise out of geometrical problems [!]. We were not able to find any interesting geometrical problem that would give rise to theorems like II 1–4. On the other hand, we found that the explanation of these theorems as arising from algebra worked well. Therefore we adopted the latter explanation."[8]

But what evidence does van der Waerden present to demonstrate that "these theorems did not arise out of geometrical problems"? The answer, he tells us, is that no "interesting geometrical problem" leads to them. How does he know? Answer: he could not find any. But the conclusion is unwarranted, since even if it is true that no interesting geometrical problems led to them, it does not follow that noninteresting geometrical problems did not lead to them either. Furthermore, what *is* an interesting geometrical problem? Van der Waerden does not say, but the answer is implicit in what follows: "we found that the explanation of these theorems as arising from algebra worked well. Therefore. ..." An interesting geometrical problem, then, seems to be a

---

4 See "On the Need to Rewrite."

5 "Defence," p. 199.

6 In this context, S. Gandz, who advances his own algebraic interpretation of ancient mathematics in "The Origin and Development of the Quadratic Equations in Babylonian, Greek and early Arabic Algebra," *Osiris*, 1938, 3:405–557, says the following: "The commentary of NEUGEBAUER ... has no foundation in the text. It only shows how far away from the truth we may err, if we try, by all means, to detect our modern school formulas in the old Babylonian text" (p. 423).

7 Otto Neugebauer, *Vorgriechische Mathematik* (Berlin: Springer-Verlag, 1934; reprinted Springer-Verlag, 1969), p. 33.

8 "Defence," pp. 203–204.

problem the assumption of which "works well" in explaining the origin of the theorems under discussion. And van der Waerden has decided arbitrarily (since he could
558 not check all possible geometrical theorems and problems) that there are no | "interesting geometrical problems" working well under the circumstances. On the other hand, what works well is the assumption of an underlying algebraic foundation to Greek geometry. What does "working well" mean, then? Again, no answer is provided, but it would clearly seem to mean something removing difficulties and enabling one to cut through to the root and thus come up with "simple," "convincing," straightforward explanations. Ultimately, then, in the paragraph under discussion, van der Waerden does say that Greek geometry (at least some important parts of it) is, taken by itself, unfathomable, puzzling, weird, and that one can get rid of these unsavory features by assuming a hidden algebraic basis to it. Therefore, "the Greeks started with algebraic problems and translated them into geometric language," Q.E.D.[9]

Leaving aside the circularity of the entire argument, and the conflation of logic and history that it involves, van der Waerden's assertions represent an unconscious but nevertheless clear-cut vindication of the argument that the real roots of the methodological position embodied in the concept "geometric algebra" lie in the modern mathematician's ability to read geometric texts algebraically without any historical qualms.

II

Why did the Greeks, according to the proponents of the idea, disguise algebra in geometrical garb? Freudenthal gives three different answers: one historical, the second philosophical (its pertinence entirely escapes me), and the third "traditional." His "historical" answer speaks of a "torturous path through foundations of mathematics,"[10] which came to an end with the Eudoxian theory of proportions. But since there was neither genuine foundational work nor a real *Grundlagenkrisis* (as Hasse and Scholz

---

[9] It is this very same approach which is involved in identifying the purely geometrical problem of the application of areas as the Greek method of solution of quadratic equations, later equated with the Babylonian method: the modern mathematician can indeed translate Greek geometry and Babylonian specific-number manipulations into the algebraic language. Thus the parabolic application of areas, in which one is asked to apply to a given straight line a rectangle *equal* to a given square, can be transcribed as $ax = b^2$, if the given line is a and the given square b^2; to apply to the given line a rectangle equal to the given square such that the applied rectangle *falls short* of the second extremity of the given line by a square (the elliptical application of areas) can be transcribed as $x + y = a$, $xy = b^2$; and, to apply to the given line a rectangle equal to the given square such that the applied rectangle *exceeds* the second extremity of the given line by a square (the hyperbolic application of areas) can be transcribed as $x - y = a$, $xy = b^2$. This mathematical possibility, however, is not a satisfactory historical justification for the claimed identity of the Greek and the algebraic procedure. Moreover, strictly speaking, it is not the case that Euclid, *Elements* I.44 corresponds exactly to the simple parabolic application of areas. The simple parabolic application does not lead (as we saw) to a quadratic equation. If anything, it corresponds to the division of a given product (area) by a given magnitude (line). Only *Elements* VI.28 and 29 lead, when transcribed algebraically, to complete quadratic equations corresponding respectively to elliptical and hyperbolic application of areas. In this context, see A. Szabo, "Zum Problem der sog. 'Geometrischen Algebra' in Euklids Elementen," completed in 1975 for a *Festschrift* in honor of Willy Hartner, p. 7 of the preprint.

[10] Freudenthal, "What Is Algebra," p. 191.

and Van der Waerden referred to it[11]) in pre-Eudoxian times,[12] speaking of "the Greek end of the torturous path through foundations of mathematics, Eudoxos' | theory" is 559
misleading and consequently Freudenthal's so-called historical answer is nonhistorical and a nonanswer. The other answers do not fare any better. For example, in the "traditional" answer Freudenthal points out that "Once canonised, the *Elements* were sacrosanct ... The mathematical community was small. To be understood within it, you had to quote Euclid and to speak his language."[13] Fine, but why did Euclid, then, adopt the very same language? In sum, Freudenthal's "three main reasons" for the alleged disguise by the Greeks of algebra under the cloak of geometry are not good reasons.

"The efficiency of a symbolism is determined by the ease with which the user can move within it, by the algorithmic autonomy it provides."[14] Granted. But to judge from the texts, neither the Babylonian nor the ancient Greek moved at all "within it," and it possessed for them no "algorithmic autonomy" whatever, since it was not yet born. "It is the virtue of symbolism that it allows us for most of the time, rather than identify object and symbol ... to forget about what the symbol means." But this task is indeed impossible in Babylonian and Greek mathematical texts, where the object is always identified, either as specific numbers or as spacial diagrams. It is impossible to forget that "14, 30" means exactly "14, 30" and "line *AB*" means exactly "line *AB*"! It is simply not true that "Almost never in the *Elements* or anywhere else in Greek mathematics does *AB* mean a line or a line segment."[15]

As Freudenthal would have it, *Elements* V is "algebra and nothing else"; it is, moreover, "a general theory of magnitude ... independent of dimension or any characteristic of specific magnitudes."[16] The problem with such a characterization is the existence of *Elements* VII, in which many of the things dealt with in Book V are repeated and applied specifically to numbers (integers). In the presence of a general theory of magnitude, such a procedure would not have been just repetitious and superfluous but outright senseless. Numbers, after all, are specific instances of magnitude, and what is true of magnitudes in general is also true of numbers. In writing Book VII, then, Euclid did not simply follow tradition (as Heath thinks),[17] thereby merely proving over again for numbers propositions which he already proved for all magnitudes, including numbers. Book V, it seems, presents a general theory of proportion applicable to all kinds of

---

11 H. Hasse and H. Scholz, "Die Grundlagenkrisis der griechischen Mathematik," *Kant Studien*, 1928, 33:4–34; B. L. van der Waerden, "Zenon und die Grundlagenkrise der griechischen Mathematik," *Mathematische Annalen*, 1940–1941, 117:141–161.

12 Wilbur R. Knorr, *The Evolution of the Euclidean Elements: A Study of the Theory of Incommensurable Magnitudes and Its Significance for Early Greek Geometry* (Dordrecht/Boston: D. Reidel, 1975), pp. 40–42, 50, 305–313; cf. also Hans Freudenthal, "Y avait-il une crise des fondements des mathématiques dans l'antiquité?" *Bulletin de la Societé Mathématique de Belgique*, 1966, 8:43–55.

13 "What Is Algebra," p. 191. Freudenthal's philosophical answer reads: "Though in daily use by laymen as well as mathematicians, fractions were taboo in highbrow mathematics, because philosophy forbade the division of the unit" (*ibid.*).

14 *Ibid.*, p. 192.

15 *Ibid.*

16 *Ibid.*, p. 193.

17 T. L. Heath, *The Thirteen Books of Euclid's Elements*, 3 vols. (Cambridge: Cambridge University Press, 1908), Vol. II, p. 113.

magnitudes but not to numbers. The reason for this is that numbers for the Greeks are not instances of a concept of general magnitude. "Magnitude, in fact, corresponds to one of the two divisions of *quantity*, *ποσόν*, namely the continuous (as a line, a surface, or a body) whereas a number is *discrete*."[18] Numbers (integers) are not illustrations of something else, they are entities in their own right, with their own distinctive features, definitions, and so forth. This is what enables Wilbur Knorr to say that "Book VII does not merely duplicate Book V. It develops a body of analogous material for *the separate class of integers* [my emphasis]; that is, it is *required* as an independent treatment, not a duplication of a special case of Book V."[19]

560 | Why should this be so? Basically, because of the Greek view that arithmetic is an independent, not a derivative, discipline and that geometry and arithmetic are different genera having their own domains, disposing of their own techniques of demonstration, and dealing with their own subject matter. Pursuing them properly means refraining from infringing upon the territory of one by means of the tools and methods of the other.[20]

If I had been aware of the existence of Euclid's *Data*, argues Freudenthal, I "would never have claimed there were no equations in Greek geometry." For Freudenthal, the *Data* is a "textbook on solving equations." He summarizes the ninety-four propositions contained therein in a succinctly and strikingly epigrammatic statement: "Given certain magnitudes a, b, c and a relation $F(a, b, c, x)$, then x, too, is given. . . ."[21] But the fact remains that Greek geometry contained no equations. One cannot find even one equation in the entire text of the *Data*. Proof (as the Hindu mathematician would say): "Look!" Unless one has at his disposal the algebraic language and the capacity to translate into it, it is impossible to sum up this little treatise of rather varied content as offhandedly as Freudenthal has done. Indeed, had Euclid at his disposal Freudenthal's functional notation, it is rather easy to infer that he would not have needed ninety-four propositions to get his point across.

Each case in Euclid's *Data* is unique, having its own method of analysis, and none is subsumable under or reducible to other cases. "Datarum magnitudinum ratio inter se data est" (Prop. I) and "Si data magnitudo ad aliam magnitudinem rationem habet datam, data est etiam illa magnitudine" (Prop. II)—to use perhaps the simplest illustration possible—are not for Euclid both instances of "Given a, b, c, and $y = F(a, b, c, x)$, x is also given," but two different problems, interesting in their own right, having their own solutions. Of course, Freudenthal's description is mathematically correct. Historically, however, it is wanting. Heath is much more to the point when he says: "The *Data* . . . are still concerned with *elementary geometry*, though forming part of the introduction to higher analysis. Their form is that of propositions proving that, if certain things *in a figure* are given (in magnitude, in species, etc.), something else is given. The subject-matter is much the same as that of the planimetrical books of the *Elements*, to which the *Data* are often supplementary."[22]

---

18 T. L. Heath, *Mathematics in Aristotle* (Oxford: Clarendon Press, 1949), p. 45.

19 Knorr, *Evolution of the Euclidean Elements*, p. 309.

20 Aristotle, *Posterior Analytics* 75a37–75b20.

21 "What Is Algebra," p. 194. By the way, references to the *Data* appear in my paper in a number of places, e.g., p. 81, n. 26, and p. 108, n. 106.

22 Heath, *Elements*, Vol. I, p. 8, my italics.

This is what the *Data* is, not a textbook on solving equations, but a treatise presenting another approach to elementary geometry (other than that of the *Elements*, that is). Neither are Archimedes' works instances of "algebraic procedure in Greek mathematics."[23] Heath's edition is "in modern notation."[24] It is faithful only to the disembodied mathematical content of the Archimedean text, but not to its form. And this is crucial. If one abandons Archimedes' form and transcribes his rhetorical statements by means of algebraic symbols, manipulating and transforming the latter, then clearly "the algebraic procedure" appears. But this procedure itself is not "in Greek mathematics." It is a result (as Freudenthal himself states it) of "replacing vernacular by artificial language, and numbering variables by cardinals, a quite recent mathematical tool."[25] Indeed! Archimedes' text is anchored securely in the *terra firma* of Greek
geometry. If one is not willing to compress wording, to replace | "vernacular" by arti- 561
ficial language, to introduce variables and number them by cardinals, and to apply all the other technical tricks which are "quite recent mathematical tools," then Archimedes' proof of Proposition X of ΠΕΡΙ ΕΛΙΚΩΝ is geometric, not algebraic. This was discerned in a curious way even by Heath, who justified his algebraic procedure and the use of the symbols A_1, A_2, ... A_n "in order to exhibit *the geometrical character of the proof.*"[26]

Dijksterhuis himself in his *Archimedes* said: "in a representation of Greek proofs in the symbolism of modern algebra it is often precisely the most characteristic qualities of the classical argument which are lost, so that the reader is not sufficiently obliged to enter into the train of thought of the original. ..."[27]

III

Both Freudenthal and van der Waerden have constructed identical operative definitions of algebra, thereby creating significant problems in their analyses of Greek geometry. Freudenthal says: "This ability to describe relations and solving procedures, and the techniques involved *in a general way*, is in my view of algebra such an important feature of algebraic thinking that I am willing to extend the name 'algebra' to it. ... But what is in a name?"[28] However, it is precisely the inability of the Babylonian mathematician "to describe relations and solving procedures, and the techniques involved *in a general way*" that warrants his disqualification as algebraist. What the Babylonian mathematician lacks is precisely the ability to dispense with specific, definite numbers, and it is this deficiency that dictates the particular form of his approach. What he can produce is recipes, not general formulas.

With respect to the Greek mathematician (geometer), on the other hand, though it is legitimate to see his approach as a general approach (the so-called theorem of Pythagoras is true of *any* right-angled triangle, etc.), the language he uses is the geometric language and the generality involved is an outgrowth of dealing with geometrical and *not* with

23 "What Is Algebra," p. 195.
24 T. L. Heath, *The Works of Archimedes* (Cambridge: Cambridge University Press, 1897), title page.
25 "What Is Algebra," p. 196.
26 Heath, *Archimedes*, p. 109, my italics.
27 E. J. Dijksterhuis, *Archimedes* (Copenhagen: Munksgaard, 1956), p. 7.
28 "What Is Algebra," pp. 193–194, my italics.

algebraic entities. Consequently, by Freudenthal's own criteria of "algebraic thinking," Babylonian and Greek mathematics are nonalgebraic.

"What's in a name?" asks Freudenthal and uses the question even as a motto for his article. The answer, clearly enough, is "it depends." Names are words, and words are important when used thoughtfully. As a matter of fact, it is possible to argue that all there is is, one way or another, in words. The *Iliad* and *Hamlet* are in words; and so is the *Magna Charta*. The Bible is in words; and so is the American Declaration of Independence. All of mathematics is in a very definite sense in words. Thought and feeling (beyond inarticulate physiological reactions) are in words. Artistic experience is in a proper sense in words, for no informed, thoughtful reaction to and communication about a work of art is possible in the absence of articulate expression, which is again in words. Our meaningful access to reality (whatever it may consist of) is always mediate: we know the world through words.[29]

562 | But words can be misused. *Mein Kampf* and the *Protocols of the Elders of Zion* are also in words; and so is *The National Enquirer*. Words are powerful weapons, and men are governed (or misgoverned) with, by, and in words. And so, "what's in a word?" As always, "it all depends." Is the word used carefully? Does the user follow the advice of Paul to the Ephesians: "Let no man deceive ... with vain words"? Are words used and understood pertinently with reference to the subject matter, according to the old legal maxim, "Verba accipienda sunt secundum subjectam materiam?" If the answer to the above questions is positive, then there is a lot in a word; if negative, the word is misleading and therefore dangerous.

The use of the word "algebra" as a term descriptive of Babylonian and Greek mathematics is a misuse of the word. When the questions enumerated above are asked in connection with that use, all the answers come out negative. The word "algebra" is used carelessly; its use is deceiving since it leads to a translation of ancient mathematical texts into a historically inappropriate language; and, if "algebra" has its proper meaning, the use of the term is unsuited to the subject matter. Words are judgments, or, as Nietzsche put it, preconceived judgments; and this is how it should be. But some judgments carry conviction while others are blatantly unjust. The word "algebra" in the context discussed belongs to the latter category.

Enthusiasts of algebraic interpretations of Greek geometry have violated one of the fundamental tenets of historical scholarship. History is the study of the present traces of past events from the standpoint of change and the particular, the idiosyncratic.[30] Although long-lasting structures, stable frameworks, and durable, quasi-constant features are legitimate topics of historical investigation, they are not what makes history what it is.[31] History is primarily, essentially interested in the event *qua* particular event,

---

[29] In a pregnant way, the Venerable Inceptor expressed this view as follows: "Si dicas: nolo loqui de vocibus sed tantum de rebus, dico quod quamvis velis loqui tantum de rebus, tamen hoc non est possibile nisi mediantibus vocibus vel conceptibus vel aliis signis" (William of Ockham, *Commentary on the Sentences*, I, dist. 2, quest. 1, in *Super quatuor libros sententiarum (In sententiarum I)*, being Vol. III (1495) of Guillelmus de Occam, O.F.M., *Opera plurima* (Lyons, 1494–1496; reprinted London: Gregg Press, 1962, in 4 vols.).

[30] See G. R. Elton, *The Practice of History* (Sidney: Sidney University Press, 1967), pp. 8–12.

[31] "... there is more to history than the study of persistent structures and the slow progress of evolution" (Fernand Braudel, *The Mediterranean and the Mediterranean World in the Age of Philip II*, 2 vols., New York: Harper & Row, 1975, Vol. II, p. 901).

in the specific happening, in change from an identifiable, individual characteristic to another identifiable, individual characteristic. History is not (or is primarily not) striving to bunch events together, to crowd them under the same heading by draining them of their individualities. On the contrary, history is the attempt at understanding each past event in its own right. The domain of history, then, is the idiosyncratic.

The historian of ideas does not discharge his obligation by showing merely the extent to which past ideas are like modern ideas. His main effort should be in the direction of showing the extent to which past ideas were unlike modern ones, irrespective of the fact that they might (or might not) have led to the modern ideas. This is a wise methodological tack, since it enables the historian to avoid reductive anachronism while channeling his historical empathy toward an understanding of the past in its own right. It is also wise to take the written documents of the past to mean precisely what they say, short of clear-cut proof to the contrary. There is no historical advantage whatever growing out of the gratuitous assumption that the men of old played tricks on us by systematically hiding their line of thought.

I shall not presume to define here what mathematics is, as that is best left to mathematicians. Besides, there are plenty of definitions available, running the gamut from Bertrand Russell's to Nicolas Bourbaki's.[32] Every reader can easily take his | pick. But 563
I can say safely what mathematics is not. It is certainly not history. The domain of mathematics is not the idiosyncratic, but, in a very real sense, the nomothetic, since what mathematicians do is to show that from certain assumptions about as yet unidentified objects some conclusions about the same objects will follow necessarily, by rule.

The history of mathematics is history not mathematics. It is the study of the idiosyncratic aspects of the activity of mathematicians who themselves are engaged in the study of the nomothetic, that is, of what is the case by law. If one is to write the history of mathematics, and not the mathematics of history, the writer must be careful not to substitute the nomothetic for the idiosyncratic, that is, not to deal with past mathematics as if mathematics had no past beyond trivial differences in the outward appearance of what is basically an unchangeable hard-core content.

In mathematics (like in anything else) form and content are not independent variables. On the contrary, they mutually condition one another and neither is immune to change. A certain form permits only a certain content, and a new content requires a new form. This is why the methodological approach which casts indiscriminately the algebraic shadow over the garden of Greek mathematics obscures precisely those features which make it *Greek* mathematics. Instead of showing the degree to which it was unlike modern, post-Renaissance mathematics, that approach, by greatly overemphasizing the similarities, prevents an understanding of Greek mathematics in its own right. It also leads in the long run to the untenable view that the Greek mathematicians did not mean what they said, but that they hid "admirably"[33] their line of thought. Coupled with this is the great danger of easily "discerning" problematic or nonexistent influences

---

32 "... mathematics may be defined as the subject in which we never know what we are talking about, nor whether what we are saying is true" (Bertrand Russell, *Mysticism and Logic and Other Essays*, New York: Barnes and Noble, 1971, pp. 59–60) and "A *mathematical theory* ... contains rules which allow us to assert that certain assemblies of signs are *terms* or relations of the theory, and other rules which allow us to assert that certain assemblies are *theorems* of the theory" (Nicolas Bourbaki, *Elements of Mathematics: Theory of Sets*, Paris/London: Addison-Wesley, 1968, p. 16).

33 B. L. van der Waerden, *Science Awakening* (Groningen: P. Noordhoff, 1954), p. 172.

between mathematical cultures a world apart, simply because when submitted to the algebraic cure all mathematical cultures look alike.

Entrenched as it is, the traditional interpretation of the history of ancient mathematics must give way to a new, more sympathetic, and historically responsive interpretation, simply because the old interpretation has outlived its usefulness and is now an obstacle on the road to a sensitive historical understanding of ancient mathematical texts. After all, like scientific theories, historical theories are tentative attempts to make sense of the past; they are provisional by their very nature, and consequently their authors should not be dreaming hopelessly of endowing them, in God-like fashion, with eternal life and immaculate beatitude.

Otto Neugebauer is right. Speaking of the fact that it was the Hindus and not the Babylonians who introduced a sign for zero to be used *always* whenever required in the writing of numbers, Neugebauer makes the following pertinent remark:

> It seems to me ... that the awareness of the arbitrariness and the purely conventional, symbolic character of all means of expression does not arise in the framework of a continuous historical development, which, of course, rests on the direct tradition from generation to generation; the awareness of all these things becomes absolutized into fixed and rigid forms which cannot be substantially changed of one's own accord, as this largely transcends the analytical capacity of mankind. Only men who are the
> 564 heirs of an alto|gether different historical tradition are able to use freely the foreign means of expression and to recognize both their limits and their potentialities.[34]

Now this seems indeed to be the case with respect to the introduction of the algebraic approach by Viète, Fermat, and Descartes, men of genius belonging to another culture than the Greek, but who managed somehow to discern in what the Greeks had done (geometry) precisely what the Greeks themselves never dreamt about when they were doing it, namely a hidden algebraic structure, which the moderns set about to extract from the Greek texts. This is the true historical origin of the concept "geometrical algebra." It is the intellectual product of foreigners, barbarians, reading Greek mathematical texts in light of their own idiosyncrasies, their own barbarian approach, and "seeing" in it what the Greeks, the autochthons, never put into it, namely, an algebraic train of thought. *Mutatis mutandis*, like the Babylonians with the general concept of zero, the Greeks never came up with a symbolic approach; it remained for the sixteenth- and seventeenth-century Europeans (playing somewhat the role of the Hindus in our comparison), the heirs and at the same time the usurpers of the Greeks, to invent the general symbolic approach and thereby to "perceive" its roots within the confines of Greek geometry.

Though what one calls a thing is, to begin with, merely a convention, once the calling (naming) has been socially accepted, departures from the standard usage without further ado are misleading and can be dangerous. Whatever *algebra* might "really" be, the term as standardly used means something definite, as do most of the words used

---

34 *Vorgriechische Mathematik*, p. 78: "Mir scheint ... dass im Rahmen einer kontinuierlichen geschichtlichen Entwicklung, die ja auf der direkten Tradition von Generation zu Generation beruht, das Bewusstsein der Willkürlichkeit und des rein konventionellen symbolischen Charakters aller Ausdrucksmittel gar nicht ensteht, dass alle diese Dinge zu aboluten und gegebenen Formen werden, die aus freien Stücken wesentlich abzuändern das analytische Vermögen der Menschen weit übersteigt. Erst Menschen die selbst einer ganz anderen geschichtlichen Tradition entstammen, sind imstande, die fremden Ausdrucksmittel frei zu gebrauchen und ihre Schranken wie ihre Möglichkeit zu erkennen."

in common parlance. This is what makes communication possible. A "table" is a table. A "chair" is a chair. Even Freudenthal agrees with that, since he says: "'algebra' has a meaning in everyday language just as 'chair' and 'table' have."[35] Calling, then, a tree "table" is misleading, in spite of the fact that trees can (and quite often do) become tables. (As a matter of fact—and this is crucial—quite often they do not.) By the same token, calling a tree "chair" is misleading. In such an arbitrary naming procedure, one substitutes one of the many potentialities of the object for its reality. This is dangerous, since trees are potentially not just tables or chairs, but also coffins or houses. Calling a tree "table," then, is misleading not only because it takes the potential for the real but also because it neglects all but one of the various potentialities of the object.

Precisely as it is only hindsight that enables one to call legitimately a certain tree "table," or "chair," or "coffin," it is only unwarranted historical hindsight that has enabled scholars to call Greek geometry "algebra," by setting up just one of the potentialities of Greek geometry into a chosen entelechy. There may exist, by divine decree, a chosen people. However, "chosen" entelechies, in the perfectly natural case of multiple potentialities, are *post factum* creations of the mind of the historian-philosopher running rampant, since the whole historical point consists exactly in the necessity to show that in the actual historical process only "the chosen entelechy" has been realized.

| It is true that names are conventions. But conventions fulfill a very important func- 565
tion, making articulate communication (i.e., intelligent life) possible. Abiding by them enables one to carry on in everyday life. Blatant transgressions against socially accepted conventions, on the other hand, prevent normal communication and can be rather troublesome. It is mere convention to kiss and embrace one's bride at the wedding. Refuse to do it, "because it is a mere convention," and you are in for some real trouble.

The name "algebra," like all names, is a convention (although it has some very definite historical roots). But it means something recognizable in common parlance. Apply it indiscriminately to what is obviously geometry and you have not merely breached a useful convention, you have thereby created a new one, less definite, sharp, and useful than the one you violated, since it substitutes potentiality for reality. And although this is possible, it is wrong historically, since history deals with reality (what happened) and not with potentiality (what could have happened logically). The approach of Freudenthal, van der Waerden, and their cohort substitutes logic for history.[36]

---

[35] "What Is Algebra," p. 193.

[36] Here, I have in mind Andre Weil's unprecedented missive to the editor of the *Archive for History of Exact Sciences*, entirely repetitive in its few *non ad hominem* passages of the arguments of van der Waerden and Freudenthal: "Who Betrayed Euclid?" *Arch. Hist. Exact Sci.*, 1978, 19:91–93. Concerning this letter, the less said the better. In adopting this position, I am guided by Simone Weil's words in her sensitive and penetrating essay on the *Iliad* (*The Iliad or the Poem of Force*, Wallingford, Pa.: Pendle Hill, n.d., pp. 3, 36): "To define force—it is that *x* that turns anybody who is subjected to it into a *thing*. Exercised to the limit, it turns man into a thing in the most literal sense: it makes a corpse out of him. Somebody was here, and the next minute there is nobody here at all;" And: "The man who does not wear the armor of the lie cannot experience force without being touched by it to the very soul. Grace can prevent this touch from corrupting him, but it cannot spare him the wound."

Boston Studies in the Philosophy of Science

Editor: Robert S. Cohen, *Boston University*

1. M.W. Wartofsky (ed.): *Proceedings of the Boston Colloquium for the Philosophy of Science, 1961/1962.* [Synthese Library 6] 1963 ISBN 90-277-0021-4
2. R.S. Cohen and M.W. Wartofsky (eds.): *Proceedings of the Boston Colloquium for the Philosophy of Science, 1962/1964.* In Honor of P. Frank. [Synthese Library 10] 1965 ISBN 90-277-9004-0
3. R.S. Cohen and M.W. Wartofsky (eds.): *Proceedings of the Boston Colloquium for the Philosophy of Science, 1964/1966.* In Memory of Norwood Russell Hanson. [Synthese Library 14] 1967 ISBN 90-277-0013-3
4. R.S. Cohen and M.W. Wartofsky (eds.): *Proceedings of the Boston Colloquium for the Philosophy of Science, 1966/1968.* [Synthese Library 18] 1969 ISBN 90-277-0014-1
5. R.S. Cohen and M.W. Wartofsky (eds.): *Proceedings of the Boston Colloquium for the Philosophy of Science, 1966/1968.* [Synthese Library 19] 1969 ISBN 90-277-0015-X
6. R.S. Cohen and R.J. Seeger (eds.): *Ernst Mach, Physicist and Philosopher.* [Synthese Library 27] 1970 ISBN 90-277-0016-8
7. M. Čapek: *Bergson and Modern Physics.* A Reinterpretation and Re-evaluation. [Synthese Library 37] 1971 ISBN 90-277-0186-5
8. R.C. Buck and R.S. Cohen (eds.): *PSA 1970.* Proceedings of the 2nd Biennial Meeting of the Philosophy and Science Association (Boston, Fall 1970). In Memory of Rudolf Carnap. [Synthese Library 39] 1971 ISBN 90-277-0187-3; Pb 90-277-0309-4
9. A.A. Zinov'ev: *Foundations of the Logical Theory of Scientific Knowledge (Complex Logic).* Translated from Russian. Revised and enlarged English Edition, with an Appendix by G.A. Smirnov, E.A. Sidorenko, A.M. Fedina and L.A. Bobrova. [Synthese Library 46] 1973 ISBN 90-277-0193-8; Pb 90-277-0324-8
10. L. Tondl: *Scientific Procedures.* A Contribution Concerning the Methodological Problems of Scientific Concepts and Scientific Explanation.Translated from Czech. [Synthese Library 47] 1973 ISBN 90-277-0147-4; Pb 90-277-0323-X
11. R.J. Seeger and R.S. Cohen (eds.): *Philosophical Foundations of Science.* Proceedings of Section L, 1969, American Association for the Advancement of Science. [Synthese Library 58] 1974 ISBN 90-277-0390-6; Pb 90-277-0376-0
12. A. Grünbaum: *Philosophical Problems of Space and Times.* 2nd enlarged ed. [Synthese Library 55] 1973 ISBN 90-277-0357-4; Pb 90-277-0358-2
13. R.S. Cohen and M.W. Wartofsky (eds.): *Logical and Epistemological Studies in Contemporary Physics.* Proceedings of the Boston Colloquium for the Philosophy of Science, 1969/72, Part I. [Synthese Library 59] 1974 ISBN 90-277-0391-4; Pb 90-277-0377-9
14. R.S. Cohen and M.W. Wartofsky (eds.): *Methodological and Historical Essays in the Natural and Social Sciences.* Proceedings of the Boston Colloquium for the Philosophy of Science, 1969/72, Part II. [Synthese Library 60] 1974 ISBN 90-277-0392-2; Pb 90-277-0378-7
15. R.S. Cohen, J.J. Stachel and M.W. Wartofsky (eds.): *For Dirk Struik.* Scientific, Historical and Political Essays in Honor of Dirk J. Struik. [Synthese Library 61] 1974 ISBN 90-277-0393-0; Pb 90-277-0379-5
16. N. Geschwind: *Selected Papers on Language and the Brains.* [Synthese Library 68] 1974 ISBN 90-277-0262-4; Pb 90-277-0263-2
17. B.G. Kuznetsov: *Reason and Being.* Translated from Russian. Edited by C.R. Fawcett and R.S. Cohen. 1987 ISBN 90-277-2181-5

18. P. Mittelstaedt: *Philosophical Problems of Modern Physics.* Translated from the revised 4th German edition by W. Riemer and edited by R.S. Cohen. [Synthese Library 95] 1976
ISBN 90-277-0285-3; Pb 90-277-0506-2
19. H. Mehlberg: *Time, Causality, and the Quantum Theory.* Studies in the Philosophy of Science. Vol. I: *Essay on the Causal Theory of Time.* Vol. II: *Time in a Quantized Universe.* Translated from French. Edited by R.S. Cohen. 1980 Vol. I: ISBN 90-277-0721-9; Pb 90-277-1074-0
Vol. II: ISBN 90-277-1075-9; Pb 90-277-1076-7
20. K.F. Schaffner and R.S. Cohen (eds.): *PSA 1972.* Proceedings of the 3rd Biennial Meeting of the Philosophy of Science Association (Lansing, Michigan, Fall 1972). [Synthese Library 64] 1974 ISBN 90-277-0408-2; Pb 90-277-0409-0
21. R.S. Cohen and J.J. Stachel (eds.): *Selected Papers of Léon Rosenfeld.* [Synthese Library 100] 1979 ISBN 90-277-0651-4; Pb 90-277-0652-2
22. M. Čapek (ed.): *The Concepts of Space and Time.* Their Structure and Their Development. [Synthese Library 74] 1976 ISBN 90-277-0355-8; Pb 90-277-0375-2
23. M. Grene: *The Understanding of Nature.* Essays in the Philosophy of Biology. [Synthese Library 66] 1974 ISBN 90-277-0462-7; Pb 90-277-0463-5
24. D. Ihde: *Technics and Praxis.* A Philosophy of Technology. [Synthese Library 130] 1979
ISBN 90-277-0953-X; Pb 90-277-0954-8
25. J. Hintikka and U. Remes: *The Method of Analysis.* Its Geometrical Origin and Its General Significance. [Synthese Library 75] 1974 ISBN 90-277-0532-1; Pb 90-277-0543-7
26. J.E. Murdoch and E.D. Sylla (eds.): *The Cultural Context of Medieval Learning.* Proceedings of the First International Colloquium on Philosophy, Science, and Theology in the Middle Ages, 1973. [Synthese Library 76] 1975 ISBN 90-277-0560-7; Pb 90-277-0587-9
27. M. Grene and E. Mendelsohn (eds.): *Topics in the Philosophy of Biology.* [Synthese Library 84] 1976 ISBN 90-277-0595-X; Pb 90-277-0596-8
28. J. Agassi: *Science in Flux.* [Synthese Library 80] 1975
ISBN 90-277-0584-4; Pb 90-277-0612-3
29. J.J. Wiatr (ed.): *Polish Essays in the Methodology of the Social Sciences.* [Synthese Library 131] 1979 ISBN 90-277-0723-5; Pb 90-277-0956-4
30. P. Janich: *Protophysics of Time.* Constructive Foundation and History of Time Measurement. Translated from German. 1985 ISBN 90-277-0724-3
31. R.S. Cohen and M.W. Wartofsky (eds.): *Language, Logic, and Method.* 1983
ISBN 90-277-0725-1
32. R.S. Cohen, C.A. Hooker, A.C. Michalos and J.W. van Evra (eds.): *PSA 1974.* Proceedings of the 4th Biennial Meeting of the Philosophy of Science Association. [Synthese Library 101] 1976 ISBN 90-277-0647-6; Pb 90-277-0648-4
33. G. Holton and W.A. Blanpied (eds.): *Science and Its Public.* The Changing Relationship. [Synthese Library 96] 1976 ISBN 90-277-0657-3; Pb 90-277-0658-1
34. M.D. Grmek, R.S. Cohen and G. Cimino (eds.): *On Scientific Discovery.* The 1977 Erice Lectures. 1981 ISBN 90-277-1122-4; Pb 90-277-1123-2
35. S. Amsterdamski: *Between Experience and Metaphysics.* Philosophical Problems of the Evolution of Science. Translated from Polish. [Synthese Library 77] 1975
ISBN 90-277-0568-2; Pb 90-277-0580-1
36. M. Marković and G. Petrović (eds.): *Praxis.* Yugoslav Essays in the Philosophy and Methodology of the Social Sciences. [Synthese Library 134] 1979
ISBN 90-277-0727-8; Pb 90-277-0968-8

37. H. von Helmholtz: *Epistemological Writings.* The Paul Hertz / Moritz Schlick Centenary Edition of 1921. Translated from German by M.F. Lowe. Edited with an Introduction and Bibliography by R.S. Cohen and Y. Elkana. [Synthese Library 79] 1977
ISBN 90-277-0290-X; Pb 90-277-0582-8
38. R.M. Martin: *Pragmatics, Truth and Language.* 1979
ISBN 90-277-0992-0; Pb 90-277-0993-9
39. R.S. Cohen, P.K. Feyerabend and M.W. Wartofsky (eds.): *Essays in Memory of Imre Lakatos.* [Synthese Library 99] 1976 ISBN 90-277-0654-9; Pb 90-277-0655-7
40. Not published.
41. Not published.
42. H.R. Maturana and F.J. Varela: *Autopoiesis and Cognition.* The Realization of the Living. With a Preface to "Autopoiesis' by S. Beer. 1980 ISBN 90-277-1015-5; Pb 90-277-1016-3
43. A. Kasher (ed.): *Language in Focus: Foundations, Methods and Systems.* Essays in Memory of Yehoshua Bar-Hillel. [Synthese Library 89] 1976
ISBN 90-277-0644-1; Pb 90-277-0645-X
44. T.D. Thao: *Investigations into the Origin of Language and Consciousness.* 1984
ISBN 90-277-0827-4
45. F.G.-I. Nagasaka (ed.): *Japanese Studies in the Philosophy of Science.* 1997
ISBN 0-7923-4781-1
46. P.L. Kapitza: *Experiment, Theory, Practice.* Articles and Addresses. Edited by R.S. Cohen. 1980 ISBN 90-277-1061-9; Pb 90-277-1062-7
47. M.L. Dalla Chiara (ed.): *Italian Studies in the Philosophy of Science.* 1981
ISBN 90-277-0735-9; Pb 90-277-1073-2
48. M.W. Wartofsky: *Models.* Representation and the Scientific Understanding. [Synthese Library 129] 1979 ISBN 90-277-0736-7; Pb 90-277-0947-5
49. T.D. Thao: *Phenomenology and Dialectical Materialism.* Edited by R.S. Cohen. 1986
ISBN 90-277-0737-5
50. Y. Fried and J. Agassi: *Paranoia.* A Study in Diagnosis. [Synthese Library 102] 1976
ISBN 90-277-0704-9; Pb 90-277-0705-7
51. K.H. Wolff: *Surrender and Cath.* Experience and Inquiry Today. [Synthese Library 105] 1976
ISBN 90-277-0758-8; Pb 90-277-0765-0
52. K. Kosík: *Dialectics of the Concrete.* A Study on Problems of Man and World. 1976
ISBN 90-277-0761-8; Pb 90-277-0764-2
53. N. Goodman: *The Structure of Appearance.* [Synthese Library 107] 1977
ISBN 90-277-0773-1; Pb 90-277-0774-X
54. H.A. Simon: *Models of Discovery* and Other Topics in the Methods of Science. [Synthese Library 114] 1977 ISBN 90-277-0812-6; Pb 90-277-0858-4
55. M. Lazerowitz: *The Language of Philosophy.* Freud and Wittgenstein. [Synthese Library 117] 1977 ISBN 90-277-0826-6; Pb 90-277-0862-2
56. T. Nickles (ed.): *Scientific Discovery, Logic, and Rationality.* 1980
ISBN 90-277-1069-4; Pb 90-277-1070-8
57. J. Margolis: *Persons and Mind.* The Prospects of Nonreductive Materialism. [Synthese Library 121] 1978 ISBN 90-277-0854-1; Pb 90-277-0863-0
58. G. Radnitzky and G. Andersson (eds.): *Progress and Rationality in Science.* [Synthese Library 125] 1978 ISBN 90-277-0921-1; Pb 90-277-0922-X
59. G. Radnitzky and G. Andersson (eds.): *The Structure and Development of Science.* [Synthese Library 136] 1979 ISBN 90-277-0994-7; Pb 90-277-0995-5

60. T. Nickles (ed.): *Scientific Discovery.* Case Studies. 1980
ISBN 90-277-1092-9; Pb 90-277-1093-7
61. M.A. Finocchiaro: *Galileo and the Art of Reasoning.* Rhetorical Foundation of Logic and Scientific Method. 1980 ISBN 90-277-1094-5; Pb 90-277-1095-3
62. W.A. Wallace: *Prelude to Galileo.* Essays on Medieval and 16th-Century Sources of Galileo's Thought. 1981 ISBN 90-277-1215-8; Pb 90-277-1216-6
63. F. Rapp: *Analytical Philosophy of Technology.* Translated from German. 1981
ISBN 90-277-1221-2; Pb 90-277-1222-0
64. R.S. Cohen and M.W. Wartofsky (eds.): *Hegel and the Sciences.* 1984 ISBN 90-277-0726-X
65. J. Agassi: *Science and Society.* Studies in the Sociology of Science. 1981
ISBN 90-277-1244-1; Pb 90-277-1245-X
66. L. Tondl: *Problems of Semantics.* A Contribution to the Analysis of the Language of Science. Translated from Czech. 1981 ISBN 90-277-0148-2; Pb 90-277-0316-7
67. J. Agassi and R.S. Cohen (eds.): *Scientific Philosophy Today.* Essays in Honor of Mario Bunge. 1982 ISBN 90-277-1262-X; Pb 90-277-1263-8
68. W. Krajewski (ed.): *Polish Essays in the Philosophy of the Natural Sciences.* Translated from Polish and edited by R.S. Cohen and C.R. Fawcett. 1982
ISBN 90-277-1286-7; Pb 90-277-1287-5
69. J.H. Fetzer: *Scientific Knowledge.* Causation, Explanation and Corroboration. 1981
ISBN 90-277-1335-9; Pb 90-277-1336-7
70. S. Grossberg: *Studies of Mind and Brain.* Neural Principles of Learning, Perception, Development, Cognition, and Motor Control. 1982 ISBN 90-277-1359-6; Pb 90-277-1360-X
71. R.S. Cohen and M.W. Wartofsky (eds.): *Epistemology, Methodology, and the Social Sciences.* 1983. ISBN 90-277-1454-1
72. K. Berka: *Measurement.* Its Concepts, Theories and Problems. Translated from Czech. 1983
ISBN 90-277-1416-9
73. G.L. Pandit: *The Structure and Growth of Scientific Knowledge.* A Study in the Methodology of Epistemic Appraisal. 1983 ISBN 90-277-1434-7
74. A.A. Zinov'ev: *Logical Physics.* Translated from Russian. Edited by R.S. Cohen. 1983
[*see also* Volume 9] ISBN 90-277-0734-0
75. G-G. Granger: *Formal Thought and the Sciences of Man.* Translated from French. With and Introduction by A. Rosenberg. 1983 ISBN 90-277-1524-6
76. R.S. Cohen and L. Laudan (eds.): *Physics, Philosophy and Psychoanalysis.* Essays in Honor of Adolf Grünbaum. 1983 ISBN 90-277-1533-5
77. G. Böhme, W. van den Daele, R. Hohlfeld, W. Krohn and W. Schäfer: *Finalization in Science.* The Social Orientation of Scientific Progress. Translated from German. Edited by W. Schäfer. 1983 ISBN 90-277-1549-1
78. D. Shapere: *Reason and the Search for Knowledge.* Investigations in the Philosophy of Science. 1984 ISBN 90-277-1551-3; Pb 90-277-1641-2
79. G. Andersson (ed.): *Rationality in Science and Politics.* Translated from German. 1984
ISBN 90-277-1575-0; Pb 90-277-1953-5
80. P.T. Durbin and F. Rapp (eds.): *Philosophy and Technology.* [*Also* Philosophy and Technology Series, Vol. 1] 1983 ISBN 90-277-1576-9
81. M. Marković: *Dialectical Theory of Meaning.* Translated from Serbo-Croat. 1984
ISBN 90-277-1596-3
82. R.S. Cohen and M.W. Wartofsky (eds.): *Physical Sciences and History of Physics.* 1984.
ISBN 90-277-1615-3

Boston Studies in the Philosophy of Science

83. É. Meyerson: *The Relativistic Deduction.* Epistemological Implications of the Theory of Relativity. Translated from French. With a Review by Albert Einstein and an Introduction by Milić Čapek. 1985 ISBN 90-277-1699-4
84. R.S. Cohen and M.W. Wartofsky (eds.): *Methodology, Metaphysics and the History of Science.* In Memory of Benjamin Nelson. 1984 ISBN 90-277-1711-7
85. G. Tamás: *The Logic of Categories.* Translated from Hungarian. Edited by R.S. Cohen. 1986 ISBN 90-277-1742-7
86. S.L. de C. Fernandes: *Foundations of Objective Knowledge.* The Relations of Popper's Theory of Knowledge to That of Kant. 1985 ISBN 90-277-1809-1
87. R.S. Cohen and T. Schnelle (eds.): *Cognition and Fact.* Materials on Ludwik Fleck. 1986 ISBN 90-277-1902-0
88. G. Freudenthal: *Atom and Individual in the Age of Newton.* On the Genesis of the Mechanistic World View. Translated from German. 1986 ISBN 90-277-1905-5
89. A. Donagan, A.N. Perovich Jr and M.V. Wedin (eds.): *Human Nature and Natural Knowledge.* Essays presented to Marjorie Grene on the Occasion of Her 75th Birthday. 1986 ISBN 90-277-1974-8
90. C. Mitcham and A. Hunning (eds.): *Philosophy and Technology II.* Information Technology and Computers in Theory and Practice. [*Also* Philosophy and Technology Series, Vol. 2] 1986 ISBN 90-277-1975-6
91. M. Grene and D. Nails (eds.): *Spinoza and the Sciences.* 1986 ISBN 90-277-1976-4
92. S.P. Turner: *The Search for a Methodology of Social Science.* Durkheim, Weber, and the 19th-Century Problem of Cause, Probability, and Action. 1986. ISBN 90-277-2067-3
93. I.C. Jarvie: *Thinking about Society.* Theory and Practice. 1986 ISBN 90-277-2068-1
94. E. Ullmann-Margalit (ed.): *The Kaleidoscope of Science.* The Israel Colloquium: Studies in History, Philosophy, and Sociology of Science, Vol. 1. 1986 ISBN 90-277-2158-0; Pb 90-277-2159-9
95. E. Ullmann-Margalit (ed.): *The Prism of Science.* The Israel Colloquium: Studies in History, Philosophy, and Sociology of Science, Vol. 2. 1986 ISBN 90-277-2160-2; Pb 90-277-2161-0
96. G. Márkus: *Language and Production.* A Critique of the Paradigms. Translated from French. 1986 ISBN 90-277-2169-6
97. F. Amrine, F.J. Zucker and H. Wheeler (eds.): *Goethe and the Sciences: A Reappraisal.* 1987 ISBN 90-277-2265-X; Pb 90-277-2400-8
98. J.C. Pitt and M. Pera (eds.): *Rational Changes in Science.* Essays on Scientific Reasoning. Translated from Italian. 1987 ISBN 90-277-2417-2
99. O. Costa de Beauregard: *Time, the Physical Magnitude.* 1987 ISBN 90-277-2444-X
100. A. Shimony and D. Nails (eds.): *Naturalistic Epistemology.* A Symposium of Two Decades. 1987 ISBN 90-277-2337-0
101. N. Rotenstreich: *Time and Meaning in History.* 1987 ISBN 90-277-2467-9
102. D.B. Zilberman: *The Birth of Meaning in Hindu Thought.* Edited by R.S. Cohen. 1988 ISBN 90-277-2497-0
103. T.F. Glick (ed.): *The Comparative Reception of Relativity.* 1987 ISBN 90-277-2498-9
104. Z. Harris, M. Gottfried, T. Ryckman, P. Mattick Jr, A. Daladier, T.N. Harris and S. Harris: *The Form of Information in Science.* Analysis of an Immunology Sublanguage. With a Preface by Hilary Putnam. 1989 ISBN 90-277-2516-0
105. F. Burwick (ed.): *Approaches to Organic Form.* Permutations in Science and Culture. 1987 ISBN 90-277-2541-1

Boston Studies in the Philosophy of Science

106. M. Almási: *The Philosophy of Appearances.* Translated from Hungarian. 1989 ISBN 90-277-2150-5
107. S. Hook, W.L. O'Neill and R. O'Toole (eds.): *Philosophy, History and Social Action.* Essays in Honor of Lewis Feuer. With an Autobiographical Essay by L. Feuer. 1988 ISBN 90-277-2644-2
108. I. Hronszky, M. Fehér and B. Dajka: *Scientific Knowledge Socialized.* Selected Proceedings of the 5th Joint International Conference on the History and Philosophy of Science organized by the IUHPS (Veszprém, Hungary, 1984). 1988 ISBN 90-277-2284-6
109. P. Tillers and E.D. Green (eds.): *Probability and Inference in the Law of Evidence.* The Uses and Limits of Bayesianism. 1988 ISBN 90-277-2689-2
110. E. Ullmann-Margalit (ed.): *Science in Reflection.* The Israel Colloquium: Studies in History, Philosophy, and Sociology of Science, Vol. 3. 1988 ISBN 90-277-2712-0; Pb 90-277-2713-9
111. K. Gavroglu, Y. Goudaroulis and P. Nicolacopoulos (eds.): *Imre Lakatos and Theories of Scientific Change.* 1989 ISBN 90-277-2766-X
112. B. Glassner and J.D. Moreno (eds.): *The Qualitative-Quantitative Distinction in the Social Sciences.* 1989 ISBN 90-277-2829-1
113. K. Arens: *Structures of Knowing.* Psychologies of the 19th Century. 1989 ISBN 0-7923-0009-2
114. A. Janik: *Style, Politics and the Future of Philosophy.* 1989 ISBN 0-7923-0056-4
115. F. Amrine (ed.): *Literature and Science as Modes of Expression.* With an Introduction by S. Weininger. 1989 ISBN 0-7923-0133-1
116. J.R. Brown and J. Mittelstrass (eds.): *An Intimate Relation.* Studies in the History and Philosophy of Science. Presented to Robert E. Butts on His 60th Birthday. 1989 ISBN 0-7923-0169-2
117. F. D'Agostino and I.C. Jarvie (eds.): *Freedom and Rationality.* Essays in Honor of John Watkins. 1989 ISBN 0-7923-0264-8
118. D. Zolo: *Reflexive Epistemology.* The Philosophical Legacy of Otto Neurath. 1989 ISBN 0-7923-0320-2
119. M. Kearn, B.S. Philips and R.S. Cohen (eds.): *Georg Simmel and Contemporary Sociology.* 1989 ISBN 0-7923-0407-1
120. T.H. Levere and W.R. Shea (eds.): *Nature, Experiment and the Science.* Essays on Galileo and the Nature of Science. In Honour of Stillman Drake. 1989 ISBN 0-7923-0420-9
121. P. Nicolacopoulos (ed.): *Greek Studies in the Philosophy and History of Science.* 1990 ISBN 0-7923-0717-8
122. R. Cooke and D. Costantini (eds.): *Statistics in Science.* The Foundations of Statistical Methods in Biology, Physics and Economics. 1990 ISBN 0-7923-0797-6
123. P. Duhem: *The Origins of Statics.* Translated from French by G.F. Leneaux, V.N. Vagliente and G.H. Wagner. With an Introduction by S.L. Jaki. 1991 ISBN 0-7923-0898-0
124. H. Kamerlingh Onnes: *Through Measurement to Knowledge.* The Selected Papers, 1853-1926. Edited and with an Introduction by K. Gavroglu and Y. Goudaroulis. 1991 ISBN 0-7923-0825-5
125. M. Čapek: *The New Aspects of Time: Its Continuity and Novelties.* Selected Papers in the Philosophy of Science. 1991 ISBN 0-7923-0911-1
126. S. Unguru (ed.): *Physics, Cosmology and Astronomy, 1300–1700.* Tension and Accommodation. 1991 ISBN 0-7923-1022-5

Boston Studies in the Philosophy of Science

127. Z. Bechler: *Newton's Physics on the Conceptual Structure of the Scientific Revolution.* 1991 ISBN 0-7923-1054-3
128. É. Meyerson: *Explanation in the Sciences.* Translated from French by M-A. Siple and D.A. Siple. 1991 ISBN 0-7923-1129-9
129. A.I. Tauber (ed.): *Organism and the Origins of Self.* 1991 ISBN 0-7923-1185-X
130. F.J. Varela and J-P. Dupuy (eds.): *Understanding Origins.* Contemporary Views on the Origin of Life, Mind and Society. 1992 ISBN 0-7923-1251-1
131. G.L. Pandit: *Methodological Variance.* Essays in Epistemological Ontology and the Methodology of Science. 1991 ISBN 0-7923-1263-5
132. G. Munévar (ed.): *Beyond Reason.* Essays on the Philosophy of Paul Feyerabend. 1991 ISBN 0-7923-1272-4
133. T.E. Uebel (ed.): *Rediscovering the Forgotten Vienna Circle.* Austrian Studies on Otto Neurath and the Vienna Circle. Partly translated from German. 1991 ISBN 0-7923-1276-7
134. W.R. Woodward and R.S. Cohen (eds.): *World Views and Scientific Discipline Formation.* Science Studies in the [former] German Democratic Republic. Partly translated from German by W.R. Woodward. 1991 ISBN 0-7923-1286-4
135. P. Zambelli: *The Speculum Astronomiae and Its Enigma.* Astrology, Theology and Science in Albertus Magnus and His Contemporaries. 1992 ISBN 0-7923-1380-1
136. P. Petitjean, C. Jami and A.M. Moulin (eds.): *Science and Empires.* Historical Studies about Scientific Development and European Expansion. ISBN 0-7923-1518-9
137. W.A. Wallace: *Galileo's Logic of Discovery and Proof.* The Background, Content, and Use of His Appropriated Treatises on Aristotle's *Posterior Analytics.* 1992 ISBN 0-7923-1577-4
138. W.A. Wallace: *Galileo's Logical Treatises.* A Translation, with Notes and Commentary, of His Appropriated Latin Questions on Aristotle's *Posterior Analytics.* 1992 ISBN 0-7923-1578-2
Set (137 + 138) ISBN 0-7923-1579-0
139. M.J. Nye, J.L. Richards and R.H. Stuewer (eds.): *The Invention of Physical Science.* Intersections of Mathematics, Theology and Natural Philosophy since the Seventeenth Century. Essays in Honor of Erwin N. Hiebert. 1992 ISBN 0-7923-1753-X
140. G. Corsi, M.L. dalla Chiara and G.C. Ghirardi (eds.): *Bridging the Gap: Philosophy, Mathematics and Physics.* Lectures on the Foundations of Science. 1992 ISBN 0-7923-1761-0
141. C.-H. Lin and D. Fu (eds.): *Philosophy and Conceptual History of Science in Taiwan.* 1992 ISBN 0-7923-1766-1
142. S. Sarkar (ed.): *The Founders of Evolutionary Genetics.* A Centenary Reappraisal. 1992 ISBN 0-7923-1777-7
143. J. Blackmore (ed.): *Ernst Mach – A Deeper Look.* Documents and New Perspectives. 1992 ISBN 0-7923-1853-6
144. P. Kroes and M. Bakker (eds.): *Technological Development and Science in the Industrial Age.* New Perspectives on the Science–Technology Relationship. 1992 ISBN 0-7923-1898-6
145. S. Amsterdamski: *Between History and Method.* Disputes about the Rationality of Science. 1992 ISBN 0-7923-1941-9
146. E. Ullmann-Margalit (ed.): *The Scientific Enterprise.* The Bar-Hillel Colloquium: Studies in History, Philosophy, and Sociology of Science, Volume 4. 1992 ISBN 0-7923-1992-3
147. L. Embree (ed.): *Metaarchaeology.* Reflections by Archaeologists and Philosophers. 1992 ISBN 0-7923-2023-9
148. S. French and H. Kamminga (eds.): *Correspondence, Invariance and Heuristics.* Essays in Honour of Heinz Post. 1993 ISBN 0-7923-2085-9
149. M. Bunzl: *The Context of Explanation.* 1993 ISBN 0-7923-2153-7

150. I.B. Cohen (ed.): *The Natural Sciences and the Social Sciences.* Some Critical and Historical Perspectives. 1994 ISBN 0-7923-2223-1
151. K. Gavroglu, Y. Christianidis and E. Nicolaidis (eds.): *Trends in the Historiography of Science.* 1994 ISBN 0-7923-2255-X
152. S. Poggi and M. Bossi (eds.): *Romanticism in Science.* Science in Europe, 1790–1840. 1994 ISBN 0-7923-2336-X
153. J. Faye and H.J. Folse (eds.): *Niels Bohr and Contemporary Philosophy.* 1994 ISBN 0-7923-2378-5
154. C.C. Gould and R.S. Cohen (eds.): *Artifacts, Representations, and Social Practice.* Essays for Marx W. Wartofsky. 1994 ISBN 0-7923-2481-1
155. R.E. Butts: *Historical Pragmatics.* Philosophical Essays. 1993 ISBN 0-7923-2498-6
156. R. Rashed: *The Development of Arabic Mathematics: Between Arithmetic and Algebra.* Translated from French by A.F.W. Armstrong. 1994 ISBN 0-7923-2565-6
157. I. Szumilewicz-Lachman (ed.): *Zygmunt Zawirski: His Life and Work.* With Selected Writings on Time, Logic and the Methodology of Science. Translations by Feliks Lachman. Ed. by R.S. Cohen, with the assistance of B. Bergo. 1994 ISBN 0-7923-2566-4
158. S.N. Haq: *Names, Natures and Things.* The Alchemist Jābir ibn Ḥayyān and His *Kitāb al-Aḥjār* (Book of Stones). 1994 ISBN 0-7923-2587-7
159. P. Plaass: *Kant's Theory of Natural Science.* Translation, Analytic Introduction and Commentary by Alfred E. and Maria G. Miller. 1994 ISBN 0-7923-2750-0
160. J. Misiek (ed.): *The Problem of Rationality in Science and its Philosophy.* On Popper vs. Polanyi. The Polish Conferences 1988–89. 1995 ISBN 0-7923-2925-2
161. I.C. Jarvie and N. Laor (eds.): *Critical Rationalism, Metaphysics and Science.* Essays for Joseph Agassi, Volume I. 1995 ISBN 0-7923-2960-0
162. I.C. Jarvie and N. Laor (eds.): *Critical Rationalism, the Social Sciences and the Humanities.* Essays for Joseph Agassi, Volume II. 1995 ISBN 0-7923-2961-9
Set (161–162) ISBN 0-7923-2962-7
163. K. Gavroglu, J. Stachel and M.W. Wartofsky (eds.): *Physics, Philosophy, and the Scientific Community.* Essays in the Philosophy and History of the Natural Sciences and Mathematics. In Honor of Robert S. Cohen. 1995 ISBN 0-7923-2988-0
164. K. Gavroglu, J. Stachel and M.W. Wartofsky (eds.): *Science, Politics and Social Practice.* Essays on Marxism and Science, Philosophy of Culture and the Social Sciences. In Honor of Robert S. Cohen. 1995 ISBN 0-7923-2989-9
165. K. Gavroglu, J. Stachel and M.W. Wartofsky (eds.): *Science, Mind and Art.* Essays on Science and the Humanistic Understanding in Art, Epistemology, Religion and Ethics. Essays in Honor of Robert S. Cohen. 1995 ISBN 0-7923-2990-2
Set (163–165) ISBN 0-7923-2991-0
166. K.H. Wolff: *Transformation in the Writing.* A Case of Surrender-and-Catch. 1995 ISBN 0-7923-3178-8
167. A.J. Kox and D.M. Siegel (eds.): *No Truth Except in the Details.* Essays in Honor of Martin J. Klein. 1995 ISBN 0-7923-3195-8
168. J. Blackmore: *Ludwig Boltzmann, His Later Life and Philosophy, 1900–1906.* Book One: A Documentary History. 1995 ISBN 0-7923-3231-8
169. R.S. Cohen, R. Hilpinen and R. Qiu (eds.): *Realism and Anti-Realism in the Philosophy of Science.* Beijing International Conference, 1992. 1996 ISBN 0-7923-3233-4
170. I. Kuçuradi and R.S. Cohen (eds.): *The Concept of Knowledge.* The Ankara Seminar. 1995 ISBN 0-7923-3241-5

171. M.A. Grodin (ed.): *Meta Medical Ethics*: The Philosophical Foundations of Bioethics. 1995 ISBN 0-7923-3344-6
172. S. Ramirez and R.S. Cohen (eds.): *Mexican Studies in the History and Philosophy of Science.* 1995 ISBN 0-7923-3462-0
173. C. Dilworth: *The Metaphysics of Science.* An Account of Modern Science in Terms of Principles, Laws and Theories. 1995 ISBN 0-7923-3693-3
174. J. Blackmore: *Ludwig Boltzmann, His Later Life and Philosophy, 1900–1906* Book Two: The Philosopher. 1995 ISBN 0-7923-3464-7
175. P. Damerow: *Abstraction and Representation.* Essays on the Cultural Evolution of Thinking. 1996 ISBN 0-7923-3816-2
176. M.S. Macrakis: *Scarcity's Ways: The Origins of Capital.* A Critical Essay on Thermodynamics, Statistical Mechanics and Economics. 1997 ISBN 0-7923-4760-9
177. M. Marion and R.S. Cohen (eds.): *Québec Studies in the Philosophy of Science.* Part I: Logic, Mathematics, Physics and History of Science. Essays in Honor of Hugues Leblanc. 1995 ISBN 0-7923-3559-7
178. M. Marion and R.S. Cohen (eds.): *Québec Studies in the Philosophy of Science.* Part II: Biology, Psychology, Cognitive Science and Economics. Essays in Honor of Hugues Leblanc. 1996 ISBN 0-7923-3560-0
Set (177–178) ISBN 0-7923-3561-9
179. Fan Dainian and R.S. Cohen (eds.): *Chinese Studies in the History and Philosophy of Science and Technology.* 1996 ISBN 0-7923-3463-9
180. P. Forman and J.M. Sánchez-Ron (eds.): *National Military Establishments and the Advancement of Science and Technology.* Studies in 20th Century History. 1996 ISBN 0-7923-3541-4
181. E.J. Post: *Quantum Reprogramming.* Ensembles and Single Systems: A Two-Tier Approach to Quantum Mechanics. 1995 ISBN 0-7923-3565-1
182. A.I. Tauber (ed.): *The Elusive Synthesis: Aesthetics and Science.* 1996 ISBN 0-7923-3904-5
183. S. Sarkar (ed.): *The Philosophy and History of Molecular Biology: New Perspectives.* 1996 ISBN 0-7923-3947-9
184. J.T. Cushing, A. Fine and S. Goldstein (eds.): *Bohmian Mechanics and Quantum Theory: An Appraisal.* 1996 ISBN 0-7923-4028-0
185. K. Michalski: *Logic and Time.* An Essay on Husserl's Theory of Meaning. 1996 ISBN 0-7923-4082-5
186. G. Munévar (ed.): *Spanish Studies in the Philosophy of Science.* 1996 ISBN 0-7923-4147-3
187. G. Schubring (ed.): *Hermann Günther Graßmann (1809–1877): Visionary Mathematician, Scientist and Neohumanist Scholar.* Papers from a Sesquicentennial Conference. 1996 ISBN 0-7923-4261-5
188. M. Bitbol: *Schrödinger's Philosophy of Quantum Mechanics.* 1996 ISBN 0-7923-4266-6
189. J. Faye, U. Scheffler and M. Urchs (eds.): *Perspectives on Time.* 1997 ISBN 0-7923-4330-1
190. K. Lehrer and J.C. Marek (eds.): *Austrian Philosophy Past and Present.* Essays in Honor of Rudolf Haller. 1996 ISBN 0-7923-4347-6
191. J.L. Lagrange: *Analytical Mechanics.* Translated and edited by Auguste Boissonade and Victor N. Vagliente. Translated from the *Mécanique Analytique, novelle édition* of 1811. 1997 ISBN 0-7923-4349-2
192. D. Ginev and R.S. Cohen (eds.): *Issues and Images in the Philosophy of Science.* Scientific and Philosophical Essays in Honour of Azarya Polikarov. 1997 ISBN 0-7923-4444-8

Boston Studies in the Philosophy of Science

193. R.S. Cohen, M. Horne and J. Stachel (eds.): *Experimental Metaphysics*. Quantum Mechanical Studies for Abner Shimony, Volume One. 1997 ISBN 0-7923-4452-9
194. R.S. Cohen, M. Horne and J. Stachel (eds.): *Potentiality, Entanglement and Passion-at-a-Distance*. Quantum Mechanical Studies for Abner Shimony, Volume Two. 1997 ISBN 0-7923-4453-7; Set 0-7923-4454-5
195. R.S. Cohen and A.I. Tauber (eds.): *Philosophies of Nature: The Human Dimension*. 1997 ISBN 0-7923-4579-7
196. M. Otte and M. Panza (eds.): *Analysis and Synthesis in Mathematics*. History and Philosophy. 1997 ISBN 0-7923-4570-3
197. A. Denkel: *The Natural Background of Meaning*. 1999 ISBN 0-7923-5331-5
198. D. Baird, R.I.G. Hughes and A. Nordmann (eds.): *Heinrich Hertz: Classical Physicist, Modern Philosopher*. 1999 ISBN 0-7923-4653-X
199. A. Franklin: *Can That be Right?* Essays on Experiment, Evidence, and Science. 1999 ISBN 0-7923-5464-8
200. D. Raven, W. Krohn and R.S. Cohen (eds.): *The Social Origins of Modern Science*. 2000 ISBN 0-7923-6457-0
201. Reserved
202. Reserved
203. B. Babich and R.S. Cohen (eds.): *Nietzsche, Theories of Knowledge, and Critical Theory*. Nietzsche and the Sciences I. 1999 ISBN 0-7923-5742-6
204. B. Babich and R.S. Cohen (eds.): *Nietzsche, Epistemology, and Philosophy of Science*. Nietzsche and the Science II. 1999 ISBN 0-7923-5743-4
205. R. Hooykaas: *Fact, Faith and Fiction in the Development of Science*. The Gifford Lectures given in the University of St Andrews 1976. 1999 ISBN 0-7923-5774-4
206. M. Fehér, O. Kiss and L. Ropolyi (eds.): *Hermeneutics and Science*. 1999 ISBN 0-7923-5798-1
207. R.M. MacLeod (ed.): *Science and the Pacific War*. Science and Survival in the Pacific, 1939-1945. 1999 ISBN 0-7923-5851-1
208. I. Hanzel: *The Concept of Scientific Law in the Philosophy of Science and Epistemology*. A Study of Theoretical Reason. 1999 ISBN 0-7923-5852-X
209. G. Helm; R.J. Deltete (ed./transl.): *The Historical Development of Energetics*. 1999 ISBN 0-7923-5874-0
210. A. Orenstein and P. Kotatko (eds.): *Knowledge, Language and Logic*. Questions for Quine. 1999 ISBN 0-7923-5986-0
211. R.S. Cohen and H. Levine (eds.): *Maimonides and the Sciences*. 2000 ISBN 0-7923-6053-2
212. H. Gourko, D.I. Williamson and A.I. Tauber (eds.): *The Evolutionary Biology Papers of Elie Metchnikoff*. 2000 ISBN 0-7923-6067-2
213. S. D'Agostino: *A History of the Ideas of Theoretical Physics*. Essays on the Nineteenth and Twentieth Century Physics. 2000 ISBN 0-7923-6094-X
214. S. Lelas: *Science and Modernity*. Toward An Integral Theory of Science. 2000 ISBN 0-7923-6303-5
215. E. Agazzi and M. Pauri (eds.): *The Reality of the Unobservable*. Observability, Unobservability and Their Impact on the Issue of Scientific Realism. 2000 ISBN 0-7923-6311-6
216. P. Hoyningen-Huene and H. Sankey (eds.): *Incommensurability and Related Matters*. 2001 ISBN 0-7923-6989-0
217. A. Nieto-Galan: *Colouring Textiles*. A History of Natural Dyestuffs in Industrial Europe. 2001 ISBN 0-7923-7022-8

Boston Studies in the Philosophy of Science

218. J. Blackmore, R. Itagaki and S. Tanaka (eds.): *Ernst Mach's Vienna 1895–1930*. Or Phenomenalism as Philosophy of Science. 2001 ISBN 0-7923-7122-4
219. R. Vihalemm (ed.): *Estonian Studies in the History and Philosophy of Science*. 2001 ISBN 0-7923-7189-5
220. W. Lefèvre (ed.): *Between Leibniz, Newton, and Kant*. Philosophy and Science in the Eighteenth Century. 2001 ISBN 0-7923-7198-4
221. T.F. Glick, M.Á. Puig-Samper and R. Ruiz (eds.): *The Reception of Darwinism in the Iberian World*. Spain, Spanish America and Brazil. 2001 ISBN 1-4020-0082-0
222. U. Klein (ed.): *Tools and Modes of Representation in the Laboratory Sciences*. 2001 ISBN 1-4020-0100-2
223. P. Duhem: *Mixture and Chemical Combination*. And Related Essays. Edited and translated, with an introduction, by Paul Needham. 2002 ISBN 1-4020-0232-7
224. J.C. Boudri: *What was Mechanical about Mechanics*. The Concept of Force Betweem Metaphysics and Mechanics from Newton to Lagrange. 2002 ISBN 1-4020-0233-5
225. B.E. Babich (ed.): *Hermeneutic Philosophy of Science, Van Gogh's Eyes, and God*. Essays in Honor of Patrick A. Heelan, S.J. 2002 ISBN 1-4020-0234-3
226. D. Davies Villemaire: *E.A. Burtt, Historian and Philosopher*. A Study of the Author of The Metaphysical Foundations of Modern Physical Science. 2002 ISBN 1-4020-0428-1
227. L.J. Cohen: *Knowledge and Language*. Selected Essays of L. Jonathan Cohen. Edited and with an introduction by James Logue. 2002 ISBN 1-4020-0474-5
228. G.E. Allen and R.M. MacLeod (eds.): *Science, History and Social Activism: A Tribute to Everett Mendelsohn*. 2002 ISBN 1-4020-0495-0
229. O. Gal: *Meanest Foundations and Nobler Superstructures*. Hooke, Newton and the "Compounding of the Celestiall Motions of the Planetts". 2002 ISBN 1-4020-0732-9
230. R. Nola: *Rescuing Reason*. A Critique of Anti-Rationalist Views of Science and Knowledge. 2003 Hb: ISBN 1-4020-1042-7; Pb ISBN 1-4020-1043-5
231. J. Agassi: *Science and Culture*. 2003 ISBN 1-4020-1156-3
232. M.C. Galavotti (ed.): *Observation and Experiment in the Natural and Social Science*. 2003 ISBN 1-4020-1251-9
233. A. Simões, A. Carneiro and M.P. Diogo (eds.): *Travels of Learning*. A Geography of Science in Europe. 2003 ISBN 1-4020-1259-4
234. A. Ashtekar, R. Cohen, D. Howard, J. Renn, S. Sarkar and A. Shimony (eds.): *Revisiting the Foundations of Relativistic Physics*. Festschrift in Honor of John Stachel. 2003 ISBN 1-4020-1284-5
235. R.P. Farell: *Feyerabend and Scientific Values*. Tightrope-Walking Rationality. 2003 ISBN 1-4020-1350-7
236. D. Ginev (ed.): *Bulgarian Studies in the Philosophy of Science*. 2003 ISBN 1-4020-1496-1
237. C. Sasaki: *Descartes Mathematical Thought*. 2003 ISBN 1-4020-1746-4
238. K. Chemla (ed.): *History of Science, History of Text*. 2004 ISBN 1-4020-2320-0
239. C.R. Palmerino and J.M.M.H. Thijssen (eds.): *The Reception of the Galilean Science of Motion in Seventeenth-Century Europe*. 2004 ISBN 1-4020-2454-1

Boston Studies in the Philosophy of Science

240. J. Christianidis (ed.): *Classics in the History of Greek Mathematics*. 2004 ISBN 1-4020-0081-2

Also of interest:
R.S. Cohen and M.W. Wartofsky (eds.): *A Portrait of Twenty-Five Years Boston Colloquia for the Philosophy of Science, 1960-1985.* 1985 ISBN Pb 90-277-1971-3

Previous volumes are still available.

KLUWER ACADEMIC PUBLISHERS – DORDRECHT / BOSTON / LONDON